Lecture Notes in Computer Science   11242

Commenced Publication in 1973
Founding and Former Series Editors:
Gerhard Goos, Juris Hartmanis, and Jan van Leeuwen

More information about this series at http://www.springer.com/series/7409

Xiaofeng Meng · Ruixuan Li
Kanliang Wang · Baoning Niu
Xin Wang · Gansen Zhao (Eds.)

Web Information Systems and Applications

15th International Conference, WISA 2018
Taiyuan, China, September 14–15, 2018
Proceedings

 Springer

Editors
Xiaofeng Meng
Renmin University of China
Beijing, China

Ruixuan Li
Huazhong University of Science
and Technology
Wuhan, China

Kanliang Wang
Renmin University of China
Beijing, China

Baoning Niu
Taiyuan University of Technology
Yuci, China

Xin Wang
Tianjin University
Tianjin, China

Gansen Zhao
South China Normal University
Guangzhou, China

ISSN 0302-9743   ISSN 1611-3349   (electronic)
Lecture Notes in Computer Science
ISBN 978-3-030-02933-3   ISBN 978-3-030-02934-0   (eBook)
https://doi.org/10.1007/978-3-030-02934-0

Library of Congress Control Number: 2018960423

LNCS Sublibrary: SL3 – Information Systems and Applications, incl. Internet/Web, and HCI

This Springer imprint is published by the registered company Springer Nature Switzerland AG
The registered company address is: Gewerbestrasse 11, 6330 Cham, Switzerland

Preface

It is our great pleasure to present the proceedings of the 15th Web Information Systems and Applications Conference (WISA 2018). WISA 2018 was organized by the China Computer Federation Technical Committee on Information Systems (CCF TCIS) and Taiyuan University of Technology, jointly sponsored by China Association for Information Systems (CNAIS) and Luoyang Normal University. WISA provides a premium forum for researchers, professionals, practitioners, and officers closely related to information systems and applications, which encourages discussions on the theme about the future intelligent information systems with big data, with a focus on difficult and critical issues, and promotes innovative technology for new application areas of information systems.

WISA 2018 was held in Taiyuan, Shanxi, China, during September 14–15, 2018. The theme of WISA 2018 was big data and intelligent information systems, focusing on intelligent cities, government information systems, intelligent medical care, fin-teches, and network security, with an emphasis on the technology to solve the difficult and critical problems in data sharing, data governance, knowledge graph, and block chains.

This year we received 103 submissions, each of which was assigned to at least three Program Committee (PC) members to review. The peer process was double-blind. The thoughtful discussion on each paper by the PC resulted in the selection of 29 full research papers (acceptance rate of 28.16%) and 16 short papers. The conference program included keynote presentations by Prof. Guoqing Chen (Tsinghua University, China), Dr. Tao Yue (Simula Research Laboratory, Oslo, Norway) Dr. Jian Wang (Alibaba). The program of WISA 2018 also included more than 20 topic-specific talks by famous experts in six areas, i.e., digital economy and information systems, knowledge graphs and information systems, smart medical and information systems, blockchain and information systems, security and privacy of information systems, and architecture and practice of information systems, to share their cutting-edge technology and views about the academic and industrial hotspots. The other events included industrial forums, a CCF TCIS salon, and a PhD forum.

We are grateful to the general chairs, Prof. Xiaofeng Meng (Renmin University of China) and Prof. Ming Li, (Taiyuan University of Technology), all the PC members, and the external reviewers who contributed their time and expertise to the paper-reviewing process. We would like to thank all the members of the Organizing Committee, and many volunteers, for their great support in the conference organization. In particular, we would also like to thank the publication chairs, Prof. Xin Wang (Tianjin University) and Prof. Gansen Zhao (South China Normal University), for their efforts with the publication of the conference proceedings. Many thanks to the authors who submitted their papers to the conference. Last but not least, special thanks go to Shanshan Yao (Taiyuan University of Technology), the PC secretary, for her contributions to the paper reviewing and proceedings publishing processes.

August 2018
Ruixuan Li
Kanliang Wang
Baoning Niu

Organization

General Chairs

Xiaofeng Meng   Renmin University of China, China
Ming Li   Taiyuan University of Technology, China

Program Committee Chairs

Ruixuan Li   Huazhong University of Science and Technology, China
Kanliang Wang   Renmin University of China, China
Baoning Niu   Taiyuan University of Technology, China

Area Chairs

Digital Economy and Information Systems

Lingyun Qiu   Peking University, China
Yan Yu   Renmin University of China, China

Knowledge Graph and Information Systems

Huajun Chen   Zejiang University, China
Haofen Wang   Shanghai Leyan Technologies Co. Ltd., China

Medical Care and Information Systems

Chunxiao Xing   Tsinghua University, China

Block Chains and Information Systems

Ge Yu   Northeast University, China

Privacy and Security in Information Systems

Weiwei Ni   Southeast University, China
Liusheng Huang   University of Science and Technology of China, China

Architectures and Practices of Information Systems

Guoqiang Ge   Enmotech, China
Song Shi   iQiyi, China

Panel Chairs

Yun Xie	Digital China, China
Huaizhong Kou	Yellow River Conservancy Commission of the Ministry of Water Resources, China

PhD Forum Chairs

Yong Qi	Xi'an Jiaotong University, China
Lei Zou	Peking University, China

Local Chairs

Yanbin Peng	Taiyuan University of Technology, China
Xiufang Feng	Taiyuan University of Technology, China
Wenjian Wang	Shanxi University, China

Publicity Chairs

Lizhen Xu	Southeast University, China
Chunhua Song	Taiyuan University of Technology, China

Publication Chairs

Xin Wang	Tianjin University, China
Gansen Zhao	South China Normal University, China
Aiwen Song	North University of China, China

Tutorial Chair

Xiaoli Hao	Taiyuan University of Technology, China

Poster/Demo Chairs

Aiping Li	Taiyuan University of Technology, China
Xingong Chang	Shanxi University of Finance and Economics, China

Website Chairs

Weifeng Zhang	Nanjing University of Post and Telecomunications, China
Xing Fan	Taiyuan University of Technology, China

Program Committee

Dezhi An	Gansu Institute of Political Science and Law, China
Xingong Chang	Shanxi University of Finance and Economics, China

Lemen Chao	Renmin University of China, China
Hong Chen	Renmin University of China, China
Huajun Chen	Zhejiang University, China
Lin Chen	Nanjing University, China
Ling Chen	Yangzhou University, China
Ting Deng	Beihang University, China
Jing Dong	Taiji, China
Markus Endres	University of Augsburg, Germany
Jing Gao	Dalian University of Technology, China
Jun Gao	Peking University, China
Kening Gao	Northeastern University, China
Yunjun Gao	Zhejiang University, China
Guoqiang Ge	Enmotech, China
Yu Gu	Northeastern University, China
Ruiqiang Guo	Hebei Normal University, China
Wenzhong Guo	Fuzhou University, China
Xixian Han	Harbin Institute of Technology, China
Yanbo Han	Institute of Computing Technology, CAS, China
Xiaoli Hao	Taiyuan University of Technology, China
Qinming He	Zhejiang University, China
Tieke He	Nanjing University, China
Yanxiang He	Wuhan University, China
XiangYan He	Wuhan University, China
Liang Hong	Wuhan University, China
Zhenting Hong	Guangxi University of Science and Technology, China
Mengxing Huang	Hainan University, China
Shujuan Jiang	China University of Mining and Technology, China
Fangjiao Jiang	Jiangsu Normal University, China
Weijin Jiang	Xiangtan University, China
Peiquan Jin	University of Science and Technology of China, China
Huaizhong Kou	Yellow River Conservancy Commission of the Ministry of Water Resources, China
Carson Leung	University of Manitoba, Canada
Bin Li	Yangzhou University, China
Bo Li	Sun Yat-sen University, China
Guoliang Li	Tsinghua University, China
Qingzhong Li	Shandong University, China
Yukun Li	Tianjin University of Technology, China
Zhenxing Li	AgileCentury, China
Aijun Li	Shanxi University of Finance and Economics, China
Aiping Li	Taiyuan University of Technology, China
Chunying Li	Guangdong Polytechnic Normal University, China
Ruixuan Li	Huazhong University of Science and Technology, China
Ziyu Lin	Xiamen University, China
Chen Liu	North China University of Technology, China
Qing Liu	Renmin University of China, China

Yongchao Liu	Ant Financial, China
Yingqi Liu	Purdue University, USA
Shan Lu	Southeast University, China
Jianming Lv	South China University of Technology, China
Youzhong Ma	Luoyang Normal University, China
Xiaofeng Meng	Renmin University of China, China
Yang-Sae Moon	Kangwon National University, Korea
Mirco Nanni	KDD-Lab ISTI-CNR Pisa, Italy
Wee Siong Ng	Institute for Infocomm Research, Singapore
Weiwei Ni	Southeast University, China
Tiezheng Nie	Northeastern University, China
Bo Ning	Dalian Maritime University, China
Baoning Niu	Taiyuan University of Technology, China
Zhenkuan Pan	Qingdao University, China
Dhaval Patel	Indian Institute of Technology, USA
Zhiyong Peng	Wuhan University, China
Hai Phan	New Jersey Institute of Technology, USA
Jianzhong Qi	The University of Melbourne, Australia
Yong Qi	Xi'an Jiaotong University, China
Tieyun Qian	Wuhan University, China
Weining Qian	East China Normal University, China
Lingyun Qiu	Peking University, China
Weiguang Qu	Nanjing Normal University, China
Jiadong Ren	Yanshan University, China
Chuitian Rong	Tianjin Polytechnic University, China
Yingxia Shao	Peking University, China
Derong Shen	Northeastern University, China
Song Shi	Meizu Technology, China
Wei Song	Wuhan University, China
Baoyan Song	Liaoning University, China
Hailong Sun	Beihang University, China
Haojun Sun	Shantou University, China
Weiwei Sun	Fudan University, China
Chih-Hua Tai	National Taiwan University, Taiwan
Yong Tang	South China Normal University, China
Xianping Tao	Nanjing University, China
Yongxin Tong	Beihang University, China
Leong Hou U	University of Macau, SAR China
Guojun Wang	Central South University, China
Haofen Wang	Shanghai Leyan Technologies Co. Ltd., China
Hongzhi Wang	Harbin Institute of Technology, China
Hua Wang	Victoria University, Australia
Junhu Wang	Griffith University, Australia
Wei Wang	The University of New South Wales, Australia
Wenjian Wang	Shanxi University, China
Xibo Wang	Shenyang University of Technology, China

Xin Wang	Tianjin University, China
Xingce Wang	Beijing Normal University, China
Yijie Wang	National University of Defence Technology, China
ZhiJie Wang	Sun Yat-Sen University, China
Kanliang Wang	Renmin University of China, China
Wei Wei	Xi'an University of Technology, China
Shengli Wu	Jiangsu University, China
Xiaoying Wu	Wuhan University, China
Feng Xia	Dalian University of Technology, China
Yanghua Xiao	Fudan University, China
Xike Xie	Aalborg University, Denmark
Hongwei Xie	Taiyuan University of Technology, China
Huarong Xu	Xiamen University of Technology, China
Jiajie Xu	Soochow University, China
Lei Xu	Nanjing University, China
Lizhen Xu	Southeast University, China
Zhuoming Xu	Hohai University, China
Baowen Xu	Nanjing University, China
Zhongmin Yan	Shandong University, China
Nan Yang	Renmin University of China, China
Dan Yin	Harbin Engineering University, China
Jian Yin	Sun Yat-sen University, China
Ge Yu	Northeastern University, China
Hong Yu	Dalian Ocean University, China
Yan Yu	Renmin University of China, China
Ziqiang Yu	University of Jinan, China
Fang Yuan	Hebei University, China
Hua Yuan	University of Science and Technology of China, China
Xiaojie Yuan	Nankai University, China
Karine Zeitouni	University of Versailles-Saint-Quentin, France
Guigang Zhang	Institute of Automation, CAS, China
Mingxin Zhang	Changshu Institute of Technology, China
Wei Zhang	East China Normal University, China
Weifeng Zhang	Nanjing University of Posts and Telecommunications, China
Xiaowang Zhang	Tianjin University, China
Yong Zhang	Tsinghua University, China
Zhiqiang Zhang	Harbin Engineering University, China
Xiaolin Zhang	Inner Mongolia University of Science and Technology, China
Ying Zhang	Nankai University, China
Feng Zhao	Huazhong University of Science and Technology, China
Gansen Zhao	South China Normal University, China
Xiangjun Zhao	Xuzhou Normal University, China
Junfeng Zhou	Donghua University, China
Junwu Zhu	Yangzhou university, China

Contents

Information Retrieval

Natural Language Processing

Data Privacy and Security

Knowledge Graphs and Social Networks

Query Processing

Recommendation Systems

Machine Learning and Data Mining

2 Many-objective Convolutional Neural Networks

In this section, a new deep neural network model many-objective CNN (MaO-CNN) is proposed. Firstly, the DET curve is extended to many-class DET (MaDET) surface and the property of MaDET surface is discussed then. Secondly, MaO-CNN is described in MaDET space and the difference of solution space between CNN and MaO-CNN will be analyzed.

2.1 The Many-Class DET Surface

The confusion matrix of many-class classifiers is shown in Table 1. In this table, we denote it as a True $c(i,i)$ ($i=1, 2 \ldots$, n), when an instance of class i is predicted as class i. When an instance of class j ($j=1, 2 \ldots$, n) is classified as class i ($i \neq j$), we denote it as a False $c(i, j)$.

Table 1. A confusion matrix of many-class classifiers.

		True labels			
		Class 1	Class 2	...	Class n
Predicted labels	Class 1	True c(1,1)	False c(1,2)	...	False c(1,n)
	Class 2	False c(2,1)	True c(2,2)	...	False c(2,n)
	...	...	...	...	...
	Class n	False c(n,1)	False c(n,2)	...	True c(c,n)

The rate of misclassification of class i is defined as Eq. (1), and the classification accuracy of class i is defined as Eq. (2). Obviously, we can find that $fc_ir + tc_ir = 1$.

$$fc_ir = \frac{\sum_{j=1}^{n} c(j,i)(j \neq i)}{\sum_{j=1}^{n} c(j,i)} \qquad (1)$$

$$tc_ir = \frac{c(i,i)}{\sum_{j=1}^{n} c(j,i)} \qquad (2)$$

Generally, the classifiers with low value of fc_ir are preferable. However, minimizing fc_ir for different classes are conflicting with each other, an improvement in the performance of an indicator may result in a reduction in the performance of another one. We define many-class DET surface to describe the trade-off among fc_ir, as it is denoted by Eq. (3).

$$\text{MaDET} \triangleq \{fc_1r, fc_2r, \cdots, fc_nr\}, \qquad (3)$$

Several points in MaDET surface are important to note. The point $(0, 0, \ldots, 0)$ represents a perfect classifier, as it means never issuing a wrong classification. Usually, such a point does not exist in reality but can be approximated as closely as possible. The point *(1, 0, ..., 0)* represents an extreme case, in which there is no instance has true label as class *1*, some instances are misclassified as class *1* and the rest instances are correctly classified. The point *(1, 0, ..., 0)* dose not appear in the real-world classification problem, which is different with general many-objective optimization problem [11]. The surface $fc_1r + fc_2r + \cdots + fc_nr = n-1$ on the MaDET surface represents the strategy of randomly guessing a class label for an instance. Here is an example of four-class classification problem, if a classifier randomly guesses the class *1* 0.20 times on average, the class *2* 0.25 times, the class *3* 0.15 times and class *4* 0.40 times. Then 20% of class *1* can be correctly classified and 80% of class *1* can be misclassified, i.e., $fc_1r = 0.80$. Similarly, it is easy for us to know that $fc_2 = 0.75$, $fc_3 = 0.85$ and $fc_4 = 0.60$. In the case of randomly guessing strategy for four-class classification we can get that $fc_1r + fc_2r + fc_3r + fc_4r = 3$.

Any classifiers on the surface $fc_1r + fc_2r + \cdots + fc_nr = n-1$ in MaDET space may be said to have no information about the classification. A classifier which produces the MaDET surface above the surface performs worse than randomly guessing. We prefer to find classifiers that appear below randomly guessing surface, i.e., $fc_1r + fc_2r + \cdots + fc_nr < n - 1$. We try to find classifiers that have low value of fc_ir simultaneously.

Every classifier can be mapped to the MaDET space. The goal of classifiers training is to find a set of parameters of a classifier that approximate the perfect point *(0, 0, ..., 0)*. The classification problem in MaDET space turns out to be a many-objective optimization problem as it is described in Eq. (4).

$$\min_{x \in \Omega} \mathbf{F}(\mathbf{x}) = \Big(fc_1r(\mathbf{x}), fc_2r(\mathbf{x}), \ldots, fc_nr(\mathbf{x}) \Big), \tag{4}$$

where $\mathbf{x}$ is the parameters of a classifier, n is the number of classes of all instances, Ω is the solution space, and $\mathbf{F}(\mathbf{x})$ is a vector function to describe the performance of classifiers in MaDET space.

Pareto dominance [15] is an important concept to compare two solutions of MaOPs. While given two solutions $\mathbf{x}^1$ and $\mathbf{x}^2$, $\mathbf{x}^1$ is said to dominate $\mathbf{x}^2$ if and only if $fc_ir(\mathbf{x}^1) \leq fc_ir(\mathbf{x}^2)$ for all $i = 1, 2, \ldots, m$, and $fc_ir(\mathbf{x}^1) \neq fc_ir(\mathbf{x}^2)$. It can be denoted as $\mathbf{x}^1 \succ \mathbf{x}^2$. A solution $\mathbf{x}^*$ is Pareto optimal if there does not exist another solution $\mathbf{x}$ that dominates it. The Pareto set (*PS*) is the set of all the Pareto optimal points, as it is denoted by Eq. (5).

$$PS \triangleq \{\mathbf{x} | \nexists \mathbf{x}^* \in \Omega, \mathbf{x}^* \succ \mathbf{x}\} \tag{5}$$

The Pareto front (*PF*) is the set of corresponding objective vectors of the *PS*, as it is denoted by Eq. (6).

$$PF \triangleq \{\mathbf{F}(\mathbf{x}) | \mathbf{x} \in PS\} \tag{6}$$

2.2 The Many-Objective CNN Model

In this part, we propose a new model called many-objective CNN (MaO-CNN), which describes the performance of classifiers in many-objective space rather than a single objective space. The structure of the CNN model does not change, only the objective function of the model is modified. Different from traditional CNN model, we try to obtain a classifier with low value of all $fc_i r$, as it is described in Eq. (7).

$$\min_{\theta \in \Omega_\theta} \text{MaO-CNN}(\theta) = \Big(fc_1 r(\theta), fc_2 r(\theta), \ldots, fc_n r(\theta) \Big), \tag{7}$$

where θ represents the parameter of the given classifier, Ω_θ is the solution space. We prefer to obtain classifiers with low value of all $fc_i r(\theta)$ simultaneously. However, these objectives are conflicting with each other, we can try to find optimal trade-off among them. By adopting the many-objective evaluation metric the solution space of CNN is reduced, as we should find solutions in the area of the intersection of several objectives. The illustration of the solution space of CNN and MaO-CNN is shown in Fig. 1. While the dataset is not enough for CNN model training, the MaO-CNN can easily find a feasible set of solutions.

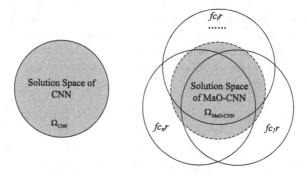

Fig. 1. Illustration of the solution space of CNN and MaO-CNN.

Generally, there is no closed form solution for the MaO-CNN. MaOEAs have been proven to be effective for many-objective optimization test problems [11]. However, the test functions of many-objective are always with less number of parameters than that of deep learning models. While dealing with many-objective problems with too many parameters, MaOEAs need much more time to find the suboptimum solution set. Usually, many researchers try to speed up the convergence of evolutionary algorithms (EAs) by combining them with local search algorithms, such as gradient based optimization methods [5]. In this paper, we proposed a hybrid framework for MaO-CNN optimization, in which MaOEA is used for global search and gradient based method is used for local search. The hybrid framework can find the optimal solutions effectively by combining the two strategies together.

3 MaO-CNN Learning

In this section, the idea of optimizing MaO-CNN by a framework of hybrid MaOEAs will be introduced. In the framework, we design a new hybrid encoding, a hybrid crossover operation and a hybrid mutation operation for MaO-CNN evolving. The hybrid encoding acts as a bridge between MaO-CNN model and evolutionary algorithms. The hybrid crossover operation improves the ability of global search and the hybrid mutation operation fasts the speed of convergence of the learning method. The details of them are discussed as follows.

3.1   The Framework of Hybrid MaOEAs

It is of great difficulty to optimize MaO-CNN model directly by using MaOEAs because of the large number of parameters [3]. In this paper, we propose a hybrid framework of MaOEAs for MaO-CNN optimization, in which MaOEAs focus on global search and gradient descent is used for local search. Inspired by the strategy of MOEA/D (multi-objective evolutionary algorithm based on decomposition) in [14], in the framework the Mao-CNN optimization problem is decomposed into a number of single objective optimization subproblems, and the single objective subproblems can be optimized by using gradient descent algorithm. Besides, a hybrid encoding is designed to describe the parameters of MaO-CNN and a novel hybrid crossover operation is adopted for population solutions evolving. The details of this framework will be described as follows.

As described above the values of all $fc_i r$ are positive, Eq. (7) is equivalent with Eq. (8).

$$\min_{\theta \in \Omega} \text{MaO-CNN}(\theta) = \left(fc_1 r(\theta)^2, fc_2 r(\theta)^2, \ldots, fc_n r(\theta)^2 \right) \tag{8}$$

Inspired by the decomposition based algorithms [9], the many-objective problem can be turned as a single objective problem by adding them together with weight, as it is denoted by Eq. (9).

$$\min_{\theta \in \Omega} J_M(\theta) = \frac{1}{2} \sum_{i=1}^{n} w_i \cdot fc_i r(\theta)^2, \tag{9}$$

where $\mathbf{w} = \{w_1, w_2, \ldots, w_C\}$ is the weight of each misclassification rate, it reflects the importance of each class for the finally classification. The value of $\mathbf{w}$ can be determined based on the distribution of the dataset.

With a given value of $\mathbf{w}$, the cost function of $J_M(\theta)$ can be minimized by gradient descent algorithm under the framework of back propagation. The partial derivative with respect to θ is denoted by Eq. (10).

$$\Delta_\theta J_M(\theta) = \sum_{i=1}^{C} w_i \cdot fc_i r(\theta) \tag{10}$$

Generally, gradient descent is susceptible to local optima, however, in practice it usually works fairly well when given a suitable weight $\mathbf{w}$. It helps to speed up the convergence of the algorithm. The value of $\mathbf{w}$ can be optimized by MaOEAs, as denoted by Eq. (11).

$$\min_{\mathbf{w} \in \Omega_{\mathbf{w}}} \text{MaO-CNN}(\mathbf{w}) = \min \Big(fc_1 r(\theta), fc_2 r(\theta), \dots, fc_n r(\theta) \Big),$$

$$subject \quad to \quad \sum_{i=1}^{n} w_i = 1, w_i > 0 \qquad (11)$$

where $\Omega_{\mathbf{w}}$ is the set of all possible solutions of $\mathbf{w}$. When given a vector of $\mathbf{w}$, the parameter θ of MaO-CNN can be approached by using hybrid evolutionary algorithm and gradient descent algorithm. The gradient descent algorithm can find the local optimal solution effectively and the hybrid evolutionary algorithm can find a good initial solution for the gradient descent algorithm. The global solutions can be found by combining these two methods together.

Algorithm 1. Learning Procedure for MaO-CNN Model

1: Initialization: Initialize the population of parameters θ and $\mathbf{w}$ randomly.
2: Local search stage: With the given $\mathbf{w}$ update θ by minimizing Eq. (10) using gradient descent algorithm with back propagation.
3: Global search stage: Update $\mathbf{w}$ using selection, crossover and mutation operation with the theory of many-objective optimization; update θ by adopting hybrid crossover operation.
4: Repeat Step 2 and Step 3 until converge.

The framework of learning procedure for MaO-CNN is described by Algorithm 1. Firstly, initialize the population of parameters θ and $\mathbf{w}$ randomly. Secondly, update each θ in the population with the given $\mathbf{w}$. Thirdly, the weight vector $\mathbf{w}$ can be optimized by MaOEAs. Then repeat Step 2 and Step 3 many times until converge. Step 2 is the local search stage, in which the gradient descent algorithm is adopted to optimize θ in the framework of backpropagation with a given $\mathbf{w}$. The parameters of neural networks can be updated to local optimal effectively. Step 3 performs global search stage, in which the adaventage of heuristic algorithms is taken. A many-objective optimization algorithm is taken to optimize the weight vector $\mathbf{w}$. By adopting the Pareto-based MaOEA not only a group of solutions can be found at the same time, but also the solutions are robust, as the solutions can provide useful information for others during the evolving stage.

3.2   MaO-CNN Encoding

In this paper, a many-objective optimization evolutionary algorithm will be applied to optimize the proposed MaO-CNN. An encoding is designed for the hybrid framework. The chromosome is used for weight vector $\mathbf{w}$ of MaO-CNN model. The real encoding is used to represent the weight vector is constituted by an array of real values in the interval $[0, 1]$. While weight vector can be updated by adopting selection operation, crossover operation and mutation operation with the theory of MaOEAs.

EMOAs are Pareto-based algorithms, which can obtain a set of solutions rather than a single solution. By adopting the hybrid MaOEAs a set of parameters of MaO-CNN can be obtained in MaDET space. The solutions obtained here are feasible solutions, rather than the optimal solution. The hybrid MaOEAs are suitable for dealing with CNN learning with not enough training dataset.

4   Experimental Results

In this section, the well-known MNIST dataset [7] and street view house numbers (SVHN) dataset [8] are selected to evaluate the performance of MaO-CNN. The many-objective optimization algorithm two_arch2 [11] is selected for model training under the framework of hybrid MaOEAs. The details of experiments are described as follows.

4.1   Dataset Description

MNIST Dataset. The MNIST dataset, which is a real-world handwritten digits data, is provided for machine learning and pattern recognition methods. It contains a training set with 60,000 examples and a testing set with 10,000 examples. The handwritten digits have been size-normalized and centered in a fixed-size image. In this paper, we use small amount examples for training, and the remains for testing. The description of MNIST is shown in the left part of Table 2.

Table 2. The details of MNIST dataset used in the experiments.

class	No. of all set	No. of testing set	No. of training set	ds4	ds5	ds6	ds7	ds8	ds9	ds10
0	6903	5923	980	*	*	*	*	*	*	*
1	7877	6742	1135	*	*	*	*	*	*	*
2	6990	5958	1032	*	*	*	*	*	*	*
3	7141	6131	1010	*	*	*	*	*	*	*
4	6824	5842	985		*	*	*	*	*	*
5	6313	5421	892			*	*	*	*	*
6	6876	5918	958				*	*	*	*
7	7293	6265	1028					*	*	*
8	6825	5851	974						*	*
9	6958	5949	1009							*

In this section, we select several sub datasets from the whole dataset, including 4, 5, 6, 7, 8, 9 and 10 classes respectively, the details are listed in the right part of Table 2. In the table, ds$n$ denotes the sub dataset that has n classes marked with * in it, for instance, ds4 has four classes images, including '0', '1', '2' and '3'.

SVHN Dataset. The street view house numbers (SVHN) dataset [8] is obtained from house numbers in Google street view images. It is a real-world image dataset for evaluation of machine learning algorithms. It consists of 99259 32×32 color digits in 10 classes, with 73257 digits for training and 26032 digits for testing. In this paper, we use small amount set of digits for training, and large amount set of digits for model evaluation instead. The details of SVHN are described in the left part of Table 3. Several sub datasets are selected from the whole dataset, including 4, 5, 6, 7, 8, 9 and 10 classes, the details are listed in the right part of Table 3.

Table 3. The details of SVHN dataset used in the experiments.

class	No. of all set	No. of testing set	No. of training set	ds4	ds5	ds6	ds7	ds8	ds9	ds10
0	6692	4948	1744	*	*	*	*	*	*	*
1	18960	13861	5099	*	*	*	*	*	*	*
2	14734	10585	4149	*	*	*	*	*	*	*
3	11379	8497	2882	*	*	*	*	*	*	*
4	9981	7458	2523		*	*	*	*	*	*
5	9266	6882	2384			*	*	*	*	*
6	7704	5727	1977				*	*	*	*
7	7614	5595	2019					*	*	*
8	6705	5045	1660						*	*
9	6254	4559	1595							*

4.2   Algorithms Involved

A many-objective optimization algorithm, i.e., two_arch2, and a single objective optimization algorithm stochastic gradient descent (SGD) [7] algorithm are tested. Experiments are performed with Matlab 2014b running on a desktop PC with an i5 3.2 GHz processor and 8 GB memory under Ubuntu14.04 LTS. A NVIDIA Quadro K2000 graphics card with 2 GB memory and GDDR5 is used in the experiments. The code implemented here are based on the package of MatConvNet [10], which is a MATLAB toolbox implementing CNNs for computer vision applications.

4.3   Evaluation Metrics

The classification accuracy (Acc) was selected to evaluate the performance of the above methods. Acc is evaluated on the test dataset, which is defined as the partition of the correctly classified samples to all samples in test dataset. We prefer to obtain CNN/MaO-CNN model with high value of classification accuracy.

4.4   Parameter Setting

In this section, the architecture of LeNet-5 is selected for MNIST and SVHN classification. In each experiment, 100 epoches are implemented for each training stage for SGD for MNIST dataset and 200 epoches are implemented for SVHN dataset. The population size is set as 10 and the maximum iteration is 20 for two_arch2 algorithm. For each local search step 10 epoches are implemented for MNIST dataset and 20 epoches are implemented for SVHN dataset. The simulated binary crossover (SBX) and polynomial bit flip mutation operators are applied for weights vector evolving in the experiments with crossover probability of $p_c = 1$, mutation probability of $p_m = 0.1$ and the hybrid crossover probability of $p_{hc} = 0.2$. For SGD algorithm, 10 independent trials are conducted, the best accuracy and the average accuracy are listed in the following. Because of it is time consuming for the hybrid MaOEA, the two_arch2 is only implemented once for each dataset, but all of the individuals of the population are compared with SGD. For each mentioned algorithm 10 results are obtained and compared in the following.

4.5   Experimental Results and Discussion

MNIST Dataset. To evaluate the performance of MaO-CNN and CNN fairly, we compare the best and mean of accuracy obtained by these methods. The best and mean of Acc (i.e., Acc_{best} and $Acc_{average}$) of MaO-CNN and traditional CNN are shown in Table 4. By comparing the results in the table we can see that not only the best accuracy but also the average accuracy obtained by MaO-CNN are higher than the best accuracy of CNN. The best and the mean results of CNN are almost the same, which means that the CNN is trapped into local optimum. The accuracy results of best and mean obtain by MaO-CNN are different, which means the MaO-CNN can obtain a set of solutions with good diversity. The results show that the traditional CNN is easily trapped into local optimum without enough training data.

Table 4. The experimental results on MNIST datasets.

dataset	MaO-CNN		CNN	
	Acc_{best}	$Acc_{average}$	Acc_{best}	$Acc_{average}$
ds4	0.9872	0.9844	0.9780	0.9780
ds5	0.9855	0.9836	0.9774	0.9774
ds6	0.9837	0.9834	0.9811	0.9811
ds7	0.9834	0.9828	0.9826	0.9826
ds8	0.9837	0.9832	0.9818	0.9818
ds9	0.9795	0.9791	0.9630	0.9630
ds10	0.9783	0.9778	0.9737	0.9737

The SGD algorithm is stable as the CNN can always converging to the same result, however, it converges to local optimal solution. The results show that the proposed MaO-CNN has better performance than CNN when the training sample is not sufficient. Since the MaO-CNN learning method is a Pareto-based searching algorithm, the proposed MaO-CNN can always obtain better solutions than CNN. Above all we can make a conclusion that the proposed MaO-CNN model is more robust than traditional CNN model.

SVHN Dataset. The best and mean of Acc (i.e., Acc_{best} and $Acc_{average}$) of MaO-CNN and traditional CNN are shown in Table 5. By comparing the results in the table we can see that not only the best accuracy but also the average accuracy obtained by MaO-CNN are better than the best accuracy of CNN. In addition, the results of MaO-CNN are robust, as the Acc_{best} and $Acc_{average}$ of MaO-CNN are almost the same. The Pareto-based method has good ability of convergence, as solution in the population can be improved by others by using evolutionary operations.

Table 5. The experimental results on SVHN datasets.

dataset	MaO-CNN		CNN	
	Acc_{best}	$Acc_{average}$	Acc_{best}	$Acc_{average}$
ds4	0.9204	0.9105	0.8647	0.8486
ds5	0.9027	0.8996	0.8526	0.8398
ds6	0.8740	0.8663	0.8415	0.8271
ds7	0.8657	0.8508	0.8082	0.7122
ds8	0.8614	0.8503	0.8234	0.8179
ds9	0.8380	0.8328	0.8036	0.7930
ds10	0.8351	0.8312	0.7992	0.7232

5   Conclusions

In this paper we proposed many-class detection error trade-off (MaDET) graph by extending DET curve to many-class classification case. Many-objective convolutional neural network (MaO-CNN) model is proposed in MaDET space. A hybrid framework of many-objective evolutionary algorithm is proposed for MaO-CNN model learning, in which the MaOEA is used for the global search and gradient based method is used for local search. The proposed framework makes it is easy to find feasible solutions by combining global search strategy with local search strategy. The new MaO-CNN can obtain better classification performance than traditional CNN with not enough training data on MNIST and SVHN datasets.

Acknowledgment. This work was partially supported by the National Key Research and Development Plan (No. 2016YFC0600908), the National Natural Science Foundation of China (No. U1610124, 61572505 and 61772530), and the National Natural Science Foundation of Jiangsu Province (No. BK20171192).

References

1. Bai, S.: Scene categorization through using objects represented by deep features. Int. J. Pattern Recogn. Artif. Intell. **31**(9), 1–21 (2017)
2. Goh, H., Thome, N., Cord, M., Lim, J.H.: Learning deep hierarchical visual feature coding. IEEE Trans. Neural Netw. Learn. Syst. **25**(12), 2212–2225 (2014)
3. Gong, M., Liu, J., Li, H., Cai, Q., Su, L.: A multiobjective sparse feature learning model for deep neural networks. IEEE Trans. Neural Netw. Learn. Syst. **26**(12), 3263–3277 (2015)
4. Goodfellow, I., Bengio, Y., Courville, A.: Deep learning (2016). http://www.deeplearningbook.org. Book in preparation for MIT Press
5. Sosa Hernández, V.A., Schütze, O., Emmerich, M.: Hypervolume maximization via set based Newton's method. In: Tantar, A.-A., et al. (eds.) EVOLVE - A Bridge between Probability, Set Oriented Numerics, and Evolutionary Computation V. AISC, vol. 288, pp. 15–28. Springer, Cham (2014). https://doi.org/10.1007/978-3-319-07494-8_2
6. Ke, Q., Zhang, J., Song, H., Wan, Y.: Big data analytics enabled by feature extraction based on partial independence. Neurocomputing **288**, 3–10 (2018). Learning System in Real-time Machine Vision
7. LeCun, Y., Bottou, L., Bengio, Y., Haffner, P.: Gradient-based learning applied to document recognition. Proc. IEEE **86**(11), 2278–2324 (1998)
8. Netzer, Y., Wang, T., Coates, A., Bissacco, A., Wu, B., Ng, A.Y.: Reading digits in natural images with unsupervised feature learning. In: NIPS Workshop on Deep Learning and Unsupervised Feature Learning (2010)
9. Trivedi, A., Srinivasan, D., Sanyal, K., Ghosh, A.: A survey of multi-objective evolutionary algorithms based on decomposition. IEEE Trans. Evol. Comput. **21**(3), 440–462 (2017)
10. Vedaldi, A., Lenc, K.: MatConvNet - convolutional neural networks for MATLAB. In: Proceeding of the ACM International Conference on Multimedia (2015)
11. Wang, H., Jiao, L., Yao, X.: Two_Arch2: an improved two-archive algorithm for many-objective optimization. IEEE Trans. Evol. Comput. **19**(4), 524–541 (2015)
12. Xia, C., Qi, F., Shi, G.: Bottom-up visual saliency estimation with deep autoencoder-based sparse reconstruction. IEEE Trans. Neural Netw. Learn. Syst. **27**(6), 1227–1240 (2016)
13. Xia, Y., Zhang, B., Coenen, F.: Face occlusion detection using deep convolutional neural networks. Int. J. Pattern Recogn. Artif. Intell. **30**(09), 1–24 (2016)
14. Zhang, Q., Li, H.: Moea/d: a multiobjective evolutionary algorithm based on decomposition. IEEE Trans. Evol. Comput. **11**(6), 712–731 (2007)
15. Zhao, J., et al.: Multiobjective optimization of classifiers by means of 3D convex-hull-based evolutionary algorithms. Inf. Sci. **367–368**, 80–104 (2016)
16. Zhao, J.: 3D fast convex-hull-based evolutionary multiobjective optimization algorithm. Appl. Soft Comput. **67**, 322–336 (2018)
17. Zhao, J.: Multiobjective sparse ensemble learning by means of evolutionary algorithms. Decis. Support Syst. **111**, 86–100 (2018)
18. Zhao, Z., Jiao, L., Zhao, J., Gu, J., Zhao, J.: Discriminant deep belief network for high-resolution SAR image classification. Pattern Recogn. **61**, 686–701 (2017)

A Research and Application Based on Gradient Boosting Decision Tree

Yun Xi, Xutian Zhuang, Xinming Wang, Ruihua Nie$^{(\boxtimes)}$, and Gansen Zhao$^{(\boxtimes)}$

School of Computer Science, South China Normal University, Guangzhou, Guangdong, China
yunxi@m.scnu.edu.cn, nrh@scnu.edu.cn

Abstract. Hand, foot, and mouth disease(HFMD) is an infectious disease of the intestines that damages people's health, severe cases could lead to cardiorespiratory failure or death.

Therefore, severe cases' identification of HFMD is important. A real-time, automatic and efficient prediction system based on multi-source data (structured and unstructured data), and gradient boosting decision tree(GBDT) is proposed in this paper for severe HFMD identification. A missing data imputation method based on GBDT model is proposed.

Experimental result shows that our model can identify severe HFMD with a reasonable area under the ROC curve (AUC) of 0.94, and which is better than that of PCIS by 17%.

Keywords: Severe HFMD · Disease identification · Missing data Machine learning

1 Introduction

Smart health-care is an active research field. In the past decades, with the rapid development of Electronic Medical Record (EMR) and other health care digital systems, more and more studies have been conducted to make health care smarter, safer and more efficient.

With the help of data mining analysis, artificial Intelligence (AI) is widely used in lots of fields, such as inferencing disease from health-related questions via deep learning. A large amount of models with machine learning methodology have shown their superiority in improving real-time identification of heart failure, chest pathology detection and other diseases.

Hand, foot, and mouth disease (HFMD) is an infectious disease of the intestines that seriously endangers people's health, especially for children under five years old. As a self-limiting disease, the most common symptoms of HFMD are fever, general malaise, and sore throat. However, in some situation the HFMD can worsen medical conditions or cause death [1].

As the outbreak of this childhood illness become more and more severe, many studies focus on disease prevention. Consequently, identifying patients with high risk of severe HFMD is crucial for disease management.

© Springer Nature Switzerland AG 2018
X. Meng et al. (Eds.): WISA 2018, LNCS 11242, pp. 15–26, 2018.
https://doi.org/10.1007/978-3-030-02934-0_2

The data were collected from the Guangzhou Women and Children Medical Center, and cover 2532 cases for children admitted between 2012 and 2015.

The identification of severe HFMD can be modeled as a binary classification problem, which consists of four steps: data preprocessing, feature extraction, feature selection, and classification. Electronic Medical Record (EMR) data includes both structured data (inspection data) and unstructured data (admission notes, physician progress notes).

In this paper we exploit XGBoost model for severe HFMD identification and missing value processing. Since the unstructured data in EMR are natural language texts recorded by human labor, it is possible that some features may not sufficiently recorded.

This may result in missing value(e.g. fever duration, body temperature), which may have a great influence on the classification decision of the model. Therefore, imputing missing value is an important problem. For missing value processing, we propose XGBoost-impute Algorithm to impute missing values of both numerical and category features, which performs better than Simple-Fill with mean of each feature values and KNN-impute method [2].

Based on such features, XGBoost is then to applied to HFMD severe case identification.

The main contributions of this paper are summarized as follow:

- We propose real-time, automatic and efficient prediction tools based on machine learning technique, for HFMD severe case identification.
- A missing data imputation method based on XGBoost model is proposed, which gets better performance than simply imputation with mean value and KNN-impute method.
- Experimental results show that our model can identify severe HFMD with a reasonable area under the ROC curve(AUC) of 0.94, and achieves 17% gain compared to the current standard of pediatric critical illness score (PCIS).

2   Related Work

Various studies have found that machine learning aided approaches produce more sophisticated and efficient results than the conventional methods currently used in clinic.

Since the emergence of deep learning and machine learning, many researchers begin to apply machine learning tools to the medical field. For example, disease diagnosis, disease prediction [3].

The outbreak of HFMD has increased in recent years, with an increasing number of cases and deaths occurring in the most Asia-Pacific Region countries [4].

In 2007, Chen et al. [5] analyzed the HFMD data from 1998 to 2005 in Taiwan manually and concluded that most cases occurred in children under age of 4 and had a higher rate of enterovirus 71 infection.

Previous studies have shown that identification of severe case is necessary, Yang et al. [6] analyzed risk factors associated with occurrence of severe HFMD through manual analysis.

In [7], Sui M et al. used laboratory parameters and logistic regression to establish a forecasting model of severe cases, reached 0.864 AUC score. In this study, we used both EMR data and inspection parameters for analysis.

Zhang et al. [8] developed a model to identify severe HFMD case with MRI-related variables. However, to collect MRI-related variables is expensive and time-consuming, which makes the model hard to be widely applied in reality. Therefore, this paper will promote a more practical, efficient and real-time prediction method. We collect variables from EMR system, which is so convenient to get that our model can be wildly used in medical field. Without the model for identification, hospital can not adopt specific medication until the patients progress to severe form of HFMD. Fortunately, our model can identify the patients who are going to develop a severe form in the very early stage and then the hospital can carries on the prevention and treatment.

3   Methodology

Disease identification problems usually use logistic regression and decision tree models. Logistic regression model cannot naturally explain the importance of features, which physicians are particularly concerned. Decision tree model often lead to overfitting, and pruning operation can't solve this problem well.

Gradient boosting decision tree(GBDT) is an additive model. It has a good strategy to prevent overfitting, and feature important on classification can easily calculated.

In this section, we provide a brief comparison between XGBoost model [9] and traditional GBDT. Then we introduce our imputation algorithm, namely, XGBoost-Impute. Next, we exploit XGBoost model to impute missing value and classification.

3.1   Gradient Boosting Decision Tree

In [10], Jerome proposed Gradient Boosting Decision Tree(GBDT) which is a veteran ensemble model. It has been considered as a benchmark in related works, and has been widely used in applications and even data analysis competitions.

In general, GBDT is a process of fitting the residuals and superimposing them on F, i.e., as shown in Eq. 1. In this process, the residuals gradually diminish, while the loss approaches to the minimum simultaneously.

$$F = \sum_{i=0}^{K} f_i. \tag{1}$$

The cost function of GBDT model can be simplified as shown in Eq. 2.

$$\mathcal{L} = \underbrace{\sum_{i=1}^{N} L(y_i, F(x_i))}_{\text{Training loss}} + \underbrace{\sum_{k=1}^{K} \Omega(f_k,)}_{\text{Regulatization}}, \tag{2}$$

where L is the training loss for the samples, including, absolute error, mean squared error and so on. Ω is the regularization function that penalizes the complexity of f_k. Comparatively, Ω is not considered in Jerome's model.

XGBoost is an extensible, end-to-end tree boosting system that is an effective implementation of gradient-boosted decision trees designed for speed and performance [9]. Compared with the traditional GBDT model, XGBoost mainly has the following advantages:

– XGBoost adds regularization term($\sum_{k=1}^{K} \Omega(f_k)$) to the cost function to control the complexity of the model and helps to smooth the final learnt weights to avoid over-fitting [9]. As shown in Eq. 3, T is the number of leaf nodes and $\sum_{j=1}^{T} w_j^2$ term is the sum of L2 modes of the score output on each leaf node.

$$\Omega(f_t) = \gamma T + \frac{1}{2}\lambda \sum_{j=1}^{T} w_j^2. \tag{3}$$

– Based on the advantage of random forests, XGBoost can reduce over-fitting and the amount of computation by using column sampling.

– XGBoost naturally accepts sparse feature format.

– The traditional GBDT uses only the first derivative information in the optimization. XGBoost performs the second-order Taylor expansion on the cost function, using both the first derivative(G_j) and the second derivative(H_j), as shown in Formula (4).

$$\tilde{\mathcal{L}}^{(t)} = \sum_{j=1}^{T} \left[G_j w_j + \frac{1}{2}(H_j + \lambda)w_j^2 \right] + \gamma T. \tag{4}$$

3.2   Impute Missing Values by XGBoost

Since the unstructured data in EMR are recorded by human labor, some features may be omitted. Therefore, imputing missing values is important. To this end, we consider this problem to be a supervised learning problem by treating numerical feature and category feature as regression and classification problems respectively. The motivation behind the proposed scheme is that XGBoost naturally accepts sparse features, as shown in Formula 2. Hence, we can apply it to impute the missing values.

Algorithm 1. XGBoost-Impute

Input: D, Extracted dataset
Input: $T = \{t_1, t_2, \cdots, t_i\}$, $t_i \subseteq \{numerical, category\}$, missing value feature
typeset
for $i = 1$ **to** n **do**
  if $T[i] == numerical$ **then**
    $new\_D[i] \leftarrow$ XGBRegression(D[i]) **if** $T[i] == category$ **then**
    $new\_D[i] \leftarrow$ XGBClassificatoin(D[i])
end
Output: Filled Dataset D_f

The detailed procedures of XGBoost-Impute are illustrated in Algorithm 1, as shown in next page. Specifically, in each step of iteration, we first select the features of a set T with missing values, i.e., the body temperature. Then, we employ XGBRegression method to predict the missing values of body temperature, i.e., T_i. If T_i is a category feature, we use XGBClassification to complete imputing.

3.3 How XGBoost Treats Missing Value?

Since XGBoost naturally accepts sparse feature format, we can directly feed data in as a sparse matrix, by using the algotithm *Sparsity-aware Split Finding* in [9] , where the missing value can be optimally imputed based on the reduction of training loss (see Algorithm 2). Specifically, XGBoost enumerate the missing values to the left and right of the splitting point, and then choose the optimal one based on the gain derived as Eq. 5.

$$Gain = \frac{G_L^2}{H_L + \lambda} + \frac{G_R^2}{H_R + \lambda} - \frac{(G_L + G_R)^2}{H_L + H_R + \lambda} - \gamma. \tag{5}$$

3.4 XGBoost Based Classification

For classification, we use the XGBoost model. Specifically, the proposed tree ensemble model is based on the decision rules of the trees with leaves for classification. As shown in Fig. 1, the final decision is obtained by summing the predicted results of both tree 1 and tree 2.

Obviously, the proposed tree ensemble classification is relatively strong since it is explicitly composed of many weak classifiers. Compared to the general model, such as logistic regression, it performs poorly if the best fit function is between quadratic function and cubic function.

4 Experiment

4.1 Experiment Setup

In experimental setup, we consider the data from inpatient EMR between March 2012 and July 2015 at Guangzhou Women's and Children's Medical Center.

Algorithm 2. Sparsity-aware Split Finding [9]

Input: I, instance set of current node
Input: $I_k = \{i \in I | x_{ik} \neq missing\}$
Input: d, feature dimension
$gain \leftarrow 0$
$G \leftarrow \sum_{i \in I} g_i, H \leftarrow \sum_{i \in I} h_i$
for $k = 1$ **to** m **do**
    $G_L \leftarrow 0, H_L \leftarrow 0, G_R \leftarrow 0, H_R \leftarrow 0$
    for j in sorted(I_k , ascent order by x_{jk}) **do**
        $G_L \leftarrow G_L + g_j, H_L \leftarrow H_L + h_j$
        $G_R \leftarrow G - G_L, H_R \leftarrow H - H_L$
        $score \leftarrow \max(score, \frac{G_L^2}{H_L+\lambda} + \frac{G_R^2}{H_R+\lambda} - \frac{G^2}{H+\lambda})$
    end
    for j in sorted(I_k, descent order by x_{jk}) **do**
        $G_R \leftarrow G_R + g_j, H_R \leftarrow H_R + h_j$
        $G_L \leftarrow G - G_R, H_L \leftarrow H - H_R$
        $score \leftarrow \max(score, \frac{G_L^2}{H_L+\lambda} + \frac{G_R^2}{H_R+\lambda} - \frac{G^2}{H+\lambda})$
    end
end
Output: Split and default directions with max gain

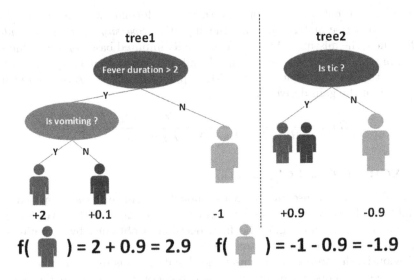

Fig. 1. The final decision for a given example is the sum of predictions from each base stump.

The data of each patient is of binary label, e.g., either mild or severe. The total number of patients is 2532, and the number of patients of severe case is 365 (14.41%). The distributions of critical numerical features that during the diagnosis(blood glucose, fever duration, platelet etc.) are shown in Fig. 2.

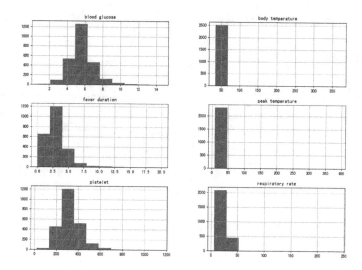

Fig. 2. Numeric data's distribution

We divide the collected data into two parts, i.e., 70% for training and 30% for testing. In the experiment, we compare the performance of the proposed XGBoost-Impute, with that of both Random Forest and Logistic Regression, by using precision, recall, F1-score and ROC curve as the metrics.

4.2   Workflow on Data Processing

Figure 3 describes the workflow of our data processing phase with raw data from EMR System. Target data is extracted from the EMR system and then filtered through data munging steps to produce useful data. After various forms of transformation(keyword extraction, key word selection, rebuild missing data, etc), the tabular data is generated.

Next, the missing values of the variables are interpolated by using three algorithms, i.e., Algorithm 1 (XGBoost-impute), Simple-Fill mean value imputation method and KNN-impute. After preprocessing, the samples in the majority groups will be processed with down-sampling to balance the data-size.

4.3   Feature Extraction

For unstructured data (mainly contains admission notes, physician progress notes), XML parser conversion is considered to generate them into medical records with natural language texts form. We extract 142 variables and 47 variables from both structured data and unstructured data respectively.

In the step of extracting features from unstructured clinical documentation(texted in Chinese), we refer to the information extraction scheme proposed

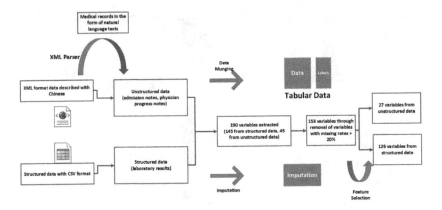

Fig. 3. Workflow on data processing

by [11]. Specifically, the approach combines medical terms extracted from Chinese context with core lexica of medical terms, an iterative bootstrapping algorithm to extract more appropriate terms [11].

By this way, the Chinese text was converted into raw tabular data including the clinical variable, the time of variable and an optional description.

4.4   Feature Selection

For the purpose of identifying patients who need more attention and resources, we first selected variables from medical examination done to admission. Then, we removed the variables with missing rates more than 20%, leaving 183 features for modeling.

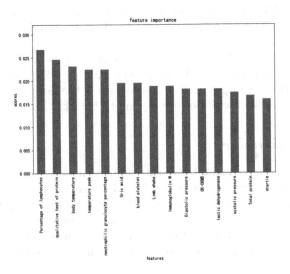

Fig. 4. Feature importance

In Fig. 4, we use XGBoost model to filter out 15 important features, to help doctors pay more attention to important illnesses. Body temperature, temperature peak, limb shake, diastolic pressure, systolic pressure come from unstructured data, and others come from structured data. This shows that EMR data and inspection data are both very important on severe HFMD case identification.

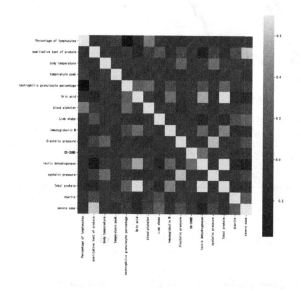

Fig. 5. Feature correlations

Figure 5 shows the correlation between top 15 features with our target. Qualitative test of protein, limb shake, and startle correlate strongly with severe HFMD case, but those features are not highly correlated among each others.

4.5   Experiment Result

According to the ROC curve as shown in Fig. 6, the three models are all based on 183 variables, the first model (XGBoost model) identifies severe HFMD with a reasonable area under the ROC curve (AUC) of 0.94, obtains a stronger performance than either the random forest(RF) model (with an AUC of 0.88) or the logistic regression(LR) model with an AUC of 0.86.

We also compare with PCIS currently used in clinic, to evaluate the performance of proposed model.

The experimental results show that our XGBoost model outperform PCIS with a 17% gain, and is superior to the conventional ones. We also list the performance comparison among the proposed XGBoost, random forest and logistic regression, in Table 1, where 'X' stands for XGBoost-Impute method, while M and K represent the Simple-Fill mean imputation and K-th Nearest Neighbor(KNN), respectively.

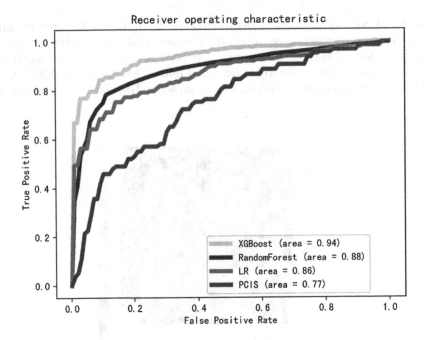

Fig. 6. Comparison of ROC with different models

For fairness of comparison, we use F1-Score as a metric by taking the precision and recall into account, to evaluate the performance of imputation method, as shown in Formula 6.

$$F1 = 2 \cdot \frac{precision \cdot recall}{precision + recall}. \tag{6}$$

According to the results from Table 1, the XGB+M model yields the best Precision(0.9096) and F1-Score(0.8691), in comparison with both LR and RF. The

Table 1. Performance of different algorithms and imputer

Algorithm	Precision	Recall	F1-score
XGB+X	**0.9096**	0.8354	**0.8691**
XGB+M	0.8931	**0.8408**	0.8652
XGB+K	0.0101	0.8104	0.8579
LR+X	0.8089	0.7812	0.7925
LR+M	0.8304	0.7563	0.7907
LR+K	0.8161	0.7592	0.7845
RF+X	0.8995	0.7757	0.8314
RF+M	0.8893	0.7393	0.8053

result is reasonable since XGBoost-impute take both feature type and correlation between different features into consideration.

5   Conclusion

In this paper, we proposed a real-time, automatic and efficient prediction tool for severe hand, foot, and mouth disease (HFMD) identification. Our model extracted features from EMR system, which is so convenient to get that our model can be wildly used in medical field. We further show that both EMR and laboratory parameters are very important for identification of severe HFMD.

In order to solve the insufficient clinical records, we presented a missing value imputation method, namely, the XGBoost-Impute method. In fact, we not only used LR,RF and XGBoost classifier, but also tried SVM, DNN which did not mention above, because of their poor performance. In the future, we will try more deep learning methods such as convolutional neural network (CNN) and so on.

The experimental results show that the proposed XGBoost model with missing value imputation outperforms the current clinic standard of PCIS, and the conventional schemes, validating the superiority of the proposed model design on the identification of severe HFMD patients.

Acknowledgements. We would like to thank Guangzhou Women and Children Medical Center, for supporting clinical data during this research.

This research is supported by national Natural Science Foundation of China (NSFC), grant No. 61471176, Pearl River Nova Program of Guangzhou, grant No. 201610010199, Science Foundation for Excellent Youth Scholars of Guangdong Province, grant No. YQ2015046, Science and Technology Planning Project of Guangdong Province, grant Nos. 2017A010101015, 2017B030308009, 2017KZ010101, Special Project for Youth Top-notch Scholars of Guangdong Province, grant No. 2016TQ03X100, and also supported by Joint Foundation of BLUEDON Information Security Technologies Co., grand No. LD20170204 and LD20170207.

References

1. Solomon, T., Lewthwaite, P., et al.: Virology, epidemiology, pathogenesis, and control of enterovirus 71. Lancet Infecti. Dis. **10**(11), 778–790 (2010)
2. Zhang, S.: Nearest neighbor selection for iteratively KNN imputation. J. Syst. Softw. **85**(11), 2541–2552 (2012)
3. Ravì, D., Wong, C., et al.: Deep learning for health informatics. IEEE J. Biomed. Health Inf. **21**(1), 4–21 (2017)
4. Xing, W., Qiaohong, L., et al.: Hand, foot, and mouth disease in china, 2008–12: an epidemiological study. Lancet Infect. Dis. **14**(4), 308–318 (2014)
5. Chen, K.T., Chang, H.L., et al.: Epidemiologic features of hand-foot-mouth disease and herpangina caused by enterovirus 71 in Taiwan 1998–2005. Pediatrics **120**(2), e244–e252 (2007)

6. Yang, T., Xu, G., et al.: A case-control study of risk factors for severe hand-foot-mouth disease among children in ningbo, China, 2010–2011. Eur. J. Pediatr. **171**(9), 1359–1364 (2012)
7. Sui, M., Huang, X., et al.: Application and comparison of laboratory parameters for forecasting severe hand-foot-mouth disease using logistic regression, discriminant analysis and decision tree. Clin. Lab. **62**(6), 1023–1031 (2016)
8. Zhang, B., Wan, X., et al.: Machine learning algorithms for risk prediction of severe hand-foot-mouth disease in children. Sci. Rep. **7**(1), 5368 (2017)
9. Chen, T., Guestrin, C.: XGBOOST: a scalable tree boosting system. In Proceedings of the 22nd ACM SIGKDD International Conference on Knowledge Discovery and Data Mining, pp. 785–794. ACM (2016)
10. Friedman, J.H.: Greedy function approximation: a gradient boosting machine. Ann. Stat. 1189–1232 (2001)
11. Xu, D., Zhang, M., et. al.: Data-driven information extraction from chinese electronic medical records. PloS one **10**(8), e0136270 (2015)

Multilingual Short Text Classification via Convolutional Neural Network

Jiao Liu, Rongyi Cui, and Yahui Zhao$^{(\boxtimes)}$

Department of Computer Science and Technology, Yanbian University,
977 Gongyuan Road, Yanji 133002, People's Republic of China
{cuirongyi,yhzhao}@ybu.edu.cn

Abstract. As multilingual text increases, the analysis of multilingual data plays a crucial role in statistical translation models, cross-language information retrieval, the construction of parallel corpus, bilingual information extraction and other fields. In this paper, we introduce convolutional neural network and propose auto-associative memory for the fusion of multilingual data to classify multilingual short text. First, the open-source tool word2vec is used to extract word vector for textual representation. Then, the auto-associative memory relationship can extract the multilingual document semantic, which need to calculate the statistical relevance of word vector between different languages. A critical problem is the domain adaptation of classifiers in different languages and we solve it by transforming multilingual text features. In order to fuse a dense combination of high-level features in multilingual text semantics, we introduce convolutional neural network into the model, and output classification prediction results. This model can process multilingual textual data well. Experiments show that convolutional neural network combined with auto-associative memory improves classification accuracy by 2 to 6% in multilingual text classification, compared to other classic models. Furthermore, the proposed model reduces the dependence of multilingual text on the parallel corpus, thus have good expansibility for multilingual data.

Keywords: Auto-associative memory
Convolutional neural network · Word embedding · Local perception

1 Introduction

With the internationalization of information communication, more and more business institutions are doing international activities. For example, government departments often require to classify documents in different languages, the international e-commerce website needs to classify and recommend the goods described in multiple languages, and digital libraries are supposed to provide multilingual information services based on multilingual classification processing for various language users. Under such circumstances, automatic classification technology of different language documents is particularly important.

© Springer Nature Switzerland AG 2018
X. Meng et al. (Eds.): WISA 2018, LNCS 11242, pp. 27–38, 2018.
https://doi.org/10.1007/978-3-030-02934-0_3

It is difficult for feature extraction of short text due to the single short text has few words and its content is scarce in semantics. Recently, multilingual short text classification has received increasing attention, and many algorithms have been proposed during the last decade. Parallel corpus-based methods [12] are usually categorized on the basis of single language documents, and then the corresponding language documents are divided into the same category. CL-ESA is an extension of parallel corpus-based methods [13,14], which represents documents by similarity vectors between documents and indexed document sets. Gliozzo *et al.* [2] categorize text across languages of English and Italian by using a comparative corpus based on latent semantic analysis and classify the document in low-dimensional projection space. Hanneman *et al.* [4] improve the accuracy of classification by constructing syntactic of full-text translation algorithm. He [6] takes advantage of a bilingual dictionary, WordNet, to translate text feature vectors, and then study the similarity between two Chinese and English texts. Tang *et al.* [15] put forward generalized vector space model for cross-lingual text clustering. Faruqui *et al.* [1,3] exploit the canonical correlation analysis for cross-lingual text analysis to find the largest correlation coefficient in the two language spaces, which aim at building a cross-language bridge. Luo *et al.* [10] improve the method of partial least squares to establish the latent intermediate semantics of multiple languages to classify the text across language in the potential space. Kim [8] proposes a convolutional neural network with multiple convolution kernels to classify texts(TextCNN).

Comparing with previous research, we introduce multilingual associative memory to extend convolutional neural network model. The model is constructed by the auto-associative memory relationship among multilingual languages, by the way counting the co-occurrence degree of multilingual words in the parallel corpus and the spatial relationships embedded in their corresponding words. We deal with word items at the document level. In order to exploit multilingual resources available, those matrixes which have the same semantic are amalgamated into single one. Local perception and weight sharing theory of convolutional neural networks could be applied to classify different documents under the combined language space.

Our work adopts TextCNN to address the issue of text characteristics, which fix convolution and learn the characteristics of multiple-word phrases from a combination of different convolution kernels of different lengths. According to the different characteristics of the deep neural network layer, the TextCNN model is extended with the superposition network layer. The experiment demonstrates that the method can merge the language space of the document, and improves the accuracy of the classification effectively.

2   Related Work

Word2Vec is a tool for computing word vectors based on a large-scale corpus, which is proposed by Mikolov [11]. It includes two structures, CBOW and Skip-Ngram, as shown in Fig. 1.

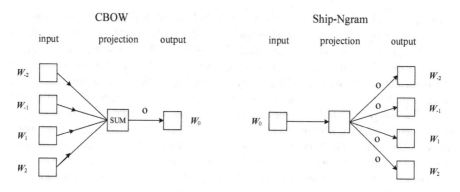

Fig. 1. CBOW and Skip-Ngram models

The model consists of three layers, including an input layer, projection layer and an output layer. For example, the input layer of the CBOW model is composed of $2k$ word vectors in the context of the current word w_0, and the projection layer vector is accumulated by these $2k$ vectors. The output vector could be corresponding with a Huffman tree, every word in the corpus is supposed to be the leaf node, which reference occurrence frequency of each word as weight. Providing that the path from root node to the leaf node is used to represent the word vector of the current word, the goal of this model is to maximize the average logarithmic likelihood function L

$$L = \frac{1}{T} \sum_{t=k}^{T-k} \log p(w_t | w_{t-k}, \cdots w_{t+k}). \tag{1}$$

Equation (1) could be regarded as the prediction of the current word w_t under the context of $w_{t-k}, \cdots, w_{t+k}$. In order to improve training efficiency, the algorithm based on Negative Sampling is proposed, which is suitable for large-scale corpus training. Note that each word is expressed as a word vector in the algorithm. It could be found that the difference between words 'France' and 'Paris' is almost the same as that obtained by 'Italy' minus 'Rome', which proved that the semantic relation between words might be represented by vector linearly.

3   Multilingual Auto-associative Memory

3.1   Co-occurrence Vocabulary Based on Word Embedding

A corpus-based approach to obtain the word co-occurrence is derived from the distribution hypothesis in [5]. In a large corpus, the distribution of words in each document can be indicated as vectors, and the degree of association between words and words can be calculated with this vector. In parallel corpus, if two words belonging to two languages appear in the same semantic document, in general, it can be deduced that the two words have high semantic relevance. By this relationship, we can find that each word in the vocabulary has a word with the greatest correlation in another language. As a result, we obtain a co-occurrence vocabulary. Not all co-occurrence word pairs can be translated into each other, notwithstanding, they are highly correlated in semantics [9], which has been proved to be suitable for cross-language document retrieval and similarity computation. The computation principle of word2vec reflects the co-occurrence relation between words and its local context when calculating word vectors. The semantic between words could be directly measured by the distance of vector space, which proves that the representation of words might be directly transformed between vectors linearly. However, in different languages, even if there is a translation relationship between source language document d_s and target language document d_t. The words have a similar distribution in the corpus of their respective languages, but they do not have contextual relations, that words in d_s and d_t could not be calculated in the same word context window. Therefore, they could only follow the semantic relations like that "v(France)-v(法国)$\approx v$(Italy)-v(意大利)". By combining the co-occurrence calculation of words and the distance of the word embedding vector, the method generating co-occurrence word pairs is presented as follows:

$$L_{s|t} = \{<x,y> | x \in V_s \wedge y \in V_t \wedge y = T - index(x)\}. \tag{2}$$

where

$$T - index(j) = \arg\max_{i \in V_t}(v_i^T * v_j + \alpha e^{m_{ij}}), j \in V_s. \tag{3}$$

V_S and V_T represent the source language and target language of word items in the document set, respectively. The value of $T - index(j)$ is the number of the word item in the target language co-occurring with the j-th words in the source language. Moreover, v_i and v_j represent the i-th and j-th word vectors of each two languages, α stands for empirical parameters and e is the base of natural logarithms. Furthermore, $m_{i,j}$ indicates the number of occurrences of the two words in the parallel corpus.

3.2 Auto-associative Memory

Auto-associative memory refers to the form or concept of two types of data related to each other have the specific form of knowledge stored in memory. According to this concept, the co-occurrence word table is used as a bridge between the data of two languages. In this paper, the auto-associative memory method is applied to the neural network, which can be described as:

$$f : R^{|v_s| \times n} \rightarrow R^{|v_t| \times n}. \tag{4}$$

$$f(v_{j|s}) = v_{T-index(j)|t}. \tag{5}$$

As shown in Eq. (4), the source language vector of the input can be associated with the correlation vector of the target language. In multilingual tasks, it is only necessary to establish a co-occurrence list among different languages, and the vectors of any language can be associated with any other language.

4 Multilingual Text Classification of Convolutional Neural Network Based on Auto-associative Memory

In the convolutional neural network, the combination of convolution kernel of different sizes can learn the expression way of phrases with a different number of words. Word2vec can generate precise word vector expression, but semantic information of a word needs to be calculated with a whole vector.

As the co-occurrence vocabulary can be regarded as the generation memory of multilingual semantics, text semantic in the source language can be associative to other target languages by mnemonic mapping. The input of the single language is extended to the input memory that contains multiple languages through the auto-associative memory. All samples in different languages space can be calculated by auto-associative memory in multidimensional space. The data generated from the associative memory relationship have a complementary semantic relationship, and the convolutional neural network can be used to extract the salient features and ignore the information that has less effect on the classifier. Therefore, this paper proposes a multilingual text categorization algorithm based on auto-associative memory and convolutional neural network.

4.1 Convolutional Multilingual Mapping Layer

According to the auto-associative memory relationship, the word has a corresponding semantic word in any other language, and each sample data of the input is supposed to be extended as shown in the following Fig. 2.

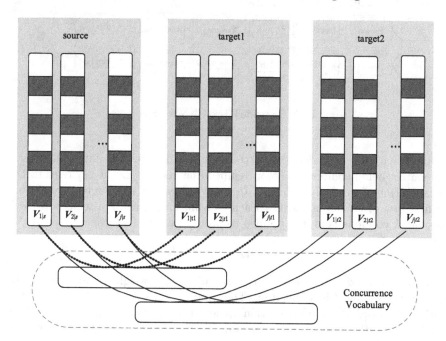

Fig. 2. Language extension based on auto-associative memory

The first frame on the right side of the graph is the output word vector of the source language text, and the co-occurrence vocabulary could find the semantic association words of each word corresponding to the target language. We look for the word vector of the associated word in the same position for the target language space. The text matrix composed of a splice source language text matrix and auto-associative memory is expressed by Eq. (6):

$$d = \begin{bmatrix} d_{i|s} \\ d_{i|t_1} \\ d_{i|t_2} \end{bmatrix}, d_{i|s}, d_{i|t_1}, d_{i|t_2} \in R^{m \times k}, d \in R^{3m \times k} \tag{6}$$

This model adapt to any language resources. The semantic mapping vector based on the input language is taken as a memory to help the model generate multilingual space, and convolutional neural network can extract the local features in that space.

4.2   Convolutional Neural Network Structure

The proposed convolutional neural network model is shown in Fig. 3:

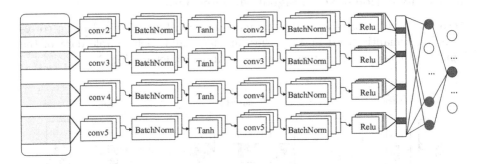

Fig. 3. Extended TextCNN models

As shown, the extended model consists of 9 layers. The input layer is a text matrix being composed of word embedding vectors. Let $d_{i:i+j}$ refer to the concatenation of words $d_i, d_{i+1}, \cdots, d_{i+j}$, a window that moving backward from the first row of the input matrix. Each convolution kernel is a window with h rows and k columns, and it is applied to produce a new feature.

$$s_i = w * d_{i:i+h-1} + b, w \in R^{h \times k}. \tag{7}$$

Among them, d represents input, w represents the weight parameter of convolution kernel, and b is a bias item. The window width k is consistent with the width of the word vector. Convolution kernel with row 2 represents it can extract phrase information composed of two words. In the same way, the convolution kernel of other lengths can also represent the extraction of phrase characteristics of the corresponding number of words.

The calculation process of the batch normalization layer is calculated as follows:

$$\mu_\beta = \frac{1}{m} \sum_{i=1}^{m} s_i, \tag{8}$$

$$\sigma_\beta^2 = \frac{1}{m} \sum_{i=1}^{m} (s_i - \mu_\beta)^2, \tag{9}$$

$$\tilde{s}_i = \frac{s_i - \mu_\beta}{\sqrt{\sigma_\beta^2 + \varepsilon}}, \tag{10}$$

$$z_i = \gamma \tilde{s}_i + \beta. \tag{11}$$

where, μ_β is the mean value of the input, and σ_β^2 is the variance of input, m is the quantity of input data. Equation (10) is used to compress the distribution range of data [7] so that results have fixed mean and variance. Neural network is a parameterized model essentially, different data distributions are supposed to be better fitted with different parameter models. When the distribution gap between training data and test data is large, the performance of the model will be greatly reduced. In addition, as the number of network layers increases, the impact of changes in lower-layer network parameters on higher-level networks will increase. Normalization of data can solve this problem, but the expression ability of network will be weakened. So Eq. (11) is used to zoom and translate normalized data.

Activation layer functions is f_{tanh} and f_{relu}.

$$f_{\tanh}(z) = \frac{e^z - e^{-z}}{e^z + e^{-z}}, \tag{12}$$

$$f_{relu}(z) = \begin{cases} z & (if z > 0) \\ 0 & (if z < 0) \end{cases}. \tag{13}$$

The output range of $f_{tanh}(z)$ is $[-1, 1]$, and the function value is saturated when the absolute value of the input data X is very large, which makes the function close to the biological neuron and can suppress or stimulate the information transmission of the neuron. And since it has a mean value of 0, it converges faster. Relu function solves the problem saturation function encountered, that is, when the function value is saturated to 0 or 1, the network layer's derivative is close to 0, so it will affect the reverse transfer of the gradient. In high layer, Relu function is suitable to ensure the transmission of gradient and alleviate the problem of gradient disappearance.

The Chunk-Max Pooling method is adopted in pooling layer, which means to divide the vectors into equal length segments. After that, only the most significant eigenvalues of each subsegment is preserved.

The last layer is the output layer of the classification result, which combined the full-connection layer with the softmax layer constitutes a softmax regression classifier. Assumed that the convolution layer, the activation layer and the pool layer can map the vector from the original input to the hidden layer feature space, then the function of the full connection layer shall map the distributed feature vector in the hidden layer space to the sample label to complete the classification task.

The formula for the softmax function is:

$$p_i = \frac{e^{y_i}}{\sum\limits_k e^{y_k}}, \tag{14}$$

where y_i represents the output of the i-th unit on the previous layer, and the value of p_i is the output of the i-th neuron on the output layer, which represents the probability that the classification label belongs to the i-th class.

5 Experiments

5.1 Datasets

The experimental data set of this paper is collected from a multilingual document management system project, including 13 categories scientific abstracts which contain more than 90,000 texts in Chinese, English, and Korean. Each language contains more than 30,000 texts, which form a translation corpus for content alignment. The data set is randomly divided into ten parts, of which are used as the training set, and the rest are used as the test set, repeat it and take the average value of the experimental results.

5.2 Experimental Setup and Baselines

We use accuracy and cross entropy to verify the performance of the proposed model. The concept of cross entropy comes from information theory. Suppose that the same sample set has two kinds of label probability distribution p and q. The cross entropy represents the average encoding length from the truly distributed p to the error distribution q, so that the cross entropy loss function can measure the similarity between the p and the q. According to the duality of entropy, when p and q are equally distributed, the following formula is minimized.

$$H = -\sum_i p_i \log_2 q_i. \tag{15}$$

For the dataset, we compare our method against state-of-the-art methods including Bilingual Word Embedding (BWE) [17], Canonical Correlation Analysis (CCA) [1], Machine Translate (MT) [4].

(1) Bilingual Word Embedding (BWE) disarranges the mixed training and use the TextCNN algorithm as the classifier to measure the effect of this algorithm.

(2) Canonical Correlation Analysis (CCA) map two languages into a single space for classification and perform correlation calculations on the word vectors between the two languages to obtain a linear transformation matrix between the word vectors. We average classification results between each two languages pairs.

(3) Based on Machine Translate (MT) model, Google translation is used to translate Chinese and Korean abstracts into English for training.

(4) In this paper, multilingual text classification of convolutional neural network based on auto-associative memory (Me-CL-CNN) has built convolution kernel set in [2, 3, 4, 5]. Proper phrase expression can give expression to semantic combination and weaken unimportant features. The depth of convolution kernel is 64, and the dropout rate of hidden nodes is equal to 0.5. According to empirical value. We choose L2 regularization method and set the regularity coefficient equal to 0.05.

5.3   Accuracy for Different Embedding Dimensions

In different Natural Language Processing tasks, there are different requirements for the length of the word vector, so the purpose of the first experiment is to determine the length of the word vector that fits the problem of short text classification. The TF-IDF algorithm is used to weigh word vectors to get text vectors [16] as input of classifier commonly used in machine learning, including SVM, KNN and RBF.

To tell the significant difference in accuracy of different embedding dimensions, We first compare the classifer in terms of classification accuracy on our Chinese datasets. The results are shown in Table 1, and the best dimension of each classifer is highlighted in bold.

Table 1. Classification accuracy (%) for different embedding dimensions

Embedding dimensions	Accuracy (%)		
	KNN	SVM	RBF
50	59.83	68.03	65.21
100	61.50	70.91	67.45
150	62.16	71.79	**69.83**
200	**62.24**	**72.03**	69.69
250	62.24	72.01	69.67

As can be seen, the classification accuracy of the text increases slowly with the increase of the word embedding dimension, which indicates that the higher dimension of the word embedding is more able to express the semantic information. When the vector dimension exceeds 200, the classification performance will no longer continue to increase. Without loss of generality, we set length of input word embedding is 200 for our experiment considering classfication performance.

5.4 Experiments and Analysis of Results for Multilingual Text Classification

We compare our method against the selected baselines in terms of classification accuracy, and the results of the experiment are shown in Table 2:

Table 2. Classification accuracy (%) and cross entropy on datasets

Method	Result evaluation indexes	
	Cross entropy	Accuracy (%)
BWE	0.97	78.53
CCA	0.64	80.8
MT	1.12	76.5
Me-CL-CNN	**0.45**	**82.41**

Generally, the bilingual transformation method based on the Machine Translation and the Canonical Correlation Analysis is similar. The former is the translation of the text, and the latter uses the relevance of the word vector itself to carry out the language transformation. In a specific field of language, the Machine Translation tool is very poor in the translation of special terms in the scientific literature, so only 76.5% of the correct rate is obtained, while CCA has an average accuracy of 80.80%. The method based on BWE is slightly worse than CCA. This is because the alignment statements of the three languages produced differences in grammar and word position during the random fusion of windows, resulting in poor training of word vectors. The width of the set context window contains three kinds of words, so it will have an obvious influence on the semantic expression of the documents.

For the single label datasets, the model we proposed shows stronger performance in classification. The sensitivity of the convolutional neural network to local features makes it more capable of complying with the multilingual semantic information based on the auto-associative memory model, thus completing the classification task well. The advantage of the model is that it does not need the aid of external tools, only needs parallel corpus to obtain the relationship between multilingual features. It has good extension ability, strong generalization ability and strong portability for each language. Moreover, the model can obtain any language text to be input, and it will help the model to generate semantic category labels by using the mapping vector of the single language semantic, so this model can counteract the effect of data imbalances due to the scarcity of language resources.

6 Conclusion

This paper proposes a way constructing the semantic auto-associative memory relationship between multilingual languages in combination with the co-occurrence degree of multilingual words in the parallel corpus and the distance of word embedding. Without language restrictions, we provide a basis for the detection and fusion of multilingual text semantics. Moreover, this paper extends the text classification model of convolutional neural network and superimpose two convolution, pooling and activation layers to extract the higher level abstract semantics. The normalization layer is added to adjust and speed up the model, which extends the input of every short text by using auto-associative memory. The experiment illustrates that the auto-associative memory model and the extended convolutional neural network can extract the deep semantic information of the multilingual feature, which can be performed in a particularly efficient way for the classification of multilingual short text.

Acknowledgment. This research was financially supported by State Language Commission of China under Grant No. YB135-76. We would like to thank editor and referee for their careful reading and valuable comments.

References

1. Faruqui, M., Dyer, C.: Improving vector space word representations using multilingual correlation. In: Proceedings of the 14th Conference of the European Chapter of the Association for Computational Linguistics, pp. 462–471 (2014)
2. Gliozzo, A., Strapparava, C.: Exploiting comparable corpora and bilingual dictionaries for cross-language text categorization. In: Proceedings of the 44th Annual Meeting of the Association for Computational Linguistics, pp. 553–560. Association for Computational Linguistics (2006)
3. Guo, J., Che, W., Yarowsky, D., Wang, H., Liu, T.: Cross-lingual dependency parsing based on distributed representations. In: Proceedings of the 53rd Annual Meeting of the Association for Computational Linguistics, Long Papers, vol. 1, pp. 1234–1244 (2015)
4. Hanneman, G., Lavie, A.: Automatic category label coarsening for syntax-based machine translation. In: Proceedings of the Fifth Workshop on Syntax, Semantics and Structure in Statistical Translation, pp. 98–106. Association for Computational Linguistics (2011)
5. Harris, Z.S.: Mathematical structures of language. In: Tracts in Pure and Applied Mathematics (1968)
6. He, W.: Research on wordNet based Chinese English cross-language text similarity measurement. Master's thesis, Shanghai Jiao Tong University (2011)
7. Ioffe, S., Szegedy, C.: Batch normalization: accelerating deep network training by reducing internal covariate shift. arXiv preprint arXiv:1502.03167 (2015)
8. Kim, Y.: Convolutional neural networks for sentence classification. arXiv preprint arXiv:1408.5882 (2014)
9. Liu, J., Cui, R.Y., Zhao, Y.H.: Cross-lingual similar documents retrieval based on co-occurrence projection. In: Proceedings of the 6th International Conference on Computer Science and Network Technology, pp. 11–15 (2017)
10. Luo, Y., Wang, M., Le, Z., Lu, X.: Bilingual latent semantic corresponding analysis and its application to cross-lingual text categorization. J. China Soc. Sci. Tech. Inf. **32**(1), 86–96 (2013)
11. Mikolov, T., Le, Q.V., Sutskever, I.: Exploiting similarities among languages for machine translation. arXiv preprint arXiv:1309.4168 (2013)

12. Peng, Z.: Research of cross-language text correlation detection technology. Master's thesis, Central South University (2014)
13. Potthast, M., Stein, B., Anderka, M.: A Wikipedia-based multilingual retrieval model. In: Macdonald, C., Ounis, I., Plachouras, V., Ruthven, I., White, R.W. (eds.) ECIR 2008. LNCS, vol. 4956, pp. 522–530. Springer, Heidelberg (2008). https://doi.org/10.1007/978-3-540-78646-7_51
14. Sorg, P., Cimiano, P.: An experimental comparison of explicit semantic analysis implementations for cross-language retrieval. In: Horacek, H., Métais, E., Muñoz, R., Wolska, M. (eds.) NLDB 2009. LNCS, vol. 5723, pp. 36–48. Springer, Heidelberg (2010). https://doi.org/10.1007/978-3-642-12550-8_4
15. Tang, G., Xia, Y., Zhang, M., Zheng, T.: Cross-lingual document clustering based on similarity space model. J. Chin. Inf. Process. **26**(2), 116–120 (2012)
16. Tang, M., Zhu, L., Zou, X.C.: Document vector representation based on word2vec. Comput. Sci. **43**(6), 214–217 (2016)
17. Vulić, I., Moens, M.F.: Monolingual and cross-lingual information retrieval models based on (bilingual) word embeddings. In: Proceedings of the 38th International ACM SIGIR Conference on Research and Development in Information Retrieval, pp. 363–372. ACM (2015)

Classifying Python Code Comments Based on Supervised Learning

Jingyi Zhang[1], Lei Xu[2(✉)], and Yanhui Li[2]

[1] School of Management and Engineering, Nanjing University,
Nanjing, Jiangsu, China
jyzhangchn@outlook.com
[2] Department of Computer Science and Technology, Nanjing University,
Nanjing, Jiangsu, China
{xlei,yanhuili}@nju.edu.cn

Abstract. Code comments can provide a great data source for understanding programmer's needs and underlying implementation. Previous work has illustrated that code comments enhance the reliability and maintainability of the code, and engineers use them to interpret their code as well as help other developers understand the code intention better. In this paper, we studied comments from 7 python open source projects and contrived a taxonomy through an iterative process. To clarify comments characteristics, we deploy an effective and automated approach using supervised learning algorithms to classify code comments according to their different intentions. With our study, we find that there does exist a pattern across different python projects: *Summary* covers about 75% of comments. Finally, we conduct an evaluation on the behaviors of two different supervised learning classifiers and find that Decision Tree classifier is more effective on accuracy and runtime than Naive Bayes classifier in our research.

Keywords: Code comments classification · Supervised learning
Python

1   Introduction

Source code documentation is vital in maintenance of a system. In contrast to external documentation, writing comments in source code is a convenient way for developers to keep documentation up to date [3]. In addition, reading documents of well-documented projects is a good way to follow engineers' ideas, as well as form a good programming style for programming language learners.

While there are previous studies concerning quality analysis of source code comments, they put an emphasis on code/comments ratio or relations between

Supported by the Natural Science Foundation of Jiangsu Province of China (Grant No. BK20140611), the Natural Science Foundation of China (Grant Nos. 61272080, 61403187).

X. Meng et al. (Eds.): WISA 2018, LNCS 11242, pp. 39–47, 2018.
https://doi.org/10.1007/978-3-030-02934-0_4

source code and comments rather than comments classification, thus the amount of research focused towards the source code classification is limited, especially for python projects (most of previous work are based on Java or C/C++ programming language [5]).

However, comment characteristics are poorly studied since studying comments has several major challenges. First, it is difficult to understand comments. As comments are written in natural language, the exact meaning and scope of comments may not be easy to classify by just reading the comments. In addition, it may require a thorough understanding of the semantics of the relevant code to determine the intention of certain comments. Moreover, unlike software bugs, which have a relatively well accepted classification, there is no unified comment taxonomy based on comment content yet.

In this paper, we focus on python code comments classification. First, we sampled several source code files to contrive a taxonomy manually. Subsequently, we extracted all the comments and applied two supervised learning classifiers to train models and classify the comments automatically. After that, we evaluated behaviors on accuracy and runtime of two classifiers and analyzed classification results according to the experiment.

```
 1  # —*— coding: utf-8 —*—
 2  """
 3  Jörgen Stenarson <jorgen.stenarson@bostream.nu>
 4  """
 5  #*********************************************************
 6  # Copyright (C) 2005 Jörgen Stenarson <jorgen.stenarson@bostream.nu>
 7  # Distributed under the terms of the BSD License.
 8  #*********************************************************
 9  import types
10  from IPython.utils.dir2 import dir2
11
12  def create_typestr2type_dicts(dont_include_in_type2typestr=["lambda"]):
13      """Return dictionaries mapping lower case typename (e.g. 'tuple') to type
14      objects from the types package, and vice versa.
15
16      TODO: Should be extended for choosing more than one type."""
17      for tname in typenamelist:
18          name = tname[:-4].lower() # Cut 'Type' off the end of the name
19      return typestr2type, type2typestr
```

Fig. 1. An example of Python files

2 Background

Comments Quality Analysis. Padioleau et al. selected three popular open source operating systems written in C and studied the comments from several dimensions including *what, whom, where* and *when*. In another study of code comments analysis, Arafati and Riehle [1] assessed the comment density of open source on a large scale and showed that the comment density of active open source projects was independent of team and project size but not of project age. Howden [6] focused on the relation between comments analysis and programming errors.

Comments Classification for Java and C/C++ Projects. Pascarella and Bacchelli [7] investigated how six diverse Java OSS projects use code comments, with the aim of understanding their purpose. They produced a taxonomy of source code comments containing 16 categories. Steidl et al. [8] provided a semi-automatic approach for quantitative and qualitative evaluation of comment quality based on comment classification for the Java and C/C++ programming language. Figure 1 shows a Python file example containing both code and comments. Code comments may include references to subroutines and descriptions of conditional processing. Specific inline comments of code may also be necessary for unusual coding.

Since most of the related work are based on Java and C/C++ programming language, we want to figure out whether their taxonomy could be applied to other programming language projects. In our work, we focus on python language, which is an interpreted language in contrast to compiled language such as Java and C/C++. We aim to devise a comprehensive classification inspired by Pascarella and Bacchelli [7].

3   Methodology

We perform comment classfication with supervised learning to differentiate between different categories.

3.1   Classification Granularity

In the process of classifying the code comments manually, we find comments can be summarized as two major forms: in block and in line. We set the automated classification to work at line level since comments in the same block may work for different purposes. For example, comments on line 13 and line 16 in Fig. 1 are in the same block, but comment on line 13 gives the summary of the function while line 16 is to remind developers for future work.

3.2   Classification Technique

As different comment categories have the same underlying syntax, no parser or complier can perform comment classification based on grammar rules [8].

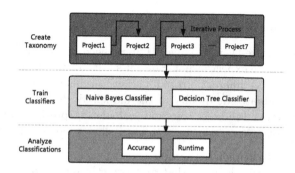

Fig. 2. Overview of our method

Hence, there's a need for heuristic approach. Since there is no simple criteria available to categorize large amount of comments, we employ supervised learning for automatic comment classification. In Particular, we choose two supervised classifiers: (1) Naive Bayes, and (2) Decision Tree. These two classifiers based on two different algorithms have different advantages and drawbacks and therefore will result in different classifications for the same data set (Fig. 2).

4   Experiment

4.1   Selection of Research Projects

This research concentrated on the programming language Python since it's one of the top 5 popular programming languages and few studies concerning code comments analysis are based on it. With respect to the samples, we select seven open-source software projects on GitHub: Pandas, Django, Pipenv, Pytorch, Ipython, Mailpile, Requests. They are all of great popularity on GitHub according to the number of stars. Details of these projects are exhibited in Table 1. To alleviate threats to the external factors such as scale and realm, these projects are of different size, contributors, stars etc.

Table 1. Details of seven open source projects

Project	Python source lines			Stars	Contributors	Commits
	Code	Comment	Ratio			
Pandas	217k	69k	32%	14k	1,168	17k
Django	208k	54k	26%	34k	1,587	26k
Pipenv	198k	71k	36%	11k	187	4k
Pytorch	136k	30k	22%	16k	640	11k
Ipython	34k	21k	61%	13k	541	23k
Mailpile	37k	6k	18%	7k	141	6k
Requests	5k	2k	38%	32k	505	5k

4.2   Classification Evaluation

When training supervised learning algorithm, we need a way to know if the model produces the results we expect or not. Since it's difficult to significantly compare the values produced by different errors functions, we instead use measures other than the loss to evaluate the quality of the model. With our study, we use the standard 10-fold cross validation method, calculating precision and recall. To define different measures we will use, we first define some notions:

* S (Sample): a set of predicted samples
* TP (True Positives): elements that are correctly retrieved by the approach under analysis

* TN (True Negatives): elements that are retrieved by the approach under analysis

One of the most common measure to evaluate the accuracy of a model is the number of samples correctly predicted divided by the total number of samples predicted. The accuracy has a range of $[0, 1]$, and a higher accuracy is better.

$$Accuracy = \frac{|TP| + |TN|}{S}$$

4.3   Categorization of Code Comments

This study concerns the role of code comments written by python software developers. Although there's similar efforts on analysis of code comments, most of the researchers put an emphasis on software development process such as maintenance and code review, ignoring some other functions of code comments when define a taxonomy. As a result, there follows our first research question:

* *RQ1. How can code comments be categorized?*

To answer this question, we conduct several iterative content analysis sessions [4]. During the first iteration, we pick 2 python files at random in Requests (reported in Table 1) and analyze all code and comments. This process resulted in the first draft taxonomy of code comments concerning some obvious categories. Subsequently, we choose 2 new python files in another project and try to categorize code comments using the previous taxonomy and improve the first draft. The following iterations are similar to the second phase: applying the newest version of taxonomy to python files in new projects and updating the current taxonomy.

Upon the categories of the code comments, we can consider the proportion of each category and investigate whether some classes of comments are of predominance in order to analyze the purpose and significance of code comments accurately and precisely. Moreover, we can discover whether there is a pattern across different projects. Therefore, the second research question is set as follows:

* *RQ2. What's the distribution of categories? Does each category distributes evenly or is there a predominant one?*

4.4   Automaticlly Categorization

Before training machine learning classifiers, we need to create training data sets as the input of classifiers first. Training data should be labeled with the classification to be learned. We randomly sampled files from seven open source projects and manually tagged 30 comments for each category corresponding to our taxonomy, thus we create 330 comments with labels in total as training data set. For implementation of Naive Bayes and Decision Tree classifiers, we use an existing python machine learning library TextBlob[1]. This naturally leads to our next two research questions:

---

[1] http://textblob.readthedocs.io/en/dev/.

* *RQ3. What are the differences between the performance of the two classifiers?*
* *RQ4. How effective are two automatic classifiers?*

5 Results and Analysis

In this section, we present our taxonomy and analyze the automatically classification results, as well as the evaluation of automated classification approach.

5.1 Taxonomy of Code Comments

To give an answer to RQ1 proposed in Sect. 4.2, we create a taxonomy containing 11 categorizes defined as follows:

```
1  """
2  >>> sps.compare('CBA')
3  False
4  In [2]: %precision 3
5  Out[2]: u'%.3f'
6  >>> [rdr.read(5), rdr.read(), rdr.read(), rdr.read()]
7  ['B', 'C', '', '']
8  >>> sps.expiration = time.time() - 5
9  True
10 """
```

Fig. 3. An example of usage comments

1. **Metadata.** The Metadata category contains two main parts: license and copyright information. This category provides the legal information about the source code and intellectual property [7]. Usually, this category only accounts for a small proportion of all comments.
2. **Summary.** This category contains a brief description of the functionability of the source code concerned. It is often used to describe the purpose or behavior of the related code.
3. **Usage.** The usage category explain how to use the code. It usually contains some examples. Figure 3 gives an example of *usage*.
4. **Parameters.** This category is used for explaining the parameters of a certain function. It's usually marked with *@param* or *:param*.
5. **Expand.** This category aims to provide more details on the code itself. Comments under this category explain in detail the purpose of a small block of code or just a line of code. It's usually a inline comment.
6. **Version.** This kind of comments identifies the applicable version of some libraries, which is of great significance for running the code.
7. **Development Notes.** This category of comments is for developers. It often covers the topics concerning ongoing work, temporary tips, and explanations of a function etc.
8. **Todo.** This type of comments regards explicit actions to be done in the future work, aiming to fix bugs or to improve the current version.

9. **Exception.** This category contains methods to handle errors and potential suggestions to prevent unwanted behaviors. It can be marked by tags such as *@throws* and *@exception*.
10. **Links.** This category refer to the linked external resources. The common tags are: *link, http://, see* etc.
11. **Noise.** Noise should be considered when classify source code comments. It contains some meaningless symbols which may be used for separation.

5.2 Behaviors of Supervised Learning Classifiers

Distribution of Categories. Figure 4 shows the distribution of the comments across the categories using two different classifiers. Numbers on x-axis correspond to the index of each category in Sect. 5.1, and y-axis indicates the proportion of different categories in each project. In this figure, red lines represent for the distribution from Naive Bayes classifier, while blue lines represent for the distribution from Decision Tree classifier. Under the same color, each line corresponds to one of the distributions of seven projects cross the categories. Since our focus here is on the differences between two classifiers, we weaken the difference among seven projects under the same classifier, and therefore use the same color to present.

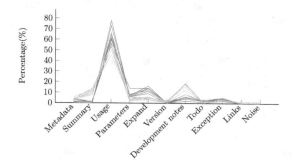

Fig. 4. Distributions of comments per project from two classifiers (Color figure online)

As for RQ2, considering the distribution of the comments under two classifiers, we see that the category *Usage* is the most predominant one which accounts for almost 80% of all code comments. This suggests that the comments in the projects are targeting end-user developers more frequently than internal developers. The second prominent category is *Summary, Expand,* and *Development Notes,* which accounts for 15% of the overall lines of comments on average. Since *Summary* and *Expand* play a significance role in code readability and maintainability, these projects are all friendly for new python learners to understand the source code.

Accuracy. To answer RQ3 and RQ4, we test two classifiers using a test data set. Figure 5 shows the accuracy of two classifiers on seven projects. Clearly the accuracy of Decision Tree classifier is higher than Naive Bayes. On average, the accuracy of Naive Bayes classifier is 80.6% while the Decision Tree classifier is up to 87.1%. The results can also be validated according to Fig. 4. Compared to blue lines from Decision Tree classifier, red lines from Naive Bayes are more concentrated, which means a minor number of comments are correctly classified by Naive Bayes classifier. For example, red lines of proportion on category *Summary* are almost 0, however, truth values are supposed to be a little bit more than that. Conversely, more comments are correctly classified by Decision Tree classifier.

Runtime. For RQ4, Fig. 6 shows classifying process runtime for 7 projects using two classifiers. It's clear that Decision Tree classifier is faster than Naive Bayes. For example, it takes 983 s (16 min) for Naive Bayes classifier to classify comments in *Pandas*, which has 69k lines of comments in total, while Decision Tree classifier only requires 61 s (1 min) to complete the process. Therefore, Decision Tree classifier is superior to Naive Bayes on runtime in this research.

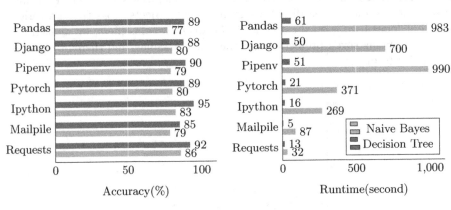

Fig. 5. Accuracy of two classifiers (Color figure online)

Fig. 6. Runtime of two classifiers (Color figure online)

6   Conclusion

This work presented a detailed approach for source code comments classification and analysis of results based on python open source projects. A supervised-learning approach for comments classification provides a foundation of comments analysis. In this study, we deployed Naive Bayes and Decision Tree classifiers to build the model and classify the code comments automatically. The results indicates that *summary* plays a prominent role in python comments, which corresponds to high maintenance and readability [2]. For those personal developers who wants to improve their coding technique or fix some bugs, these python projects are also good choices. Furthermore, we assessed behaviors of two classifiers and found that Decision Tree classifier is superior to Naive Bayes classifier in accuracy and runtime. The rationale of the differences will be discussed in our future work.

References

1. Arafati, O., Riehle, D.: The comment density of open source software code. In: 2009 31st International Conference on Software Engineering - Companion Volume, pp. 195–198, May 2009. https://doi.org/10.1109/ICSE-COMPANION.2009.5070980
2. Fjeldstad, R.K., Hamlen, W.T.: Application program maintenance study: report to our respondents. In: Proceedings GUIDE, vol. 48, April 1983
3. Fluri, B., Wursch, M., Gall, H.C.: Do code and comments co-evolve? On the relation between source code and comment changes. In: Proceedings of the 14th Working Conference on Reverse Engineering, WCRE 2007, pp. 70–79. IEEE Computer Society, Washington, DC (2007)
4. Lidwell, W., Holden, K., Butler, J.: Universal Principles of Design, Revised and Updated: 125 Ways to Enhance Usability, Influence Perception, Increase Appeal, Make Better Design Decisions. Rockport Publishers, Beverly (2010)
5. Nurvitadhi, E., Leung, W.W., Cook, C.: Do class comments aid Java program understanding? In: 33rd Annual Frontiers in Education, FIE 2003, vol. 1, pp. T3C-13–T3C-17, November 2003. https://doi.org/10.1109/FIE.2003.1263332
6. Howden, W.E.: Comments analysis and programming errors. IEEE Trans. Softw. Eng. **16**(1), 72–81 (1990)
7. Pascarella, L., Bacchelli, A.: Classifying code comments in Java open-source software systems. In: 2017 IEEE/ACM 14th International Conference on Mining Software Repositories (MSR), pp. 227–237, May 2017
8. Steidl, D., Hummel, B., Juergens, E.: Quality analysis of source code comments. In: 2013 21st International Conference on Program Comprehension (ICPC), pp. 83–92, May 2013. https://doi.org/10.1109/ICPC.2013.6613836

A Frequent Sequential Pattern Based Approach for Discovering Event Correlations

Yunmeng Cao[1,2(✉)], Chen Liu[1,2], and Yanbo Han[1,2]

[1] Beijing Key Laboratory on Integration and Analysis of Large-Scale Stream Data, North China University of Technology, Beijing, China
m849434653@163.com, {liuchen,hanyanbo}@ncut.edu.cn
[2] Data Engineering Institute, North China University of Technology, Beijing, China

Abstract. In an IoT environment, event correlation becomes more complex as events usually span over many interrelated sensors. This paper refines event correlations in an IoT environment, and proposes an algorithm to discover event correlations. We transform the event correlation discovery problem into a time-constrained frequent sequence mining problem. Moreover, we apply our approach in anomaly warning in a coal power plant. We have made extensive experiments to verify the effectiveness of our approach.

Keywords: Event correlation · Sensor event · Anomaly warning

1 Introduction

IoT (Internet of Things) allows industry devices to be sensed and controlled remotely, resulting in better efficiency, accuracy and economic benefit. Nowadays, sensors are widely deployed in industrial environments to monitor devices' status in real-time. A sensor continuously generates sensor events and sensor events are correlated with each other. These correlations can be dynamically interwoven, and data-driven analysis would be of help.

We will examine a partial but typical scenario with the example of anomaly warning in a coal power plant. In a coal power plant, there are hundreds of machines running continuously and thousands of sensors have been deployed. Events from different sensors correlate with each other in multiple ways. As Sect. 2 shows, such correlations uncover possible propagation paths of anomalies among different devices. It is very helpful to explain the root cause of an observable device anomaly/failure and make warnings in advance.

In recent years, event correlation discovery problem has received notable attentions [1–8]. A lot of research aims at identifying the event correlations from event logs. The semantic relationship is the most common type of correlations among events. It can be widely applied in the discovery, monitoring and analysis of business processes [1–4]. Motivated by the anomaly warning scenario, in this paper, we put our focus on a statistical correlation among sensor events. Such correlations provide clues for us to find possible anomaly propagation paths.

© Springer Nature Switzerland AG 2018
X. Meng et al. (Eds.): WISA 2018, LNCS 11242, pp. 48–59, 2018.
https://doi.org/10.1007/978-3-030-02934-0_5

This paper proposes an algorithm to discover event correlations in a set of sensor events. We transform the correlation discovery problem into a time-constrained frequent sequence mining problem and update traditional frequent sequence mining algorithm. Then, in a real application, we use the event correlations to make anomaly warnings in a coal power plant. Furthermore, a lot of experiments are done to show the effectiveness of our approach based on a real dataset from the power plant.

2 A Case Study

Actually, an anomaly/failure is not always isolated. Owing to the obscure physical interactions, a trivial anomaly will propagate among different sensors and devices and gradually transform itself into a severe one. Such propagation paths can be observed from the correlations among events from different sensors.

Figure 1 shows a real case about how a trivial anomaly propagates and gradually evolves into a severe failure. The events are from three sensors on different devices: coal feeder B device (CF-B device), coal mill B device (CM-B device) and primary air fan B device (PAF-B device). As a maintenance record shows, a severe failure happens at the PAF-B device (id: 939) from 2014-10-13 00:12:00 to 2014-10-13 08:00:00. The description of the failure is "#1B bearing vibration is too slow on primary air fan B device". As Fig. 1 shows, we can find the maintenance record actually doesn't denote the root cause of this failure. Actually, this failure is initially caused by the value decrease of coal feed sensor event on the CF-B device. Such value decrease will trigger the value decrease of electricity sensor event on the CM-B device after almost 9 s. Then, the value decrease of electricity sensor event will finally cause the above failure in the maintenance record after 13 s.

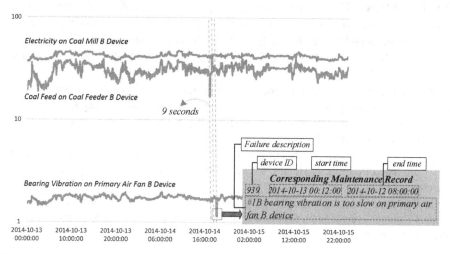

Fig. 1. Propagation path of a trivial anomaly evolving into a server failure in a primary air fan device: a real case.

This case illustrates how an anomaly propagates itself among different devices and finally evolves into a severe one. But mining such propagation paths is a challenging problem as we cannot clarify and depict complicated physical interactions among these devices. However, the correlations among sensor events show us lots of clues. From the statistical view, we can easily observe that the above three sensor event sequences in Fig. 1 are obviously correlated. If we can find such correlations, we have chances to splice them to form propagation paths. With these paths, we can make warnings before an anomaly evolves into a severe one. Furthermore, we can also explain an anomaly and find its root causes.

3   Definitions

In an industrial environment, there are thousands of sensors have been deployed. Each sensor continuously generates sensor events to reflect the devices' running status. The formal definition of sensor event is listed below.

Definition 1 (Sensor Event): A sensor event, also refers to a sensor event instance, is a 4-tuple: $e = (timestamp, eventid, type, val)$, in which *timestamp* is the generation time of e; *eventid* is the unique identifier of e; *type* is the type of e, which is represented by a sensor id; *val* is the value of e.

Definition 2 (Sensor Event Sequence): Given a sensor s_i, a sensor event sequence is a time-ordered list of sensor events $E_i = \langle e_{i,1}, e_{i,2}, \ldots, e_{i,n} \rangle$ from s_i.

Based on the case in Fig. 1, we observe that event correlation is a relationship between two sensor event sequences. It implies that a sensor event sequence always happens earlier than another one within a short enough time period.

Definition 3 (Event Correlation): Let $E_i = \langle e_{i,1}, e_{i,2}, \ldots, e_{i,n} \rangle, E_j = \langle e_{j,1}, e_{j,2}, \ldots, e_{j,n} \rangle$ be two sensor event sequences $(n \neq 0)$ generated by two sensors respectively. We denote the event correlation between E_i and E_j as $(E_i, E_j, \Delta t, conf)$, where E_i is the source, E_j is the target, Δt is the time E_j delayed to E_i, and *conf* is a measure of relationship between E_i and E_j.

To reduce the computation complexity of discovering event correlations among sensor event, we firstly symbolize the sensor event sequences by learning from a classic symbolic representation algorithm, called SAX [9]. SAX algorithm allows a numeric sequence of length n to be reduced to a symbolic sequence of length m $(m \ll n)$ composed of k different symbols. The symbolization of sensor event sequences in Fig. 1 are listed in Table 1. Herein, we take Table 1 as a running example to describe our event correlation discovery algorithm.

In this paper, our main goal is how to discover event correlations from a set of symbolic sensor events. Such event correlations can facilitate IoT applications, such as anomaly warning.

Table 1. A sample of a symbolic event sequence set (running example).

SID	t_1	t_2	t_3	t_4	t_5	t_6	t_7	t_8	t_9	t_{10}	t_{11}	t_{12}	t_{13}	t_{14}	t_{15}	t_{16}	t_{17}	t_{18}	t_{19}	t_{20}	t_{21}	t_{22}	t_{23}	t_{24}	...	t_{77}
CF		c	c	c	c	c	d	g				f	g	g	g	f	f	f	e	e	g	g	g	f	...	d
E	...	c	c	c	c	c	d		e			e	f	g	g	f	f	f	e	d	g	g	g		...	d
BV	...	...	c	c	c	c	c	d				g	g	f	g	g	f	f	f	e	f	g	g		...	b

CF: Coal Feed, E: Electricity, BV: Bearing Vibration.

4 Discovery of Sensor Event Correlations

4.1 The Overview of Our Solution

This section introduces the overview of our solution. Physical sensors deployed on devices generate sensor events continuously. Collected sensor events are symbolized by SAX so that we transform numerical sensor events into symbolic ones. This is responsible for discretizing and extracting features of input sensor events. Next, from symbolic events we discover event correlations. In this paper, the problem is transformed into time-constrained frequent sequence mining. Finally, such event correlations are applied in anomaly warning in a coal power plant.

Our main idea is to discover event correlations from a set of symbolic event sequences. To do this, we transform event correlation discovery into a frequent sequence mining problem. Essentially, symbolization is a coarse-grained description since each symbol corresponds to a segment of the original sequence. In this manner, if a sequence correlates with another one, there probably exists a frequent sequence between their symbolized sequences [9]. It inspires us to use the frequent sequence to measure event correlation. In other words, if two symbolized sequences have a long enough frequent sequence, there is a relationship between them.

One challenge is how to identify the time delay between two related event sequences shown in Fig. 1. It actually reflects how long that a sensor will be affected by its related sensor. Unfortunately, traditional frequent sequence mining algorithms cannot directly solve such problem. They only focused on the occurrence frequency of a sequence in a sequence set [10, 11]. Hence, we try to design an algorithm. It can discover a frequent sequence. The occurrences of each element in the discovered sequence are within a short time period, i.e., time delay Δt in Definition 3.

Another challenge is how to determine the target and source by a frequent sequence. If two symbolic sequences have a long enough above frequent sequence, the original sequences of them, denoted as E_i and E_j, have an event correlation $(E_i, E_j, \Delta t, conf)$. $conf$ can be computed as the ratio of the frequent sequence length to the length of S_i. Such a frequent sequence is called as a time-constrained frequent sequence in this paper.

4.2 Event Correlation Discovery

Frequent Sequence Mining. We list some related concepts about frequent sequence mining. A sequence in a sequence set $\mathcal{D}$ is associated with an identifier, called a *SID*.

A support of a sequence is the number contained in $\mathcal{D}$. A sequence becomes frequent if its support exceeds a pre-specified minimum support threshold in $\mathcal{D}$. A frequent sequence with length l is called a l-frequent sequence. It becomes closed if there is no super-sequence of it with the same support in $\mathcal{D}$. A projection database of sequence S in $\mathcal{D}$ is defined as $\mathcal{D}_S = \{\alpha | \eta \in \mathcal{D}, \eta = \beta \Diamond \alpha\}$ (β is the minimum prefix of η containing S).

Projection-based algorithms are a category of traditional algorithms in frequent sequence mining [11]. They adopt a divide-and-conquer strategy to discover frequent sequences by building projection databases. These algorithms firstly generate 1-frequent sequences F_1, where $F_1 = \{s_1 : sup_1, s_2 : sup_2, \dots, s_n : sup_n\}$, s_i is a 1-frequent sequence and sup_i is its support. This step is followed by the construction of projection database for each 1-frequent sequence. In each projection database above, they generate 1-frequent sequences F_2 and projection database of each element in F_2. The process is repeated until there is no 1-frequent sequence.

Time-Constrained Frequent Sequence Mining. In this section, we explain what a time-constrained frequent sequence is, what the differences between time-constrained frequent sequence mining and traditional frequent sequence mining are, and how to mine time-constrained frequent sequences.

Time-Constrained Frequent Sequence. Traditionally, a frequent sequence with length 1 is called 1-frequent sequence. In this paper, a 1-frequent sequence f_1 becomes a **1-Δt frequent sequence** if any two occurrence timestamps of f_1 span no more than the time threshold Δt. The concept is formalized as follows: if the occurrence timestamps of a 1-frequent sequence f_1 in a sequence set is $t_1, t_2, \dots, t_k$ respectively, and $\max_{t_i, t_j \in \{t_1, t_2, \dots, t_k\}} \{|t_i - t_j|\} \leq \Delta t$, f_1 is called a 1-Δt frequent sequence. A frequent sequence with length l, each element of which is a 1-Δt frequent sequence, is called a l-Δt **frequent sequence**. A l-Δt frequent sequence ($l \geq 1$) is the **time-constrained frequent sequence** this paper focuses on. To avoid redundant computation, our goal to discover all closed l-Δt frequent sequences.

Some examples of above concepts are shown in Table 1. Three symbolic events c with slashes in Table 1 forms a 1–60 s frequent sequence with support 3. The symbolic event sequences $ccccccd$ in bold is a 7–60 s frequent sequence with support 3.

Differences Between Time-Constrained Frequent Sequence Mining and Traditional Frequent Sequence Mining. Our approach to discovering time-constrained frequent sequences derives from traditional projection-based frequent sequence mining approaches. However, it is significantly different from them. Firstly, the ways to mine 1-frequent sequences and 1-Δt frequent sequences are quite different. In traditional algorithms, each 1-frequent sequence has a unique item, i.e., a symbol in this paper. However, our approach has to generate multiple 1-Δt frequent sequences with same symbol. The reason is they may cause different results. The above example has shown a 7–60 s frequent sequence $ccccccd$. In its projection database, our approach will generate the 1-Δt frequent sequence f with horizontal lines and the one with vertical lines both, which are presented in Table 1. The former one will generate a time-constrained frequent sequence $ccccccdff$ with support 2. The latter one will generate a result of $ccccccdf$ with support 3.

Another difference is that, a 1-Δt frequent sequence is required to identify sources and targets among sequences. To achieve this goal, we have to specify a sequence as a source beforehand and find its all targets (or specify a sequence as a target and find its all sources). As a result, once our approach generates a 1-Δt frequent sequence, its occurrence timestamp in the source sequence should be the earliest.

Time-Constrained Frequent Sequence Mining. We try to mine time-constrained frequent sequences by generating 1-Δt frequent sequences and their projection database recursively as traditional algorithms do. However, the differences between time-constrained frequent sequence mining and traditional one increase redundant computation greatly. We have to make some improvements to prevent such redundant computation.

For one thing, different 1-Δt frequent sequences may generate repeated projection database. For example, in the projection database of c with slashes in Table 1, each 1–60 s frequent sequence in bold (e.g., c, c, c, c, c and d) will be appended to c to generate projection database recursively. It will generate $2^6 - 1 = 63$ results such as cc, cd, ccd, and only $ccccccd$ is the closed one. In this case, our approach generates 1-Δt frequent sequences chronologically in a projection database. Each one corresponds to a set of SID. A 1-Δt frequent sequence will be kept, only if its SID set is not contained by a previous one.

For another, if the SID set of a kept 1-Δt frequent sequence contains that of a previous one, their projection databases may has a large repeated portion. For example, in the projection of sequence $ccccccd$ in bold in Table 1, our approach will generate three 1-Δt frequent sequences, including f with horizontal lines, g with grid, and f with vertical lines. Obviously, most part of the projection database of $ccccccdg$ (the g with grid) is contained by that of $ccccccdf$ (the f with vertical lines). Hence, our approach prevents redundant results by avoiding mining repeated portion of two projection databases twice. Our approach only extends $ccccccdg$ into $ccccccdgf$. The rest part can be extracted from the results of mining projection database of $ccccccdf$ (the f with vertical lines).

The Implementation of Our Algorithm. In this section, we illustrate the details of our algorithm on discovering event correlations. Our algorithm learns from the traditional divide-and-conquer strategy, which is a recursion of generating 1-Δt frequent sequences and their projection database. Figure 2 shows an example of the main steps. Furthermore, there are still some skills to reduce redundant computation.

The input sequence of our algorithm is regarded as a source sequence. It discovers all 1-Δt frequent sequences chronologically. In each 1-Δt frequent sequence, the item in the source sequence occurs earliest. A 1-Δt frequent sequence will be removed, if its SID set is contained by a previous one.

For each kept 1-Δt frequent sequence, our algorithm builds a projection database. Herein, if the SID set of a 1-Δt frequent sequence f is contained by a later one f', it handles f in the sequence set until f'. The rest part can be extracted from the results of handling f'.

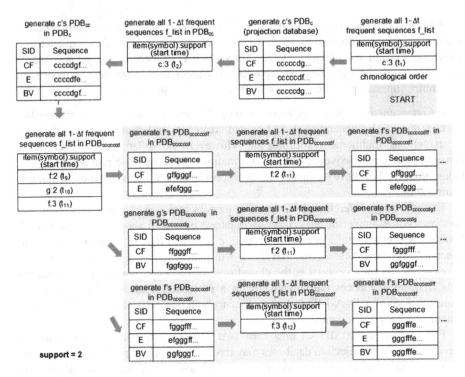

Fig. 2. An example of our algorithm by using data scenario in Table 1.

In each projection database, our algorithm recursively finds 1-Δt frequent sequences then builds projection databases in the same manner. The recursion continues until there is no 1-Δt frequent sequence. In this way, our algorithm can finds all closed l-Δt frequent sequence so that it can list all target sequences of the specified sequence. Finally, if the *conf* between the source sequence and a target sequence is no less than an input parameter *min_conf*, our algorithm outputs the event correlation.

4.3 Application of Our Approach for Anomaly Warning

An event correlation $\left(E_i, E_j, \Delta t, conf\right)$ between E_i and E_j indicates that the sensor event sequence E_i can propagates to E_j. Taking the coal feed sensor, electricity sensor, and bearing vibration sensor as example. The coal feed sensor event sequence in Fig. 1 can propagate to electricity sensor event sequence, and finally to bearing vibration sensor event sequence. And when bearing vibration sensor event sequence in Fig. 1 occurs, a failure occurs in PAF-B device. In this way, we can make an early warning when the coal feed sensor event sequence in Fig. 1 occurs.

As the above example shows, if we can learn how sensor events propagate among sensors, we can make early warnings of device failures. Splicing the event correlations is an effective way to learn the propagation paths. Each maintenance record is a 4-tuple, denoted as $r = \langle rid, start\_time, end\_time, failure\_desc \rangle$. Each maintenance record may correspond to a sensor event sequence E_0 specified by the start time, end time and failure description. Compute event correlations related to E_0. For each related event correlation $(E_i, E_0, \Delta t, conf)$, recursively find event correlations related to E_i in the same manner. The process is stopped if there are no corresponding event correlations. Then, we splice the event correlations into a propagation path. For maintenance records with same anomaly/failure type, we get the maximum common sub path as the anomaly propagation path of this type of anomaly/failure.

We denote such a common path as $P = \langle Vertices, Edges \rangle$, where *Vertices* is the set of sensors in the path, and *Edges* is the event correlations set between two vertices in *Vertices*. In fact, a propagation path P is a directed graph.

Once a sensor event sequence E_i occurs, matches it with all sensor event sequences in each path learned above. If there is any sequence matches up with E_i, make a warning and propagate E_i along with the corresponding event correlation. If the target sensor event sequence in time period Δt occurs, process it in the same way.

5   Experiments

5.1   Experiment Setup

Datasets: The following experiments use a real sensor event set from a coal power plant. The set contains sensor events from 2015-07-01 00:00:00 to 2016-10-01 00:00:00. Totally 157 sensors deployed on 5 devices are involved and each sensor generates one event per second. Firstly, we test the effects of our algorithm on some one-day sets. We analyze that how parameters affect the event correlation number. Secondly, we apply our solution in anomaly warning in a coal power plant. We divide the set into two parts. The training set is responsible for learning propagation paths. The testing set is used for making early warnings. We use real maintenance records of the plant power from 2015-07-01 00:00:00 to 2016-10-01 00:00:00 to verify our warnings. Notably, in this paper, we only consider the records with failures occurring both in training set and testing set.

Environments: The experiments are done on a PC with four Intel Core i5-2400 CPUs 3.10 GHz and 4.00 GB RAM. The operating system is Windows 7 Ultimate. All the algorithms are implemented in Java with JDK 1.8.0.

5.2   Experiment Results

Effects of Our Service Hyperlink Generation Algorithm. In this experiment, we try to verify how key parameters affect discovered service hyperlink number. Parameter *conf* threshold *min_conf* is set to be 0.8. Time threshold Δt is the significant parameter we concentrate on.

We randomly select 45 days from the whole event set, which spans one year. The selected sets are around 10% of the whole set. For each one-day set, we get 157 sensor event sequences. Each sensor event sequence is transformed into a symbolic event sequence. Then we invoke our algorithm under different values of Δt to generate service hyperlinks from 157 symbolic event sequences. We record the average number for the 45 one-day event sets in Fig. 3. Figure 3 shows that the average number of service hyperlinks increases linearly with the rise of Δt.

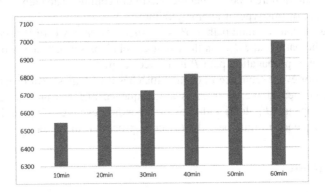

Fig. 3. Average number of service hyperlinks under different values of Δt (*min_conf = 0.8*).

Comparative Effectiveness of Different Anomaly Warning Approaches. In this experiment, sensor event sequences in the training set are input into our algorithm to compute event correlations. For each maintenance record during the time period of the training set, we compute propagation paths as follows.

Each sensor event sequence in the testing set is simulated as a stream and input into a sliding window. Our algorithm judges whether it should make a warning and propagate the sequence in the current window. After all streams are processed, we count the results to analyze the effects. Details of the process can be found in Sect. 4.3.

To measure the effects, we use the following indicators. **Precision** is the number of correct results divided by the number of all results. **Recall** is the number of correct results divided by the number of results that should have been returned.

In this part, we compare the results between our approach and other three typical approaches: LOF (Local Outlier Factor), COF (Connectivity based Outlier Factor), KNN (K-Nearest Neighbor). We compare the precision and recall of the four approaches on training datasets with different scale, and the final results are listed in Table 2.

Table 2. The precision and recall of different methods

	3-month dataset		6-month dataset		9-month dataset		12-month dataset	
	Precision	Recall	Precision	Recall	Precision	Recall	Precision	Recall
LOF	0.51	0.75	0.65	0.78	0.69	0.83	0.70	0.84
COF	0.71	0.76	0.73	0.76	0.74	0.83	0.79	0.85
KNN	0.63	0.78	0.64	0.79	0.66	0.81	0.68	0.86
Our approach	0.61	0.85	0.75	0.88	0.85	0.90	0.88	0.91

From the Table 2, our precision and recall both show a growing trend with the rise of training dataset scale. It indicates that our approach makes more correct warnings on a larger dataset. The reason is larger training set will generate more service hyperlinks, which will improve the precision and recall. On the other hand, our approach shows lower precision and recall than two typical approaches on the 3-month dataset. But it has a higher recall than the three approaches. The reason is the 3-month dataset generates small service hyperlinks and a few faults, which provide little clues for make warnings for our approach. Another reason is the three typical approaches can only detect the anomalies, which are the sensor data deviating from most one. They cannot identify faults formed by several anomalies.

6   Related Works

Event correlation discovery is a hot topic [1–8]. It can be used in various areas like process discovery [1–4], anomaly detection [5, 6], healthcare monitoring [7, 8] and so on. In the field of business process discovery, event correlation challenge is well known as the difficulty to relate events that belong to the same case. Pourmirza et al. proposed a technique called correlation miner, to facilitates discovery of business process models when events are not associated with a case identifier [1, 3]. Cheng et al. proposed a new algorithm called RF-GraP, which provides a more efficient way to discover correlation over distributed systems [2]. Reguieg et al. regarded event correlation as correlation condition, which is a predicate over the attributes of events that can verify which sets of events belong to the same instance of a process [4]. Some studies used event correlation to detect anomalies. Friedberg et al. proposed a novel anomaly detection approach. It keeps track of system events, their dependencies and occurrences. Thus it learns the normal system behavior over time and reports all actions that differ from the created system model [5]. Fu et al. focused on temporal correlation and spatial correlation among failure events. They developed a model to quantify the temporal correlation and characterize spatial correlation. Failure events are clustered by correlations to predict their future occurrences [6]. Other works applied event correlation in healthcare monitoring. Forkan et al. concentrated on vital signs, which are used to monitor a patient's physiological functions of health. The authors proposed a probabilistic model. The model is used to make predictions of future clinical events of an unknown patient in real-time using temporal correlations of multiple vital signs from many similar patients [7, 8].

Recently, some researchers focus on event dependencies. Song et al. mined activity dependencies (i.e., control dependency and data dependency) to discover process instances when event logs cannot meet the completeness criteria [12]. In this paper, the control dependency indicates the execution order and the data dependency indicates the input/output dependency in service dependency. A dependency graph is utilized to mine process instances. In fact, the authors do not consider the dependency among events. Plantevit et al. presented a new approach to mine temporal dependencies between streams of interval-based events [13]. Two events have a temporal dependency if the intervals of one are repeatedly followed by the appearance of the intervals of the other, in a certain time delay.

7 Conclusion

In this paper, we focus on mining event correlations in an IoT environment. By the event correlations, we can mine the anomaly propagation among sensors and devices in the software layer. This paper applies our approach to make anomaly warnings in a coal power plant scenario. Experiments show that, our approach can make warning of anomalies before they happen.

Acknowledgement. National Key R&D Plan (No. 2017YFC0804406); National Natural Science Foundation of China (No. 61672042); The Program for Youth Backbone Individual, supported by Beijing Municipal Party Committee Organization Department, "Research of Instant Fusion of Multi-Source and Large-Scale Sensor Data".

References

1. Pourmirza, S., Dijkman, R., Grefen, P.: Correlation miner: mining business process models and event correlations without case identifiers. Int. J. Coop. Inf. Syst. **26**(2), 1–32 (2017)
2. Cheng, L., Van Dongen, B.F., Van Der Aalst, W.M.P.: Efficient event correlation over distributed systems. In: 17th IEEE/ACM International Symposium on Cluster, Cloud and Grid Computing, pp. 1–10. Institute of Electrical and Electronics Engineers Inc., Madrid (2017)
3. Pourmirza, S., Dijkman, R., Grefen, P.: Correlation mining: mining process orchestrations without case identifiers. In: Barros, A., Grigori, D., Narendra, N.C., Dam, H.K. (eds.) ICSOC 2015. LNCS, vol. 9435, pp. 237–252. Springer, Heidelberg (2015). https://doi.org/10.1007/978-3-662-48616-0_15
4. Reguieg, H., Benatallah, B., Nezhad, H.R.M., Toumani, F.: Event correlation analytics: scaling process mining using mapreduce-aware event correlation discovery techniques. IEEE Trans. Serv. Comput. **8**(6), 847–860 (2015)
5. Friedberg, I., Skopik, F., Settanni, G., Fiedler, R.: Combating advanced persistent threats: from network event correlation to incident detection. Comput. Secur. **48**, 35–57 (2015)
6. Fu, S., Xu, C.: Quantifying event correlations for proactive failure management in networked computing systems. J. Parallel Distrib. Comput. **70**(11), 1100–1109 (2010)
7. Forkan, A.R.M., Khalil, I.: PEACE-home: probabilistic estimation of abnormal clinical events using vital sign correlations for reliable home-based monitoring. Pervasive Mob. Comput. **38**, 296–311 (2017)
8. Forkan, A.R.M., Khalil, I.: A probabilistic model for early prediction of abnormal clinical events using vital sign correlations in home-based monitoring. In: 14th IEEE International Conference on Pervasive Computing and Communications, pp. 1–9. Institute of Electrical and Electronics Engineers Inc., Sydney (2016)
9. Lin, J., Keogh, E., Lonardi, S., Chiu, B.: A symbolic representation of time series, with implications for streaming algorithms. In: 8th ACM SIGMOD Workshop on Research Issues in Data Mining and Knowledge Discovery, pp. 2–11. Association for Computing Machinery, San Diego (2003)
10. Pei, J., Han, J., Wang, W.: Constraint-based sequential pattern mining: the pattern-growth methods. J. Intell. Inf. Syst. **28**(2), 133–160 (2007)

11. Mooney, C.H., Roddick, J.F.: Sequential pattern mining: approaches and algorithms. ACM Comput. Surv. **45**(2), 1–39 (2013)
12. Song, W., Jacobsen, H.A., Ye, C., Ma, X.: Process discovery from dependence-complete event logs. IEEE Trans. Serv. Comput. **9**(5), 714–727 (2016)
13. Plantevit, M., Robardet, C., Scuturici, V.M.: Graph dependency construction based on interval-event dependencies detection in data streams. Intell. Data Anal. **20**(2), 223–256 (2016)

Using a LSTM-RNN Based Deep Learning Framework for ICU Mortality Prediction

Hanzhong Zheng[1] and Dejia Shi[2(✉)]

[1] Department of Computer Science, University of Pittsburgh, Pittsburgh,
PA 15213, USA
haz78@pitt.edu

[2] Key Laboratory of Hunan Province for New Retail Virtual Reality Technology,
Hunan University of Commerce, Changsha 410205, China
shidejia@126.com

Abstract. In Intensive Care Units (ICU), the machine learning technique has been widely used in ICU patient data. A mortality risky model can provide assessment on patients' current and when the disease may worsen. The prediction of mortality outcomes even intervenes doctor's decision making on patient's treatment. Based on the patient's condition, a timely intervention treatment is adopted to prevent the patient's condition gets worse. However, the common major challenges in ICU patient data are irregular data sampling and missing variables values. In this paper, we used a statistical approach to preprocess the data. We introduced a data imputation method based on Gaussian process and proposed a deep learning technology using LSTM-RUN that emphasizes on long time dependency relation inside the patient data records to predict the probability of patient's mortality in ICU. The experiment results show that LSTM improved the mortality prediction accuracy than base RNN using the new statistical imputation method for handling missing data problem.

Keywords: Deep learning · Recurrent neural network
Mortality prediction · Gaussian process

1 Introduction

Intensive Care Unit, also referred as ICU, is the important unit in modern hospital for saving patients with serious diseases. In the past several decades, the number of ICUs has dramatically increased by 50%. As populations in many countries age, doctors who can work in emergency and ICU could become increasingly pressed for time. For example, by the end of 2015, the number of people in China who is above 60 years old is approximately 222 million, which is 16.1% of the total population. Among them, 143.86 million people aged 65 or above, accounting for 10.5% of the total population. Now, the number of people who

© Springer Nature Switzerland AG 2018
X. Meng et al. (Eds.): WISA 2018, LNCS 11242, pp. 60–67, 2018.
https://doi.org/10.1007/978-3-030-02934-0_6

is above 85 years old in U.S. is 3 million. This number is estimated to be 9 million in 2030, which will bring great pressure to ICU. Automation may be an important solution to this problem. Under the background of rapid development of machine learning, many researchers try to use data mining and deep learning approach to study the mortality prediction problem for ICU patients.

Nowadays, machine learning techniques have been widely used in medical fields, such as the diagnosis procedure [3], gentic information extraction [6], etc. Continuous monitoring patients in ICUs can easily generate sufficient amount of medical records, which provide large enough amount of medical data to build a risk assessment model for ICU patients. This model can be used to evaluate the current patient's condition and predict the mortality probabilities at each timestamp to prevent the circumstance of patient worsen. The prediction of ICU outcomes is essential to underpin critical care quality improvement programs.

Deep learning neural network has also applied in the area of medical research: classifying the bio-medical text, disease symptoms identification and visualization, bio-medical images analysis, etc. However, Electronic Health Records (EMR) is another source of information that can be used to provide the assistance on disease diagnosis or evaluation on caring procedure for patients. However, EMR is very different comparing with other medical data resources. EMR has the time dependency inside the data. Deep learning neural network is a forward-feeding neural network that is not suitable for modeling the time dependent data. In this paper, we used a Recurrent Neural Network (RNN) to model the time-series data. The Recurrence in the RNN allows it to remember the information from previous calculation and the previous information will influence the calculation on current input. In addition to base RNN, we also experimented LSTM-RNN, which is a variation of RNN. Comparing with base RNN, the LSTM-RNN has the long term memory that can memorize the information from the calculations in the much further time stamps. For the data that crosses over a long time interval, LSTM-RNN is more suitable than base RNN.

In this paper, we studied the problem inside the data set: irregular sampling and missing values and built two deep learning neural network models using base-RNN and LSTM-RNN. We used the supervised learning method to train and test our models. Then, we compared the test results of RNN, LSTM-RNN and other machine learning algorithms to evaluate their performance on real hospital data.

2   Related Work

The irregular data sampling in medical records is a very common problem. Many researches have done using the LSTM to solve the time irregularities. Inci Baytas et al. [1] proposed a novel LSTM framework called Time-Aware LSTM, also referred as T-LSTM. In their approach, they modified the sigmoid layer of the LSTM cell, which enables time decay to adjust the memory content in the cell. Their experiment results indicate the T-LSTM architecture is able to cluster the patients into clinical subtypes. Che et al. [2] studied the task of pattern recognition and feature extraction in clinical time series data. They used a differ-ent variation of recurrent neural network so called GRU-RNN that can also uses the prior knowledge. Che evaluated their model on two real-word hospital data sets and showed their neural nets can learn interpretable and clinical relevant features form the data set. Harutyunyan [4] also used deep learning framework to make predictions on clinical times series data. In their work, they studied multi-ple tasks involving modeling risk of mortality, forecasting length of stay in ICU, detecting physiological decline, and phenotype classification. They built a RNN model to explore the correlations between those multiple tasks. However, they only explored the traditional data imputation method that fills the missing data using the summary statistics. The tradition data imputation method ignores the correlation between variables. For example, the variable Temperature and Heart Rate may be highly correlated. High Temperature value also could also indicate a high Heart Rate value. However, if we impute the low mean Heart Rate value under a high Temperature condition, it could influence the prediction accuracy.

In this paper, we focused on developing a new data imputation approach using Gaussian process and propose a deep learning framework to predict the probability of mortality in ICU on real hospital patient data. For this prediction task, we built and compared the performance of base-RNN and LSTM-RNN model, especially on false positive errors made by these two models.

3   Data Imputation and Multivariate Data Modelling

In this paper, we used ICU data set: *The PhysioNet*, it includes over 4000 patient records. Each record maintains the 36 variables measurement at least once dur-ing the first 48 h after admission to the ICU. Each patient has a result variable: *In-hospital death* is a binary value (0: survivor, or 1: died in-hospital). However, there are three major problems existing in this data set: (1) missing value prob-lem: not all variables are available in all cases. For example, at time stamp t_i, there could be only 7 values out of 36 variables. (2) Irregular sampling: The each record was measured at irregular time stamp. Patient's measurements were tak-ing at different time stamp. The interval between two measurements are not the same. These two problems require the data pre-processing before using the data to train our model. (3) "Imbalanced" data sets: the number of dead patients only contains a very small proportion of the data set.

We use the time window and statistical summary imputation method to manually fill the missing values. To be more specific, we divide each patient's record into equal length window and the length of the window can be 2 h, 5 h, 10 h, etc. For each of the variable, we use the 5 summary statistics min, max, mean, median, standard deviation. Using the summary statistic of a time interval can solve the problem of missing data at a specific time stamp. However, this dataset has the problem of serious missing data. Many time interval still does not have the values.

At each time stamp t_i, we use a tensor that contains the value of each variable for each data entry. We proposed to use statistical model to study overall data form for each tensor. For each missing interval value, we introduced a Gaussian process that estimates the mean and variance from the recorded measurements. Let $\chi = \{x_1, x_2, x_3, ..., x_j\}$ be the collection of tensors from the patient record with j number of known measurements, in particularly, we can denoted it as $\{f(x_i) : x_i \in \chi\}$, where which is drawn from a Gaussian process with a mean function $m(\cdot)$ and kernel function $k(\cdot, \cdot)$. Then, the distribution of the set χ is denoted as,

$$
\begin{bmatrix} f(x_1) \\ f(x_2) \\ \vdots \\ f(x_j) \end{bmatrix} \sim N \left(\begin{bmatrix} m(x_1) \\ m(x_2) \\ \vdots \\ m(x_j) \end{bmatrix} \begin{bmatrix} k(x_1, x_2) & \dots & k(x_i, x_j) \\ k(x_2, x_1) & \dots & k(x_2, x_j) \\ \vdots & \ddots & \vdots \\ k(x_j, x_1) & \dots & k(x_j, x_j) \end{bmatrix} \right)
$$

or $f(\cdot) \sim gp(m(\cdot), k(\cdot, \cdot))$. The purpose of kernel function is to transform to a valid covariance matrix corresponding to some multivariate Gaussian distribution. For a kernel transformation, the kernel function must satisfy the Mercer's condition (illustrated in Definition 1). In Mercer's condition, the function needs to be square-integrable (illustrate in Definition 2) Therefore, we choose the squared exponential kernel, defined in Eq. (1), where parameter τ determines the smoothness of the Gaussian process prior with $k_{SE}(\cdot, \cdot)$.

$$
k_{SE}(x_i, x_j) = exp(-\frac{1}{2\tau^2} \|x_i - x_j\|^2) \tag{1}
$$

Definition 1. *A real-valued kernel function $K(x, y)$ satisfies Mercer's condition if $\int \int K(x, y)g(x)g(y)dxdy \geq 0$ for all square-integrable functions $g(\cdot)$.*

Definition 2. *A function $g(x)$ is square-integrable if $\int_{-\infty}^{+\infty} |g(x)|^2 \, dx < \infty$*

Then each patient record can be modelled through multivariate Gaussian distribution, illustrated in Eq. (2), where $\mu = m(\cdot)$ and $\Sigma = k(\cdot, \cdot)$.

$$
f(\mathbf{x}) = \frac{1}{(2\pi)^{d/2} |\Sigma|^{1/2}} exp \left| -\frac{1}{2}(\mathbf{x} - \boldsymbol{\mu})^T \Sigma^{-1} (\mathbf{x} - \boldsymbol{\mu}) \right| \tag{2}
$$

4   RNN and LSTM-RNN

Comparing with feedforward network, the recurrent neural network takes the current input and it also takes the what they previously perceived. The information from previous inputs can be kept into the hidden layers in the RNN, which will influence the calculation of the current input. The main difference between the recurrent network and feedforward is the feedback look connected to their past decisions.

$$
h_t = \Phi(W * x_t + U * h_{t-1}) \tag{3}
$$

Equation (3) shows the mathematical expression of updating the hidden layers in RNN. It takes the previous hidden layer h_{t-1} and current input x_t to calculate the current hidden layer output. However, the main disadvantage of the RNN is the "long term memorization". To be more explicit, if there are two data inputs d_i and d_j across a long time interval, the RNN cannot remember the information from the input d_i when it does the calculation on current input d_j. From the PhysioNet, each patient record contains the information more than 48 h in ICU. Using the RNN may not be able to "remember" the patients' information many hours ago. The loss of information in the neural network is referred as the "vanishing gradient" problem.

A variation of recurrent neural network, so called Long Short-Term Memory Unit (LSTM), was proposed by the German researchers Sepp Hochreiter and Juergen Schmidhuber as a solution to the vanishing gradient problem [5]. The architecture of the LSTM can be viewed as a gated cell. The cell decides which information will be remembered, or forgot through gate opening and closing. By maintaining this gate switch, it allows LSTM to continue to learn over a long time interval.

$$f_t = \sigma(W_f * [h_{t-1}, x_t] + b_f) \tag{4}$$

$$i_t = \sigma(W_i * [h_{t-1}, x_t] + b_i) \tag{5}$$

$$\tilde{C}_t = tanh(W_c * [h_{t-1}, x_t] + b_c) \tag{6}$$

$$C_t = f_t * C_{t-1} + i_t * \tilde{C}_t \tag{7}$$

$$o_t = \sigma(W_o[h_{t-1}, x_t] + b_o) \tag{8}$$

$$h_t = o_t * tanh(C_t) \tag{9}$$

The above 6 equations illustrate the update procedure for a layer of memory cell between time stamp $t-1$ and t. The sigmoid layer, also called "forget gate layer" decides what information will be dropped out from the cell, illustrated in Eq. (4). Equations (5) and (6) refer to the "input gate layer", which contains two parts: one sigmoid layer and one tanh layers. This sigmoid layer decides what information the cell will update and the tanh layer controls the new information will be stored into the cell. The Eq. (5) illustrates the process of forgetting information and updating information. Eventually, the LSTM cell will generate outputs using Eqs. (7), (8), and (9), where h_t is the output of the hidden layer and C_t is the output of the cell, which represented as a tensor with 2 dimensions. Since the LSTM decides to drop up some information at each time stamp, it is able to store the information from longer time stamp, when comparing with base-RNN. Then, we defined the softmax layer that maps the outputs generated by the LSTM cell into the probability representation using Eq. (10), where $f(C_{t_i})$ denotes as the probability of class i.

$$f(C_{t_i}) = \frac{exp^{C_{t_i}}}{\sum_j^{|C_t|} exp^{C_{t_j}}} \tag{10}$$

5   Results and Discussion

We built two neural networks RNN and LSTM-RNN with the same structure: 36 feature inputs, 1296 hidden units with 2 layers. We split the 4000 data samples into the training group and testing group. In order to resolve the "imbalanced" number of dead patients records and survival patients. We randomly selected 400 survival patients and 400 dead patients as the training set and 200 survival patients and 200 dead patients as the testing set.

The output of the model is two probabilities: $[Prob(survival), Prob(dead)]$, denoted as $(v1), p(v2)]$ If $p(v1) > p(v2)$, then we classify the patient as dead (0), otherwise, we classify the patient as survival (1). We used the mean squared error as the loss measurement of the model. The mean squared error is measured by the sum of the variance of the model and the squared bias of the model. If the patient outcome is survival (0), then the target variable is $[1,0]$, where can be interpreted as $[p(v1) = 1, p(v2) = 0]$. If the patient outcome is dead (1), then the target variable is $[0,1]$.

$$MSE = \frac{1}{N} * \sum^{N} [(p(v_1) - \widehat{p(v_1)})^2 + (p(v_2) - \widehat{p(v_2)})^2]$$ (11)

Equation (11) is the mathematical expression of the Mean Squared Error measurement of our model, where N is the batch size. The model uses the loss during the learning phase to gradually adjust the model until there is no improvement or very small improvement.

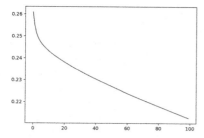

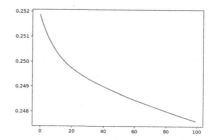

Fig. 1. The loss of the RNN: epoch = 100, learning rate = 0.01, batch size = 800, window size = 10 h.

Fig. 2. The loss of the LSTM: epoch = 100, learning rate = 0.01, batch size = 800, window size = 10 h

The epoch actually represents the learning phase of RNN and LSTM. Figures 1 and 2 show the loss of RNN and LSTM at each epoch. The loss is calculated by MSE mathematical function and it indicates the model's learning outcomes. Both RNN and LSTM can reduce their loss through each epoch. At the initial, the loss of LSTM is lower than RNN; then during the beginning, both of RNN and LSTM can rapidly reduce their loss. However, at the end, RNN received lower loss than LSTM. The window size could be the reason because the model

Table 1. Confusion Matrix of RNN and LSTM on testing data: batch size = 400 (200 survival and 200 dead)

	RNN: survival	RNN: dead	LSTM: survival	LSTM: dead
Target: survival	137	63	170	30
Target: dead	90	110	55	145

Table 2. Evaluation statistics of RNN and LSTM

	RNN	LSTM
Specificity	60.35%	82.86%
Sensitivity	63.58%	75.56%

does not need to remember too many previous information when we have a large window size.

From the Tables 1 and 2 of RNN and LSTM, even though the RNN can achieve a lower loss than LSTM, the testing result shows that LSTM did a better job than RNN for time-series data that has a long term dependency. For ICU mortality prediction, the most important error is the false positive. LSTM had lower errors than RNN. For specificity and sensitivity, the LSTM also achieves higher values than RNN.

Figure 3 also shows that the ROC curve of LSTM is always higher than the RNN. In Table 3, we compared several different machine learning algorithms. Th

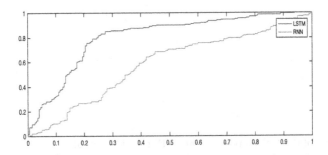

Fig. 3. The ROC curve of RNN and LSTM

Table 3. The comparison of the AUC score of different machine learning algorithms

Algorithm	AUC score	Algorithm	AUC score
SVM	0.563	LDA	0.608
QDA	0.673	LSTM	0.8025
RNN	0.581	RF	0.642
LR	0.602		

AUC value of LSTM is the highest among the results obtained from different algorithms. Therefore, for the task of modeling time-series data, especially for the long term data, LSTM can produce an improved prediction results using Gaussian data imputation method among different algorithms.

6   Conclusion and Future Work

The major problems of Electronic Health Record (EMR) are irregular data sampling and missing values. In the paper, we imputed the missing variable values by 5 summary statistics. The mean and stand deviation values were modeled by multivariate Gaussian distribution through kernalization of Gaussian process, which ensures the correlation between variables is considered into the imputation process. The recurrent neural network emphasizes on the time dependency relationship in the data. The experimental results indicate that LSTM produces higher accuracy than RNN on modeling time series data that has the long term dependency.

For the future work, we plan to use a Convolutional Neural Network based LSTM (CNN-LSTM). We can consider the each patient record as an image and use the CNN to automatically extract useful features. In addition, we also need to consider that whether the missing values are also informative. Seriously ill patients normally has less missing variable values than less ill patients. Therefore, we can use indicator variables for each value. For example, if the variable's value is missing, the indicator sets to 0, otherwise, the indicator sets to 1. Then, the indicator variables would also be the input of the LSTM.

References

1. Baytas, I.M., Xiao, C., Zhang, X., Wang, F., Jain, A.K., Zhou, J.: Patient subtyping via time-aware LSTM networks. In: Proceedings of the 23rd ACM SIGKDD International Conference on Knowledge Discovery and Data Mining, KDD 2017, pp. 65–74. ACM, New York (2017). https://doi.org/10.1145/3097983.3097997
2. Che, Z., Kale, D., Li, W., Bahadori, M.T., Liu, Y.: Deep computational phenotyping. In: Proceedings of the 21th ACM SIGKDD International Conference on Knowledge Discovery and Data Mining, KDD 2015, pp. 507–516. ACM, New York (2015). https://doi.org/10.1145/2783258.2783365
3. Deligiannis, P., Loidl, H.W., Kouidi, E.: Improving the diagnosis of mild hypertrophic cardiomyopathy with mapreduce. In: Proceedings of Third International Workshop on MapReduce and Its Applications Date, pp. 41–48, MapReduce 2012. ACM, New York, NY, USA (2012). https://doi.org/10.1145/2287016.2287025
4. Harutyunyan, H., Khachatrian, H., Kale, D., Galstyan, A.: Multitask learning and benchmarking with clinical time series data, March 2017
5. Hochreiter, S., Schmidhuber, J.: Long short-term memory. Neural Comput. **9**(8), 1735–1780 (1997). https://doi.org/10.1162/neco.1997.9.8.1735
6. Hussain, A.: Machine learning approaches for extracting genetic medical data information. In: Proceedings of the Second International Conference on Internet of Things, Data and Cloud Computing, ICC 2017, p. 1. ACM, New York (2017). https://doi.org/10.1145/3018896.3066906

Modularized and Attention-Based Recurrent Convolutional Neural Network for Automatic Academic Paper Aspect Scoring

Feng Qiao[⊠], Lizhen Xu, and Xiaowei Han

School of Computer Science and Engineering,
Southeast University Nanjing, Nanjing, China
qf_sun@seu.edu.cn

Abstract. Thousands of academic papers are submitted at top venues each year. Manual audits are time-consuming and laborious. And the result may be influenced by human factors. This paper investigates a modularized and attention-based recurrent convolutional network model to represent academic paper and predict aspect scores. This model treats input text as module-document hierarchies, uses attention pooling CNN and LSTM to represent text, and outputs prediction with a linear layer. Empirical results on PeerRead data show that this model give the best performance among the baseline models.

Keywords: Aspect scoring · Convolutional neural network · Deep learning
LSTM · Attention pooling

1 Introduction

Every year prestigious scientific venues receive thousands of academic papers to their conferences and journals. Because rated by humans, it is exhausting and affected by personal factors of reviewers. In this paper, a method on how to rating and aspect scoring academic papers based on their source files and meta information which called automatic academic paper aspect scoring (AAPAS) will be proposed. This helps reviewers get rid of unqualified papers, and helps authors obtain suggestions for improvement.

A number of researchers have done some work on consistency, prejudice, author response and quality of general reviews (e.g., Greaves et al. 2006; Ragone et al. 2011; De Silva and Vance 2017). The organizer of the Nips 2014 took out 10%-part presentation of the papers and measured the consistency of the reviews. Find that the number of controversial papers is over one-fourth (Langford and Guzdial 2015). The recently launched dataset PeerRead consists of 14,700 papers and 10,700 peer reviews from the ARXIV, ICLR and NIPS conferences, which include the aspect scores of appropriateness, clarity, originality, empirical soundness/correctness, theoretical soundness/correctness, meaningful comparison, substance, impact of ideas or results, impact of accompanying software, impact of accompanying dataset, recommendation and reviewer confidence (https://github.com/allenai/PeerRead). All aspect scores are

X. Meng et al. (Eds.): WISA 2018, LNCS 11242, pp. 68–76, 2018.
https://doi.org/10.1007/978-3-030-02934-0_7

integers from 1 to 5 and some are strongly artificial subjective. The goal of this paper is to predict the aspect scores for academic papers.

2   Related Work

The task of AAPAS is usually treated as a supervised learning problem. The main purposes can be listed in three categories: text classification (Larkey 1998; Rudner and Liang 2002; Yannakoudakis et al. 2011; Chen and He 2013), score regression (Attali and Burstein 2004; Phandi et al. 2015; Zesch et al. 2015) and preference ranking (Yannakoudakis et al. 2011; Phandi et al. 2015). However, there is a need for human-specific characteristics in the process. Alikaniotis and Yannakoudakis (2016) employed a LSTM model to learn features for AES (Automatic Easy Scoring) without any predefined features. Taghipour and Ng (2016) used a CNN model followed by a recurrent layer to extract local features and model sequence dependencies. And a two-layer CNN model was proposed by Dong and Zhang (2016) to obtain higher level and more abstract information. Dong et al. (2017) further proposed to add attention-based mechanism to the pooling layer to automatically decide which part is more important in determining the quality of the paper.

The work in this paper is also related to building modularized hierarchical level representations of documents. Li et al. (2015) built a hierarchical LSTM auto-encoder for documents. And Tang et al. (2015) used a hierarchical Gated RNN for sentiment classification. This paper proposes a modularized hierarchical and attention-based CNN-LSTM model for aspect scoring, which is a regression task.

3   Model

3.1   Module Representation

A complete resource paper r is divided into modules according to the general structure of academic paper (abstract, keywords, title, authors, introduction, related work, methods, conclusion, reference and appendices). For initialization, the word is embedded into the vector with its one-hot representation through an embedding matrix. In each module, attention-based CNN is used with word-level embeddings to get the representation of the i-th module.

In addition, the presentation of the author module is somewhat different, and different authors have different effects on the author module. The Eq. (1) shows method of constructing author module:

$$m_{authors} = \sum_{i=1}^{A} \vartheta_i \alpha_i \tag{1}$$

where $\vartheta = (\vartheta_1, \ldots, \vartheta_A)^T$ is weight parameter, and α_i is embedding vector of the i-th author of paper. It is randomly initialized and will be learned in the training program. A is the max length of all the author list.

3.2 Details of Modularized Hierarchical CNN

Attention-based CNN consists of a convolution layer and an attention pooling layer. Convolution layer is used to extract local features, while pooling layer automatically determines the connection between words, sentences, and modules.

Module Representation. After obtaining word embeddings $x_i, i = 1, 2, \ldots, n$, a convolutional layer is employed on each module:

$$z_i = f\left(W_z \cdot \left[x_i^j : x_i^{j+h_w-1}\right] + b_z\right) \tag{2}$$

where W_z, b_z, are weight matrix and bias vector, h_w is window size in the convolutional layer and z_i is abstract feature representation. Following the convolutional layer, an attention polling layer is used to acquire a module representation. The model is described in Fig. 1.

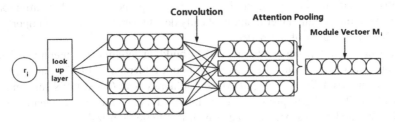

Fig. 1. Module representation using ConvNet and attention pooling.

Attentive Pooling Layer. The equations of convolutional and attention pooling layers are defined as follows:

$$s_i = \tanh(W_s \cdot z_i + b_s) \tag{3}$$

$$u_i = \frac{e^{w_u \cdot s_i}}{\sum e^{w_u \cdot s_j}} \tag{4}$$

$$m = \sum u_i z_i \tag{5}$$

where W_s, w_u, are weight matrix and vector, respectively, b_s is bias vector, s_i and u_i are attention vector and attention weight respectively for i-th word. And m is the final module representation, which is the weighted sum of all word vectors.

3.3 Text Representation

A recurrent layer is used to build the representation of entire document which is similar to the models of Alikaniotis et al. (2016) and Taghipour and Ng (2016). But the main difference is that this model treats paper as sequence of modules rather than sequence of sentences. Alikaniotis et al. (2016) use score-specific word embeddings as word features and take the last hidden state of Long short-term memory (LSTM) as text representation. Taghipour and Ng (2016) take the average value over all hidden states of LSTM as text representation. This paper uses LSTM to learn from module sequences and attention pooling on hidden states to obtain the contribution of each module to the final text representation. The structure of text representation using LSTM is shown in Fig. 2.

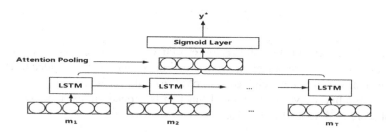

Fig. 2. Text representation using LSTM and attention

LSTM Model. Hochreiter and Schmidhuber (1997) and Pascanu et al. (2013) modified LSTM units into recurrent units which used to handle the problem of vanishing gradients effectively. Gates are used to control information flow, preserving or forgetting for each cell units in LSTM. To control the flow of information while processing vector sequences, an input gate, a forget gate and an output gate are employed to decide the passing of information at each time step. Assuming that a paper consists of T modules, $s_1, s_2, \ldots, s_T$ with s_t being the feature representation of t-th word s_t. LSTM cell units are addressed in the following equations:

$$i_t = \sigma(W_i \cdot s_t + U_i \cdot h_{t-1} + b_i) \tag{6}$$

$$f_t = \sigma\left(W_f \cdot s_t + U_f \cdot h_{t-1} + b_f\right) \tag{7}$$

$$\tilde{c}_t = \tanh(W_c \cdot s_t + U_c \cdot h_{t-1} + b_c) \tag{8}$$

$$c_t = i_t \odot \tilde{c}_t + f_t \odot c_{t-1} \tag{9}$$

$$o_t = \sigma(W_o \cdot s_t + U_o \cdot h_{t-1} + b_o) \tag{10}$$

$$h_t = o_t \odot \tanh(c_t) \tag{11}$$

where s_t and h_t are input module and output module vectors at time t, respectively. W_i, W_f, W_c, W_o, U_i, U_f, and U_o are weight matrices and b_i, b_f, b_c, b_o are bias vectors. The symbol $\odot$ means element-wise multiplication and σ represents the sigmoid function.

Attentive Pooling Layer. After obtaining hidden layer state of LSTM, $h_1, h_2, \ldots, h_T$, another attention pooling layer over modules is used to learn the final text representation. This helps to acquire the weight of module's contribution to final quality of the text. The attention pooling over modules is addressed as:

$$a_i = \tanh(W_a \cdot h_t + b_a) \tag{12}$$

$$\alpha_i = \frac{e^{w_\alpha \cdot a_i}}{\sum e^{w_\alpha \cdot a_j}} \tag{13}$$

$$o = \sum \alpha_i h_i \tag{14}$$

where W_a, w_α are weight matrix and vector respectively, b_a is bias, a_i is attention vector for i-th module, and α_i is attention weight of i-th module. o is considered as final representation, which is the weighted sum of all module vectors.

Finally, one linear layer with sigmoid function is applied on text representation to get final score as described in Eq. 15.

$$y = \text{sigmoid}(w_y o + b_y) \tag{15}$$

4 Experiments

4.1 Objective

Mean square error (MSE) is widely used in regression tasks. It measures average value of square error between gold standard score y_i^* and prediction scores y_i. Given N papers, calculate MSE according to the equation as follows:

$$\text{mse}(y, y^*) = \frac{1}{N} \sum_{i=1}^{N} (y_i - y_i^*)^2 \tag{16}$$

The module is trained in a fixed number of epochs and evaluated on development set at every epoch. The details of model hyper-parameters are listed in Table 1.

Table 1. Hyper-parameters

Layer	Parameter	Value
Lookup	Word embedding dim	50
CNN	Windows size/number of filters	5/100
LSTM	Hidden units	100
Dropout	Rate	0.5
	Epochs/batch size/initial learning rate η/momentum	50/10/0.001/0.9

4.2 Word Embeddings

Stanford's publicly available GloVe 50-dimensional embeddings are used as word pretrained embeddings, which are trained on 6 billion words from Wikipedia and web text (Pennington et al. 2014). During training process, word embeddings are fine-tuned.

4.3 Optimization

RMSprop (Dauphin et al. 2015) are used as optimizer to train the whole model. The initial learning rate η is set to 0.001 and momentum is set to 0.9. Dropout regularization is used to avoid overfitting and drop rate is 0.5.

Table 2. The PeerRead dataset. Asp. indicates whether the data contains the aspect scores. Acc/Rej indicates the number of papers accepted and rejected.

Section	Papers	Reviews	Asp.	Acc/Rej
NIPS2013–2017	2420	9152	×	2420/0
ICLR2017	427	1304	√	172/255
ACL2017	137	275	√	88/49
CoNLL2016	22	39	√	11/11
Arxiv2007–2017	11778	-	-	2891/8887
Total	14784	10770		

4.4 Data

The PeerRead dataset is used as evaluation data of our system. It consists of five section of papers which listed in Table 2. 60% of data is used for training, 20% is used for development, and the rest 20% is used for test. All the aspect scores are scaled to range [0, 1]. During evaluation phase, the scaled scores are rescaled to original integer scores, which are used to calculate evaluation metric quadratic weighted kappa (QWK) values.

4.5 Baseline Models

LSTM with Mean-over-Time Pooling (LSTM-MoT) (Taghipour and Ng 2016) and hierarchical CNN (CNN-CNN-MoT) (Dong and Zhang 2016).

LSTM-MoT uses one layer of LSTM over word embedding sequences and takes average pooling over all time-step states which is the current state-of-the-art neural model on the text-level. CNN-CNN-MoT uses two layers of CNN which is a state-of-the-art model on the sentence-level. One layer operates on sentences and the other is stacked above. And mean-over-time pooling gets the final text representation.

Table 3. Comparison between different models and pooling methods on test data.

Aspects	LSTM-MoT	CNN-CNN-MoT	M-LSTM-CNN-att
Appropriateness	0.664	0.620	0.712
Clarity	0.751	0.721	0.719
Originality	0.694	0.640	0.723
Empirical soundness/correctness	0.697	0.680	0.702
Theoretical soundness/correctness	0.678	0.632	0.690
Meaningful comparison	0.680	0.650	0.729
Substance	0.671	0.662	0.672
Impact of ideas or results	0.630	0.647	0.720
Impact of accompanying software	0.641	0.642	0.706
Impact of accompanying dataset	0.632	0.621	0.634
Recommendation	0.691	0.624	0.713
Reviewer confidence	0.631	0.621	0.681
Avg.	0.672	0.647	**0.700**

4.6   Results

The results are listed in Table 3. The model modularized and attention-based recurrent CNN (M-LSTM-CNN-att) outperform LSTM-MoT by 4.2%, CNN-CNN-MoT by 8.3% on average quadratic weighted kappa.

4.7   Analysis

Several experiments are performed to verify the effectiveness of M-LSTM-CNN-att model. As is shown in the Table 3, there are minor improvements in every aspect, except for clarity which reduced by 4.3% compared to LSTM-MoT and 0.2% compared to CNN-CNN-MoT. This is because M-LSTM-CNN-att is based on module-level while the other two is based on sentence-level, so that the baseline models could catch more details about structure of paper. And M-LSTM-CNN-att model has barely improved in impact of accompanying dataset. This is because the description of the data itself is relatively less descriptive in the paper. And according to experience and earlier research about PeerRead, the more important parts about the academic paper are author, abstract and conclusion (Kang et al. 2018). While the data only occupies a very small part. However, the improvements of other aspects show that the modularized hierarchical structure of the model is of great help to obtain better representations by incorporating knowledge of the structure of the source paper.

5   Conclusion

This paper investigates a modularized and attention-based recurrent convolutional network model to represent academic paper and predict aspect scores. This model treats input text as module-document hierarchies, uses attention pooling CNN and LSTM to represent text, and output the prediction with a linear layer. Although in a few aspects it is not as good as the baseline models, the overall effect is better than the baselines.

References

Alikaniotis, D., Yannakoudakis, H., Rei, M.: Automatic text scoring using neural networks. arXiv preprint arXiv:1606.04289 (2016)

Dong, F., Zhang, Y.: Automatic features for essay scoring – an empirical study. In: Proceedings of the 2016 Conference on Empirical Methods in Natural Language Processing, pp. 1072–1077 (2016)

Taghipour, K., Ng, H.T.: A neural approach to automated essay scoring. In: Proceedings of the 2016 Conference on Empirical Methods in Natural Language Processing, pp. 1882–1891 (2016)

De Silva, P.U.K., Vance, C.K.: Preserving the quality of scientific research: peer review of research articles. In: De Silva, P.U.K., Vance, C.K. (eds.) Scientific Scholarly Communication. FLS, pp. 73–99. Springer, Cham (2017). https://doi.org/10.1007/978-3-319-50627-2_6

Langford, J., Guzdial, M.: The arbitrariness of reviews, and advice for school administrators. Commun. ACM Blog **58**(4), 12–13 (2015)

Larkey, L.S.: Automatic essay grading using text categorization techniques. In: Proceedings of the 21st Annual International ACM SIGIR Conference on Research and Development in Information Retrieval, pp. 90–95. ACM (1998)

Rudner, L.M., Liang, T.: Automated essay scoring using bayes' theorem. J. Technol. Learn. Assess. **1**(2), 3 (2002)

Chen, H., He, B.: Automated essay scoring by maximizing human-machine agreement. In: Proceedings of the 2013 Conference on Empirical Methods in Natural Language Processing, pp. 1741–1752 (2013)

Yannakoudakis, H., Briscoe, T., Medlock, B.: A new dataset and method for automatically grading ESOL texts. In: Proceedings of the 49th Annual Meeting of the Association for Computational Linguistics: Human Language Technologies, vol. 1, pp. 180–189. Association for Computational Linguistics (2011)

Attali, Y., Burstein, J.: Automated essay scoring with e-rater R v. 2.0. ETS Res. Rep. Ser. **2**, i–21 (2004)

Phandi, P., Chai, K.M.A., Ng, H.T.: Flexible domain adaptation for automated essay scoring using correlated linear regression (2015)

Zesch, T., Wojatzki, M., Akoun, D.S.: Task-independent features for automated essay grading. In: Proceedings of the Tenth Workshop on Innovative Use of NLP for Building Educational Applications, pp. 224–232 (2015)

Dong, F., Zhang, Y., Yang, J.: Attention based recurrent convolutional neural network for automatic essay scoring. In: Proceedings of the 21st Conference on Computational Natural Language Learning (CoNLL 2017), pp. 153–162 (2017)

Li, J., Luong, M.-T., Jurafsky, D.: A hierarchical neural autoencoder for paragraphs and documents. arXiv preprint arXiv:1506.01057 (2015)

Tang, D., Qin, B., Liu, T.: Document modeling with gated recurrent neural network for sentiment classification. In: EMNLP, pp. 1422–1432 (2015)

Kang, D., et al.: A Dataset of Peer Reviews (PeerRead): Collection, Insights and NLP Applications. arXiv:1804.09635v1 (2018)

Cloud Computing and Big Data

Fast Homomorphic Encryption Based on CPU-4GPUs Hybrid System in Cloud

Jing Xia[✉], Zhong Ma, Xinfa Dai, and Jianping Xu

Wuhan Digital Engineering Institute, No. 1 Canglong Bei Road, Jiangxia,
Wuhan 430205, People's Republic of China
673718032@qq.com

Abstract. Security is an ever-present consideration for applications and data in the cloud computing environment. As an important method of performing computations directly on encrypted data without any need of decryption and compromising privacy, homomorphic encryption is an increasingly popular topic of protecting the privacy of data in cloud security research. However, as high computational complexity of the homomorphic encryption, it will be a heavy workload for computing resources in the cloud computing paradigm. Motivated by this observation, this paper proposes a fast parallel scheme with DGHV algorithm based on CPU-4GPUs hybrid system. Our main contribution of this paper is to present a parallel processing stream scheme for large-scale data encryption based on CPU-4GPUs hybrid system as fast as possible. Particularly, the proposed method applies CPU-4GPUs parallel implementation to accelerating encryption operation with DGHV algorithm to reduce the time duration and provide a comparative performance study. We also make further efforts to design a pipeline architecture of processing stream in CPU-4GPUs hybrid system to accelerate encryption for DGHV algorithm. The experiment results show that our method gains more than 91% improvement (run time) and 70% improvement compared to the serial addition and multiplication operation with DGHV algorithm respectively.

Keywords: Homomorphic encryption · Cloud computing · Data security
GPU · Privacy · Parallel

1 Introduction

Cloud computing platform has become the popular way for different customers to store, compute, and analyze their large-scale data on the third-party cloud providers generally, which have the risk of private data leakage.

Standard encryption algorithms help protect sensitive data from outside attackers, however they cannot be utilized to compute ciphertext directly. Homomorphic Encryption presents a very useful algorithm that can compute on encrypted data without the need to decrypt it. Nevertheless, the computational complexity of homomorphic algorithm is huge. This is the motivation for us to build an accelerated computing platform for execution of homomorphic algorithm. In cloud environment, the clients are given the possibility of encrypting their sensitive information before

© Springer Nature Switzerland AG 2018
X. Meng et al. (Eds.): WISA 2018, LNCS 11242, pp. 79–90, 2018.
https://doi.org/10.1007/978-3-030-02934-0_8

sending it to the cloud server. The cloud server then compute over their encrypted data without the need for the decryption key. For example, user's privacy is well protected in the fields of medicine and finance by using homomorphic algorithm. Therefore, it is meaningful to apply homomorphic cryptosystem for ensuring security and privacy of data in such platform.

Along with the rise of concept of full homomorphic encryption (FHE) first proposed by Rivest et al. [1], the FHE gradually has become one of the most important topic in the field of cryptosystem. Nevertheless, satisfactory results were not obtained for a long time before 2009. Only some semi-homomorphic or only finite-step full-homomorphic encryption schemes have been obtained, such as the RSA algorithm which only supports the multiplication homomorphism. Until 2009, Gentry [2] has proposed the realization of the FHE using bootstrapping. Subsequently, Gentry invented a fully homomorphic encryption scheme based on an ideal lattice [3] in the same year. In addition, the application value of homomorphic encryption has been started to receive attentions by scholars and entrepreneurs. A lot of variants based on Gentry's approach have emerged from that time.

However, the FHE scheme leads to high workload for computational resources and therefore making the cryptosystem computationally inefficient. In 2010, Dijk et al. [4] proposed another concise and easy-to-understand full-homomorphic encryption scheme, have been called integer (ring)-based full homomorphic encryption scheme (DGHV). The DGHV performs homomorphic encryption over integers and not on lattices. Accordingly, the DGHV is suitable for computer operations in theory. In contrast, the reality is that the DGHV is impractical for use, especially in real time applications. For example, the application of homomorphic encryption in a simple plaintext search will increase the amount of computing in trillions of times. Thus, on the basis of the homomorphic encryption principle, we need to put forward a high efficiency solution so that it can be better applied to the field of cloud computing.

Much parallel computing architecture has been designed successfully in recent years. Under the background, we combine the homomorphic encryption and the parallel framework to create this paper's idea. The purpose of this paper is to identify that the proposed CPU-4GPUs parallel scheme can benefit the improvement of the efficiency for the DGHV quantitatively. Firstly, we analyze the sequential processing stream and defined the processing unit in order to find the noninterference subtasks. Secondly, we analyze the computing characteristics of the DGHV cryptosystem. Then based on the analysis we design a cluster of data structures for the process which are suitable for the parallel computing environment. Because the response time is the key consideration for the cryptosystem, we employ four GPUs to accelerate encryption operation in order to obtain a preferable response time. Besides, a pipeline framework has been designed for stream processing as a optimization measure in our study. To verify the effectiveness of our method, we aim at test the performance of the proposed system by carrying out a set of experiments.

In order to find out the bottleneck of the cryptosystem, we can analyze the serial homomorphic encryption based on the only CPU system which is referred to as Serial Homomorphic Encryption (SHE). In this paper, we make use of the only one GPU to execute the encryption for DGHV, which is referred to as Somewhat Parallel Homomorphic Encryption (SPHE). By contrast, we make use of the hybrid CPU-4GPUs system to accelerate the encryption for DGHV, which is referred to as a Fast Parallel Homomorphic Encryption (FPHE). We further do some comparison experiments in order to quantitatively test the improvement of the performance and economical efficiency.

The main contributions of this paper are the following. (1) The CPU-4GPUs hybrid system is explored as a parallel scheme to improve the performance of DGHV algorithm. (2) We also make further efforts to design a pipeline architecture of processing stream in CPU-4GPUs hybrid system as an assisted acceleration of DGHV algorithm. (3) Extensive experiments are conducted to demonstrate the effectiveness of the proposed method in optimization of DGHV algorithm.

The rest of the paper is organized as follows. Section 2 reviews related work on homomorphic cryptosystem using various parallel systems. The core of our paper, Sect. 3, proposes a method that is used to parallel DGHV computation. Section 4 shows the experimental results. Finally, we draw some conclusions and discussions in the Sect. 5.

2 Related Work

The reliability and security of data remain major concerns in cloud computing environment. Fugkeaw et al. [5] propose a hybrid access control model to support fine-grained access control for big data outsourced in cloud storage systems. In order to check the cloud data reliability of the user's uploads the data in server, Swathi et al. [6] propose an approach using certificate less public auditing scheme which is used to generate key value. Jiang et al. [7] propose a novel cloud security storage solution which is based on autonomous data storage, management, and access control. Moreover, due to loss of ownership of user's data, the reliability and security of data stored in the outsourcing environment give rise to a lot of concerns. Therefore, protecting user's information privacy in outsourcing environment becoming increasingly important. For example, Shu et al. [8] propose a privacy-preserving task recommendation scheme which is used to encrypt information about tasks and workers before being outsourced to the crowdsourcing platform. Zhang et al. [9] providing a strong privacy protection to safeguard against privacy disclosure and information tampering. These techniques are utilized to protect privacy of user's data.

However, in cloud computing environment, due to the demand of computing ciphertext, many common encryption algorithms are not applicable. Therefore, homomorphic encryption algorithm attract scholar's attention because of the feature of algebraic homomorphism. But the high computational complexity of homomorphic algorithm hinder the practical application. Therefore, many scholars have studied how to improve the efficiency of homomorphic encryption algorithm for sake of popularizing the homomorphism application in cloud computing environment in recent years. Jayapandian et al. [10] have tried to present an efficient algorithm which combines the characteristics of both probabilistic and homomorphic encryption techniques. This is to improve execution efficiency from the point of homomorphic algorithm itself. Nevertheless, the time complexity has not been reduced much. Consequently, we gave up improving homomorphic algorithm itself to promote efficiency.

Sethi et al. [11] proposed a model to parallel homomorphic encryption based on MapReduce framework. Similarly, another parallel homomorphic encryption scheme is proposed based on MapReduce environment with 16 cores and 4 nodes by Min et al. [12]. The evidence of the above two studies illustrate that a good parallel scheme may improve the performance of homomorphic encryption. This is the most motivation for us to apply hybrid CPU-4GPUs hybrid system to homomorphic cryptosystem.

Because of the fact is that the privacy of sensitive personal information is an increasingly important topic as a result of the increased availability of cloud services, especially in the medical and financial field. Khedr et al. [13] propose a secure medical computation using GPU-Accelerated homomorphic encryption scheme. Accelerating somewhat homomorphic cryptography is also researched by many scholars. For example, Tian et al. [14] design a GPU-assisted homomorphic cryptograph for matrix multiplication operation. Moayedfard et al. [15] present the parallel implementations of somewhat homomorphic encryption scheme over integers using Open-MP and CUDA programming to reduce the time duration. Nevertheless, many applications require both addition and multiplication operations, so we study the improvement of the executing efficiency for the DGHV using CPU-4GPUs.

However, to the best of our knowledge, there is no research trying to quantitatively research the acceleration improvement of multiple GPUs compared to a single GPU. Moreover, this paper present a parallel processing stream scheme for large-scale data encryption based on CPU-4GPUs hybrid system as fast as possible, which is also different from the existing research.

3 Proposed Method

3.1 Background Knowledge of DGHV

In the homomorphic application scenario, when the cloud server cannot decrypt the data for security reasons, but the server needs to process and respond to the user's query request based on performing computation on the ciphertext which is uploaded by users. The key point is that the output of the ciphertext is equivalent to the result of performing the same operation on the plaintext and then encrypting it. Specifically, different homomorphic encryption algorithms can perform different homomorphic calculations on ciphertext.

In our study, the DGHV both have the additive homomorphism and the multiplicative homomorphism. Generally, the homomorphic encryption algorithm has the following homomorphic properties.

Additive Homomorphism. If there is an effective algorithm that makes the plaintext x, y satisfies the equation as $E(x+y) = E(x) \oplus E(y)$ or $x+y = D(E(x) \oplus E(y))$, and do not reveal the plaintext x, y.

Multiplicative Homomorphism. If there is an effective algorithm that makes the plaintext x, y satisfies the equation as $E(x \times y) = E(x) \otimes E(y)$ or $x \times y = D(E(x) \otimes E(y))$, and do not reveal the plaintext x, y.

A homomorphic encryption scheme consists of four part, namely homomorphic key generation (KeyGen), homomorphic encryption (Enc), homomorphic decryption (Dec), and homomorphic assignment (Evaluate). All of the four algorithms are probabilistic polynomials time complexity. The DGHV encryption scheme supports both addition and multiplication in the integer range, which is using only basic modular arithmetic, instead of using the ideal lattice concept in polynomial rings. The specific process is as follows.

KeyGen. The encryption select a random positive prime as the secret p and an n bit odd integer from an interval $p \in (2Z+1) \cap [2^{n-1}, 2^n)$.

Enc. For every bit data, we define q and r as random positive integers, where q is a large positive integer and r satisfy the equation as $|2r| < p/2$. To encrypt a bit of plaintext m belong to the binary set, which describe as $m \in \{0, 1\}$. In order to generate ciphertext c in the DGHV algorithm is defined as $c = pq + 2r + m$, where q is a large random number and r is a small random number. In practical application, for a composite message denoted by $M = m_1 m_2 \ldots m_n$, where $n \in N$(the set of natural numbers). Subsequently, the ciphertext C is defined as equation one and equation two. In our study, the encryption step is executed by using four GPUs.

$$C = E(p, M) = E(p, m_1)E(p, m_2) \ldots E(p, m_n) = c_1 c_2 \ldots c_n \qquad (1)$$

$$c_i = E(p, m_i) = pq_i + 2r_i + m_i \qquad (2)$$

Dec. The decryption algorithm use the secret key and ciphertext to decrypt the ciphertext, which is described as the equation three.

$$m = D(p, c) = (c \bmod p) \bmod 2 \qquad (3)$$

Evaluate. Since the public key is open to the public, we can obtain the $c - pq$ by the equation $c - pq = m + 2r$. We call the right part of the equation denoted by $m + 2r$ as the noise. While performing the decryption, it is necessary to satisfy the condition denoted by $c \bmod p = m + 2r < p/2$. In detail, only satisfy the condition, the mode operation of noise m + 2r can perform the correct decryption process described by $(m + 2r) \bmod 2 = m$. As a conclusion, the noise $m + 2r$ is the key to perform the correct decryption. The re-encryption technology provides a feasible way to reduce the noise. Then Evaluate(pk_2, Dec, s', c') is performed, and the obtained result is a new ciphertext. The ciphertext is obtained by performing the encryption over the plaintext information using the secret key pk_2. This is one of the most important characteristics of the DGHV's idea, which is the decryption of the homomorphism. Therefore, every time before we carry out the computational operation of the ciphertext, we apply the re-encryption operation over the ciphertext. Then the noise will remain in the controllable range.

3.2 Serial Homomorphic Encryption (SHE)

To parallelize the serial algorithm, we should find out the parallelizable part. As there is noninterference with each other among bit encryption, the process of computing the data encryption can be parallelized. It is sensible that the process of plaintext encryption can be divided into smaller data encryption units. These units can be sent to many process units to execute. The Amdahl's law indicates that a parallel algorithm can obtain speedup ratio compared with the serial version as equation nine.

$$S(m) = \frac{m}{1 + (m - 1)f} \tag{4}$$

Where S is the speedup ratio, m is the number of processing units, and f indicates the proportion of serial parts in the parallel algorithm. This law states that the speedup ratio can be up to 1/f when $m \to \infty$. Theoretically, in our algorithm, the factor f depends on the time cost by the following steps: initialization, data partition, task distribution, synchronization, data encryption, data merging, result outputting. The executing flow of serial homomorphic encryption task based on CPU is as shown in Fig. 1.

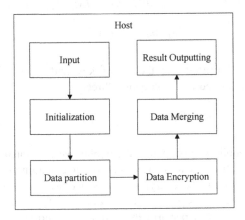

Fig. 1. The executing flow of the serial homomorphic encryption based on CPU

By analyzing the executing flow, it is found that except the process of data encryption, other steps are hard to be parallelized. As is shown in the Fig. 1, we can analyze the serial homomorphic encryption based on the only CPU and find out the bottleneck of the executing flow. When CPU execute the data encryption, other tasks may be stay waiting state. Therefore, data encryption is the main bottleneck step in DGHV algorithm. In fact, one of the reasons for the high computational complexity of the DGHV is to encrypt data by bits, which is referred to data encryption in our study.

3.3 Somewhat Parallel Homomorphic Encryption (SPHE)

In the study of parallel computing methods, we found that using GPU-assisted to accelerate is applicable and popular in many applications from the hardware perspective. Moreover, with continuous architectural improvements, GPUs have evolved into a massively parallel, multithreaded, many-core processor system with tremendous computational power. Many scholars utilize GPU to accelerate efficiency of the homomorphic encryption. For example, Wang et al. present the only one GPU implementation of a full homomorphic encryption scheme [16], which referred to the SPHE in this paper. However, the experimental results show that the speedup factors of 7.68, 7.4 and 6.59 for encryption, decryption and evaluation respectively, when compared with the existing CPU implementation. By analyzing the experimental results, the executing flow of DGHV encryption based on CPU-GPU system as shown in the Fig. 2, and it is found that the rate of speedup is not high.

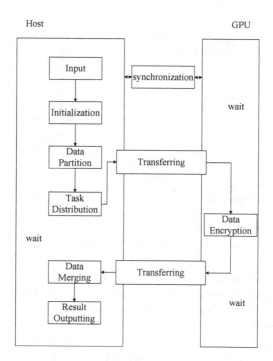

Fig. 2. The executing flow of the DGHV based on CPU-GPU system

Figure 2 shows the status of host, system bus and GPU in a task of the homomorphic encryption based on CPU-GPU system. Although the data encryption has been paralleled in GPU, but only one of the three parts of the hybrid system, including host, system bus and GPU, is in working state while others are waiting. Therefore, there is a considerable waste of time in this algorithm. Maybe, it is not enough to apply only one GPU to accelerate the homomorphic encryption for actual application.

3.4   Fast Parallel Homomorphic Encryption (FPHE)

To solve this problem, we design a hybrid CPU-4GPUs system shown in the Fig. 3 and a pipeline system shown in the Fig. 4, which transforms the above process into three procedures for accelerating DGHV algorithm. In order to quantitatively prove the acceleration of multiple GPUs compared to a single GPU, and maximize the acceleration for homomorphic algorithm in our experiment. Furthermore, on account of our lab having 4 GPUs, our study adopts the CPU-4GPUs accelerated architecture.

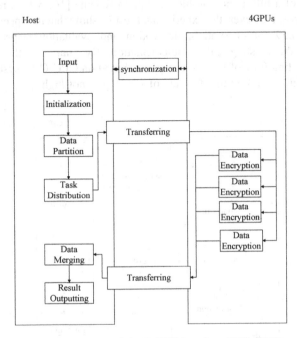

Fig. 3. The executing flow of the DGHV based on CPU-4GPUs system

Preprocessing. This step mainly includes three subtasks, that is, initialization, data partition, and task distribution. First, the host gets the input data which contains many data blocks. In the next step, data partition task mainly splits the data block into smaller data unit for the step of preparing data encryption. In special, the task distribution mainly assign data unit to the four GPUs in order to perform encryption tasks in parallelism. These three steps are all processed in CPU and host memory. We combine them to a procedure called preprocessing.

Transferring and Encrypting. Four GPUs can be used by s single host process at the same time. Transferring input data from host to the 4GPUs and transferring the ciphertext back from the 4GPUs to host relay on the system bus. Therefore, we need to consider the 4GPUs device and the system bus as the critical resources. In detail, while the 4GPUs execute encryption in parallel, the host can still perform preprocessing step and the third step. This is the mainly reason for acceleration in our study. Moreover, the 4GPUs have been stay working state all the time.

Merging and Outputting. Finally, the host will take the tasks that merge the ciphertext of data unit into the ciphertext data block as output. Based on the above abstraction, we present a pipeline architecture for CPU-4GPUs hybrid system, as shown in Fig. 4. The preprocessing is handled by a host process unit, which is called Host_Proc1. The transferring and encrypting procedure is handled by the system bus (with CPU instructions) and 4GPUs, which is called Host_Proc2_GPU. The merging and outputting procedure is handled by a host process unit called Host_Proc3. In Fig. 4, the process unit is a batch of data encryption, and the granularity of process unit is very important to the efficiency.

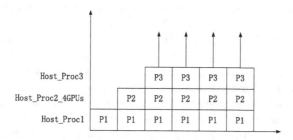

Fig. 4. Pipeline architecture of processing stream in CPU-4GPUs hybrid system

4 Experimental Results

4.1 Experiment Environment and Dataset

We have used a system with Intel Core i5-5200U CPU having clock speed of 2.20 GHz with 2 cores and 4 logical processors, 12 GB RAM, and Windows 10 Operating System. And we make use of the NVIDIA Tesla C2050s, each of which has 3 GB memory running at 1150 MHz. All the codes are implemented in C++ and CUDA, which is an SDK for NVIDIA GPU, and which is an integrated complier for C++ and CUDA. We took a data set containing data block with size of 128 MiB. In detail, We have generated plaintext as a series of random integers for homomorphic computation and used encrypted 64-bit representation for integer. Thus, our data sets ranged from 8 MiB to 64 MiB of plaintexts. In this section, we will evaluate our method in performance and compare the FPHE with the SHE and the SPHE respectively.

4.2 Granularity of Process Unit

In our study, the granularity of process unit indicates the maximum number of data units, which are processed by the 4 GPUs. This is a major factor that influences the performance of the proposed method. As it determines whether the processing stream can flow smoothly or not. We have generated plaintext as a series of random integers for encryption and used encrypted 64-bit representation for integer in our experiment.

In specific, we mainly tested the throughput in a various process unit granularities, and the test result with different number of data units was shown in Fig. 5. In this test, we kept the queue of waiting tasks full, which contained the tasks to be processed, so that the throughput would not be influenced by input. It has been numerically shown that the throughput of the system is low when the number of number of data units is small. This is because the GPUs resources cannot be effectively utilized, and the most of time is wasted on waiting and communications between host and GPUs. With the increase of the number of data units, the throughput also rises. It means that, with the growth of the number of data units, the GPU resources can be utilized more effectively. However, we can see the throughput dropping down when the number of data units is more than 55. The reason is that, when the number of data units goes beyond the optimum value, a large-size process unit will cost more time in GPU, while the host must spend more time on waiting.

Figure 6 shows the response time of a data unit encryption. It has been numerically shown that the response time increases with the growth of the number of data units. By analyzing, the reason is that the larger data size cost more time in the transferring and data encryption procedure. Nevertheless, the slope of the curve becomes larger and tends to be constant. This is because the number of data units is small, time is mostly consumed in data transferring step, and also when the number becomes large enough, the response time will have an approximate linear increase with the data size.

In detail, compared with a serial approach, there are many steps added in our method, including data distribution and transmission, interaction within and between the processing units. These additional parts will affect the efficiency of the system, especially when the system is under low workload. Therefore, relatively large size data are utilized in the experiment and the data are representative for the workload in the parallelism. Therefore, we estimate the performance for different granularities from throughput and response time, which is also called window size. By regulating the sliding window size, finally we adopt 55 as the value of granularity. Because our main purpose is to improve the throughput of system and the throughput achieves maximum while the response time in this granularity is also preferable. As a compromise solution, we choose the case of window size w = 55 for our implementation, which is referred as granularity of process unit.

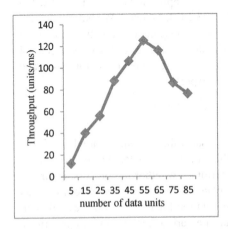

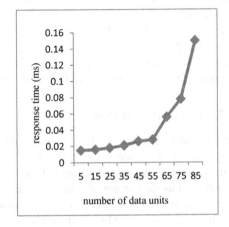

Fig. 5. Throughput under different load levels

Fig. 6. Response time under different load levels

4.3   Evaluation of the Proposed Method

In order to evaluate how well the proposed method has practical application significance, we quantitatively test the improvement of the run time by comparing the FPHE
with the SHE and the SPHE based on the add and multiplication functions over
integers. In detail, the data block is divided into different data units to test, with the size
of data blocks being 8MiB, 16MiB, 32MiB, and 64MiB.

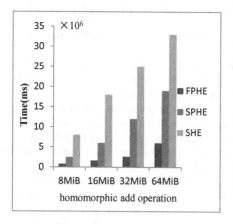

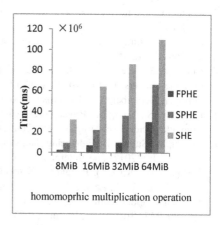

Fig. 7. Run time of homomorphic add
operation

Fig. 8. Run time of homomorphic multiplication operation

Figures 7 and 8 shows run time of applying different size of data block by comparison of experimental results between the three methods with homomorphic add
operation and multiplication operation respectively. Our results show that the FPHE
provides more than 91% improvement (run time) over the SHE and more than 70%
improvement over the SPHE with homomorphic add operation. Also the results show
that the FPHE provides more 90% improvement over the SHE and more than 71%
improvement over the SPHE with homomorphic multiplication operation. As shown in
the Figs. 7 and 8, run time of the FPHE has been parallelized using the 4GPUs is
shorter than the serial mode and the somewhat parallel mode.

5   Conclusions and Discussions

In this research, we use hybrid CPU-4GPUs as a parallel coprocessor to execute DGHV
homomorphic encryption. We further adopt the pipeline architecture of processing
stream in CPU-4GPUs hybrid system to accelerate the method in our study. The
experimental results showed that our proposed method gains more than 91%
improvement (run time) and 71% improvement compared to the SHE and the SPHE
respectively. However, to further illustrate the effectiveness of the proposed method,

decryption and evaluation algorithms of DGHV will be implemented on the hybrid CPU-4GPUs system.

References

1. Rivest, R.L., Adleman, L., Dertouzos, M.L.: On data banks and privacy homeomorphisms. In: Foundations of Secure Computation, pp. 169–177 (1978)
2. Gentry, C.: A fully homomorphic encryption scheme. Ph.D. thesis, Stanford University (2009)
3. Gentry, C.: Fully homomrphic encryption using ideal lattices. In: ACM STOC 2009, pp. 169–178 (2009)
4. van Dijk, M., Gentry, C., Halevi, S., Vaikuntanathan, V.: Fully homomorphic encryption over the integers. In: Gilbert, H. (ed.) EUROCRYPT 2010. LNCS, vol. 6110, pp. 24–43. Springer, Heidelberg (2010). https://doi.org/10.1007/978-3-642-13190-5_2
5. Fugkeaw, S., Sato, H.: Privacy-preserving access control model for big data cloud. In: Computer Science and Engineering Conference, pp. 1–6. IEEE Press, Changchun (2016)
6. Swathi, R., Subha, T.: Enhancing data storage security in cloud using certificateless public auditing. In: 2017 2nd International Conference on Computing and Communications Technologies (ICCCT), Bangkok, Thailand, pp. 348–352 (2017)
7. Jiang, W., Zhao, Z., Laat, C.D.: An autonomous security storage solution for data-intensive cooperative cloud computing. In: 2013 IEEE 9th International Conference on e-Science, pp. 369–372. IEEE Press, Beijing (2013)
8. Shu, J., Jia, X., Yang, K., et al.: Privacy-preserving task recommendation services for crowdsourcing. In: IEEE Transactions on Services Computing (2018)
9. Zhang, J., Li, H., Liu, X., et al.: On efficient and robust anonymization for privacy protection on massive streaming categorical information. In: IEEE Transactions on Dependable and Secure Computing, vol. 14, pp. 507–520. IEEE (2017)
10. Jayapandian, N., Rahman, A.M.J.M.Z.: Secure and efficient online data storage and sharing over cloud environment using probabilistic with homomorphic encryption. Clust. Comput. **20**, 1561–1573 (2017)
11. Sethi, K., Majumdar, A., Bera, P.: A novel implementation of parallel homomorphic encryption for secure data storage in cloud. In: 2017 International Conference on Cyber Security and Protection of Digital Services (Cyber Security), pp. 1–7. IEEE Press, San Francisco (2017)
12. Min, Z., Yang, G., Shi, J.: A privacy-preserving parallel and homomorphic encryption scheme. Open Phys. **15**, 135–142 (2017). De Gruyter Open
13. Khedr, A., Gulak, G.: SecureMed: secure medical computation using GPU-accelerated homomorphic encryption scheme. IEEE J. Biomed. Health Inform. **22**, 597–606 (2017)
14. Tian, Y., Al-Rodhaan, M., Song, B., et al.: Somewhat homomorphic cryptography for matrix multiplication using GPU acceleration. In: 2014 International Symposium on Biometrics and Security Technologies (ISBAST), pp. 166–170. IEEE Press, Kuala Lumpur (2015)
15. Moayedfard, M., Molahosseini, A.S.: Parallel implementations of somewhat homomorphic encryption based on open-MP and CUDA. In: 2015 International Congress on Technology, Communication and Knowledge (ICTCK), pp. 186–190. IEEE Press, Mashhad (2015)
16. Wang, W., Hu, Y., Chen, L., et al.: Accelerating fully homomorphic encryption using GPU. In: 2012 IEEE Conference on High Performance Extreme Computing, pp. 1–5. IEEE Press, Waltham (2012)

A Method of Component Discovery in Cloud Migration

Jian-tao Zhou, Ting Wu, Yan Chen, and Jun-feng Zhao[✉]

College of Computer Science, Inner Mongolia University, Hohhot, China
310221433@qq.com, cswt@mail.imu.edu.cn, hillstone369@sina.com,
cszjf@imu.edu.cn

Abstract. Cloud migration is an important means of software development on the cloud. The identification of reusable components of legacy systems directly determines the quality of cloud migration. The existed clustering algorithms do not consider the factor of relation types between classes, which affects the accuracy of clustering result. In this paper, the relation type information between classes is introduced in software clustering. Multi-objective genetic algorithm is used to cluster the module dependency graph with the relationship types (R-MDG). The experimental results show that the above method can effectively improve the quality of reusable components.

Keywords: Reusable component · Software clustering
Relation type · Genetic algorithm

1 Introduction

With the increasing scalability and complexity of software, software development has become increasingly difficult so that the software quality might decline [1]. Software reengineering improves the efficiency of software development by using the original parts of software or legacy systems [2]. Software migration is one of the common ways to implement software reengineering. In the process of software migration, the most key and difficult work is to analyze the legacy systems and try to find the appropriate reusable components for the system developers to reuse [3]. Software clustering is a useful method for reusable component discovery, which can not only reflect the relationship between modules or components but also make the result of clustering more accord with the actual system.

Especially after the concept of cloud computing was proposed, cloud-oriented software development is becoming a hot topic in the area of software engineering. Reengineering the existing software or the legacy systems to cloud can reduce the workload of developers [4,5], rather than developing the similar software systems from the beginning in cloud. Cloud migration is the process of migrating all or part of a software's data, applications or services to the cloud to enable information on-demand services. [6] divided the migration into three strategies: migration to IaaS, migration to PaaS, and SaaS-related migration.

X. Meng et al. (Eds.): WISA 2018, LNCS 11242, pp. 91–102, 2018.
https://doi.org/10.1007/978-3-030-02934-0_9

In the field of software migration, existing clustering methods do not take into account the specific relationship types between modules and only consider the calling relationship, which is more depends on the user's input. In addition, existing software clustering methods are not combined with cloud computing, so as to fully utilize the advantages of cloud computing.

This paper is a research on the algorithm of reusable components discovery in cloud migration, which proposes a method of software clustering based on the relationship between classes. The rest of this paper is organized as follows. Section 2 introduces the related work in the field of software clustering. Section 3 presents a method of software clustering. The experiment is presented in Sect. 4 and conclusions are drawn in Sect. 5.

2   Related Work

Through clustering algorithm, complex system can be divided into many smaller subsystems, which separately complete some special functions, or have complete meaning [7]. In the process of software clustering, using different clustering algorithm can obtain different clustering results [8,9]. [10–12] introduce different ways to implement software module clustering.

In [13] and [14], an extended hierarchical clustering algorithm LIMBO is proposed, which allows for weighting schemes that reflect the importance of various attributes to be applied. [15] uses directed graph, which called module dependency graphs (MDG), to make the complex structure of software system easier to understand. In MDG, nodes represent the system's modules, and directed edges represent the relationship between nodes. In this way, software clustering is transformed into the division of the graph.

A graph has many partitions. In the process of software clustering, the clusters are expected to have higher cohesion and lower coupling. [15] uses modularization quality (MQ) to evaluate clusters. Figure 1 is an example of MDG.

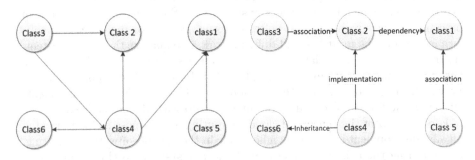

Fig. 1. Example of MDG.   **Fig. 2.** Example of R-MDG.

Praditwong et al. proposed the use of Pareto multi-objective genetic algorithm for software clustering [16]. However, the method just takes into account the call relationship in the source code of software system. The number of the relations usually more depends on the user's input, which is relatively subjective.

The existing software clustering methods do not take into account the specific relation types between classes such as inheritance, implementation and so on, which may benefit to the accuracy and efficiency of component discovery. In addition, when dealing with large-scale data, the traditional computing model is inefficient and cannot obtain the software clustering results efficiently.

Based on the above analysis, this paper proposes the idea of mapping relation types between classes to dependency weights. Combining with the above mentioned Pareto multi-objective genetic algorithm to achieve software clustering, the revised method can get reusable components more reasonably. In addition, the algorithm can be deployed into cloud platform to improve the efficiency of reusable component discovery. Through experimental analysis, the clustering result based on MDG with the relationship types (R-MDG) is more in line with the real structure of software system.

3   Software Clustering Based on the R-MDG

In order to enable system developers to use more reusable components reasonably in the software reusing process, the proposed clustering method takes advantage of the relation type information in the process of clustering.

3.1   Description of Software Realization

In the process of analyzing the software by DependencyFinder [17], the relation types between classes should be extracted. The different relation types reflect the different degrees of cohesion and coupling. In the process of software clustering, R-MDG instead of MDG will be used as input.

The nodes in R-MDG represent the classes or components in object-oriented software, and the directed edges in the graph represent the relationship between the classes or components, and the weight on the directed edge represents the specific relation type.

For an object-oriented system, the relationship between the classes exists in the following four types:

- Inheritance. Express the relationship between general and special classes.
- Implementation. Specify the relationship between interface and implementation class.
- Association. Represent a structure connection between two classes.
- Dependency. Describe the business dependency between two classes.

According to the relationship specified above, these different types are mapped to the different weights on the edges between the nodes in R-MDG.

When the relationship between the classes is relatively strong, a higher weight will be assigned to the edge. Different weights help to improve the accuracy of clustering algorithm.

3.2   Definition of R-MDG

According to the description above, R-MDG will be constructed as a directed graph with weight $G = (V, E)$. R-MDG can be divided into n clusters by clustering ($\prod_G = \{G_1, G_2, G_3, \cdots, G_{n-1}, G_n\}$), and each $G_i((1 \leqslant i \leqslant n) \wedge (n \leqslant |V|))$ represents a cluster, which is formalized as follows:

$$
\begin{gathered}
G = (V, E), \\
\prod_G = \bigcup_{i=1}^{n} G_i, G_i = (V_i, E_i), V = \bigcup_{i=1}^{n} V_i, \\
E = \{\langle V_i, V_j, R\rangle | V_i, V_j \in V\}, \\
\forall ((1 \leqslant i, j \leqslant n) \wedge (i \neq j)), V_i \cap V_j = \phi, \\
R = \{inheritance, implementation, association, dependency\}
\end{gathered}
\tag{1}
$$

In R-MDG, if class A has relation with class B, there will exist a directed edge <A, B> or <B, A>, which is assigned the corresponding attribute. Figure 2 is an example of R-MDG. In clustering, different relation types are mapped to different weights.

3.3   Evaluation of R-MDG Partition in Software Clustering

Before clustering, DependencyFinder can be used to analyze the source code of software and discover the relationship between classes in software. Based on the classes and their relations, R-MDG will be established, and then the problem of software clustering is transformed into R-MDG partition.

A R-MDG has many different clustering partitions. How to evaluate the partitions is the key. Modularization quality (MQ) is used to evaluate the quality of MDG partitions [18]. The higher the cohesion degree and the lower the coupling degree to one clustering partition is, the larger the value of the MQ will be.

The relations between the classes within same clusters determine the degree of cohesion. μ_{inner} represents the cohesion degree of the ith cluster in one partition. The relations between the clusters determine the coupling degree. μ_{outer} represents the coupling degree of the ith cluster of the R-MDG.

Thus, for a partition, the R-MQ is calculated as follows:

$$
R - MQ = \sum_{i-1}^{k} CF_i,
$$

$$
CF_i = \begin{cases} 0 & \mu_{inner} = 0 \\ \dfrac{2\mu_{inner}}{2\mu_{inner} + \mu_{outer}} & otherwise \end{cases},
$$

$$
\mu_{inner} = W_E * E_\mu + W_I * I_\mu + W_A * A_\mu + W_D * D_\mu,
$$

$$
\mu_{outer} = \sum_{j=1, j \neq i}^{n} \begin{pmatrix} W_E * (E_{i,j} + E_{j,i}) + W_I * (I_{i,j} + I_{j,i}) + \\ W_A * (A_{i,j} + A_{j,i}) + W_D * (D_{i,j} + D_{j,i}) \end{pmatrix}
\tag{2}
$$

CF_i represents the modular quality of the ith cluster. W_E, W_I, W_A and W_D represent the weight value of relation inheritance, implementation, association and dependency. E_μ, I_μ, A_μ, and D_μ represent the number of edges with the relations in the ith cluster. $E_{i,j}$, $I_{i,j}$, $A_{i,j}$, and $D_{i,j}$ denote the number of the edges with the relations from the ith cluster to the jth cluster.

According to the difference of objective functions in the process of clustering with method of Pareto multi-objective genetic algorithm, there exist two different multi-objective evaluation methods, which are Maximal-Size Cluster Approach (R-MCA) and Equal-Size Cluster Approach (R-ECA).

In R-MCA, the following five objective functions are used:

- Weighted cohesion degree
- Weighted coupling degree
- Number of clusters
- R-MQ
- Number of isolated nodes

In R-ECA, the following five objective functions are used:

- Weighted cohesion
- Weighted coupling
- Number of clusters
- R-MQ
- The difference between the maximum and minimum number of nodes in the clusters

3.4   Description of the Software Clustering Algorithm

According to the process of software clustering, the R-MDG is partitioned by genetic algorithm. The reusable components can be obtained by finding the optimal partition of the R-MDG. The process of software clustering using Pareto multi-objective genetic algorithm is shown in Algorithm 1.

Algorithm 1. Pareto Software Clustering by GA

Input: $R - MDG$, the number of iterations T;
Output: optimal *Partition*;
1 **Initialize** n $R - MDG$ partitions;
2 $Pop \leftarrow encode(n\ Partitions)$;
3 $EvaluateMultiObjective(Pop)$;
4 **for** $j = 1 \rightarrow T$ **do**
5   **while** $Pop.size() \neq n$ **do**
6   *Randomly Generate Partitions*;
7        $Pop \leftarrow encode(Partitions)$;
8   **end**
9      $Nondominant\ individual \leftarrow Sort(Pop)$;
10      $Selection(Pop)$;
11      $Crossover(Pop)$;
12      $Mutation(Pop)$;
13      $newPop \leftarrow Pop$;
14      $EvaluateMultiObjective(newPop)$;
15      $j \leftarrow j + 1$;
16 **end**

According to the definition of the Pareto optimal, Algorithm 1 uses multi-objective to evaluate the partition, then sort the partitions to find the non-dominated partitions. Through the implementation of genetic algorithms to find optimal partitions.

3.5   Realization of the Clustering Method

In the process of Pareto multi-objective genetic algorithm, it is necessary to calculate the value of each Pareto objective function for each cluster in each generation. Due to the large scale of clusters in each generation population, the computation is time-consuming. Migrating the traditional algorithm to the cloud platform and using parallel computing model, the objective functions can be evaluated efficiently.

MapReduce can deal with the large-scale data in parallel [19], so the clusters in each generation can be divided into several data sets. In each data set, the values of the individual objective functions for each partition will be calculated. In this way, the calculations of the objective functions in each generation can be distributed to different computing nodes. Summarizing the calculation results of each node, the evaluation value can be obtained for each generation.

4   Experiment and Analysis

A tool is implemented to analyze the reusable components in Java software system by clustering. In order to verify the rationality of the clustering process, three packages in JUnit framework are selected to be analyzed.

4.1   The Construction of R-MDG

The classes in the JUnit kernel package are converted to the nodes of the R-MDG, and the relations between the classes are converted into the edges of the R-MDG. The R-MDG represented by the GraphML [20].

GraphML is a kind of XML-based file format to describe graph. To the description of the edge element, in addition to the source and target attributes which represent the source node and the destination node of the edge, the attribute relationType represents the relation type between nodes. The GraphML is used as the input of the clustering process.

4.2   The Evaluation of Clustering Results

MoJo is a distance matrix which compares the clustering result to the standard clustering result through calculating the minimum number of operations required for adjusting the produced result to the standard result [21]. In this way, the difference between the experimental results and the reference standard can be more visualized.

[6] gives a reference partition for JUnit as shown in Fig. 3. The columns of the matrix represent the classes, and the rows represent the clusters. The number 1 indicates that some class is partitioned into some cluster.

$$\begin{bmatrix} 1\,1\,0\,0\,0\,0\,0\,0\,0\,0\,0\,0\,0\,0\,0\,0\,0\,0\,0\,0\,0\,0\,0\,0\,0 \\ 0\,0\,0\,1\,1\,0\,0\,0\,0\,1\,0\,0\,0\,0\,0\,1\,0\,0\,0\,0\,0\,0\,0\,0\,0 \\ 0\,0\,0\,0\,0\,1\,1\,0\,1\,0\,0\,0\,0\,0\,0\,0\,0\,0\,0\,0\,0\,0\,0\,0\,0 \\ 0\,0\,0\,0\,0\,0\,0\,1\,0\,0\,0\,0\,0\,1\,0\,1\,0\,0\,0\,0\,1\,0\,0\,0 \\ 0\,0\,0\,0\,0\,0\,0\,0\,0\,0\,1\,0\,0\,0\,0\,0\,1\,1\,0\,0\,0\,0\,1\,0 \\ 0\,0\,0\,0\,0\,0\,0\,0\,0\,0\,0\,1\,1\,1\,0\,0\,0\,0\,0\,0\,0\,0\,0\,0 \\ 0\,0\,0\,0\,0\,0\,0\,0\,0\,0\,0\,0\,0\,0\,0\,0\,0\,1\,1\,0\,1\,0\,1 \end{bmatrix}$$

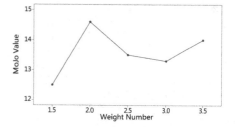

Fig. 3. A reference partition of JUnit.

Fig. 4. MoJo values of different values of parent relationship.

4.3 The Determination of the Weights of Specific Relation Types

In Java, the cohesion degrees of generalization and implementation relation are similar, and the cohesion degrees of dependency and association relation are close. Therefore, when constructing R-MDG, the relations with generalization and implementation between classes are unified into parent relationship. Because the cohesion degrees of dependency and association relation are relatively weak, the weights of them are set to 1.0, and the weights of the generalization and implementation relations are determined by experiment.

The experiment sets the weight of the generalization and implementation to a value between 1.5 and 3.5. In order to highlight the comparison, the weights are at 0.5 intervals and each configuration is run 10 times. The result of the clustering is compared with the reference partition to obtain the average MoJo, and the average MoJo value is shown in Fig. 4.

It can be seen from the figure that the average of MoJo is 12.5 when the weight of the generalization and realization relation is set to 1.5, and the mean value of MoJo varies between 13.3 and 14.6 when the weight of generalization and realization is greater than 2. Therefore, the clustering result is better when the weight of the generalization and realization relation is set to 1.5.

4.4 The Comparison of the Clustering Results Between Introducing Relation Type and Not Introducing Relation Type

According to the above software clustering process, two different parameters are set in the genetic algorithm to conduct the experiment. When calculating R-MQ, 1.5, 1.5, 1.0, and 1.0 are selected as the edge weights in the R-MDG that represent inheritance, implementation, association, and dependency. Two group of different parameters are assigned to genetic algorithm, and every clustering process is run 10 times. The result of each execution is compared with the reference partition, and the values of MoJo are shown in Figs. 5 and 6. The two group of different parameters are listed as follows.

In Fig. 5, the number of genetic iterations is 200, the population size is 100, the crossover probability is 0.8, and the mutation probability is 0.001. In Fig. 6, the number of genetic iterations is 100, the population size is 200, the crossover probability and the mutation probability are same with Fig. 5.

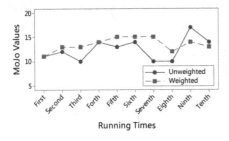
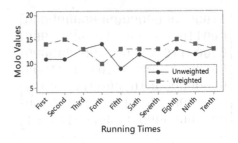

Fig. 5. One group of MoJo values considering relationship type or not.

Fig. 6. Another group of MoJo values considering relationship type or not.

From the comparison of the data in the figure, it can be find in Fig. 1, the average MoJo obtained from weighted experiments is 12.5, and the average MoJo obtained from unweighted experiments is 13.5. However in Fig. 2, the average MoJo obtained from weighted experiments is 11.8, and the average MoJo obtained from unweighted experiments is 13.3. Therefore, the introduction of the relation types between classes into the software clustering process can effectively improve the accuracy of software clustering, and clustering results are closer to the actual structure of software system.

4.5   The Comparison of the Clustering Results Between Single-Objective and Multi-objective Evaluation

According to the process of software clustering using Pareto multi-objective genetic algorithm, the crossover probability of genetic algorithm is set to 0.8, the mutation probability is 0.001, the number of genetic iterations is 200, and the population size is 100. The weights of the generalization and realization relation are set to 1.5, and the weights of dependency and association are set to 1.0. According to the selection of different objective function, the R-MCA and the R-ECA are used to cluster the logical classes in JUnit 3.8. The average MoJo value of single-objective and multi-objective obtained by running 10 times is shown in Figs. 7 and 8.

(1) R-MCA

It can be seen from the experiment that there are five average MoJo values of multi-objective clustering lower than that of single-objective clustering in ten times. Therefore, the results obtained by single-objective software clustering method are similar to R-MCA software clustering method. The results of Pareto multi-objective software clustering using R-MCA are analyzed as follows. In the results of the fifth software clustering, 48 non-dominated solutions are generated, and 10 non-dominated solutions are shown in Fig. 9. It also can be seen R-MQ value can not determine the quality of the partition in the multi-objective clustering method. When the cohesion is low, the value of R-MQ may be large, such as the 7th clustering result.

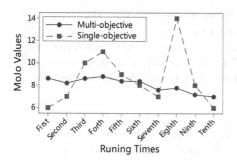

Fig. 7. Comparison of the MoJo values of single-objective results and R-MCA results.

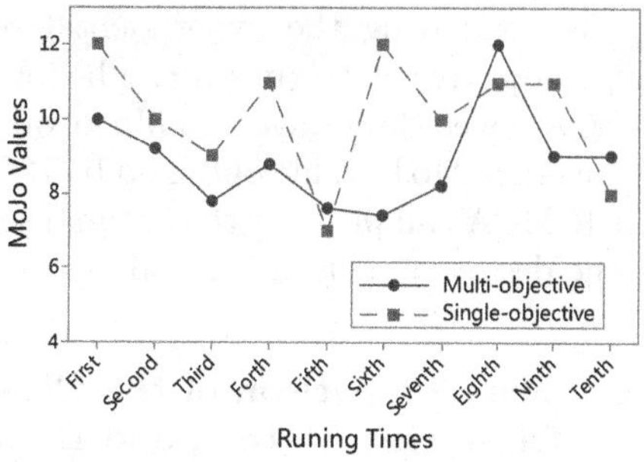

Fig. 8. Comparison of the MoJo values of single-objective results and R-ECA results.

(2) R-ECA

It can be seen from the Fig. 8 that there are seven average MoJo values of multi-objective clustering lower than that of single-objective clustering in ten times. Therefore, compared with the clustering results obtained by the single-objective software clustering method, the result of R-ECA software clustering method is closer to standard partition.

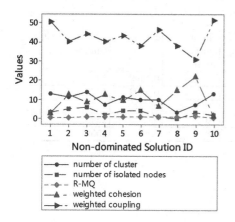

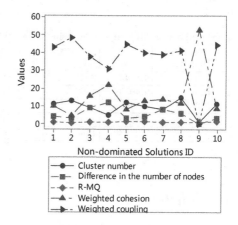

Fig. 9. Objective function values of R-MCA.

Fig. 10. Objective function values of R-ECA.

The results of Pareto multi-objective clustering using R-ECA are analyzed as follows. In the results of the first software clustering, 35 non-dominated solutions are generated, and 10 non-dominated solutions are shown in Fig. 10.

As can be seen from Fig. 10, if the number of cluster partitions required is large, each node should be distributed evenly into each cluster partition as much as possible.

By comparing the experimental results of single-objective clustering and multi-objective clustering, it can be found that the clustering results obtained by R-ECA have a lower average MoJo value than single-objective results. However, the average MoJo value obtained by R-MCA is similar to that in single-objective, but R-MCA can produce the optimal solution when the number of isolated nodes in the desired partition is small.

4.6 The Comparison of the Clustering Results Between Improved Clustering Method and Bunch Method

Bunch clustering method [15] is a traditional method to solve clustering problems which uses MDG for clustering. Using GraphML file as input, each method executes 5 times. Tables 1 and 2 show the clustering result of Bunch method and improved clustering method. It can be found that the results of the improved method are more stable than the Bunch. The MQ values in Table 1 are much lower than the values of MQ in Table 2. The reason is that the improved method averages the MQ in the original method when calculating MQ. At the same time, since the partial redundancy calculation in the Bunch method is avoided, the execution time of the improved clustering algorithm is smaller than that of the Bunch.

Table 1. Clustering result obtained by Bunch

Run	Number of clusters in result	MQ value	Clustering time (millisecond)
1	10	4.3044	2031
2	10	4.4069	1766
3	2	1.5406	922
4	8	4.2988	2203
5	5	2.9583	1140

Table 2. Clustering result obtained by improved method

Run	Number of clusters in result	MQ value	Clustering time (millisecond)
1	9	0.5409	1275
2	11	0.5402	1473
3	9	0.5229	1335
4	10	0.5289	1245
5	8	0.5578	1559

5 Conclusion

In the process of software development, in order to improve development efficiency and avoid duplication of work, discovering the proper size of the reusable components has become the key. A revised clustering algorithm is proposed in this paper which uses genetic algorithm for software clustering and considers the relationship between classes. It can be seen from the experimental results that the clustering result of introducing the relationship types between classes is obviously better than the result of no introduction. Two multi-objective evaluation methods are used to evaluate the results of software clustering, which is convenient for system developers to choose richer partitions in software reuse.

This method analyzes the explicit and implicit relations between the classes. Through clustering, the system developers can select the appropriate components to reuse according to user needs. However, due to the size and complexity of legacy systems, the computation task of software clustering is large and complex. As a result, the efficiency of the component discovery process can be improved in a parallel manner. Clustering can be efficiently realized, and reusable components can be discovered efficiently by using cloud computing.

Acknowledgment. The authors wish to thank Natural Science Foundation of China under Grant No. 61462066, 61662054, Natural Science Foundation of Inner Mongolia under Grand No. 2015MS0608, Inner Mongolia Science and Technology Innovation Team of Cloud Computing and Software Engineering and Inner Mongolia Application Technology Research and Development Funding Project "Mutual Creation Service Platform Research and Development Based on Service Optimizing and Operation Integrating". Inner Mongolia Engineering Lab of Cloud Computing and Service Software and Inner Mongolia Engineering Lab of Big Data Analysis Technology.

References

1. Ekabua, O.O., Isong, B.E.: On choosing program refactoring and slicing reengineering practice towards software quality, April 2012
2. Masiero, P.C., Braga, R.T.V.: Legacy systems reengineering using software patterns. In: Proceedings of the XIX International Conference of the Chilean Computer Science Society, SCCC 1999, pp. 160–169 (1999)
3. Zheng, Y.L., Hu, H.P.: Making software using reusable component. Comput. Appl. **20**(2), 35–38 (2000)
4. Jamshidi, P., Ahmad, A., Pahl, C.: Cloud migration research: a systematic review. IEEE Trans. Cloud Comput. **1**(2), 142–157 (2014)
5. Fowley, F., Elango, D.M., Magar, H., Pahl, C.: Software system migration to cloud-native architectures for SME-sized software vendors. In: Steffen, B., Baier, C., van den Brand, M., Eder, J., Hinchey, M., Margaria, T. (eds.) SOFSEM 2017. LNCS, vol. 10139, pp. 498–509. Springer, Cham (2017). https://doi.org/10.1007/978-3-319-51963-0_39
6. Zhao, J.F.: Research on component reuse of legacy system in cloud migration. PhD thesis, Inner Mongolia University (2015)
7. Wu, J., Hassan, A.E., Holt, R.C.: Comparison of clustering algorithms in the context of software evolution. In: IEEE International Conference on Software Maintenance, pp. 525–535 (2005)
8. Anquetil, N., Fourrier, C., Lethbridge, T.C.: Experiments with clustering as a software remodularization method. In: Proceedings of the Sixth Working Conference on Reverse Engineering, pp. 235–255 (1999)
9. Maqbool, O., Babri, H.: Hierarchical clustering for software architecture recovery. IEEE Trans. Softw. Eng. **33**(11), 759–780 (2007)
10. Kumari, A.C., Srinivas, K.: Hyper-heuristic approach for multi-objective software module clustering. J. Syst. Softw. **117**, 384–401 (2016)
11. Jeet, K., Dhir, R.: Software module clustering using hybrid socioevolutionary algorithms. **8**, 43–53 (2016)

12. Zhong, L., Xue, L., Zhang, N., Xia, J., Chen, J.: A tool to support software clustering using the software evolution information. In: IEEE International Conference on Software Engineering and Service Science, pp. 304–307 (2017)
13. Andritsos, P., Tzerpos, V.: Information-theoretic software clustering. IEEE Trans. Softw. Eng. **31**(2), 150–165 (2005)
14. Wang, Y., Liu, P., Guo, H., Han, L., Chen, X.: Improved hierarchical clustering algorithm for software architecture recovery, 247–250 (2010)
15. Doval, D., Mancoridis, S., Mitchell, B.S.: Automatic clustering of software systems using a genetic algorithm. In: Software Technology and Engineering Practice, p. 73 (1999)
16. Praditwong, K., Harman, M., Yao, X.: Software module clustering as a multi-objective search problem. IEEE Trans. Softw. Eng. **37**(2), 264–282 (2010)
17. Dependencyfinder. https://github.com/Laumania/DependencyFinder
18. Mancoridis, S., Mitchell, B.S., Chen, Y., Gansner, E.R.: Bunch: a clustering tool for the recovery and maintenance of software system structures. In: IEEE International Conference on Software Maintenance, pp. 50–59 (1999)
19. Jingbo, X.I.A., Zekun, W.E.I., Kai, F.U., Zhen, C.H.E.N.: Review of research and application on hadoop in cloud computing. Comput. Sci. **43**(11), 6–11 (2016)
20. The graphml file format. http://www.graphml.graphdrawing.org/
21. Tzerpos, V., Holt, R.C.: MoJo: a distance metric for software clusterings. In: Working Conference on Reverse Engineering, p. 187 (1999)

Online Aggregation: A Review

Yun Li[2], Yanlong Wen[2]([✉]), and Xiaojie Yuan[1,2]

[1] College of Cyberspace Security, Nankai University, Tianjin, China
[2] College of Computer Science, Nankai University, Tianjin, China
{liyun,wenyanlong,yuanxiaojie}@dbis.nankai.edu.cn

Abstract. Recent demands for querying big data have revealed various shortcomings of traditional database systems. This, in turn, has led to the emergency of a new kind of query mode, approximate query.Online aggregation is a sample-based technology for approximate querying. It becomes quite indispensable in the era of information explosion today. Online aggregation continuously gives an approximate result with some error estimation (usually confidence interval) until all data are processed. This survey mainly aims at elucidating the most critical two steps for online aggregation: sampling mechanism and error estimation methods. As the development of MapReduce, researchers try to implement online aggregation in MapReduce framework. We will also briefly introduce some implementations of online aggregation in MapReduce and evaluate their features, strength, and drawbacks. Finally, we disclose some existing challenges in online aggregation, which needs attention of the research community and application designers.

Keywords: Online aggregation · Big data · Approximate query
MapReduce

1 Introduction

Nowadays, with the explosive growth of big data in the fields of government management, medical services, retailing, manufacturing, and location services involving individuals, the enormous social value and industrial space it brings cannot be ignored. The huge commercial value, scientific research value, social management and public service value contained in big data and the value of supporting scientific decision-making are being recognized and developed. More and more organizations tend to excavate the valuable information and knowledge contained in large amounts of data according to actual need, and ultimately achieve the goal of fully utilizing the value of the data. On the one hand, it can reduce the cost of enterprises and improve the efficiency of the industry. On the other hand, it can promote enterprise innovation, and it may create strategic emerging industries such as data services and data manufacturing. Under the above background, our society is shifting to a data-driven society and government management, macro management, industrial policy, education, finance,

© Springer Nature Switzerland AG 2018
X. Meng et al. (Eds.): WISA 2018, LNCS 11242, pp. 103–114, 2018.
https://doi.org/10.1007/978-3-030-02934-0_10

operations and other management activities all exhibit high-frequency real-time characteristics, which also requires much more rapid and effective data analysis methods. The initial stage of data analysis usually uses a data exploration phase to grasp some of the key features of the data set, such as mean, sum, maximum/minimum, and so on. In a traditional relational database, these operations can be implemented by an aggregate function. However, in the era of big data, the time spent on a complete data scan is enormous. In many cases, it is not necessary for the user to spend such a huge calculation cost to get an accurate result. An estimate with some kind of accuracy guarantee is acceptable to the user. Online aggregation is such a method to solve this problem. Let's consider a query like this:

SELECT $attr$, $AGG(expression)$
FROM $R_1, R_2, ..., R_k$
WHERE $predicate$
GROUP BY $attr$;

Where AGG represents aggregate operations including COUNT, SUM, AVG, VARIANCE and STDDEV, $expression$ is what it means in a sql query, and $predicate$ also involves join conditions on the k relations. Let $\nu(i)$ be the aggregate value of tuple i, set $u(i) = 1$ if tuple i satisfies $predicate$ and 0 otherwise. **Online aggregation** is a sampling-based technology to answer aggregation queries immediately with an estimate result after the user submits the query, and the confidence interval gets tighter over time, the result is continuously refined during the increase of the sampled data. Aggregation operator contains COUNT, SUM, AVG, VARIANCE and STDEV. In practical applications, this approximate result can meet the need of subsequent in-depth analysis.

Online aggregation was first proposed by Hellerstein in 1997 [1], it is groundbreaking because it breaks the traditional batch mode of aggregation in a database system. Aggregate queries are usually used to get some characteristics of data. In traditional database, you have to run the entire dataset to get a result, meaning waste of a lot of time. In realistic applications, a long response time will give the user a bad experience. Well, online aggregation can offer a way of interactive query:

(1) The result is continuously refined with the increase of samples. Users can stop the query as soon as the running query is sufficiently precise, then the user can proceed to the next query. User does not have to wait for a long time to see the final result until the query is completed.The interactive query greatly improves query efficiency and saves both time and labor. Meanwhile, it provides a better user experience than the traditional query.

(2) For many big data applications, rather than expending much time and computing resources to get a completely precise result, to obtain a quick estimate result with accuracy assurance is more meaningful and valuable. It meets the need of real-time processing and helps to quickly extract high-value-density information from huge amounts of data.

Online aggregation can be divided into two phases. The first phase is sampling. A high quality sample means the sample is sufficiently random. The second phase is estimation. We can get an approximate result on the sample. The accuracy of the approximate result depends on the variance and bias of samples. We will summary the work from the above two aspects. With the popularity of the prominent parallel data processing tool MapReduce [2], researchers begin to implement online aggregation in MapReduce framework. We will also summarize the work in this area.

2   Sampling Mechanism

The essential core process in online aggregation is sampling. The main idea of online aggregation is estimating the final result by sampled data through a specific method. Therefore, sampling quality is of great importance to statistical guarantees on accuracy. From the process of data sampling, the randomness and unbiasedness of the sample directly affect the accuracy of the query result and the convergence speed of the confidence interval. The initial research work often assumes that the data is stored in a random order or assumes there exists a random queue so that a sequential scan from head to tail can achieve random results. However, in practical applications, the data storage order is often related to a certain attribute. How to perform random sampling from such non-randomly distributed data is a key issue in the online aggregation process. The estimation is meaningful only if the samples are random. We cannot determine the way data is stored on disk, we have to generate a stream of random samples regardless of the characteristics of dataset. As it was, well sampling is half done. As is shown in Fig. 1, we summarize the main technology used in sampling.

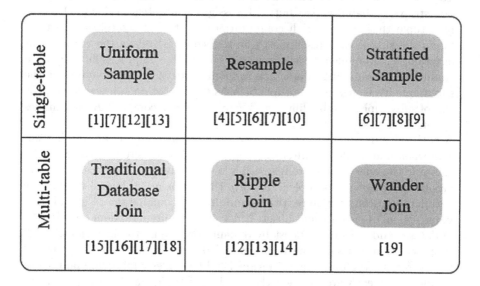

Fig. 1. Sample methods in online aggregation

2.1   Single-Table Sampling

In the initial studies, researchers focused on single table's online aggregation, the problem becomes random sampling data from a relational table. The sampling methods used in online aggregation can be divided into three types.

- **Uniform sample.** Uniform sample ensures that each tuple is sampled with the same probability. The preliminary work of online aggregation uses Olken's method [3] to sample from a single table. Hass uses heap scans and index scans to guarantee the records are retrieved in random order [1]. But it works badly if the order of a leap file is correlated with the values of some attributes of the records or the indexed attributes are not the same as those being aggregated in the query. To this end, [4] uses an iterative scrambler to generate a random-like dataset so that sequential scanning can obtain one full page of random samples from a page rather than several random samples thus reduce the I/O cost. The more random the sample, the more accurate the estimation will be. Uniform sample cost too much to get random samples and assume the dataset has good randomness or randomize the dataset in advance. Unfortunately, the assumption does not hold in most scenarios.

- **Resample.** Resampling is a sampling method represented by bootstrap. Accessing the database multiple times to draw samples causes large I/O, Bootstrap works like this: (1) draw s tuples from the original data D, denoted as S, (2) take K trials on S instead of D, we sample $|S|$ tuples i.i.d from S with replacement. We denote the $i - th$ trail as S^i, $|S^i| = |S|$. Bootstrap use K resamples to simulate the distribution of the original data. Bootstrap method repeatedly sample from sample thus causes huge computational cost. To address this problem, instead resampling continuously, [5,6] simulates the bootstrap's probability distribution by assigning an integer value to the sampled tuple thus reduces high computational cost. Bootstrap-based method limits the size of each resample to be the same, [7] relaxes the requirement of resample and named it variational subsample thus improve efficiency than traditional subsampling. It assigns a random S_{id} to each tuple to represent the subsample it belongs to, one tuple belongs to at most on subsample and the size of subsamples can be different. Therefore we can complete K subsamples through one-pass.

- **Stratified sample.** Estimation based on uniform sampling are often accurate when queries do not filter or group data. When there is a GROUP BY clause in the query, some groups with a low selectivity which is called "small groups" may be undersampled. To solve this problem, [8] logically splits the data into non-conjunctive partitions and linearly combine the results of individual calculations for each subset. It does not require a random distribution of dataset and reduces I/O cost by dynamically adjust the partitions. [9,10] proposed stratified sampling technology, [10] sample the "same" number of tuples from each group, "same" means BlinkDB takes a minimum number of samples from each group when the number of tuples in the group is insufficient, BlinkDB takes the whole group as sample which means the result is accurate and therefore does not need to estimate. All these work above guarantee the accuracy of each group, but ignore to preserve the correct order between groups. [11] proposes IFOCUS algorithm whose basic idea is to maintain confidence interval for each group, keep drawing samples when the group's confidence interval overlaps with other groups', stop sampling when there are no other groups overlap with itself. The algorithm has been proven to give correct ordering results with certain accuracy.

2.2 Multi-table Sampling

Actually, queries on a table are usually simple and account for a small proportion of the total query operations. Complex queries often involve multiple tables in which joins are of great concern. During the entire query execution, joins cost the most. Online aggregation was first proposed for a single table, then Hass extended this work on multiple tables. He proposed ripple join [12,13] which was designed based on nested-loop join and hash join, aimed to obtain an estimate result as soon as possible for the premise of accuracy. The main idea is to continuously draw samples from each table, iteratively join till the confidence interval is accurate enough. It overcomes the non-uniform distribution of samples in traditional join query of two uniform random samples. Ripple join performs badly when memory overflows. On this basis, [14] raised parallel hash ripple join algorithm, it extended ripple join via simultaneously sampling and simultaneously executing the merge-sort join on multiple tables. It maintains good performances when memory overflows. There are also some improved join algorithms based on the traditional database join algorithm [15–18]. All the above algorithm is based on ripple join which is still expensive and assumes data is stored randomly or use some technology to generate a stream of random samples, this is contrary to query optimization algorithms such as sorting and indexing. Ripple join is uniform, however, not independent. To solve the problem, [19] proposed a new algorithm–wander join. Wander join converts join algorithm into random walks over the join graph where a node represents a tuple and an edge between them if they can join, thus does not have to make any assumptions about data. Wander join has improved a lot efficiency in comparison with ripple join both in memory and memory overflows. But wander join is not uniform.

Maybe you have found that the above works are all for a single query, when a series of queries arrives, something can be shared. Wu et al organizes multiple queries in a DAG(Dynamic Acyclic Graph) to discover the intersection between multiple query results [4], taking advantage of the intermediate results to get higher efficiency. [20] uses "sample buffer" in the first level to store sampling results for sharing and save partial statistics calculation in the seconde level to reduce the computational cost. Sharing reduces the repetitive sampling or calculation process, but needs extra space to store intermediate results. Anyway, it gives us space for tade-offs.

3   Estimation Methods

With high-quality samples in hand, the question now is how to estimate the ture query result from the samples. In other words, we have to give a set of standards that measure the accuracy of the estimation. Aggregation queries can be divided into two categories: one category contains SUM/COUNT/AVG/VARIANCE which measure some aspects of the overall data, the other contains MAX/MIN which reflect the endpoints of the data set distribution or the characteristics of extreme values. There are few studies on the latter, so we mainly discuss the former category of aggregate functions.

3.1   Closed-Form Estimation

Closed-form estimation ensures the result falls into the confidence interval with a certain probability. The confidence interval refers to a possible range of the final result. Let s be the number of samples, Y_s be the aggregate result on the sample, μ be the true result. The confidence p is determined in advance (95% is usually considered a reasonable value), our goal is to calculate ε_s to produce confidence interval $[Y_s - \varepsilon_s, Y_s + \varepsilon_s]$. Hellers divides confidence interval into two categories: conservative confidence interval and large-sample confidence interval [1].

Conservative confidence interval calculation is based on Hoffing inequality, it promises the probability that μ falls into $[Y_s - \varepsilon_s, Y_s + \varepsilon_s]$ is greater or equal to p(p is defined in advance). Hoffing inequality is as follows:

$$P\{|Y_s - \mu| < \varepsilon_s\} > 1 - 2e^{-2N\varepsilon_s^2} \tag{1}$$

Set the right side of the inequality to p, we can get ε_s. Conservative confidence interval can be applied when s 1, it is usually too loose to achieve the required accuracy, therefore is not often used in online aggregation.

Large-sample confidence interval calculation uses central limit theory, it guarantees the probability of μ falling into $[Y_s - \varepsilon_s, Y_s + \varepsilon_s]$ is equal to p. Taking s samples from any population whose mean is μ and variance is δ^2, when n is sufficiently large, the mean of the sample follows a normal distribution whose mean is μ and variance is $\delta^2 \setminus n$, i.e $Y_s \sim N(\mu, \delta^2)$, note δ^2, μ is the variance and mean of the original data. Since the variance of the original data is unknown, we use sample's variance instead, $T_{s,2}(v) = \sum_{i=1}^{s}(v(i) - Y_s)^2/s$. Thus,

$$P\{|Y_s - \mu| < \varepsilon_s\} = P\{|\frac{\sqrt{s}(Y_s - \mu)}{T_{s,2}^{1/2}(v)}| < \frac{\varepsilon_s\sqrt{s}}{T_{s,2}^{1/2}(v)}\} \approx 2\phi(\frac{\varepsilon_s\sqrt{s}}{T_{s,2}^{1/2}(v)}) - 1 \tag{2}$$

set the equality to 1 and we can get ε_s. When the number of samples is large enough, samples follow a normal distribution and calculate the confidence interval according to the Central Limit Theory. Large-sample confidence interval gives a tighter interval than the conservative confidence interval and does not require original data follow any distribution. It is often used in online aggregation.

3.2   Bootstrap Estimation

When sample size is really small, the above estimation method does not work because it gives a very loose confidence interval. Bootstrap [21] "resamples" on sample and therefore its accuracy is not affected by sample size. There are three widely-used bootstrap methods: the standard bootstrap; the percentile bootstrap; the t-percentile bootstrap [21];

Standard Bootstrap. Standard bootstrap first takes a random and independent sample of size s from the original data D, denoted as $x_1, x_2, x_3, ..., x_s$. Then take B "resamples" of size s from the sample with replacement. We can get a result on each resample, which is an estimation of the true result running. The estimation result distribution is the same as the exact result. Let $\tilde{x}$ be the estimation of x, $\tilde{x}_{(i)}$ be the estimation based on the $i - th$ resample, then:

$$\tilde{x} = \frac{1}{B} \sum_{i=1}^{B} \tilde{x}_{(i)}, Var(\tilde{x}) = \frac{1}{B-1} \sum_{i=1}^{B} \{\tilde{x}_{(i)} - \tilde{x}\}^2 \tag{3}$$

$Var(x)$ is the variance of the resample. Thus the $(1 - \alpha)\%$ confidence interval of the ture result is:

$$[\tilde{x} - \mu_{1-\alpha/2} Var(x), \tilde{x} + \mu_{1-\alpha/2} Var(x)] \tag{4}$$

Percentile Bootstrap. The $\alpha/2\,th$ quantile and $(1 - \alpha/2)th$ quantile of bootstrap empirical distribution are the upper and loer bounds of statistical confidence interval with $(1 - \alpha)$ confidence. Percentile bootstrap uses this principle to sort the B resamples by $\tilde{x}$, the sorted $\tilde{x}$ denoted as $\tilde{x}^*$ is also called order statistics. The ture result lies in the confidence interval:

$$[\tilde{x}^*_{\frac{\alpha}{2}B}, \tilde{x}^*_{((1-\frac{\alpha}{2})B)}] \tag{5}$$

Percentile gives a tighter confidence interval than the standard bootstrap.

t-Percentile Bootstrap. t-Percentile bootstrap is based on percentile bootstrap, it computes t statistics for each resample: $t_{(i)} = \frac{\tilde{x}_{(i)} - \tilde{x}}{\sqrt{Var(x)}}$. We get the order statistics by sorting $t_{(i)}$. When the significance level amounts to α, the $\alpha/2\,th$ quantile and $(1 - \alpha/2)th$ quantile is $t^*_{(\frac{\alpha}{2}B)}$ and $t^*_{((1-\frac{alpha}{2})B)}$ respectively. The true result falls into confidence interval with probability $1 - \alpha$:

$$[\tilde{x} - t^*_{(\frac{\alpha}{2}B)} * \sqrt{Var(x)}, \tilde{x} - t^*_{((1-\frac{\alpha}{2})B)} * \sqrt{Var(x)}] \tag{6}$$

t-percentile bootstrap gives a tighter confidence interval than the percentile bootstrap.

4   OA in Distributed Environment

MapReduce is a popular cluster-based framework for parallel computing of large-scale datasets. It is first developed by Google in 2003 [22], and the corresponding Google Distributed File System (GFS) is published in 2004 [23]. Google has widely applied it to many large-scale data processing problems [2]. It is natural to think of applying it to online aggregation to improve performance. However, there are many challenges to adapt online aggregation to MapReduce. MapReduce cares about shortening the total execution time rather than providing intermediate estimation result before the computation is completed. By taking into account their various design decisions, we have selected a collection of representative academic online aggregation systems in order to exemplify and crystallize the implementations of different strategies, their strengths, limitations, and some suggested improvements.

- **HOP** [24,25] modifies MapReduce architecture to support pipeline intermediate data between operators. The architecture of HOP is shown as Fig. 2, job J_2 depends on job J_1's output. It allows pipelining within a job and pipelining between jobs. Each mapper stores its output in an in-memory buffer, when the buffer grows to a threshold size α, it will be sent to a reducer. Also, the reducer can apply the reduce function to the currently received data. HOP does build a good framework for interactive queries on MapReduce, but there are still many limitations. First, HOP makes a sequential reading to get a sample because it assumes data is randomly distributed, but this is not true in most cases. Besides that it only provides query progress other than any guarantee of accuracy. User cannot judge whether the accuracy meets the

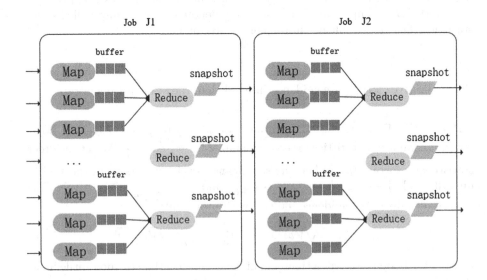

Fig. 2. Architecture of HOP

demand according to the query process. Therefore, [26] thinks it is a partial aggregation system rather than an online aggregation system.

- **OA on Hyracks** [27] solve the problem of the missing estimation methods in HOP. The proposed system works well both in the case where the data in fully randomly stored and the case data skew. The reason why it can apply to both cases is that it uses a Bayesian framework to produce estimates and confidence bounds. The system organizes data into a single queue of *blocks*. It takes a snapshot for each block when the system starts statistical analysis. A snapshot includes *state, aggregatevalue* if the block is done, *IPaddress* of the mapper, *scheduletime* and *processingtime*. The system divides the data into "observed" data denoted as X and "unobserved" data denoted as Θ. They infer the distribution $P(\Theta|X)$ using Bayesian framework. $P(\Theta|X)$ is the posterior distribution for Θ. Previous studies only focuses on the sample itself and estimates the final result by the sample, this work takes in head in making a prediction for the aggregate results of the "unobserved" data and refining the result as the increase of "observed" data, making a trade-off between the accuracy and efficiency.

- **OATS** [20] is built on HOP and focuses on the sharing issue of multiple queries. The system organizes the data blocks according to content, i.e data in the same block is continuous within a certain range while the block size can be different. Given a set of continuous queries, the system first groups the map tasks of those queries which process the same block and have identical *predicates* into a map task. Secondly it collects samples by one-pass disk scan and builds a sample buffer in the main memory so that samples can be shared. In addition to having the same *predicates*, when the aggregate type of the queries is also identical, the computation can be shared by reusing the intermediate result and perform incremental calculation on the new-coming blocks.

- **G-OLA** [28] focuses on non-monotonic queries, i.e whose result cannot be updated by only looking at the new input tuples. It first randomly partitions the dataset into multiple mini batches of uniform size, and process through the data by taking a single mini-batch at a time. It partitions the interme- diate output into the *deterministic* and the *uncertain* sets, when dealing with a new mini-batch, it runs a delta update on the uncertain sets and the

Table 1. Online aggregation in distributed system

System	Platform	Sample method	Data skew	Multi-table	Continuous queries
HOP	Hadoop	Sequential reading	No	No	Yes
G-OLA	FluoDB	Uniform sampling	No	Yes	Yes
COLA	Hadoop	Bootstrap	Yes	Yes	Yes
OATS	HOP	Bootstrap	Yes	Yes	Yes
PF-OLA	Hadoop	Parallel sampling	Yes	Yes	Yes
HOLA	Hadoop	BlockSample	Yes	No	Yes

new-coming mini-batch. This is an incremental calculation process and has prospects in real-time processing.

There are many other work that considers Online Aggregation in the context of MapReduce [10,26,29–33]. Although much technical progress has been made in academic, we are yet to see its impact on practical industrial production (Table 1).

5 Summary and Future Work

Online aggregation has been studied for more than twenty years, howerver, it is still in the academic research stage. There are many reasons why it cannot be applied in industry. First sampling is not an easy thing. Data from various industries have different structures. Each organization has its own way of storing and managing data. It is difficult to find a common sampling method for these heterogeneous data. In addition, academic researches often implement online aggregation by changing the database kernel. This is too expensive because it may change the way users manage data and the data interface that connects to the upper application may also be affected. We're happy to see Y Park et al build a universe approximate query processing system called VerdictDB [7], it serves as a middleware and does not have to modify the backend database, greatly speeding up the query process.

Reviewing the development of online aggregation over 20 years, it still needs future research in the following aspects:

- **Proper sampling methods.** The variance and bias of sample has a great influence on the final estimation. We can sample without bias by random sampling, but the variance cannot be avoided. In order to meet these requirements, the cost of sampling will not be very low. Some researchers try to share a sample with multiple queries. In fact, shared sampling has serious side effects. How to make a better trade-off is a matter worth considering.
- **Aggregate query on extreme values.** A lot of achievements have been made in online aggregation, but most of the research focuses on the aggregate function such as Count, Sum, Average and Stdev. The other two most common aggregate functions, Max and Min, are rarely studied. Max and Min reflect the endpoints of the data distribution.
- **Real-time query processing.** Since online aggregation is a sample-based technology, data needs to be pre-processed or sampling should be done in advance in order to reduce query processing time. This is in contrast to the requirement for real-time processing. So how to perform incremental online aggregation with the increase of real-time data is an issue.
- **Applicable to emerging platforms.** With the emergence of various big data processing platforms, each organization uses a specific data processing platform due to many factors. Online aggregation is an attractive technology, however, developers are put off by the extra overhead which is brought by migrating to a new platform or adding new features to the existing platform.

Acknowledgements. This work is supported by the National Natural Science Foundation of China under Grant No. 61772289 and the Fundamental Research Funds for the Central Universities.

References

1. Hellerstein, J.M., Haas, P.J., Wang, H.J.: Online aggregation. In: ACM SIGMOD Record, vol. 26, pp. 171–182. ACM (1997)
2. Aarnio, T.: Parallel data processing with MapReduce. In: TKK T-110.5190, Seminar on Internetworking (2009)
3. Olken, F.: Random sampling from databases. Ph.D. thesis, University of California, Berkeley (1993)
4. Wu, S., Ooi, B.C., Tan, K.L.: Continuous sampling for online aggregation over multiple queries. In: Proceedings of the 2010 ACM SIGMOD International Conference on Management of Data, pp. 651–662. ACM (2010)
5. Agarwal, S., et al.: Knowing when you're wrong: building fast and reliable approximate query processing systems. In: Proceedings of the 2014 ACM SIGMOD International Conference on Management of Data, pp. 481–492. ACM (2014)
6. Zeng, K., Gao, S., Mozafari, B., Zaniolo, C.: The analytical bootstrap: a new method for fast error estimation in approximate query processing. In: Proceedings of the 2014 ACM SIGMOD International Conference on Management of Data, pp. 277–288. ACM (2014)
7. Park, Y., Mozafari, B., Sorenson, J., Wang, J.: VerdictDB: universalizing approximate query processing. arXiv preprint arXiv:1804.00770 (2018)
8. An, M., Sun, X., Ninghui, S.: Dynamic data partitioned online aggregation. J. Comput. Res. Dev. (2010)
9. Joshi, S., Jermaine, C.: Robust stratified sampling plans for low selectivity queries. In: IEEE 24th International Conference on Data Engineering, ICDE 2008, pp. 199–208. IEEE (2008)
10. Agarwal, S., Mozafari, B., Panda, A., Milner, H., Madden, S., Stoica, I.: BlinkDB: queries with bounded errors and bounded response times on very large data. In: Proceedings of the 8th ACM European Conference on Computer Systems, pp. 29–42. ACM (2013)
11. Kim, A., Blais, E., Parameswaran, A., Indyk, P., Madden, S., Rubinfeld, R.: Rapid sampling for visualizations with ordering guarantees. Proc. VLDB Endow. **8**(5), 521–532 (2015)
12. Haas, P.J., Hellerstein, J.M.: Ripple joins for online aggregation. ACM SIGMOD Rec. **28**(2), 287–298 (1999)
13. Haas, P.J.: Large-sample and deterministic confidence intervals for online aggregation. In: Proceedings of Ninth International Conference on Scientific and Statistical Database Management, pp. 51–62. IEEE (1997)
14. Luo, G., Ellmann, C.J., Haas, P.J., Naughton, J.F.: A scalable hash ripple join algorithm. In: Proceedings of the 2002 ACM SIGMOD International Conference on Management of Data, pp. 252–262. ACM (2002)
15. Dittrich, J.P., Seeger, B., Taylor, D.S., Widmayer, P.: Progressive merge join: a generic and non-blocking sort-based join algorithm** this work has been supported by grant no. se 553/2-2 from DFG. In: VLDB 2002: Proceedings of the 28th International Conference on Very Large Databases, pp. 299–310. Elsevier (2002)
16. Jermaine, C., Dobra, A., Arumugam, S., Joshi, S., Pol, A.: The sort-merge-shrink join. ACM Trans. Database Syst. (TODS) **31**(4), 1382–1416 (2006)

17. Jermaine, C., Dobra, A., Arumugam, S., Joshi, S., Pol, A.: A disk-based join with probabilistic guarantees. In: Proceedings of the 2005 ACM SIGMOD International Conference on Management of Data, pp. 563–574. ACM (2005)

18. Jermaine, C., Dobra, A., Pol, A., Joshi, S.: Online estimation for subset-based SQL queries. In: Proceedings of the 31st International Conference on Very Large Data Bases, pp. 745–756. VLDB Endowment (2005)

19. Li, F., Wu, B., Yi, K., Zhao, Z.: Wander join: online aggregation via random walks. In: Proceedings of the 2016 International Conference on Management of Data, pp. 615–629. ACM (2016)

20. Wang, Y., Luo, J., Song, A., Dong, F.: Oats: online aggregation with two-level sharing strategy in cloud. Distrib. Parallel Databases 32(4), 467–505 (2014)

21. Efron, B.: Bootstrap methods: another look at the jackknife. In: Kotz, S., Johnson, N.L. (eds.) Breakthroughs in Statistics, pp. 569–593. Springer, New York (1992). https://doi.org/10.1007/978-1-4612-4380-9_41

22. Ghemawat, S., Gobioff, H., Leung, S.T.: The Google file system, vol. 37. ACM (2003)

23. Dean, J., Ghemawat, S.: MapReduce: simplified data processing on large clusters. Commun. ACM 51, 107–113 (2008)

24. Condie, T., et al.: Online aggregation and continuous query support in MapReduce. In: ACM SIGMOD International Conference on Management of Data, pp. 1115–1118 (2010)

25. Condie, T., Conway, N., Alvaro, P., Hellerstein, J.M., Elmeleegy, K., Sears, R.: MapReduce online. In: NSDI, vol. 10, p. 20 (2010)

26. Qin, C., Rusu, F.: PF-OLA: a high-performance framework for parallel online aggregation. Distrib. Parallel Databases 32(3), 337–375 (2014)

27. Pansare, N., Borkar, V.R., Jermaine, C., Condie, T.: Online aggregation for large MapReduce jobs. Proc. VLDB Endow. 4(11), 1135–1145 (2011)

28. Agarwal, S., Agarwal, S., Armbrust, M., Armbrust, M., Stoica, I.: G-OLA: generalized on-line aggregation for interactive analysis on big data. In: ACM SIGMOD International Conference on Management of Data, pp. 913–918 (2015)

29. Zeng, K., Gao, S., Gu, J., Mozafari, B., Zaniolo, C.: ABS: a system for scalable approximate queries with accuracy guarantees. In: Proceedings of the 2014 ACM SIGMOD International Conference on Management of Data, pp. 1067–1070. ACM (2014)

30. Zhang, Z., Hu, J., Xie, X., Pan, H., Feng, X.: An online approximate aggregation query processing method based on hadoop. In: 2016 IEEE 20th International Conference on Computer Supported Cooperative Work in Design (CSCWD), pp. 117–122. IEEE (2016)

31. Cheng, Y., Zhao, W., Rusu, F.: Bi-level online aggregation on raw data. In: Proceedings of the 29th International Conference on Scientific and Statistical Database Management, p. 10. ACM (2017)

32. Shi, Y., Meng, X., Wang, F., Gan, Y.: You can stop early with cola: online processing of aggregate queries in the cloud. In: Proceedings of the 21st ACM International Conference on Information and Knowledge Management, pp. 1223–1232. ACM (2012)

33. Gan, Y., Meng, X., Shi, Y.: COLA: a cloud-based system for online aggregation. In: 2013 IEEE 29th International Conference on Data Engineering (ICDE), pp. 1368–1371. IEEE (2013)

Collaborative Caching in P2P Streaming Systems

Guoqiang Gao[1(✉)] and Ruixuan Li[2]

[1] Engineering Research Center of Hubei Province for Clothing Information,
Wuhan Textile University, Wuhan 430073, China
ggao@wtu.edu.cn
[2] Intelligent and Distributed Computing Laboratory, School of Computer
Science and Technology, Huazhong University of Science and Technology,
Wuhan 430074, China
rxli@hust.edu.cn

Abstract. In the past decade, Peer-to-Peer (P2P) Systems achieved great successes. Its fascinating characteristics, such as decentralized control, self-governance, fault tolerance and load balancing, make it the default infrastructure for file sharing applications. Today, P2P system is one of the largest Internet bandwidth consumers. In order to relieve the burden on Internet backbone and improve the user access experience, efficient caching strategies should be applied. In this paper, we propose a novel collaborative caching in P2P streaming systems. This strategy first calculates the factor of each cached file through their caching value, and then replaces the file with the lowest factor. The simulation experiments show that our strategy has higher cache hits and lower system load.

Keywords: P2P · Streaming · Caching

1 Introduction

The Internet has been tremendous growth in the last ten years. According to the report of eMarketer [1], more than 50% of the world's population will use the internet regularly in 2019. In the reference [14], Cisco forecasts that global IP traffic will reach 1.1 ZB (zettabyte: 1000 exabytes) per year in 2016 and will exceed 2.3 ZB by 2021. The explosion was mainly due to popular distributed applications such as P2P file sharing, distributed computing, Internet phone service, Internet streaming and online social networks. Among all these applications, Internet streaming and P2P file sharing systems are the killing applications which occupy the largest percentage of Internet traffic. Similar to P2P file sharing systems, Internet streaming applications also carry video traffic but have more stringent time constraints for data delivery. Viewers will experience severe watching disruption if data chunks cannot be received before the deadline. Consequently, they demand higher Internet bandwidth.

Nowadays, there are many commercial P2P streaming applications, such as PPTV [2], IQIY [3], UUSee [12], SopCast [13] and etc. They achieved great successes and each of them has millions of registered users. The practices of these systems proved

X. Meng et al. (Eds.): WISA 2018, LNCS 11242, pp. 115–122, 2018.
https://doi.org/10.1007/978-3-030-02934-0_11

that P2P is an effective solution to delivery video contents to millions of simultaneous viewers across the world in real time. P2P streaming applications are still in their infant era. The biggest problem is how to reduce the network overhead of these P2P streaming systems. For distributed applications, caching is an effective mechanism to reduce the amount of data transmission on the Internet backbone and improve user access experience. In P2P streaming systems, the client will cache the downloaded files to a local folder. These cached files can be used as data sources downloaded by other peers, which can improve system efficiency and reduce network overhead. However, the caching space is limited. Therefore, P2P streaming systems need to replace some of the content when the caching is full. In order to design efficient caching scheme in P2P streaming systems, we measure some existing P2P streaming systems firstly, such as PPTV and iQIY. The experimental results show that the two measured P2P streaming systems in China all use the interval cache which only caches a part of the adjacent contents. This strategy has a lower caching hit ratio when a peer interacts with other peers. In this paper, we propose a collaborative caching. This strategy first calculates the factor of each cached file through their chunk value that is the minimum unit of the video data requested by peers, and then replaces the file with the lowest factor. We will discuss the chunk value in detail in the following sections. Our main goal is to keep those hard-won files because they will bring the greatest benefits to the system.

The rest of the paper is organized as follows. In Sect. 2, we first briefly describe the caching of P2P streaming applications, and then present related works in recent years. In Sect. 3, we introduce our caching strategy, including file caching value and replacement algorithm. In Sect. 4, We discuss the experiments we conducted and analyze the results. Finally, in Sect. 5, we present our conclusion and the future works.

2   Related Works

Most commercial P2P streaming applications, such as PPTV, IQIY, UUSee and SopCast, and etc., use mesh-based P2P infrastructure. In these systems, all the viewers of a channel form an overlay network. The content of each channel is divided into multiple sub-streams and further divided into chunks. A small number of dedicated video servers are deployed for initialization of video broadcasting. Users who are interested in a video stream often have broadband connections which are high enough for them to act as relay points and forward the video clips to other users. Peers serve each other by exchanging chunks of data periodically.

To relieve the burden imposed by P2P traffic, design and implement an effective caching infrastructure in P2P systems attracted great interests from both industry and academia [4, 5, 15]. However, it is difficult due to the unique features such as self-governing, and dynamic membership, large number of peers, and even larger amount of shared files in P2P applications.

Huang et al. study the unique issue of caching fairness in edge environment in reference [6]. Due to distinct ownership of peer devices, caching load balance is critical. They consider fairness metrics and formulate an integer linear programming problem. They develop a distributed algorithm where devices exchange data reachability and identify popular candidates as caching peers. Their experiments show that their algorithms significantly improve data caching fairness while keeping the contention induced latency similar to the best existing algorithms. However, They only consider a constant number of data chunks. When the number of chunks increases, their algorithm might become inefficient. In [11], Hefeeda et al. proposed similar idea by deploying caches at or near the border of the ASs (autonomous systems), pCache will intercept P2P traffics go through the AS, and try to cache the most popular contents. The cache size is relatively small, and the objects in P2P applications are very big, so the effectiveness of this approach is doubtable. Furthermore, pCache itself became a bottleneck and a single point of failure which affect its efficiency.

SECC [7] introduces a novel index structure called Signature Augment Tree (SAT) to implement collaborative caching. Simulation results show that SECC schemes significantly outperform other collaborative caching methods which are based on existing spatial caching schemes in a number of metrics, including traffic volume, query latency and power consumption. However SECC is only suitable for mobile P2P networks. In PECAN [8], each peer adjusts its cache capacity adaptively to meet the server's upload bandwidth constraint and achieve the fairness. Dai et al. [9] have developed a theoretical framework to analyze representative cache resource allocation schemes within the design space of collaborative caching, with a particular focus on minimizing costly inter-ISP traffic. Their researches not only help themselves and other researchers understand the traffic characteristics of existing P2P systems in light of realistic elements, but also offer fundamental insights into designing collaborative ISP caching mechanisms.

3 Caching Algorithm

In our paper, we will propose a novel caching replacement algorithm which will use chunk value to optimize system performance. If the caching space is full, one or several cached files have to be evicted from the current cache to make room for the coming file. Here, cache replacement algorithms make the decision which files should be replaced.

3.1 Chunk Value

We define the chunk value to provide reference for cache replacement. If a chunk is never requested, it will be of no value and is not necessary to be kept in the caching space. In this paper, we propose a concept called chunk value (CV). If a chunk with higher CV, we keep it longer in the cache because the other peers may be more need to download it. However, to accurately answer which chunk is more valuable is very difficult in a distributed environment as well. In P2P networks, if a chunk is very popular (that is many peers own it), a peer can quickly gain it from nearby peers. On the contrary, if a chunk is unpopular, it is difficult that a peer gets it, perhaps to pass through great distances because only few peers own it. By some inspiration, we use the communication delay as a chunk's CV. The higher the communication delay, the higher the value of the file being accessed. Our basic point is if a file is access by a lot of remote peers, then can be considered it is useful. A chunk c's CV Vc at time t in Eq. 1 is refined to

$$V_c(t+1) = \mu V_c(t) + \eta \sum_{i \in t} d_i(c) \tag{1}$$

where, $0 < \mu, \eta < 1$, $\mu + \eta = 1$. The larger the value of μ, the more important history requested information on the chunk C. On the contrary, the larger the value of η, the more important recent requested information on the chunk C. The value $d_i(c)$ is the delay of the i-th request for chunk C in the period t. If a chunk has more requests or the delay of its request is high, then it will have a higher CV value. The value of the data block is calculated separately by each node, which can be used as an important reference factors in cache replacement mechanism.

3.2 Replacement Algorithm

In P2P streaming systems, a big movie file is converted into many small blocks with sequence number. The client downloads the current blocks based on their sequence number from dedicated server and other peers when the program is playing. The adjacent blocks are stored in a file whose size is 20 M in PPTV, and then the file is putted into the cache. If the cache is full, the system will evict a file from the cache for the new coming file. As for iQIY, it caches its downloading data into a big file whose structure is more complex than PPTV. We understand its structure after a lot of analysis and write a shell to obtain caching information from it just like PPTV. The value of a cached file f is defined as follows

$$V_f = \frac{1}{n} \sum_{t \in T} V_{c,c \in f}(t) \tag{2}$$

where T is the length of f's caching time, and $1/n$ represents the average value of the sum. The smallest unit of t can be adjusted according to the application. We use 5 s in the experiment, that is, the value of a chunk is calculated once every 5 s. In our strategy, the system calculates V_f for each f in the cache, and evicts the file f with the lowest V_f when the cache is full. The detail of our cache replacement algorithm is depicted as Algorithm 1.

Algorithm 1
where playing program **do**
    convert the download blocks into file f_{new}
    if f_{new} should be putted into the cache **then**
        if the cache is full **then**
            calculate each cached file f's V_f
            evict the cached file f with the lowest V_f
        end if
        put f_{new} into the cache
    end if
end where

4   Evaluation

In this section, we first simply introduce our experiment methodology, and then we analyze the efficiency of our strategy.

4.1   Experiment Methodology

To save time and increase efficiency, we use PeerSim [10] as the simulation-driven kernel. PeerSim has been designed to cope with P2P networks. It can support simulations on a very dynamic environment. It has great scalability and can support thousands of peers. In our experiments, we implement P2P streaming protocol in Peersim. Although we can use Peersim to simulate our strategy, the close-source P2P streaming applications cannot be simulated because we do not know their mechanism. We develop a program to collect the caching data of PPTV and IQIY to estimate their caching strategies for the comparative analysis. Our method is to observe the change of the cache, and record this change. To obtain the request time of the cached files, we use Wireshark to measure the communicating packets. We can judge which files are downloaded by other peers through observing the sequence number in packets.

4.2   Cache Hits

In order to evaluate the cache efficiency for requesting files, we define cache hits of a peer as the number of downloading files from this peer by other peers. In P2P streaming system, a peer actually accesses their cached data when it downloads data from another peers. Obviously, the higher the cache hits, the better cache utilization an algorithm can achieve. We run PPTV and iQIY 24 h, and use Wireshark to collect their communicating data. After cleaning and analyzing the collected data, we can obtain the cache hits of PPTV and iQIY. As for our strategy, we can get similar information from the simulated experiment.

Figure 1 depicts the cache hits in one peer for three strategies, where the y-axis shows the cumulative number of caching hits. Our strategy (Proposal) has the best performance, in that the cached files totally have been accessed 732 times within 24 h. The two real systems have similar performance, and the total number of requesting cached files does not exceed 500. However, the amount of access to a peer in the network is very large. This is to show that, for a node, most of the data is obtained from other peers and less data is provided to others. Form the figure, we can that there is less access in the middle of the night. Therefore, the curve of cache hits is flat between 10 and 16. Based on our measurement data, we can estimate that these P2P applications have low cache utilization. Our strategy has not such rule because the data requests are random in the simulation experiment.

4.3   System Load

In order to understand the caching strategy of P2P streaming systems, we design an experiment to measure the replacement strategy of the caching. In this experiment, we set PPTV cache size to 1024 M, and collect the residence time of video files in PPTV

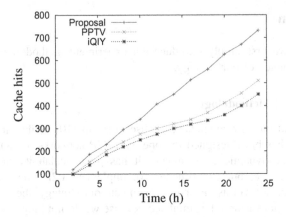

Fig. 1. The cache hits of three strategies in 24 h

cache directory. As for iQIY, its cache size is set to 1024 M as well, and the keep time of video files is obtained from its cache file. Based on the collected data, we estimate the caching replacement strategies of PPTV and iQIY, the results are shown in Figs. 2. To clearly describe the results, we only choose the popular channels to present, the unpopular channels have similar performance. The original X-axis is the number of the sorted video files. Since the length of each channel is different, we normalize them to 0–100, and used it to express the entire contents of the channels. The Y-axis is the residence time of the corresponding part of channels. For both applications, the preservation time of video files has the tooth-like curve. This strategy seems to be the interval cache which only caches a part of the adjacent contents. From the discussion about peer collaboration in the previous section, we can find that this caching strategy in PPTV and iQIY is not very good. Therefore, it is urgent that P2P streaming applications need new caching strategy to improve performance.

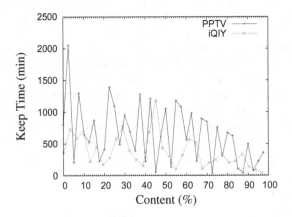

Fig. 2. Cache replacement

For the real P2P streaming systems, we cannot get the system load of their networks. We use simulation experiments to analyze it for PPTV and iQIY in this paper. According to the previous measurement experiment, we can roughly estimate that PPTV and iQIY use alternate caching mechanism (interval cache) to replace the cached data. For simplicity, we consider that the real P2P streaming systems use interval cache as their caching strategies. To analyze our strategy compared to the existing applications, we use Peersim to simulate the system with interval cache. In the simulation experiments, we set two nodes as dedicated servers, and collect the communication traffic among peers under different the network sizes. As for our simulations, there are no dedicated servers. As can be seen from Fig. 3, our strategy can reduce the traffic load of the dedicated server compared to the systems with interval cache, where the traffic messages is collected in one simulation cycle. The proposed algorithm can reduce about 38% of the traffic when the network size is 5000 peers.

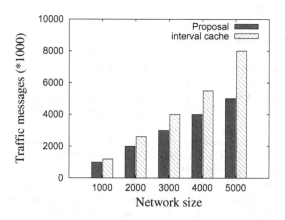

Fig. 3. System load for different strategies

5   Evaluation

Caching techniques are widely used to boost the performance in large-scale distributed applications. However, as one of the most bandwidth consuming applications on the Internet, not enough efforts have been conducted on P2P caching. In this paper, we propose novel and effective cache replacement algorithms for P2P streaming systems. We use the chunk value as the factor to keep more valuable cached files for relieving network load and improving system performance. The simulation experiments show that our strategy has higher cache hits and lower system load. In the future, we plan to apply our algorithm on other P2P based application such as online streaming services. Moreover, in order to obtain more realistic results, we are working to carry out the simulation based trace-driven.

Acknowledgements. This work is supported by Major Projects of the National Social Science Foundation (No. 16ZDA092), National Natural Science Foundation of China under Grants

61572221, U1401258, 61433006, 61300222, 61370230 and 61173170, Innovation Fund of Huazhong University of Science and Technology under Grants 2015TS069 and 2015TS071, Science and Technology Support Program of Hubei Province under Grants 2015AAA013 and 2014BCH270, Science and Technology Program of Guangdong Province under Grant 2014B010111007, and Excellent Young and Middle-aged Science and Technology Innovation Team Plan of Hubei Higher Education (T201807).

References

1. eMarketer: Internet users and penetration worldwide, 2016–2021 (2018)
2. PPTV. http://www.pptv.com. Accessed 1 May 2018
3. IQIY. http://www.iqiyi.com. Accessed 1 May 2018
4. Moeini, H., Yen, I.-L., Bastani, F.B.: Efficient caching for peer-to-peer service discovery in internet of things. In: 2017 IEEE International Conference on Web Services, ICWS 2017, Honolulu, HI, USA, 25–30 June 2017
5. Das, S.K., Naor, Z., Raj, M.: Popularity-based caching for IPTV services over P2P networks. Peer-to-Peer Netw. Appl. **10**(1), 156–169 (2017)
6. Huang, Y., Song, X., Ye, F., Yang, Y., Li, X.: Fair caching algorithms for peer data sharing in pervasive edge computing environments. In: 37th IEEE International Conference on Distributed Computing Systems, ICDCS 2017, Atlanta, GA, USA, 5–8 June 2017
7. Zhu, Q., Lee, D.L., Lee, W.-C.: Collaborative caching for spatial queries in mobile P2P networks. In: Proceedings of the 27th International Conference on Data Engineering, ICDE 2011, Hannover, Germany, 11–16 April 2011
8. Kim, J., Im, H., Bahk, S.: Adaptive peer caching for P2P video-on-demand streaming. In: Proceedings of the Global Communications Conference 2010. GLOBECOM 2010, Miami, Florida, USA, 6–10 December 2010
9. Dai, J., Li, B., Liu, F., Li, B., Jin, H.: On the efficiency of collaborative caching in ISP-aware P2P networks. In: 30th IEEE International Conference on Computer Communications, Joint Conference of the IEEE Computer and Communications Societies, Shanghai, China, 10–15 April 2011
10. Montresor, A., Jelasity, M.: PeerSim: a scalable P2P simulator. In: Proceedings of the 9th International Conference on Peer-to-Peer (P2P 2009), Seattle, WA, USA, September 2009
11. Hefeeda, M., Hsu, C.-H., Mokhtarian, K.: pCache: a proxy cache for peer-to-peer traffic. In: Proceedings of the ACM SIGCOMM 2008 Conference on Applications, Technologies, Architectures, and Protocols for Computer Communications, Seattle, WA, USA, 17–22 August 2008
12. UUSee. http://www.uusee.com. Accessed 1 May 2018
13. SopCast. http://www.sopcast.com. Accessed 1 May 2018
14. Cisco Inc.: Cisco visual networking index: forecast and methodology, 2016–2021 (2018)
15. Pacifici, V., Lehrieder, F., Dan, G.: Cache bandwidth allocation for P2P file-sharing systems to minimize inter-ISP traffic. IEEE/ACM Trans. Netw. **24**(1), 437–448 (2016)

A Distributed Rule Engine for Streaming Big Data

Debo Cai[1(✉)], Di Hou[1], Yong Qi[1], Jinpei Yan[1], and Yu Lu[2]

[1] Department of Computer Science and Technology, Xi'an Jiaotong University,
Xi'an, China
yushisx@163.com
[2] Troops 69064 of PLA, Xinjiang, China

Abstract. The rules engine has been widely used in industry and academia, because it can separate the rules from the execution logic and incorporate the features of expert knowledge. With the advent of big data era, the amount of data has grown at an unprecedented rate. However, traditional rule engines based on PCs or servers are hard to handle streaming big data owing to limitation of hardware performance. The structured streaming computing framework can provide new solutions for these challenges. In this paper, we design a distributed rule engine based on Kafka and Structured Streaming (KSSRE), and propose a rule-fact matching strategy using the Spark SQL engine to support a large number of event stream inferences. KSSRE uses DataFrame to store data and inherits the load balancing, scalability and fault-tolerance mechanisms of Spark2.x. In addition, in order to remove the possible repetitive rules and optimize the matching process, we use the ternary grid model [1] for representing rules and design a scheduling model to improve the memory sharing in the matching process. The evaluation shows that KSSRE has a better performance, scalability and fault tolerance based on DBLP data sets.

Keywords: Rule engine · Spark2.x · Event stream

1 Introduction

The rules engine simulates the decision process of a human expert and handles events and triggers corresponding actions based on prior knowledge in the pre-set rule base. Because the rules engine separates the rules from the execution logic, and the interface with expert experience is friendly, it has been successfully applied in insurance and insurance claims, bank credit and many other areas. With the development of information technology, big data has become one of the main themes of the information age. For example, Mobike, which is based on the Internet of Things (IoT), officially announces that the average amount of data generated per minute is close to 1G. How to perform multi-dimensional analysis and processing of a large number of data streams in real time and accurately will be a serious challenge for the rule engine to adapt to development.

In order to solve the above problems, many researchers have designed a distributed rule engine based on big data processing frameworks such as Hadoop and Spark to

© Springer Nature Switzerland AG 2018
X. Meng et al. (Eds.): WISA 2018, LNCS 11242, pp. 123–130, 2018.
https://doi.org/10.1007/978-3-030-02934-0_12

improve the matching efficiency. However, these solutions also have their own imperfections. Referring to these scenarios, based on the Kafka and Structured Streaming computing framework, we designed and implemented a distributed rules engine (KSSRE) to support a large number of event flow inference. The purpose is to improve the matching efficiency of the rules engine and achieve better load balancing and fault tolerance. Using the Kafka clustering feature to decouple the event flow, a relatively efficient rule-fact matching strategy is designed and implemented on the Spark SQL engine. At the same time, in order to improve the calculation rate, use DataFrame/DataSet which is better than RDD in both time and space to store data. In order to remove the possible repetitive rules and optimize the matching process, we improved the ternary grid model for representing rules, and designed a scheduling model to improve the memory sharing in the matching process. In addition, because KSSRE is based on Structured Streaming, it inherits the load balancing, scalability, and fault-tolerance mechanisms of Spark 2.x.

The rest of the paper is organized as follows. Section 2 provides some background information and explains related work. Section 3 elaborates on the design and implementation of KSSRE. In Sect. 4, we use the DBLP data set to conduct an experimental analysis of the KSSRE. Section 5 concludes the paper and discusses future work.

2   Background and Related Work

2.1   Rule Engine

The rule engine usually consists of three parts, namely rule base, fact collection, and inference engine. The fact is that there is a multiple relationship between objects and their attributes. Rules are inferential sentences that consist of conditions and conclusions. When facts meet the conditions, the corresponding conclusions are activated. The general form of the rule is as follows:

> Rule_1: /* Rule Name*/
> Attributes /* Rule-Attributes*/
> LHS /* conditions*/ => RHS /* actions*/

The LHS is a condition and consists of several conditions. It is a generalized form of known facts and a fact that it has not been instantiated. The RHS is a conclusion and consists of several actions.

2.2   Apache Kafka and Structured Streaming

Apache Kafka [2] is a distributed streaming platform, which consists of Producer, Kafka cluster and consumer. Producer publishes the message to the specified topic according to the set policy. After receiving the message from the Producer, the Kafka cluster stores it on the hard disk. The Consumer pulls the data from the Kafka cluster and uses the offset to record the location of the consumption. Kafka guarantees high processing speed while guaranteeing low latency and zero loss in data processing. Even with terabytes of data, it can guarantee stable performance.

Structured Streaming [3] is a real-time computational framework for Spark 2.x. It uses DataFrame to abstract data. DataFrame is a collection of Row objects (each Row object represents a row of records) and contains detailed structural information (patterns). Spark clearly knows the structure and boundaries of the dataset, so that it is easy to implement the exactly-once of the data at the framework level. In particular, Structured Streaming re-uses its Catalyst engine to optimize SQL operations, which improves computational efficiency.

2.3 Related Work

With the rise of big data and the IoT, some researchers based on the Hadoop framework to decompose the rules and map the matching tasks into the Map and Reduce processes in the cluster and obtain the matching results [4, 5]. Zhou and other researchers use the message passing model to transform the matching process of rules into messages between processes, and implement parallel and distributed reasoning [6]. Researchers such as Chen and others used Spark 1.x's stream data calculation framework to map rules and facts to Dstream operations for event stream processing [7]. Researchers such as Zhang and others used Spark's process and relational API to map the matching process of rules and facts to the operation of an enhanced RDD, which is DataFrame, and achieved parallel distribution rule matching [8]. Referring to these scenarios, we have designed and implemented a distributed rule engine based on Kafka and Structured Streaming for reasoning on a large number of event streams.

3 Implementation and Optimization of the KSSRE

3.1 Overall Design

The overall design of the KSSRE is shown in Fig. 1 and consists of three parts.

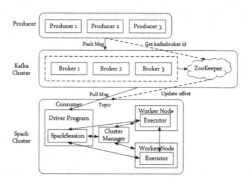

Fig. 1. KSSRE architecture

Producer is the source of real-time data generation. The Kafka cluster receives real-time events from Producer, which decouples these real-time data and performs pre-processing. The Spark cluster pulls data from the Kafka cluster and processes it to generate inference results.

3.2 Inference Process

KSSRE decoupled and preprocess event flows through Kafka clusters. Based on the Structured Streaming real-time computing framework, we use DataFrame/DataSet which is better than RDD in both time and space to store data. We designed and implemented an effective rule-fact matching strategy, converting the rules to SQL operations and using the Catalyst engine to optimize the SQL operations, ultimately achieving inference.

As shown in Fig. 2, KSSRE divides the inference process into four stages of "Hash-Filter-Trigger-Select", and implements inference by periodically cycling through four stages.

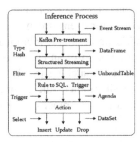

Fig. 2. Inference process.

- The first layer is the Kafka data preprocessing layer that implements asynchronous processing of data producers and consumers.
- The second layer is the Structured Streaming data filter layer, which implements the matching of the LHS part of the rules and the facts.
- The third layer is the rule preprocessing layer, which implements conversion query and conflict resolution from rules to SQL statements.
- The fourth layer is the SQL execution layer, which executes all SQL statements in Agenda and produces a Result Table.

3.3 Optimization

In the actual experiment, the efficiency of the above algorithm is low. There are two main reasons for this: First, there may be duplication of rules in the rule base. In addition, in the Filter stage, all rules need to be filtered in turn, and for rules that have mutually exclusive conditions, there will be a lot of redundancy. Second, the use frequencies of conditions in the LHS part of the rules are different. The degree of memory sharing corresponding to different execution orders has a great influence on the matching efficiency. For these two reasons, we directionally optimize the algorithm.

On the one hand, the rules are handled in advance. As shown in Fig. 3, we have designed a new rule storage style from the ternary grid [1]:

Fig. 3. The representation of a rule.

R_i stands for the rule and i is the rule ID. F_{ij} indicates the conditions contained in the rule. Status uses "$-1/1/0/-2/2$" to indicate the current status of the rule. "0" indicates unused, "1" indicates that F_{ij} is a sub-condition in the LHS part of the rule R_i. "-1" indicates that F_{ij} is a negative form of the sub-condition with the number j. "2" indicates that F_{ij} is a sub-statement in the RHS part of the rule R_i. "-2" indicates that F_{ij} is a negative form of the sub-sequence number j. The Queue Pointer is a pointer to a rule ID queue in which there is an exclusive sub-condition with the rule R_i. The ternary grid is mainly to convert rules into rule matrices to eliminate duplicate rules and meaningless rules. The improved model not only pre-processes the rules, but also reduces the number of rules and facts in the matching process through the Mutex Queue.

```
Algorithm 1 The Inference Engine Algorithm
Input: Event Stream
Output: DataSet
 1: function PRETREATMENT(R₁, R₂,...Rₙ)
 2:     L is the length of all facts
 3:     F[1][L] ← R₁.LHS.split
 4:     F[2][L] ← R₂.LHS.split
 5:     ...
 6:     F[n][L]Rₙ.Rₙ.LHS.split
 7:     for i ← 1 to n do
 8:         for j ← (i + 1) to (n − 1) do
 9:             if F[i][L] and F[j][L] are mutually inverse then
10:                 Rᵢ.Mutexqueue ← Rᵢ.Mutexqueue.append(Rⱼ)
11:                 Rⱼ.Mutexqueue ← Rⱼ.Mutexqueue.append(Rᵢ)
12:             end if
13:         end for
14:     end for
15: end function
16:
17: procedure INFERENCE(DataSet, Rules)
18:     rules ← Rules.sortBy(priority)
19:     for i ← 1 to rules.length do
20:         if DataSet.filter(rules[i]) is not null then
21:             rules ← rules.drop(rules[i].inverse)
22:             ConflictSet ← rules[i]
23:         end if
24:     end for
25:     Agenda ← ConflictSet.sortBy(priority)
26:     for j ← 1 to Agenda.length do
27:         DataSet.Select(Agenda[j])
28:     end for
29: end procedure
```

The second is that for the problem that the rule LHS partial use frequency influences efficiency differently, we can directly set the priority of the rule through the conditional use frequency. But this is a simple optimization method in the ideal case where the LHS part of the rule contains only one condition. To this end, we have established a scheduling model. Trigger the execution sequence by changing the rules to maximize the reuse of existing matching results. The scheduling model of the rule is:

$$R_p = \frac{1}{3} \sum_{i=1}^{n} (C_i + C_m + n) \tag{1}$$

Among them, C_i is the frequency of use of each condition. n is the number of conditions contained in the LHS. C_m is the most frequently used condition of all LHS conditions. 1/3 indicates that these three factors each occupy the weight of the scheduling model. R_p reflects the frequency of use of the LHS part of rule i in the rule base. Finally, we optimize the Filter process based on the scheduling model. The pseudo code of the optimization algorithm is shown in Algorithm 1.

4 Experiments

In this section, we refer to the OpenRuleBench [9] and use the DBLP [10] (Digital Bibliography & Library Project) data set to perform an experimental analysis of the KSSRE in terms of both performance and scalability. DBLP is an English literature database in the field of computers. As of January 1, 2018, more than 6.6 million papers and more than 2.1 million scholars were included. The rules use the four parts of the DBLP filter, negative filter, and join operations to generate a total of 20 valid rules.

Query (Id, T, A, Y, M):
- att (Id, title, T), - att (Id, year, Y)
-att (Id, author, A), -att (Id, month, M)

The experiment was based on three servers. The CPU of each server is Inter E3-1231*8 and the memory is 8 G. The servers were interconnected via Gigabit Ethernet and the operating system was Centos 7.2. Each server runs three virtual machines totaling nine nodes. In addition, considering that 90% of the time in the production system is used for matching [11], we use match time as an indicator of performance testing.

4.1 Performance Testing

Performance testing is mainly to compare KSSRE with Drools [12]. Deploying KSSRE on 3 virtual machines on 1 server is compared with Drools 6.5 deployed on another server.

As shown in Fig. 4, Drools has better processing performance than KSSRE when the number of facts is less than 1 million. Drools and KSSRE has similar performance when the amount of data increases to 1 million. When the amount of data increased to 2 million, KSSRE began to provide slightly better performance. When the number of facts continues to increase to 5 million, Drools cannot handle it and KSSRE finishes processing in about 80 s. Although we can make Drools continue to work with large amount of data in a way that improves hardware performance, it is clear that Drools' memory model is less flexible in terms of garbage collection, which makes it impossible to handle large data sets.

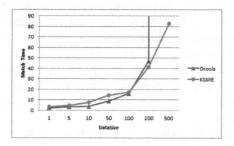

Fig. 4. Performance testing.

4.2   Extensibility Test

Extensibility testing extends the Spark cluster within the same facts, rules, and time intervals to change the number of cluster nodes and record the processing time.

As shown in Fig. 5, matching times for different scales of facts are shown. It is obvious that the matching time decreases as the number of nodes increases. With the same order of magnitude of the fact, the matching time decreases as the number of cluster nodes increases. And on a larger scale of fact, the effect is even more pronounced. Therefore, we can improve the matching efficiency by simply scaling the cluster. In addition, the graph can reflect that when the cluster increases from 3 to 5 nodes, the KSSRE matching time decreases the most. The decrease in the matching time for the change in the number of other nodes is reduced. This is because when the number of nodes of the cluster is 3, the data communication is the internal communication of the server, and when the number of the nodes is more than 3, the cost of the network communication between the nodes needs to be considered.

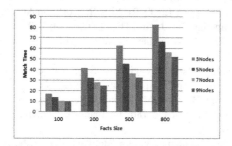

Fig. 5. Comparison of match time.

5 Conclusion

Based on Kafka and Structured Streaming, we extended the general algorithm of the inference engine and implemented a prototype system of distributed rule engine, which is suitable for streaming big data. According to the experimental situation, the rule storage trinomial mesh model is improved and a scheduling model is designed to remove repetitive rules and optimize the matching process. Most importantly, KSSRE is characterized by the reliability, scalability, and fault-tolerance of Structured Streaming. Finally, experimental results show that KSSRE not only supports large-scale factual data inference, but also can achieve better performance by extending the number of cluster nodes.

In the future, we plan to improve the performance of the rules engine from the aspects of optimizing conflict resolution strategies and improving the degree of memory sharing during matching. Moreover, the reasoning model of the current KSSRE is relatively simple and does not consider the time of data generation (Event-time in the data). In addition, we also plan to write inference results to storage or display in real time.

Acknowledgment. This work is partially supported by the National Key Research and Development Program of China under Grant No. 2016YFB1000600.

References

1. Erdani, Y.: Developing algorithms of ternary grid technique for optimizing expert system's knowledge base. In: 2006 Seminar Nasional Aplikasi Teknologi Informasi (2006)
2. Apache Kafka. http://kafka.apache.org/. Accessed May 2018
3. Structured Streaming. http://spark.apache.org. Accessed May 2018
4. Dean, J., Ghemawat, S.: MapReduce: simplified data processing on large clusters. Commun. ACM **51**, 107–113 (2008)
5. Cao, B., Yin, J., Zhang, Q., Ye, Y.: A MapReduce-based architecture for rule matching in production system, pp. 790–795. IEEE (2010)
6. Zhou, R., Wang, G., Wang, J., Li, J.: RUNES II: a distributed rule engine based on rete network in cloud computing. Int. J. Grid Distrib. Comput. **7**, 91–110 (2014)
7. Chen, Y., Bordbar, B.: DRESS: a rule engine on spark for event stream processing, pp. 46–51. ACM (2016)
8. Zhang, J., Yang, J., Li, J.: When rule engine meets big data: design and implementation of a distributed rule engine using spark, pp. 41–49. IEEE (2017)
9. Liang, S., Fodor, P., Wan, H., Kifer, M.: OpenRuleBench: an analysis of the performance of rule engines. In: Proceedings of the 18th International Conference on World Wide Web, pp. 601–610. ACM (2009)
10. DBLP: computer science bibliography. http://dblp.uni-trier.de/db/. Accessed May 2018
11. Forgy, C.L.: Rete: a fast algorithm for the many pattern/many object pattern match problem. Artif. Intell. **19**, 17–37 (1982)
12. Drools. https://www.drools.org/. Accessed May 2018

Optimization of Depth from Defocus Based on Iterative Shrinkage Thresholding Algorithm

Mingxin Zhang[1(✉)], Qiuyu Wu[2], Yongjun Liu[1], and Jinlong Zheng[1]

[1] School of Computer Science and Engineering,
Changshu Institute of Technology, Changshu, China
mxzhang163@163.com, zjinlong@163.com,
yongjun1981@126.com
[2] School of Computer Science and Technology,
China University of Mining and Technology, Xuzhou, China
wuqiuyu_cumt@163.com

Abstract. In solving the dynamic optimization of depth from defocus with the iterative shrinkage threshold algorithm (ISTA), the fixed iteration step decelerated the convergence efficiency of the algorithm, which led to inaccuracy of reconstructed microscopic 3D shape. Aiming at the above problems, an optimization of ISTA algorithm based on gradient estimation of acceleration operator and secant linear search (FL-ISTA) was proposed. Firstly, the acceleration operator, which consists of the linear combination of the current and previous points, was introduced to estimate the gradient and update the iteration point. Secondly, the secant linear search was used to dynamically determine the optimal iteration step, which accelerated the convergence rate of solving the dynamic optimization of depth from defocus. Experimental results of standard 500 nm grid show that compared with ISTA、FISTA and MFISTA algorithms, the efficiency of FL-ISTA algorithm was great improved and the depth from defocus decreased by 10 percentage points, which close to the scale of 500 nm grid. The experimental results indicate that FL-ISTA algorithm can effectively improve the convergence rate of solving dynamic optimization of depth from defocus and the accuracy of the reconstructed microscopic 3D shape.

Keywords: Microscopic 3D reconstruction · Depth from defocus
Iterative shrinkage threshold algorithm
Gradient estimation of acceleration operator · Secant linear search

1 Introduction

In micro-nano computer vision, microscopic 3D shape reconstruction based on vision is of great significance for a more comprehensive understanding of sample characteristics and evaluation of operational processes. The more commonly used microscopic 3D reconstruction methods [1] mainly include a volume recovery method, depth from stereo, depth from focus and depth from defocus. Depth from defocus was first proposed by Pentland [3]. This method uses the mapping relationship between the defocus

© Springer Nature Switzerland AG 2018
X. Meng et al. (Eds.): WISA 2018, LNCS 11242, pp. 131–144, 2018.
https://doi.org/10.1007/978-3-030-02934-0_13

degree feature of two-dimensional image and the depth of the scene, which resolves the three-dimensional depth information of the scene [4, 5].

In the depth from defocus of the scene, it is necessary to first obtain the defocus images with different degrees of scenery, which leads to change the camera parameters. However, in the observation of micro-nano images, the observation space is very limited and the camera with high magnification is used, so the camera's imaging model will change as the camera parameters change [6]. Considering the limitations of micro-nano observations, Wei et al. [7] proposed a new global depth from defocus with fixed camera parameters. This method uses the relative ambiguity [7, 8] and the diffusion equation [7, 9] to solve the depth information of the scene, and transforms the solution of depth information into a dynamic optimization problem to solve.

However, in solving the dynamic optimization problem of depth from defocus, the document [7] uses the classic iterative shrinkage threshold algorithm [10, 11] (Iterative Shrinkage Threshold Algorithm, ISTA) to solve. Because the ISTA algorithm is an extension of the gradient descent method, the iterative process only considers the information of the current point to update the iteration point of the gradient estimation. The optimization process takes the form of zigzag [12–14] approaching the minimum point, and the convergence speed is slow. In addition, the algorithm adopts a fixed step in the iterative process. When the iterative process is close to the minimum point, the fixed step size may be larger than the actual iterative step size, resulting in a poor iteration efficiency of the algorithm, which led to inaccuracy of the reconstruction of microscopic 3D shape.

In order to solve the problem of ISTA algorithm for dynamic optimization of depth from defocus, this paper proposed a method based on the gradient estimation of acceleration operator and the secant line search to optimize the ISTA algorithm. The acceleration operator consisted of a linear combination of the current point and the previous point, and then update the iteration point. The secant linear search was used to determine the optimal step dynamically. Finally, the FL-ISTA algorithm was applied to the reconstruction of depth information on a standard grid of 500 nm scale. The experimental results show that compared with ISTA algorithm, Fast Iterative Shrinkage Threshold Algorithm (FISTA) and Monotone Fast Iterative Threshold Algorithm (MFISTA), the convergence efficiency of FL-ISTA algorithm was faster, the depth information of the reconstruction decreased faster, which close to the scale of standard 500 nm grid. Experimental results indicate that FL-ISTA algorithm can effectively improve the convergence rate of solving the dynamic optimization of depth from defocus and elevate the accuracy of reconstructed microscopic 3D shape.

2   Global Depth from Defocus

2.1   Defocus Imaging Model

In image processing, image blurring can usually be expressed as a convolution.

$$I_2(x, y) = I_1(x, y) * H(x, y) \tag{1}$$

Where $I_1(x, y)$, $I_2(x, y)$ and $H(x, y)$ represent clear and blurred images and point spread functions respectively. "$*$"represents a convolution. According to the principle of point spread, the point spread function can be approximated by a two-dimensional Gaussian function, and the fuzzy diffusion parameter therein σ indicates the degree of blurring of the image. Since the equation of heat radiation is isotropic, Eq. (1) can be expressed as

$$\begin{cases} \dot{z}(x, y, t) = c\nabla z(x, y, t) & c \in [0, \infty)\, t \in (0, \infty) \\ z(x, y, 0) = g(x, y) \end{cases} \tag{2}$$

Where "c" represents fuzz diffusion parameter, $\dot{z} = \partial z / \partial t$, "$\nabla$" represents laplacian

$$\nabla z = \frac{\partial^2 z(x, y, t)}{\partial x^2} + \frac{\partial^2 z(x, y, t)}{\partial y^2}$$

$$\sigma^2 = 2tc \tag{3}$$

If the depth map is an ideal plane, then "c" is a constant, otherwise it can be expressed as

$$\begin{cases} \dot{z}(x, y, t) = \nabla \cdot (c(x, y)\nabla z(x, y, t)) & t \in (0, \infty) \\ z(x, y, 0) = g(x, y) \end{cases} \tag{4}$$

Where "∇"and"$\nabla \cdot$" represent gradient and differential operators

$$\nabla = \begin{bmatrix} \frac{\partial}{\partial x} & \frac{\partial}{\partial y} \end{bmatrix}^T \quad \nabla \cdot = \frac{\partial}{\partial x} + \frac{\partial}{\partial y}$$

From the above equation, we firstly need to know the clear image to solve the heat radiation equation, but this is a complicated process. Therefore, Favaro [8] proposed the concept of relatively ambiguity.

Suppose there are two different degrees of blurred images $I_1(x, y)$ and $I_2(x, y)$, the fuzzy diffusion coefficients are σ_1 and σ_2, $\sigma_1 < \sigma_2$, so $I_2(x, y)$ can be expressed as

$$\begin{aligned} I_2(x, y) &= \iint \frac{1}{2\pi\sigma_2^2} \exp\left(-\frac{(x - u)^2 - (y - v)^2}{2\sigma_2^2}\right) g(u, v) du dv \\ &= \iint \frac{1}{2\pi\Delta\sigma^2} \exp\left(-\frac{(x - u)^2 + (y - v)^2}{2\Delta\sigma^2}\right) I_1(u, v) du dv \end{aligned} \tag{5}$$

Where $\Delta\sigma^2 = \sigma_2^2 - \sigma_1^2$ represents relatively ambiguity, Eq. (2) can be expressed as

$$\begin{cases} \dot{z}(x,y,t) = c\nabla z(x,y,t) & c \in (0,\infty) \quad t \in (0,\infty) \\ z(x,y,0) = I_1(x,y) \end{cases} \tag{6}$$

Equation (4) can be expressed as

$$\begin{cases} \dot{z}(x,y,t) = \nabla \cdot (c(x,y)\nabla z(x,y,t)) & t \in (0,\infty) \\ z(x,y,0) = I_1(x,y) \end{cases} \tag{7}$$

And at the moment of Δt, $u(x,y,t) = I_2(x,y)$, Δt can be defined as

$$\Delta\sigma^2 = 2(t_2 - t_1)c \doteq 2\Delta tc \tag{8}$$

In addition

$$\Delta\sigma^2 = \lambda^2(\eta_2^2 - \eta_1^2) \tag{9}$$

Where $\eta_i(i = 1,2)$ represent fuzzy circle radius, λ represents the constant between the ambiguity and the radius of the fuzzy circle.

$$\eta = \frac{Dv}{2}\left|\frac{1}{f} - \frac{1}{v} - \frac{1}{s}\right| \tag{10}$$

Where v, f, s are the camera's image distance, focal length and object distance respectively, and D is the convex lens radius.

2.2   Depth Information Reconstruction

Assume that the image $I_1(x,y)$ is a defocus image before depth variation, its depth information is $S_1(x,y)$, and the image $I_2(x,y)$ is a defocus image after the depth variation, its depth information is $S_2(x,y)$, s_0 is known as the ideal focal length. In this section, the depth information of the image is obtained by changing the object distance. The schematic diagram is shown in Fig. 1.

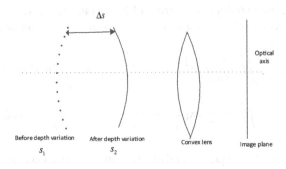

Fig. 1. Illustration of depth from defocus

First, establishing the heat radiation equation from the above condition

$$\begin{cases} \dot{z}(x,y,t) = \nabla \cdot (c(x,y)\nabla z(x,y,t)) & t \in (0,\infty) \\ z(x,y,0) = I_1(x,y) \\ z(x,y,\Delta t) = I_2(x,y) \end{cases} \qquad (11)$$

Where relatively ambiguity can be expressed as

$$\begin{aligned} \Delta\sigma^2(x,y) &= \lambda^2 \left(\eta_2^2(x,y) - \eta_1^2(x,y) \right) \\ &= \frac{\lambda^2 D^2 v^2}{4} \left[\left(\frac{1}{f} - \frac{1}{v} - \frac{1}{s_2(x,y)} \right)^2 - \left(\frac{1}{f} - \frac{1}{v} - \frac{1}{s_1(x,y)} \right)^2 \right] \\ &= \frac{\lambda^2 D^2 v^2}{4} \left[\left(\frac{1}{s_0} - \frac{1}{s_2(x,y)} \right)^2 - \left(\frac{1}{s_0} - \frac{1}{s_1(x,y)} \right)^2 \right] \end{aligned} \qquad (12)$$

It defines in this paper

$$b = \frac{4\Delta\sigma^2}{\lambda^2 D^2 v^2} + \left(\frac{1}{s_0} - \frac{1}{s_1(x,y)} \right)^2 \qquad (13)$$

Therefore, the final depth information is

$$s_2(x,y) = \frac{1}{((1/s_0 \pm \sqrt{b}))} \qquad (14)$$

In order to solve the dynamic optimization problem of depth from defocus, the following objective function is established to calculate the thermal radiation equation

$$\bar{s} = \arg\min \iint (z(x,y,\Delta t) - I_2(x,y))^2 dx dy \qquad (15)$$

Since the above solution process may be ill-conditioned, a Tikhonov regularization term is added to the end of the objective function, expressed as

$$\bar{s} = \arg\min \iint (z(x,y,\Delta t) - I_2(x,y))^2 dx dy + \rho \|\nabla s_2(x,y)\|^2 + \rho l \|s_2(x,y)\|^2 \qquad (16)$$

The third term of the above formula is about the smoothness constraint of the depth map, and the fourth is the guarantee of the boundedness of the depth map, in fact, the energy coefficient $\rho > 0$, $l > 0$. Lost energy is expressed as

$$E(s) = \iint (z(x,y,\Delta t) - I_2(x,y))^2 dx dy + \rho \|\nabla s\|^2 + \rho l \|s\|^2 \qquad (17)$$

Therefore, the scene depth information can be solved by solving the following optimization problem.

$$\tilde{s} = \arg\min E(s)$$
$$s.t (11) 和 (14) \tag{18}$$

3 FL-ISTA Algorithm

3.1 Using ISTA to Solve the Dynamic Optimization of Depth from Defocus

For the dynamic optimization problem of depth from defocus in Sect. 2.2, the solution is based on the ISTA algorithm

$$E_1(s) = \iint (z(x, y, \Delta t) - I_2(x, y))^2 dxdy \tag{19}$$

So, for unconstrained optimization problem

$$E_1(s) = \iint (z(x, y, \Delta t) - I_2(x, y))^2 dxdy$$

The solution to the above gradient algorithm is

$$s_j = s_{j-1} - t_j \nabla E_1(s_{j-1}) \tag{20}$$

Where t_j represents iteration step, the second-order estimation model of the above formula can represent as

$$s_j = \arg\min \left\{ E_1(s_{j-1}) + \nabla E_1(s_{j-1})(s - s_{j-1}) + \frac{1}{2t_j} \|s - s_{j-1}\|_2^2 \right\} \tag{21}$$

When formula (19) is added to the Tikhonov regularization term

$$E(s) = E_1(s) + \rho \|\nabla s\|^2 + \rho l \|s\|^2$$

It can get the following iteration formula

$$s_j = \arg\min \left\{ E_1(s_{j-1}) + \nabla E_1(s_{j-1})(s - s_{j-1}) + \frac{1}{2t_j} \|s - s_{j-1}\|_2^2 + \rho \|\nabla s\|^2 + \rho l \|s\|^2 \right\} \tag{22}$$

Ignore the constant term, you can get

$$s_j = \arg\min \left\{ \frac{1}{2t_j} \|s - (s_{j-1} - t_j \nabla E_1(s_{j-1}))\|_2^2 + \rho \|\nabla s\|^2 + \rho l \|s\|^2 \right\} \tag{23}$$

Since Tikhonov regularization terms are separable, the calculations can become a one-dimensional problem to solve each element, and its iteration formula is

$$s_j = T_{\rho t_j}\left(s_{j-1} - t_j \nabla E_1\left(s_{j-1}\right)\right) \tag{24}$$

Among them, $T_\rho : R^n \to R^n$ represents contraction operator, defined as

$$T_\rho(y_i) = \begin{cases} 0, & |y_i| \le \rho \\ y_i - \rho sign(y_i), & otherwise \end{cases} \tag{25}$$

3.2 Gradient Estimation of Acceleration Operator

The FL-ISTA algorithm is introduced based on the ISTA algorithm. First, the acceleration operator consists of a linear combination of the current point and the previous point. Then, the acceleration operator is used to estimate the gradient to find the next iteration point. The acceleration operator is represented as

$$y_{j+1} = s_j + a_j\left(s_j - s_{j-1}\right) \tag{26}$$

Where s_j, s_{j-1} represent current depth information value and previous depth information value. $s_j - s_{j-1}$ represents search direction. a_j represents the starting current depth information s_j, the step factor required to iterate through the constructed search directions $s_j - s_{j-1}$. y_{j+1} represents the starting current depth s_j, following the resulting search direction $s_j - s_{j-1}$ for the resulting depth information. The schematic diagram of using gradient operator to evaluate the gradient is shown in Fig. 2.

Fig. 2. Illustration of gradient estimation of acceleration operator

where $a_j = n_j - 1/n_j$, $n_{j+1} = \left(\alpha_1\alpha_k + \sqrt{\alpha_1\alpha_k + 4n_j^2}\right)/2n_1 = \alpha_1 = 1, 0 < \alpha_k < 1$.

Based on the ISTA algorithm, an acceleration operator is introduced to perform gradient estimation again to find the next iteration point, so Eq. (24) can be written in the following form

$$s_{j+1} = T_{\rho t_j}\left(y_{j+1} - t_j \nabla E_1\left(y_{j+1}\right)\right) \tag{27}$$

3.3   Secant Linear Search to Find Optimal Iteration Step

Because the ISTA algorithm adopts fixed steps in the iterative process and the optimization process is close to the minimum point, the fixed step size may be larger than the actual iteration step length, resulting in poor efficiency of the algorithm. Therefore, a linear search is used to change the limit of fixed steps.

The secant formula of the secant method is expressed as

$$x^{(k+1)} = x^{(k)} - \frac{\left(x^{(k)} - x^{(k-1)}\right) f\left(x^{(k)}\right)}{f'\left(x^{(k)}\right) - f'\left(x^{(k-1)}\right)} \tag{28}$$

Where $x^{(k)}, x^{(k-1)}$ are the two initial points obtained after the iteration. The iterative process is shown in Fig. 3.

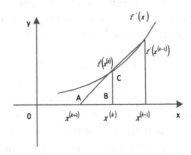

Fig. 3. Illustration of secant linear search

In order to solve the optimal iterative step, the following unary function on the step size is established.

$$t^* = \arg \min \psi(t) \tag{29}$$

$$\psi(t) = \iint (z(x, y, \Delta t) - I_2(x, y))^2 dx dy + \rho \|\nabla(s + td)\|^2 + \rho l \|s + td\|^2 \tag{30}$$

Where $\psi(t)$ is the one-variable function about t, t is the iteration step size, d is the iteration direction, and t^* is the optimal iteration step, so one-dimensional linear search method can be used to find the minimum value.

$$t_{j+1} = t_j - \psi'(t_j) \left[\frac{t_j - t_{j-1}}{\psi'(t_j) - \psi'(t_{j-1})} \right] \tag{31}$$

The optimal iterative step calculated using secant linear search, which can be written in the following form

$$s_{j+1} = T_{\rho t_j^*}\left(y_{j+1} - t_j^* \nabla E_1\left(y_{j+1}\right)\right) \tag{32}$$

4   The Flow Chart of the FL-ISTA Algorithm

The FL-ISTA algorithm steps are shown as follow:

Step 1: Give camera parameters focal length f, distance v, ideal object distance s_0, convex lens radius D, constant λ between ambiguity and fuzzy circle, two blurred images I_1, I_2, energy threshold ε, energy coefficient ρ, iterative step size t_0, $y_1 = s_0, j = 1, n_1 = \alpha_1 = 1, 0 < \alpha_k < 1$;

Step 2: The initialization depth information is s_0;

Step 3: The Eq. (12) is calculated to obtain relatively ambiguity $\Delta\sigma^2$;

Step 4: The Eq. (11) is calculated to obtain the solution of the diffusion equation $z(x, y, \Delta t)$;

Step 5: Using the above formula calculates the energy formula. If the energy is less than the threshold ε, stop it. The depth information at this time is the required value; otherwise, go to step 6.

Step 6: Calculate $n_{j+1} = \left(\alpha_1\alpha_k + \sqrt{\alpha_1\alpha_k + 4n_j^2}\right)\Big/2$;

Step 7: Calculate $y_{j+1} = s_j + \left(n_j - 1/n_{j+1}\right)\left(s_j - s_{j-1}\right)$;

Step 8: Bringing the above equation into Eq. (27);

Step 9: Using the secant line search to find the optimal iteration step, and bring it into Eq. (32) and find the s_{j+1};

Step 10: Calculate s_{j+1}, update the depth, return to step 3, continue iterating (Fig. 4).

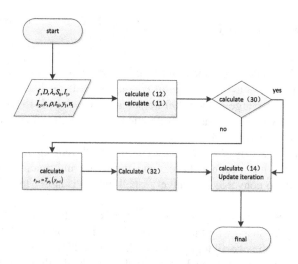

Fig. 4. FL-ISTA algorithm flow chart

5 The Simulation

In order to verify the correctness and effectiveness of the algorithm proposed in this paper, the ISTA, FISTA and MFISTA and the proposed FL-ISTA algorithm were used to perform depth information reconstruction on the standard 500 nm grid. The high and wide of the standard nanoscale grid is 500 nm and 1500 nm respectively. The HIROX-7700 microscope was used in the experiment and the magnification was 7000 times. The focal length is 0.357 mm, ideal object distance is 3.4 mm and aperture size is 2.

The results of the standard 500 nm scale grid are shown below. Among them, Fig. 5 is the two defocus images of the grid, Fig. 6 is the standard 500 nm scale grid. Figures 8, 9, 10 and 11 are the 3D shape reconstructed by ISTA, FISTA, MFISTA and FL-ISTA algorithm respectively. In the figure, the unit of the plane coordinate is pixels, and the size of the pixel is 115.36 nm*115.36 nm.

Fig. 5. Two defocus images of grid

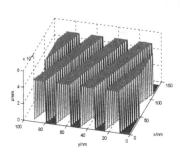

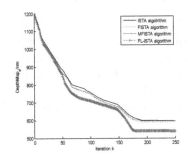

Fig. 6. True 3D shape of grid

Fig. 7. Convergence curve of depth from defocus

First, according to the experiment in [7], the energy threshold ε is set to 2, and the energy coefficient ρ is set to 0.2. Under this condition, the ISTA,FISTA and MISTSTA algorithms tend to converge around 200 iterations, while the FL-ISTA algorithm has already converged within less than 200 iterations. The depth information value convergence curve is shown in Fig. 7.

From the Fig. 7, it can be clearly seen that the depth information obtained by the FL-ISTA algorithm has declined faster, and the depth information obtained when convergence tends to converge is relatively small, which is closer to the real standard 500 nm.

From Figs. 8, 9, 10 and 11, although the error is slightly larger at the edges from the reconstructed 3D shape with the four algorithms, the reconstructed 3D topography roughly conforms to the overall topography of the standard 500 nm scale grid. It can be clearly seen from the reconstructed 3D topography that the depth information values in

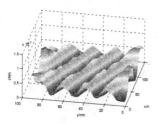

Fig. 8. Reconstructed 3D shape by ISTA

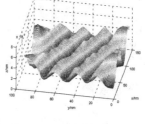

Fig. 9. Reconstructed 3D shape by FISTA

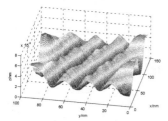

Fig. 10. Reconstructed 3D shape by MFISTA

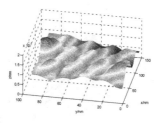

Fig. 11. Reconstructed 3D shape by FL-ISTA

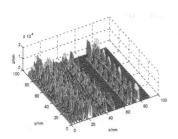

Fig. 12. Error map of ISTA algorithm

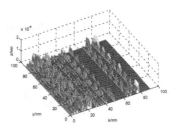

Fig. 13. Error map of FISTA algorithm

the 3D topographies reconstructed by the first three algorithms are almost the same, while the depth information values in the FL-ISTA algorithm are significantly decreased. The information value is about 540 nm, which is closer to the standard 500 nm grid scale, making the restored 3D appearance more accurate.

To further test the accuracy of the algorithm, the relative error surface between the estimated depth and the true depth value is calculated using Eq. (33). Then, using Eq. (34) calculates the mean square error. formula (35) calculates the average error of the entire plane.

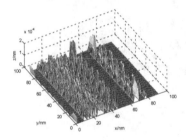

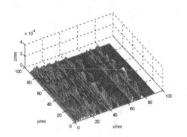

Fig. 14. Error map of MFISTA algorithm

Fig. 15. Error map of FL-ISTA algorithm

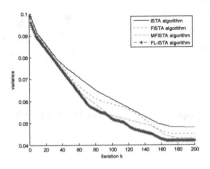

Fig. 16. Convergence curve of mean square error

$$\phi = \tilde{s}/s - 1 \tag{33}$$

$$\psi = \sqrt{E\left[(\tilde{s}/s - 1)^2\right]} \tag{34}$$

$$E = \frac{1}{n}\sum_{k=1}^{n}\left|H_k - \tilde{H}_k\right| \tag{35}$$

Where n is the number of sample points, H_k is the true height of the kth point of the grid, and $\tilde{H}_k$ is the estimated height of the kth point.

As can be seen from Figs. 12, 13, 14 and 15, the relative error surfaces obtained by the ISTA, FISTA and MFISTA algorithms do not change much, and the maximum relative error is about 80 nm, 70 nm and 70 nm. but the relative error surface obtained from the FL-ISTA calculation is only 20 nm, which elevated the accuracy of reconstructed 3D shape.

As can be seen from Fig. 16, the variances of the ISTA, FISTA and MFISTA algorithms are 0.05, 0.048 and 0.045 respectively, but the FL-ISTA algorithm has a variance of 0.041. As is known from Eq. (35), the average error of 3D topographies reconstructed by ISTA, FISTA and MFISTA algorithms are 161 nm,160 nm and

158 nm respectively, while the average error of FL-ISTA algorithm is only 96 nm, which improved the accuracy of reconstructed 3D shape.

6 Conclusion

The FL-ISTA algorithm proposed in this paper is to solve the poor convergence rate of the dynamic optimization of depth from defocus with the ISTA algorithm, which makes the reconstructed microscope 3D shape more accurate. Using the acceleration operation estimates the gradient and updates the iteration points, and secant linear search finds the optimal iteration step, which improves efficiency of the algorithm. The experimental results based on the standard 500 nm scale grid show that compared with the ISTA, FISTA and MFISTA algorithms, the convergence of FL-ISTA algorithm is faster, and the reconstructed 3D shape by the FL-ISTA algorithm decreased faster, which close to the scale of standard 500 nm grid. The experimental results show that FL-ISTA algorithm can effectively improve the efficiency of solving the dynamic optimization of depth from defocus and the accuracy of microscopic 3D reconstruction. Since it takes time to find the optimization iteration step by adopting the secant line search in the iterative process, the nonlinear search method will be explored to improve the efficiency of the algorithm, the specific method needs further exploration

References

1. Li, C., Su, S., Matsushita, Y., et al.: Bayesian depth-from-defocus with shading constraints. IEEE Trans. Image Process. **25**(2), 589–600 (2016)
2. Tao, M.W., Srinivasan, P.P., Hadap, S., et al.: Shape estimation from shaping, defocus, and correspondence using light-field angular coherence. IEEE Trans. Pattern Anal. Mach. Intell. **39**(3), 546–560 (2017)
3. Pentland, A.P.: A new sense for depth of field. IEEE Trans. Pattern Anal. Mach. Intell. **9**(4), 523–531 (1987)
4. Nayar, S.K., Watanabe, M., Noguchi, M.: Real time focus range sensor. IEEE Trans. Pattern Anal. Mach. Intell. **18**(12), 1186–1198 (1996)
5. Subbarao, M., Surya, G.: Depth from defocus: a spatial domain approach. Int. J. Comput. Vis. **13**(3), 271–294 (1994)
6. Yang, J., Tian, C.P., Zhong, G.S.: Stochastic optical reconstruction microscopy and its application. J. Opt. **37**(3), 44–56 (2017)
7. Wei, Y.J., Dong, Z.L., Wu, C.D.: Global shape reconstruction with fixed camera parameters. J. Image Graph. **15**(12), 1811–1817 (2010)
8. Favaro, P., Soatto, S., Burger, M., et al.: Shape from defocus via diffusion. IEEE Trans. Pattern Anal. Mach. Intell. **30**(3), 518–531 (2008)
9. Favaro, P., Mennucci, A., Soatto, S.: Observing shape from defocused images. Int. J. Comput. Vis. **52**(1), 25–43 (2003)
10. Beck, A., Teboulle, M.: A fast iterative shrinkage-thresholding algorithm for linear inverse problems. SIAM J. Imaging Sci. **2**(1), 183–202 (2009)
11. Zibetti, M.V.W., Helou, E.S., Pipa, D.R.: Accelerating overrelaxed and monotone fast iterative shrinkage-thresholding algorithms with line search for sparse reconstructions. IEEE Trans. Image Process. **26**(7), 3569–3578 (2017)

12. Zibetti, M.V.W., Pipa, D.R., De Pierro, A.R.: Fast and exact unidimensional L2–L1 optimization as an accelerator for iterative reconstruction algorithms. Digit. Signal Process. **48**(5), 178–187 (2016)
13. Zibulevsky, M., Elad, M.: L1-L2 optimization in signal and image processing. IEEE Signal Process. Mag. **27**(3), 76–88 (2010)
14. Beck, A., Teboulle, M.: Fast gradient-based algorithms for constrained total variation image denoising and deblurring problems. IEEE Trans. Image Process. **18**(11), 2419–2434 (2009)

Information Retrieval

LDA-Based Resource Selection for Results Diversification in Federated Search

Liang Li, Zhongmin Zhang, and Shengli Wu[⊠]

School of Computer Science, Jiangsu University, Zhenjiang 212013, China
swu@ujs.edu.cn

Abstract. Resource selection is an important step in federated search environment, especially for search result diversification. Most of prior work on resource selection in federated search only considered relevance of the resource to the information need, and very few considered both relevance and diversification of the information inside them. In this paper, we propose a method that uses the Latent Dirichlet Allocation (LDA) model to discover underlying topics in each resource by sampling a number of documents from it. Thus the vector representation of each resource can be used to calculate the similarity between different resources and to decide the diversity of them. Using a group of diversity-related metrics, we find that the LDA-based resource selection method is more effective than other state-of-the-art methods in the same category.

Keywords: Resource selection · Latent Dirichlet Allocation model
Results diversification · Federated web search · Information retrieval

1 Introduction

Federated search, also known as distributed information retrieval (DIR), is focused on retrieving multiple distributed and independent information resources through a broker [1–3]. In the federated search environment, queries are submitted to the search interface. Then based on a certain criterion, the broker chooses a given number of most suitable information resources, forwards the query to those selected resources and collects the resultant lists from them. Finally, the resultant lists are merged into a final list of results and returned to the user. Federated search is especially favorable for obtaining information from the hidden, deep web. Some search engines, such as LinkedIn[1] and WorldWideScience[2], are typical examples that take this approach.

Prior research on federated search is mainly focus on resource selection and results merging when relevance of the document to the query is the only concern [1–3]. Results diversification has been identified as necessary for web search, especially for the ad hoc task [4–6]. Its purpose is to return documents that are able to cover various sub-topics of the query. In most cases, especially for those ambiguous queries with a

---

[1] LinkedIn is a business- and employment-oriented service that operates via websites and mobile apps. Its website is located at https://www.linkedin.com/.

[2] WorldWideScience is hosted by the U.S. Department of Energy's Office of Scientific and Technical Information, and it is composed of more than 40 information sources, several of which are federated search portals themselves. Its website is located at https://worldwidescience.org/.

© Springer Nature Switzerland AG 2018
X. Meng et al. (Eds.): WISA 2018, LNCS 11242, pp. 147–156, 2018.
https://doi.org/10.1007/978-3-030-02934-0_14

small number of query words, users' intents for the search are often not very clear, and different users may have different information requirements for the same query. For example, the query "Lincoln" could refer to the 16th US President Abraham Lincoln, a luxury car brand, or a Brazilian professional football player. Therefore, this query may be related to history, cars, sports and some other aspects. If the search engine only considers the relevance of the documents to the query generally but without special treatment for these sub-topics, then it may disappoint some of the users.

In order to support resource selection and results merging effectively, each resource needs to be represented properly at the broker level. In a non-cooperative environment, usually the broker needs to collect a group of sample documents through the end-user interface of each resource and keeps them locally. The sample documents from all available resources are put together to form a central sample document base. Its content is used as the representation of each resource, resource selection methods such as CVV [2], ReDDE [3], CORI [7] and others are able to select a given number of resources by using collections of such sample documents.

When both relevance and diversity are considered, the broker may have two types of strategies for resource selection. They are related to the re-ranking algorithms of results diversification: explicit and implicit [5, 8, 15]. The explicit strategy needs to know the sub-topics involved for the query while the implicit strategy does not have such a requirement. Categorized as explicit methods, Hong and Si [4] proposed two resource selection methods DivD and DivS. Both of them are based on the diversity-oriented re-arranging method PM-2 [8], combined with the relevance-oriented resource selection algorithm ReDDE [3]. Assuming that all the sub-topics of the query are known, these two algorithms are effective. However, such a condition is strong and it might not always be possible to satisfy it. In contrast, there is no such a requirement for the implicit strategy. Therefore, some scholars studied the implicit resource selection strategy that combines different technologies. Naini et al. proposed GLS that combines MMR with clustering [6]; while Ghansah and Wu proposed MnStD [9] that combines ReDDE [3] with the portfolio theory in modern economics.

In this paper we propose a LDA-based implicit resource selection method. To our knowledge, Latent Dirichlet Allocation (LDA) has not been used for such a purpose before. Our experiments with the TREC data set ClueWeb12B demonstrate both of them are effective.

2   The Latent Dirichlet Allocation Topic Model

Latent Dirichlet Allocation (LDA) is a generative statistical model that is developed in natural language processing [10]. In LDA, Each document is viewed as a mixture of various topics and each topic only uses a small set of words frequently.

As shown in Fig. 1, the dependencies among different variables can be captured concisely. The boxes represent replicates. The bottom outer box represents documents, the bottom inner box represents the repeated choice of topics and words within a document, while the top box represents topics. M denotes the number of documents, N the number of words in a document. K denotes the number of topics and $\varphi_1, \ldots, \varphi_K$ are v-dimensional vectors storing the parameters of the Dirichlet-distributed topic-word

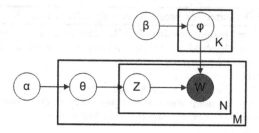

Fig. 1. The LDA model

distributions (V is the number of words in the vocabulary). The words w_{ij} are the only observable variables, and the other variables are latent variables. As proposed in [10], a sparse Dirichlet prior can be put over the topic-word distribution. Thus:

- α is the parameter of the Dirichlet prior on the per-document topic distributions,
- β is the parameter of the Dirichlet prior on the per-topic word distribution,
- θ_m is the topic distribution for document m,
- φ_k is the word distribution for topic k,
- z_{mn} is the topic for the n-th word in document m, and
- w_{mn} is the specific word.

Learning the various distributions (the set of topics, their associated word probabilities, the topic of each word, and the particular topic mixture of each document) is a problem of Bayesian inference [11, 12]. Gibbs sampling [16] and expectation propagation [17] are two typical approaches for this learning process. Gibbs sampling is used in this work.

3 The Proposed Algorithm

In this section we present the LDA-based resource selection algorithm for both relevance and diversity. Assume that we are able to collect a number of sample documents from each of all available resources. Thus all the documents are put together to form a Central Document Base (CDB) and those from the same resource are preventatives of that resource. The primary idea is: for a given query, we retrieve a group of results from the CDB by using a traditional search engine such as Indri. At this stage only relevance is considered. Then LDA is applied to a number of top-ranked documents retrieved to discover the topics involved. The next step is to use those leant topics to re-rank the results and to make them as diversified as possible by covering all the topics discovered in the last step. Finally, we rank all the resources according to the number and ranking positions of their documents appeared in the last step. Figure 2 presents the LDA-based Resource Selection algorithm (LDA-RS) for results diversification.

In Algorithm 1, each document is represented as vector $\theta(d) = (a_1, a_2, \ldots, a_K)$ after applying the LDA model with K topics, in which a_i indicates the probability that d belongs to the i-th topic. GoodnessDiv$\left(d_i, Q, D_{\eta,div}, \lambda\right)$ is a function which evaluates

---

Algorithm 1. a LDA-based resource selection algorithm LDA-RS

Input: Query Q, Central Document Base D, LDA topic model of each document in D, parameter for relevance threshold η, balance parameter λ, optimal objective function GoodnessDiv, document ranking scoring function G

Output: Ranked list of resources C

/* The first part is to build topic model for each retrieved document in D*/

1.   Search D with a given information search engine and query Q and obtain a ranked list of documents From D, choose a group of top-ranked documents $D_{\eta,rel}$ whose relevance score is greater than η

2.   Apply the LDA model to all the documents in $D_{\eta,rel}$ to obtain K topics on $D_{\eta,rel}$ and the probability distributions of topics for any document in $D_{\eta,rel}$; each document is represented as a vector $\theta(d)=(a_1,a_2,...,a_k)$

   /* The second part is to generate a diversified list of documents */

3. $D_{\eta,div} \leftarrow \emptyset$ //$D_{\eta,div}$ is a diversified list of documents to be generated

4.   while $|D_{\eta,rel}| > 0$ do

5.   for $d \in D_{\alpha,rel}$ do

6. $v(d) \leftarrow$ GoodnessDiv$(d, Q, D_{\eta,div}, \lambda)$ //see below for the explanation

7.   end for

8. $d^* \leftarrow$ argmax$\{ v(d) \}$, $D_{\alpha,rel} \leftarrow D_{\alpha,rel}/ \{d^*\}, D_{\eta,div} \leftarrow D_{\eta,div} \cup \{d^*\}$

9.   end while

   /* The third part is to rank resources */

10. $S(c_1), S(c_2), ... \leftarrow 0$ //$C = \{c_1, c_2, ...\}$ are all the resources

11.   for $d \in D_{\eta,div}$ and $d \in c_i$ do

12. $S(c_i) \leftarrow S(c_i) + G(r_d)$ // $G(r_d)$ is a function of the ranking position of d, or r_d

13.   end for

14.   Rank all the resources in C in ascending order of $s(c_i)$ values

15.   return ranked list of resources C

Fig. 2. LDA-based resource selection algorithm (LDA-RS) for results diversification

the goodness of the ranked list of $D_{\eta,div}$ followed by d_i, or $D_{\eta,div} \cup \{d_i\}$. More specifically, we use the following equation

$$\text{GoodnessDiv}(d_i, Q, D_{\alpha,div}, \lambda) = \lambda \text{Rel}(d_i, Q) - (1 - \lambda) \max_{d_j \in D_{\alpha,div}} \text{Sim}(d_i, d_j) \quad (1)$$

for this purpose. It has two parts. The first part concerns relevance and the second concerns diversity. $\text{Sim}(d_i, d_j)$ is calculated by the cosine similarity of their vector representation $\theta(d_i)$ and $\theta(d_j)$. The values of $\text{Rel}(d_i, Q)$ for all the documents retrieved are obtained from Indri. λ is used to control the weights of each part.

G(r) is a ranking function, which is defined as follows

$$G(r) = 1/(r+60) \quad (2)$$

In this way, the scores decrease with rank. This method has been used for data fusion with good performance [18]. We find it is good for our purpose as well.

4 Experimental Settings

4.1 Data Set and Baselines

ClueWeb12B[3] is a document collection used by TREC for three successive years from 2013 to 2015. It was used for diversity-related search tasks. It includes 52,343,021 documents. Some treatment is required for using this collection in our experiments.

First we sample 1% of the documents randomly from ClueWeb12B and use the K-means clustering algorithm to cluster them into 100 initial clusters. A KL divergence based method is used to calculate the similarity of two documents [19]. For these 100 clusters, their centroids are calculated. After that, we add all the rest into these 100 clusters and each document goes to the closest cluster available. Table 1 summarizes the statistics of the data set for our federated search experiments and Fig. 3 shows the number of documents in all the resources.

Table 1. Statistics of the federated search testbed based on Clueweb12

Number of resources	Total number of documents	Smallest collection	Largest collection	Average per collection
100	52,343,021	17,835	6,653,521	523,430

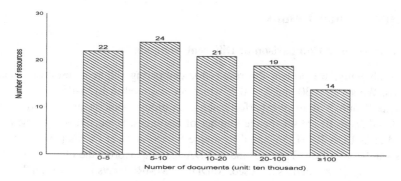

Fig. 3. Number of documents in all the resources

In order to support resource selection, we sample 1% documents from each resource to generate the Central Document Base. Documents are indexed using Indri[4] with the BM25 model. We use the 100 queries that were used in the TREC's Web tracks of 2013 and 2014 [13, 14].

---

[3] ClueWeb12B is a subset of ClueWeb12. Both ClueWeb12 and ClueWeb12B were used in TREC. ClueWeb12 consists of 733,019,372 English web pages. The major consideration for that is to encourage more research groups, especially those having limited computing facilities, to participate in the event.

[4] Its web page is located at: http://www.lemurproject.org/indri/

Apart from the resource selection algorithm proposed by us, we also test some other methods including GLS [6], MnStD [9], and ReDDE + MMR [9]. These are up-to-date implicit resource selection methods.

4.2 Evaluation

As in [3, 9], we also use the R-based diversification metric, which has been used in resource selection for federated search for some time. It is defined as

$$R_M(S, m) = \frac{M(\text{all the documents in } S)}{M(\text{all the documents in all the resources})} \quad (3)$$

where S represents the union of the resources selected by the resource selection algorithm. For all the documents either in all selected resources or all available resources, we index them centrally using Indri. The same query is sent to both search systems and results are collected. Then we apply MMR to diversify both of them and evaluate them using the same metric m. In this experiment we use seven metrics including α-nDCG@20, ERR-IA@20, nERR-IA@20, MAP-IA@20, NRBP@20, nNRBP@20, and P-IA@20. In this way, we can evaluate the performance of our resource selection algorithm without a results merging method. It is a good strategy because any results merging method may be biased to certain aspects of a resource selection algorithm.

5 Experimental Results

5.1 Performance Comparison of Different Methods

In this subsection we present the results for comparing different resource selection methods. We set K = 30 and λ = 0.5, which are required in LDA-RS.

Tables 2 and 3 show the performance of various methods in R metrics when $N_c = 3$ and $N_c = 10$ are set for the number of resources, respectively. Our method is referred to as LDA-RS (LDA-based Resource Selection) later in this paper.

As can be seen from Tables 2 and 3, LDA-RS outperforms the others most of the time when any of the metrics are used. ReDDE + MMR always performs badly, while GLS and MnStD are close. When three sources are selected, we can obtain over 70% of the performance, while 10 resources are selected, we can obtain over 80% of the performance for any of the metrics used.

Table 2. Performance Comparison of different methods ($N_c = 3$, the figures in bold are the highest for a given metric)

Method	R(ERR-IA@20)	R(nERR-IA@20)	R(α-ndcg@20)	R (NBRP)	R (nNBRP)
LDA-RS	**0.7510**	**0.7242**	**0.7504**	**0.7512**	0.7129
GLS	0.7059	0.6988	0.7037	0.7137	0.7023
MnStD	0.7071	0.6985	0.6830	0.7338	**0.7204**
ReDDE + MMR	0.3995	0.3861	0.3809	0.4095	0.3903

Table 3. Performance Comparison of different methods ($N_c = 10$, the figures in bold are the *highest for a given metric*)

Method	R(ERR-IA@20)	R(nERR-IA@20)	R(α-ndcg@20)	R (NBRP)	R (nNBRP)
LDA-RS	**0.8659**	**0.8537**	**0.8711**	**0.8576**	**0.8405**
GLS	0.7464	0.7414	0.7670	0.7470	0.7341
MnStD	0.7452	0.7342	0.7652	0.7395	0.7244
ReDDE + MMR	0.6826	0.6682	0.6920	0.6774	0.6569

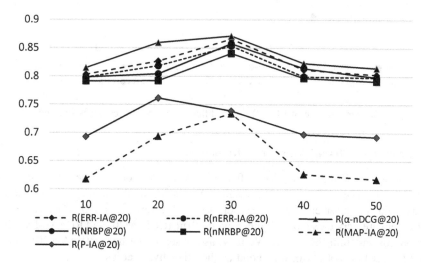

Fig. 4. Performance of LDA-RS with different values for number of topics

5.2 Selecting Different Number of Topics

The LDA topic model is used in our method. In this subsection we try to find the impact of the number of topics generated by LDA on LDA-RS.

In this experiment, we set the selected number of resources $N_c = 10$, and let the number of topics k vary from 10, 20,..., to 50. The performance of LDA-RS under various conditions is shown in Fig. 4. As can be seen from the figure, $K = 30$ is likely the best point for most of these metrics, $K = 20$ is in the second place, while K = 10, 40, or 50 are generally not good. If we look at these metrics closely, then we may find that five metrics including R(α-nDCG@20), R(nERR-IA@20), R(NRBP@20), R (nNRBP@20), and R(P-IA@20) does not change much, while we observe more fluctuations for the other two metrics R(P-IA@20) and R(MAP-IA@20) when different K values are taken. It shows that both R(P-IA@20) and R(MAP-IA@20) are hard metrics for resource selection methods to achieve good performance, while the other five are easier.

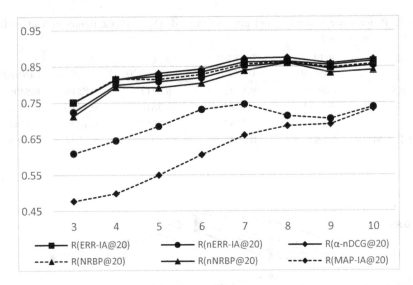

Fig. 5. Performance of LDA-RS with different number of resources selected

5.3   Selecting Different Number of Resources

This time we set $K = 30$ but select different number of resources: the range of N_c is
{3, 4, ..., 10}. Figure 5 shows the results. As shown in the figure, all the curves
increase with the number of resources selected with a few exceptions. The values of
five of them are very close to each other, while R(nERR-IA@20) and R(MAP-IA@20)
are more distant from the others. We may also observe that both R(P-IA@20) and R
(MAP-IA@20) do not perform as good as the other five metrics.

Considering both experimental results in 5.2 and 5.3, we observe that it with
different combinations of K and N_c values, the performance of resource selection is
good and stable if measured by any of the five metrics R(α-nDCG@20), R(nERR-
IA@20), R(NRBP@20), R(nNRBP@20), and R(P-IA@20). However, it is harder to
obtain good performance for resource selection if either R(P-IA@20) or R(MAP-
IA@20) is used. This is understandable because impact of the number of relevant
documents is quite different on these two groups of metrics.

5.4   Efficiency of the LDA-RS Algorithm

For the LDA-RS algorithm, to generate topic models by LDA is the most time-
consuming part. Therefore, the time required varies with the number of topics generated, as shown in Table 4. The experiment is carried out on a desktop computer with
Intel Core i7-3770 quad-core CPU and 6 GB of RAM. When we set 10 topics, it takes
over 6 s; when we set 50 topics, it takes more than 16 s. The efficiency of this
algorithm may be an issue that needs to be addressed when applying it for practice.

Table 4. Time requirement for the LDA-RS algorithm

Number of topics	Time (in seconds)
10	6.765
20	8.216
30	10.832
40	13.470
50	16.230

6 Conclusions and Future Directions

Results diversity has been investigated in many areas including information search. In this paper, we have presented a LDA-based approach for resource selection to support results diversification in a federated search system. Experiments with TREC data demonstrate the effectiveness of our method when a variety of diversity-related metrics are used for evaluation. It outperforms some other state-of-the-art methods in the same category. We also find that some commonly used diversity-related metrics are harder than some others to achieve good performance.

As our future work, we would consider some other approaches, such as deep-learning based approaches, to deal with the resource selection problem. For example, word embedding can be used as a new way to calculate the similarity between documents or queries. It may be helpful for implicit resource selection approaches. Another direction is to investigate the results merging problem in the same framework. One of the key factors is how to assign weights for those selected resources. We hypothesize that the scores obtained in the resource selection stage is still useful for that as well.

References

1. Nguyen, D., Demeester, T., Trieschnigg, D., Hiemstra, D.: Federated search in the wild: the combined power of over a hundred search engines. In: CIKM 2012, pp. 1874–1878 (2012)
2. Yuwono, B., Lee, D.L.: Server ranking for distributed text retrieval systems on the internet. In: DASFAA, pp. 41–50 (1997)
3. Si, L., Callan, J.P.: Relevant document distribution estimation method for resource selection. In: SIGIR 2003, pp. 298–305 (2003)
4. Hong, D., Si, L.: Search result diversification in resource selection for federated search. In: SIGIR 2013, pp. 613–622 (2013)
5. Carbonell, J.G., Goldstein, J.: The use of MMR, diversity-based reranking for reordering documents and producing summaries. In: SIGIR 1998, pp. 335–336 (1998)
6. Naini, K.D., Altingovde, I.S., Siberski, W.: Scalable and efficient web search result diversification. TWEB 10(3), 15:1–15:30 (2016)
7. Thomas, P., Shokouhi, M.: SUSHI: scoring scaled samples for server selection. In: SIGIR 2009, pp. 419–426 (2009)
8. Dang, V., Croft, W.B.: Diversity by proportionality: an election-based approach to search result diversification. In: SIGIR 2012, pp. 65–74 (2012)

9. Ghansah, B., Shengli, W.: A mean-variance analysis based approach for search result diversification in federated search. Int. J. Uncert. Fuzz. Knowl. Based Syst. **24**(2), 195–212 (2016)

10. Blei, D.M., Ng, A.Y., Jordan, M.I.: Latent Dirichlet allocation. J. Mach. Learn. Res. **3**, 993–1022 (2003)

11. Geman, S., Geman, D.: Stochastic relaxation, Gibbs distributions, and the Bayesian restoration of images. IEEE Trans. Pattern Anal. Mach. Intell. **6**(6), 721–741 (1984)

12. Metzler, D., Croft, W.B.: Combining the language model and inference network approaches to retrieval. Inf. Process. Manag. **40**(5), 735–750 (2004)

13. Demeester, T., Trieschnigg, D., Nguyen, D., Hiemstra, D.: Overview of the TREC 2013 federated web search track. In: TREC 2013 (2013)

14. Collins-Thompson, K., Macdonald, C., Bennett, P.N., Diaz, F., Voorhees, E.M.: TREC 2014 web track overview. In: TREC 2014 (2014)

15. Krestel, R., Fankhauser, P.: Reranking web search results for diversity. Inf. Retr. **15**(5), 458–477 (2012)

16. Griffiths, T.L., Steyvers, M.: Finding scientific topics. Proc. Natl. Acad. Sci. **101**(Suppl. 1), 5228–5235 (2004). https://doi.org/10.1073/pnas.0307752101

17. Minka, T.P., Lafferty, J.D.: Expectation-propagation for the generative aspect model. In: UAI 2002, pp. 352–359 (2002)

18. Cormack, G.V., Clarke, C.L.A., Büttcher, S.: Reciprocal rank fusion outperforms Condorcet and individual rank learning methods. In: SIGIR 2009, pp. 758–759 (2009)

19. Khudyak Kozorovitzky, A., Kurland, O.: Cluster-based fusion of retrieved lists. In: SIGIR 2011, pp. 893–902 (2011)

Extracting 5W from Baidu Hot News Search Words for Societal Risk Events Analysis

Nuo Xu[1,2] and Xijin Tang[1,2(✉)]

[1] Academy of Mathematics and Systems Science,
Chinese Academy of Sciences, Beijing 100190, China
xunuo1991@amss.ac.cn, xjtang@iss.ac.cn
[2] University of Chinese Academy of Sciences, Beijing 100049, China

Abstract. Nowadays risk events occur more frequently than ever in China during the critical periods of social and economic transformation and spread rapidly via a variety of social media, which have impacts on social stability. Online societal risk perception is acquired by mapping online community concerns into respective societal risks. What we concern is how to recognize and describe societal risk events in a formal and structured way. So to get a structured view of those risk events, we propose an event extraction framework on HNSW including 5 elements, namely where, who, when, why, and what (5W). The task for extracting 5W of risk events is converted into different machine learning tasks. Three methods are explored to tackle the extraction tasks. The framework of 5W extraction on the basis of online concerns can not only timely access to societal risk perception but also expose the events by a structured image, which is of great help for social management to monitor online public opinion timely and efficiently.

Keywords: 5W · Societal risk perception · Baidu hot news search words
Conditional random fields · TextRank · Risk-labeled keywords

1 Introduction

Modern China has stepped into a new era with the transformation of the principal contradiction of Chinese society, i.e. the contradiction between the people's ever-growing needs and the unbalanced and inadequate development, widely seen in economic, social, and ecological fields. Societal risk events are those events involving major hazards and possible harm, and raising the whole concern of society [1]. They occur more frequently at this new critical time, timely expose on the Internet, especially in Web 2.0 era, and pose threats to social stability. Societal risk perception is the subjective evaluation of public to societal risk events, which helps shape their attitudes toward public policy on societal issues [2]. Traditional research on societal risk perception was studied from social psychology based on the psychometric paradigm and questionnaires [3] which is time-consuming to be performed and unable to analyze the emerging risk events timely.

Search engines, as one of the most common tools to access information, not only meet search requirements but also record foci of netizens. Baidu is the largest Chinese

© Springer Nature Switzerland AG 2018
X. Meng et al. (Eds.): WISA 2018, LNCS 11242, pp. 157–169, 2018.
https://doi.org/10.1007/978-3-030-02934-0_15

search engine. Baidu hot news search words (HNSW) released at Baidu News Portal are based on real-time search behaviors of Internet users, reflecting the current public concerns. As those user-generated contents are actively expressed by Internet users, and can be regarded as one kind of effective way to perceive societal risk. Tang proposed to map HNSW into either risk-free events or one event with risk label from one of 7 risk categories including national security, economy/finance, public morals, daily life, social stability, government management, and resources/environment [4]. The 7 risk categories came from socio-psychological study results [2]. The effects of HNSW-based derived societal risk perception on China stock market volatility were studied by Granger causality analysis [5].

In this paper, what we concern is how to recognize and describe societal risk events in a formal and structured way in this year. To get a structured view of those risky events, we propose an event extraction framework on HNSW including 5 elements, namely where, who, when, why, and what (5W). 5W extraction is specifically for societal risk events, thus the elements "why" and "what" are defined as the risk category the event is labeled and the contents respectively, as the risk category and the contents respectively describe why the event is being public concerned and what happened to the risky event. The task for extracting 5W of risk events are converted into different machine learning tasks, which are: named entity recognition, geographical name recognition, risk category classification, and keyword extraction. In this paper, we first illustrate challenges for societal risk annotation performed on HNSW. Then, to tackle the problems of each W extraction component, our approaches are introduced. An effective model conditional random fields (CRFs) is applied for the risk category classification. And for the task of geographical name recognition, we try to propose an algorithm through encoding the place names based on area number of National Bureau of Statistics of China. We adopt an unsupervised TextRank-based method that automatically extracts keywords for HNSW. Next, the testing results of 5W extraction toward HNSW are presented. Finally, conclusions and future work are given. The framework of 5W extraction allows to extract structured information from the wicked societal problems and provide a new perspective to conduct societal risk studies.

2   Challenges for Societal Risk Event Extraction

Online societal risk perception is acquired based on HNSW which serve as an instantaneous corpus to maintain a view of netizens' empathic feedback for social hot spots, etc. Each of HNSW corresponds to 1–20 news whose URLs are at the first page of hot words search results. By labeling those HNSW and corresponding hot news with societal risk categories, not only a rough image toward China's societal risk is acquired but the risk levels of different hazards along the time are studied in details as well [5]. The 5W extraction of societal risk event is an important task, which is more complicated and challenging related to socio-psychology.

The extraction process for "when" and "who" is straightforward. We tag "when" that refers to the date when a societal risk event occurred and "who" that refers to personal names which appear in HNSW. Person names are recognized based on named entity recognition. The main extraction challenges include societal risk classification, geographical name recognition and keywords extraction which are addressed in this section

2.1 Societal Risk Classification

The challenges on societal risk classification are summarized as the following 3 points. First, societal risk events are manually annotated to 7 risk categories. However, different individuals may have the different subjective perception of risks, which may lead the inconsistent annotation results. Second, as the search is real-time behavior with diverse topics, new words with risk are emerging and the same words may have different labels [6]. Thus, new words and multiple risk labels for one word will affect the classification results. Third, the distribution of HNSW labeled with risks is seriously unbalanced. More than 50% HNSW during one year are labeled as "risk-free". Moreover, HNSW are short texts with no punctuations and spaces, which makes it more difficult to deal with. Relevant news texts are crawled and extracted simultaneously to provide corpus for machine learning. Experiments are conducted which have carried out societal risk classifications on news contents, while accuracy is barely needed to be improved [7]. More recently, deep learning based methods have drawn much attention in the field of natural language processing (NLP). Models such as recurrent neural network [8] and convolutional neural network [9] have excellent performances in text classification. However, the length of the short text still seriously affects the performance of the models [9].

To tackle the problems, CRFs model is applied to societal risk classification for the first time directly dealing with HNSW (short texts) without news texts compared with previous studies. We regard the risk classification as a sequence labeling problem and use CRFs model to capture the relations among terms in hot words. Furthermore, the feature terms with greater weights generated by the state features of CRFs model are picked out as risk factors. The feasibility and validity of CRFs model on societal risk classification will be verified in Sect. 3.

2.2 Geographical Name Recognition

As to the characteristics of geographical name in HNSW, there is a phenomenon that multiple geographical names appear in one piece of HNSW. For instance, the HNSW "Sichuan Gansu Earthquake" corresponds to two province names. In such a case, two separate province names are extracted as "Sichuan" and "Gansu". Another example, for the HNSW "Shandong Laizhou Earthquake", when we conduct the geographical name recognition, two places "Shangdong" and "Laizhou" will individually be annotated. However, the town "Laizhou" is affiliated with Yantai City, Shandong Province. Different from the 1st case, the standard extraction results of the geographical name for this should be "Laizhou, Yantai, Shandong". Obviously, two examples correspond to two kinds of geographical name recognition policies respectively.

In order to differentiate the affiliations among the different geographical names, an algorithm is proposed through encoding the place names based on area number of National Bureau of Statistics of China to make them unified to the provincial level.

2.3   What Extraction by Feature Words

The task of keywords extraction for societal risk event is aimed at extracting risk-labeled keywords as the minimal set of words which represent not only the contents but also the corresponding risk category. One hot search word and its corresponding news texts are both about one event. To get the comprehensive description to a societal risk event, we perform keywords extraction on news texts. Given the challenges and drawbacks of supervised methods, we try an unsupervised TextRank-based method for feature words extraction through extracting from each piece of news text combining with risk factors generated by CRFs model mentioned in Subsect. 2.1.

In summary, the existing event extraction methods [10, 11] in NLP are inappropriate directly applied to societal risk events. Next, our approaches for these three problems are given.

3   Approaches

The societal risk event extraction framework, as shown in Fig. 1, includes societal risk classification, geographical name recognition, and unsupervised keywords extraction. First the approach applied on societal risk classification is introduced.

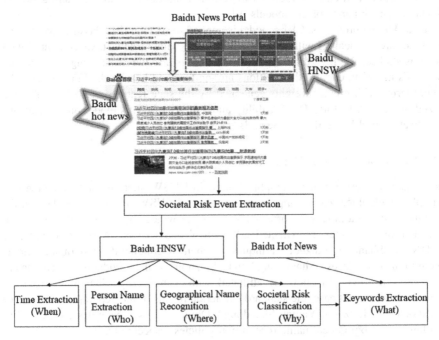

Fig. 1. The societal risk event extraction framework

3.1   Societal Risk Classification Method

3.1.1   CRFs Model

Actually, HNSW are short texts, often composed of fewer words. Traditional classification methods face the challenges of feature sparseness. CRFs model is adopted to deal with this problem. CRFs model is an undirected graphical model used to calculate the conditional probability of a set of labels given a set of input variables [12]. It has better performance in most natural language processing (NLP) applications, such as sequence labeling, part-of-speech tagging, syntactic parsing, and so on. The CRFs model has previously been used for short text classification and sentiment classification [13, 14]. The results has proved that CRFs outperformed the traditional methods like SVM and MaxEnt [15, 16]. In this paper, we utilize CRFs model for societal risk classification. Original societal risk classification is treated as a sequence labeling problem.

Linear Chain Conditional Random Field (LCCRF) is the most simple and commonly used form of CRFs model. In this paper, we use LCCRF to carry out societal risk classification. In view of the risk classification process, $X = \{x_1, x_2, \ldots x_n\}$ is a set of input random variables. $Y = \{y_1, y_2, \ldots y_n\}$ is a set of random labels. HNSW and their corresponding risk categories are respectively represented as the observed sequences and state sequences. For example, the hot search word "京广高铁票价高" (Beijing-Guangzhou high-speed rail fares are high) can be expressed as the observation sequence and state sequence. The observation sequence is $X=$ { 京,广,高铁,票价,高}. The state sequence is $Y = $ {daily life, daily life, …, daily life}.

In a given observation sequence X, the probability distribution of generating the output sequence can be described as follows:

$$p_w(Y|X) = \frac{exp(w \cdot F(Y,X))}{\sum_Y exp(w \cdot F(Y,X))} \tag{1}$$

Here, $F(Y, X) = (f_1(Y, X), f_2(Y, X), \ldots, f_K(Y, X))^T$ is the feature vector, where K is the number of features, and $f_i(Y, X)$ is a binary indicator function, with $f_i(Y, X) = 1$ when both feature and label are presented in a hot word and 0 otherwise; w is a learned weight for each feature function as well as the main parameter to be optimized.

It is necessary to define feature template to train LCCRFs model. We use the example above to illustrate the process of feature generation. When assuming the current token is "高铁", the feature templates and corresponding features are defined as Table 1.

We define two variables, namely L and N. L represents the number of categories including 7 risk categories and risk-free category, N represents the number of features generated by the feature template. Here, $L = 8$ and $N = 19$. There are $L * N$ potential functions, that is to say, there are 152 potential functions. The training of CRFs is based on Maximum Likelihood Principle. The log likelihood function is

$$L(w) = \sum_{Y,X} \left[\tilde{P}(Y,X)w \cdot F(Y,X) - \tilde{P}(Y,X)log \sum_Y exp(w \cdot F(Y,X)) \right] \tag{2}$$

Table 1. Feature template and corresponding features

Template	Implication	Feature
U00: %x[−2, 0]	The second term before current token	京
U01: %x[−1, 0]	The previous term	广
U02: %x[0, 0]	Current token	高铁
U03: %x[1, 0]	The next term	票价
U04: %x[2, 0]	The second term after current token	高

3.1.2 Experiments and Results

We perform risk multi-classification on respectively Baidu HNSW collected from November 1, 2011 to December 31, 2016 and their corresponding news texts. Table 2 lists the number distribution of each risk category on hot search words and hot news.

In the implementation, we first choose Jieba[1] as the segmentation tool to split one piece of HNSW (short texts) into segmented words. Then remove stop words and only reserve verbs, nouns, adjectives and adverbs. Each piece of segmented words is represented as a label sequence. The feature template defined in Table 1 is chosen for feature extraction. Hot words from November 1, 2011 to December 31, 2015 are chosen as the training set, while hot words in the whole 2016 as the testing set. L-BFGS algorithm is introduced to optimize the objective function.

We use *sklearn_crfsuite*[2] package for CRFs model. The iteration number is set to 100 in the training process. SVM model based on bag-of-words (BOW) is adopted in previous study for HNSW classification [7]. We utilize accuracy as the evaluation metrics to evaluate the overall performance of each model. Precision, recall and F-measure are used for performance measurement of individual societal risk classification represented in Table 3. The accuracy of CRFs model and SVM based on BOW are 0.78 and 0.74 respectively. CRFs have achieved better performance.

As to the task of societal risk multi-classification, it is essential to find out risk words as many as possible. In other words, we pay more attention to recall for evaluation. The recalls of CRFs model on risk category "national security", "economy/finance" and "public morals" are respectively 0.43, 0.34 and 0.23 and that of SVM based on BOW are in turn increased by 0.48, 1.10 and 0.64. In other words, CRFs model tends to capture risk hot words. To sum up, CRFs method achieves better performances apparently and shows the discriminatory power of predictive models in societal risk multi-classification.

3.1.3 Societal Risk Factor Extraction

State features are obtained when CRFs model completes parameters learning on the training set. The state features can be expressed as the distribution of terms' weights in each risk category. The feature terms with greater weights generated by the state features of CRFs model are picked out as risk factors. The multi-classification result of

---

[1] https://pypi.python.org/pypi/jieba/.

[2] https://pypi.python.org/pypi/sklearn-crfsuite/0.3.3/.

Table 2. Descriptive statistics of hot words and hot news

Risk category	Training dataset		Testing dataset	
	Hot words#	Hot news#	Hot words#	Hot news#
National security	2258	18472	178	1568
Economy/finance	1222	8403	119	1205
Public morals	3368	25004	399	3440
Daily life	4920	32037	656	5870
Social stability	5342	58890	364	3087
Government management	5552	52748	339	3428
Resources/environment	1716	14653	358	3156
Risk-free	24587	274669	4855	42978

Table 3. Comparison results of two models

Risk category	Precision		Recall		F-measure	
	BOW	CRFs	BOW	CRFs	BOW	CRFs
National security	0.56	0.66	0.29	0.43	0.38	0.52
Economy/finance	0.58	0.68	0.16	0.34	0.25	0.46
Public morals	0.43	0.54	0.14	0.23	0.21	0.32
Daily life	0.63	0.7	0.25	0.51	0.36	0.59
Social stability	0.47	0.56	0.49	0.51	0.48	0.54
Government management	0.49	0.58	0.35	0.56	0.41	0.57
Resources/environment	0.82	0.93	0.54	0.56	0.65	0.7
Risk-free	0.78	0.81	0.94	0.93	0.85	0.87

CRFs model has manifested the feasibility of extraction of risk factors due to its promising performance. We now perform risk factor extraction on the corpus chosen from November 1, 2011 to December 31, 2016 including 56,233 hot news search words. When the training process is completed, the state features are expressed as the distribution of terms' weights in each risk category. For example, the occurrence frequency of word "房价" (house price) under "daily life", "economy/finance", and "risk-free" are respectively 139, 2, 23. And the corresponding weights are 4.88, −0.82, 0.32. The significance of these weights can be explained from CRFs formula. Since $\sum_Y \exp(w \cdot F(Y, X))$ is the normalization factor, values of $p_w(y|x)$ depend on value of $exp(w \cdot F(Y, X))$. Take "house price" for example, we assume that the sequence only has one phrase "house price" for simplicity.

$$p_{w_1}(daily\,life|house\,price) > p_{w_3}(risk - free|house\,price)$$

$$> p_{w_2}(economy/finance|house\,price)$$

As is seen, weights show the contribution to annotation of risk category. The higher the weights of terms in one risk category are, the greater the contributions of terms to conditional probability are. As a result, terms with greater weights could be selected to construct the domain-specific dictionaries for societal risk. Feature terms with higher weights are chosen as the risk factors. The weights of all terms are ranging from −3.56 to 6.41. The set of risk factors is screened out through setting different weights. Therefore, we choose feature terms whose weights are non-negative. We respectively construct 7 dictionaries for each risk category. The number of terms each risk category dictionary contains is listed in Table 4.

Table 4. The number of terms in risk dictionaries

Risk category	Terms#
National security	1774
Economy/finance	1283
Public morals	3537
Daily life	3867
Social stability	3766
Government management	3933
Resources/environment	1446

3.2   Geographical Name Recognition

For the geographical name recognition, we propose an algorithm by encoding the place names appearing in HNSW based on area number of National Bureau of Statistics of China[3] including 3209 place names. Then, fill the geographical names to the provincial level. Take the city "Laizhou" for example, its area number is "370683", while that of "Yantai" is "370600" and that of "Shandong" is "370000". As is seen, the first two digits of the code represent the province, while the middle two digits represent the city and the last two represent the county. Therefore, the algorithm is designed as follows according to the coding rule of area number to complement geographical names to their respective provinces. Jieba is employed to do the Chinese hot news search words segmentation and part-of-speech tagging. Then terms tagged "ns" (place name) are selected for the geographical name recognition.

According to the algorithm, we randomly selected 1000 hot news search words for geographical name recognition from November 1, 2011 to December 31, 2016 with a total number of 56,233. We achieve an accuracy of 99%, proving the effectiveness of the method we proposed.

---

[3] http://www.stats.gov.cn/tjsj/tjbz/xzqhdm/201703/t20170310_1471429.html.

Algorithm: geographical name recognition

Input: place_name: geographical name
        place_code: corresponding area number of place_name.
        area_number_dictionary: {place_code:place_name}
Ouput: place_name_standard.
(1)   province_code ⟵ place_code/10000*10000;
(2)   if (place_code%10000=0)then
        {
(3)   province_name= area_number_dictionary[province_code];
(4)   place_name_standard = place_name;
(5)   **output** place_name_standard;
(6)   }
(7)   else{city_code ⟵ place/100*100;
(8)   if (place_code%100=0)then{
(9)   city_name = area_number_dictionary[city_code];
(10)   place_name_standard = province_name+city_name;
(11)   **output** place_name_standard;
(12)   }
(13)   else{
(14)   place_name_standard=
         province_name+city_name+place_name;
(15)   **output** place_name_standard;
(16)   }
(17)   }

3.3 Keywords Extraction

Considering the specificity of "What" extraction of societal risk event, we explore a TextRank-based method that automatically identifies risk-labeled keywords for societal risk event. TextRank [17] is an adaptation of PageRank algorithm for natural language texts where a text document is considered as a lexical or semantic network of words. TextRank has found its applications in different information retrieval tasks such as keyword and key phase extraction, extractive summarization, semantic disambiguation and other tasks involving graph-based term weighting [18, 19].

When we calculate the TextRank value of each piece of HNSW, the risk factors are taken into consideration to recalculate the weights of the terms. The steps of keywords extraction are as follows:

Given a set of HNSW $W = \{w_1, w_2, \ldots, w_l\}$ where l indicates the number of HNSW and one hot word w_i with its corresponding news texts set $D_i = \{d_{i1}, d_{i2}, \ldots, d_{im}\}$ where m indicates the number of news texts, the algorithm for keywords extraction is defined as follows:

(1) As to a news text $d_{ij}(j \in \{1, \ldots, m\})$, we do the Chinese news text segmentation, then remove the stop words. The set of candidate words $V_{ij} = \{cw_1, cw_2, \ldots, cw_p\}$ where p is the number of segmented words is acquired;

(2) Construct the undirected and unweighted graph $G = (V, E)$, where V is the set of vertices and E is set of edges which are added between lexical units that co-occur within a window of 5 words;

(3) Set the score associated with each vertex to an initial value of 1, and the ranking algorithm is run on the graph for several iterations until it converges;

(4) Once a final score is obtained for each vertex in the graph, vertices are sorted in reversed order of their score, and the top 20 vertices in the ranking are selected as the set of candidate keywords $C_{ij} = \{kw_1, kw_2, \ldots, kw_q\}$ where $q = 20$;

(5) Construct the sets of candidate keywords $C_i = \{kw_1', kw_2', \ldots, kw_n'\}$ for all the news text corresponding to hot word w_i, here, n is the total number of keywords of hot word w_i. Sort in reversed order of frequency values, retain maximum frequency value $Max_{fre} = max(fre(kw_h'))$ $(h \in \{1, \ldots, n\})$;

(6) Given the risk factors set of the risk category that hot word w_i belongs to $R_i = \{rw_1, rw_2, \ldots, rw_s\}$ where s indicates the number of risk factor, the weights of risk factors $RW = Max_{fre} + 1$, for $kw_h' \in R_i$, set $FW = fre(kw_h') * RW$, otherwise $FW = fre(kw_h')$;

(7) Terms in the set of candidate keywords $C_i = \{kw_1', kw_2', \ldots, kw_n'\}$ are sorted in reversed order of their new scores, and the top 20 terms in the ranking are selected as the final set of risk-labeled keywords, which represent what the event occurred in a certain risk category.

4 5W Extraction Results

Each societal risk event can be represented by these 5W (where, who, when, what, why) based on extraction methods above. We conduct the extraction of 5W elements on hot search words from November 1, 2011 to December 31, 2016 with a total number of 56,233 hot news search words and 549,608 corresponding hot news. Our methods for 5W extraction of societal risk event work real well. Here are some examples of the results (no order).

(1) HNSW: The orange alarm for heavy pollution in Beijing

> Where: Beijing,
> When: December 2, 2016,
> Who: None,
> Why: resources/environment
> What: pollution, orange, Beijing, the air, early warning, temperature, concentration, observatory, weather, emergency, requirements, release, air quality, Beijing-Tianjin-Hebei, building, deploying, overweight, enterprise, starting, city

(2) HNSW: Terrorist attacks

> Where: None,
> When: August 3, 2013,

Who: None,

Why: national security

What: the United States, threats, warnings, off, occur, base, organization, middle east, attack, anti-terrorism, president, global, coping, consulate, people, terrorism, media

(3) HNSW: Xi Jinping pursed the responsibility of Tianjin explosion

Where: Tianjin,

When: August 21, 2015,

Who: Xi Jinping,

Why: social stability

What: accident, accident, Tianjin, thorough investigation, disposition, charge, request, rescue, person, make response, occurrence, production, emergency, warehouse, emphasis, instruction, response, mystery

(4) HNSW: Liu Zhijun was prosecuted

Where: None,

When: April 11, 2013,

Who: Liu Zhijun,

Why: government management

What: ministers, bribe-taking, state, railway, interest, prosecution, system, discipline, investigation, lawyer, bribery case, cadres, project, accept, official, video, allegations, duties, involved, sack

(5) HNSW: Hong Kong college entrance examination cheating

Where: Hong Kong,

When: July 16, 2013,

Who: None,

Why: daily life

What: college entrance examination, cheat, performance, examination, education, high school, admission, college, teacher, the end, release, rubber, examination room, a lot, college entrance exam ace

The framework of 5W extraction can not only timely access to societal risk perception but also effectively and formally describe societal risk events. It provides a novel approach to extract structured information from the wicked problems when dealing with societal issues. Besides, the 5W extraction on societal risk event plays an important role in constructing knowledge graph on societal risk. It generates a more open, richer and efficient social network structure as well.

5 Conclusions

Societal risk perception is one of important subjects in social management and public opinion warning, which attracts a large number of scholars' attention. An event extraction framework 5W for societal risk events is proposed. The challenges for 5W extraction are highlighted and approaches are proposed to tackle them. The major contributions of this paper are summarized as follows:

(1) An automatic extraction of 5W for societal risk events, a general yet effective framework, is proposed to extract structured information for societal issues;
(2) CRFs model is applied to societal risk classification for the first time, which tackles challenges of feature sparseness. Besides, terms with greater weight values generated by the state features are picked out as risk factors;
(3) An effective approach is put forward to extract geographical name by assigning the place names appearing in each hot word to provincial level.
(4) We adopt a TextRank-based method that automatically identifies risk labeled keywords which represent what the event occurred in a certain risk category.

Lots of work need to be improved. Once the data of 5W of societal risk events are obtained, the construction of knowledge graph on societal risk will be further studied.

Acknowledgement. This research is supported by National Key Research and Development Program of China (2016YFB1000902) and National Natural Science Foundation of China (61473284 & 71731002).

References

1. Ball, D.J., Boehmer-Christiansen, S.: Societal concerns and risk decisions. J. Hazard. Mater. **144**, 556–563 (2007)
2. Zheng, R., Shi, K., Li, S.: The influence factors and mechanism of societal risk perception. In: Zhou, J. (ed.) Complex 2009. LNICST, vol. 5, pp. 2266–2275. Springer, Heidelberg (2009). https://doi.org/10.1007/978-3-642-02469-6_104
3. Dong, Y.H., Chen, H., Tang, X.J., Qian, W.N., Zhou, A.Y.: Prediction of social mood on Chinese societal risk perception. In: Proceedings of International Conference on Behavioral, Economic and Socio-Cultural Computing, pp. 102–108 (2015)
4. Tang, X.J.: Applying search words and BBS posts to societal risk perception and harmonious society measurement. In: Proceedings of IEEE International Conference on Systems, Man, and Cybernetics, pp. 2191–2196 (2013)
5. Xu, N., Tang, X.: Societal risk and stock market volatility in china: a causality analysis. In: Chen, J., Theeramunkong, T., Supnithi, T., Tang, X. (eds.) KSS 2017. CCIS, vol. 780, pp. 175–185. Springer, Singapore (2017). https://doi.org/10.1007/978-981-10-6989-5_15
6. Tang, X.J.: Exploring on-line societal risk perception for harmonious society measurement. J. Syst. Sci. Syst. Eng. **22**, 469–486 (2013)
7. Hu, Y., Tang, X.: Using support vector machine for classification of Baidu hot word. In: Wang, M. (ed.) KSEM 2013. LNCS (LNAI), vol. 8041, pp. 580–590. Springer, Heidelberg (2013). https://doi.org/10.1007/978-3-642-39787-5_49
8. Zhou, Y.J., Xu, B., Xu, J.M., Yang, L., Li, C.L.: Compositional recurrent neural networks for Chinese short text classification. In: International Conference on Web Intelligence, pp. 137–144 (2017)
9. Wang, P., Xu, B., Xu, J.M., et al.: Semantic expansion using word embedding clustering and convolutional neural network for improving short text classification. Neurocomputing **174**, 806–814 (2016)

10. Li, H., Ji, H., Zhao, L.: Social event extraction: task, challenges and techniques. In: Proceedings of International Conference on Advances in Social Networks Analysis and Mining, pp. 526–532. IEEE (2015)
11. Zhao, Y.Y., Qin, B., Che, W.X., Liu, T.: Research on Chinese event extraction. J. Chin. Inf. Process. **22**, 3–8 (2008). (in Chinese)
12. Lafferty, J., McCallum, A., Pereira, F.: Conditional random fields: probabilistic models for segmenting and labeling sequence data. In: Proceedings of the Eighteenth International Conference on Machine Learning, ICML, vol. 1, pp. 282–289 (2001)
13. Zhang, C.Y.: Text categorization model based on conditional random fields. Comput. Technol. Dev. **21**, 77–80 (2011)
14. Zhao, J., Liu, K., Wang, G.: Adding redundant features for CRFs-based sentence sentiment classification. In: Proceedings of Conference on Empirical Methods in Natural Language, pp. 117–126 (2008)
15. Sudhof, M., Goméz Emilsson, A., Maas, A.L., Potts, C.: Sentiment expression conditioned by affective transitions and social forces. In: Proceedings of the 20th ACM SIGKDD International Conference on Knowledge Discovery and Data Mining, pp. 1136–1145 (2014)
16. Li, T.T., Ji, D.H.: Sentiment analysis of micro-blog based on SVM and CRF using various combinations of features. Appl. Res. Comput. **32**, 978–981 (2015)
17. Mihalcea, R., Tarau, P.: TextRank: bringing order into text. In: EMNLP, pp. 404–411(2004)
18. Liang, W., Huang, C.N.: Extracting keyphrases from Chinese news articles using textrank and query log knowledge? In: Proceedings of the 23rd Asia Conference on Language, Information and Computation, pp. 733–740 (2009)
19. Rahman, M.M., Roy, C.K.: TextRank based search term identification for software change tasks. In: Proceedings of International Conference on Software Analysis, Evolution and Reengineering, pp. 540–544. IEEE (2015)

Jointly Trained Convolutional Neural Networks for Online News Emotion Analysis

Xue Zhao, Ying Zhang[✉], Wenya Guo, and Xiaojie Yuan

College of Computer Science, Nankai University, Tianjin, China
{zhaoxue,zhangying,guowenya,yuanxiaojie}@dbis.nankai.edu.cn

Abstract. Emotion analysis, as a sub topic of sentiment analysis, crosses many fields so as philosophy, education, and psychology. Grasping the possible emotions of the public can help government develop their policies and help many businesses build their developing strategies properly. Online news services have attracted millions of web users to explicitly discuss their opinions and express their feelings towards the news. Most of the existing works are based on emotion lexicons. However, same word may trigger different emotions under different context, which makes lexicon-based methods less effective. Some works focus on predefined features for classification, which can be very labor intensive. In this paper, we build a convolutional neural network (CNN) based model to extract features that can represent both local and global information automatically. Additionally, due to the fact that most of online news share the similar word distributions and similar emotion categories, we train the neural networks on two data sets simultaneously so that the model can learn the knowledge from both dataset and benefit the classification on both data sets. In this paper, we elaborate our jointly trained CNN based model and prove its effectiveness by comparing with strong baselines.

Keywords: Emotion analysis · CNN · Joint training

1 Introduction

Emotions have been widely studied in psychology and behavior sciences, as they are an important element of human nature. The emotion analysis has attracted increasing interest in many research and industry fields, for example, natural language processing (NLP) and social media. With the explosive growth of various social media services, more and more web users have a chance to express their opinions and share their thoughts freely. Identifying their opinions and sentiments towards a specific object, such as news, hot topic, product, government policy, is of great help for corresponding organizations to understand the public needs and expectations. Sentiment analysis is to study coarse-grained sentiments, which tells if something is positive, negative or neutral. Emotion

© Springer Nature Switzerland AG 2018
X. Meng et al. (Eds.): WISA 2018, LNCS 11242, pp. 170–181, 2018.
https://doi.org/10.1007/978-3-030-02934-0_16

analysis, by contrast, entails a fine-grained emotion, which deep dives into the themes associated with each emotion and allows businesses to discover customers experience by their emotion response. The study goes beyond positive-negative dimension to discrete emotion categories like happiness, sadness, anger, etc. In this paper, we focus on emotion analysis in NLP area, which is to automatically identify emotions in text.

Fig. 1. An example of emotion labels voted by 13,428 users for a *Sina News* article.

Nowadays, many online news sites dedicate to know readers' responses after reading news and collect readers' comments towards the news. Some of them display several emoji at the end of news articles, as is shown in Fig. 1, to indicate different emotions, so the readers can click one of them to show their emotions after reading the news. Online news sites possess a large amount of texts of news and each text naturally has user-contributed emotion vote results. This opens up the research on online news emotion analysis, which can help news publishers and many other organizations to improve their services. Additionally, this research result can predict and monitor the readers emotions, which will help organizations develop strategies more appropriately.

In this paper, we use adequate online news articles with their emotion votes from general web users, to analyze the most possible emotion of readers that will be triggered by the news content. There are numerous of supervised machine learning models and their variants suitable for emotion analysis on text, some of them require extensive labor work to extract features from the text, while most of deel learning models, such as CNN, are able to automatically learn features and even the structural relationship among features, owning to the fact that the trained neurons in the networks are conveying useful information in the text. They are connected in many different forms and controlled by different operations and functions. After properly trained, the networks can generate representations of text to classify texts automatically, without any feature engineering works. CNN is one of deep learning models that has achieved impressive result in image recognition several years ago. It has proved remarkable results in NLP area recent years, for the ability to extract features from global information with convolution operation and from local information with the help of filters [1]. In addition, neural networks are good at modeling complicated and implicit relationships among features, syntactically and structurally. Due to the implicit associations between emotions [2], emotion-related features should also be correlated which makes neural networks very suitable for the task. Therefore, in this paper, we adopt CNN for our online news emotion analysis task.

More importantly, we found that news from different period and even different data sources actually have many things in common, such as word distribution and emotion categories, which inspires the idea to jointly train a model on two different data sets. Although the word distribution cannot be exactly the same, the emotion categories of data sets may also have minor difference, the implicit features among lines of texts from different data sets can actually make the features more detailed and complete. For example, an emotion-embedded word 'furious' scarcely exists in data set A but appears many times in data set B, by training jointly, the model can learn the emotion conveyed in word 'furious' better in data set A with the knowledge from data set B. In addition, the enlarged data size can also boost the classification performance of the model. For these reasons, a jointly trained classification model on two different data sets can technically enhance the classification performance on both data sets.

Accordingly, we trained a jointly trained CNN-based neural network for task of emotion analysis for online news. We collected more than $76,633$ pieces of real-world online news which happen to have two different sets of emotion categories, $40,897$ of them have 8 emotion categories: touched, sympathetic, bored, angry, amused, sad, novel and warm. The rest of the data containing $35,736$ news with 6 emotions: touched, shocked, amused, sad, surprised, and angry. We conducted extensive experiments on these data sets and compared our model with strong baselines. The result proves the effectiveness of the proposed model. Furthermore, the experiment also illustrates the jointly training strategy is capable of handling data sets with different emotion categories.

The paper is organized as follow. In Sect. 2 we give a brief review of emotion analysis and deep learning models in NLP field. We introduce our model in Sect. 3 and describe our experiment setups, baselines, results and discussion about the experiment in Sect. 4. Section 5 is the conclusion of this paper.

2   Related Work

2.1   Online News Emotion Analysis

There is a large body of research on sentiment analysis. Emotion analysis, as the sub topic of sentiment analysis, initially relies on emotion lexicons. Mohammad [3,4] had built Word-Emotion Lexicon which contains a list of English words and their associations with eight basic emotions (anger, fear, anticipation, trust, surprise, sadness, joy, and disgust) and two sentiments (negative and positive). The annotations were manually done by crowdsourcing. [5] proposed an efficient algorithm to build a word-level emotional dictionary for social emotions. [6] presented a novel approach to build a high-coverage and high-precision lexicon of roughly 37 thousand terms annotated with emotion scores, called DepecheMood. [7–9] also made contributions for building emotion lexicons. Aother class of approaches are machine learning based methods. [5,10–12] built topic models with the intention to bridge the gab between social media materials and readers' emotions by introducing an intermediate layer. The models can also generate a social emotion lexicon. [13,14] proposed a joint emotion-topic model by augmenting Latent Dirichlet Allocation with an additional layer for emotion modeling.

Among these works, [5,10–16] are conducted on online news for emotion classification. Works of emotion classification for online news originates from the SemEval tasks in 2007 [17], which introduced the task of "Affective Text" to explore the connection between news headlines and the evoked emotions of readers. However, the lexicon-based methods have their limited utility: the lexicons cannot differentiate the emotions of the words in different context circumstances. The machine learning based methods ask for sufficient and effective features to train machine learning models, which mainly rely on manually extracting features or using complicated feature engineering algorithms. In this paper, we take advantage of neural networks to automatically extract features. Specifically, we use convolutional neural network which is able to encode both local and global context information, so that it can generate better document representations.

2.2 Convolutional Neural Networks for Text Classification

CNN have achieved excellent results in many types of NLP tasks, for example, semantic parsing, search query retrieval, sentence modeling, etc. Work [18] reported a series of experiments with CNN trained on top of pre-trained, task-specific and static word vectors, respectively, for sentence-level classification tasks. In work [19], a recurrent covolutional neural network for text classification had been designed. They applied a recurrent structure to capture contextual information as far as possible for learning word vectors, which followed by a max-pooling layer for automatically identifying the key components in texts. The model outperforms the baseline methods especially on document-level data sets. [20] explored the use of character-level CNN for text classification, compared to traditional models such as bag of words, n-grams and TFIDF variants, and deel learning models such as word-based CNN, the character-level CNN can generally achieve competitive results. [21–23] also provide good ideas to use CNN for text classification.

2.3 Jointly Trained Neural Networks in NLP

It is worth mentioning that jointly training have nothing to do with multi-task learning. Joint learning is to train a model at one time but it can only perform one task, whereas the multi-task training is supposed to handle a problem with more than one purposes. In this paper, we only discuss jointly training because we only have one task, that is emotion classification. Although the model will be applied on two different data sets, in fact, we only allow the model to have one output, which makes it generate one result for only one task. It is very rare to happen that the predicted emotion for news from one data set falls out of its own emotion categories, we will ignore such instances. The reason that such instances can be ignored during classification lies in the fact where most data share the same common emotions, the number of instances from noncommon emotion categories are less than the major common emotions, when they are divided to training, validation and testing data sets, they become much less important.

Multi-task learning can be found in many tasks in NLP [24–26]. However, to the best of our knowledge, this is the first work that applies one model on two similar data sets to boost the classification performance on both data sets.

3 Proposed Model

In this section, we describe our jointly trained convolutional neural networks for emotion analysis, which is shown in Fig. 2. In particular, the system has a shared CNN structure for document representation learning and task-specific layers for emotion representation modeling. The shared CNN can benefit from joint training due to the similar text content and common emotion categories.

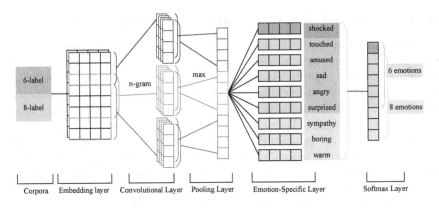

Fig. 2. An example of emotion labels voted by 13,428 users for a *Sina News* article.

3.1 Shared Convolutional Neural Network

The idea of joint training is to learn knowledge from both data sets in order to extract more complete feature set. Since two data sets have similar text contents (news articles), word-level features from two data sets are expected to have a very large overlap. Intuitively, we first use a shared *embedding layer* as the input layer, with which we train the word embedding by converting the words indices into vectors. Each dimension is randomly initialized by sampling from a uniform distribution. The *embedding layer* will be trained together with other layers and fine-tuned by input instances. Because the news corpora we have are big enough to optimize the word vectors to adapt to the news domain and emotion analysis task very well, the need to pre-train embeddings can be omitted without hurting the model performance.

Therefore, the input of the system is a index sequence of words in news s with length $|s|$, represented as $[w_1, \ldots, w_{|s|}]$, and each word w is drawn from the vocabulary V. The output from the embedding layer is a word vector matrix $\mathbf{S} \in \mathbb{R}^{|s| \times d}$, where ith row represents a d-dimensional word vector of the ith word in the document.

To capture and compose features from word representation into a concrete semantic document representation, we apply a series of transformations to the input document matrix $\mathbf{S}$ using convolution and pooling operations. Formally, the convolution operation between document matrix $\mathbf{S} \in \mathbb{R}^{|s| \times d}$ and a filter $\mathbf{f} \in \mathbb{R}^{m \times d}$ with size m results in a vector $\mathbf{c} \in \mathbb{R}^{|s| - m + 1}$ where each component is as follows:

$$c_i = f(\mathbf{f} \cdot \mathbf{S}_{i:i+m-1} + b). \tag{1}$$

where $b \in \mathbb{R}$ is a bias term and f is a non-linear function, here we use hyperbolic tangent. Therefore, we can obtain a feature map which is a vector $\mathbf{c} \in \mathbb{R}^{|s| - m + 1}$:

$$\mathbf{c} = [c_1, c_2, \ldots, c_{|s| - m + 1}] \tag{2}$$

To capture the most salient semantics of a document, we apply the max pooling operation [24] over the feature map which will take $\hat{c} = max(\mathbf{c})$ as the feature for the corresponding filter $\mathbf{f}$. For now, we have obtained one feature extracted from one filter. More convolutional filters of different width can be applied to capture the semantics of n-grams of various granularities. The features from different filters will be concatenated to generate a document representation.

For regularization we employ dropout on the concatenated document representation with a constraint on l_2-norms of the weight vectors, as suggested in [18] to avoid overfitting.

3.2 Emotion-Specific Layers

For the reason that two data sets share several emotion categories, the higher-level features can also be partially modeled with a shared group of neurons in networks. Specifically, the document representation will be further fed to the *emotion-specific layers*, which contains nine parallel regular layers connected to the pooling layer. Each emotion-specific layer receives the same document vectors but can interpret into different emotion representations, which can be defined as:

$$\mathbf{e}_k = h(\mathbf{w} \cdot \hat{\mathbf{c}} + b) \tag{3}$$

where $\mathbf{w}$ is the weights in k-th emotion-specific layer, $\hat{\mathbf{c}}$ is the result of concatenating all features $\hat{c}$, b and n denote the bias and the number of filters in CNN structure, respectively. h is a non-linear activation function, such as hyperbolic tangent. Therefore, the design of emotion-specific layers is to convert the discriminative document features into emotion-specific features via nine emotion layers. Besides, the separation of emotion-specific layers provides great convenience for the assignment of emotions to the corresponding task. We can simply concatenate 6 and 8 output vectors of 9 layers, where five of them will be integrated into both tasks simultaneously. The generated vectors can be viewed as the emotion representation converted from the same document representation. As a result, the classification task on two data sets becomes one, which can be modeled by a

3.3   Emotional Classification

The generated emotion representations for two data sets can be naturally regarded as emotion-specific features of documents for emotion classification without any engineering work. Specifically, we add a *softmax layer* to convert each emotion representation to conditional probabilities, which is calculated as follows.

$$P_i = \frac{exp(x_i)}{\sum_{i'}^{C} exp(x_{i'})} \tag{4}$$

where C is the class number for all emotion categories, which is 9 in our case.

During training, the data from two data sets will flow through the shared low-level word embedding layer and the shared CNN structure, as well as five emotion-specific layers. The model will learn from both data sets at the same time so that it can effectively improve the classification performance and generalization ability. The forward prorogation starts from the input word sequence from either 6- or 8-category corpus, and bifurcate at the point right after generating the emotion representations for two data sets. The independent *softmax layers* serve as the outputs of the whole system, which will generate the predicted emotion distribution over 9 emotion categories. We selected the most possible emotion as the predicted emotion.

Before training, two corpora will be integrated and shuffled, so the input news can be from either of news corpora. We relabel the news by mapping the 6 and 8 emotions to overall 9 emotions. As a result, the model will be trained on both data sets jointly. The backpropagation is also simple and straightforward that the parameters in the system will be updated at the same time to improve the overall performance, by minimizing the loss in forward prorogation. We use the cross-entropy loss between gold emotion distribution $P^g(d)$ and predicted emotion distribution $P(d)$ as the loss function. For task t, we have:

$$loss_t = -\sum_{d \in D} \sum_{i=1}^{C} P_{ti}^{(g)}(d) \cdot log(P_{ti}(d)) \tag{5}$$

where d, D, and C denote a document, the training data, and the number of classes, respectively. We treat these two tasks with equal importance and add the loss in two tasks, so we can get the final loss $Loss$ as:

$$Loss = -\sum_{t \in T} loss_t \tag{6}$$

where T represents all the tasks we have. The parameters will be updated with Adadelta algorithm. We set the dimension of word vectors generated from embedding layer as 100, the widths of three convolutional filters as 3, 4, and 5 as recommended in [18]. The output length of convolutional filter and emotion-specific layer is 100. The dropout rate is set as 0.2 in our experiment.

4 Experiments

In this section, we describe our experiment setups, introduce relative strong baselines and display our experiment results.

4.1 Experiment Setting

To test the effectiveness of the proposed approach, two large Chinese news datasets extracted from *Sina News* have been employed, *Sina-6* and *Sina-8*. *Sina-8*, utilized by [5], contains 40,897 news articles published from August 2009 to April 2012, with the total number of ratings for each news article over 8 emotions larger than 0, inlcuding *touched, sympathetic, bored, angry, amused, sad, surprised* and *heartwarming*. *Sina-6* is the news published from January 2013 to November 2015, which has 35,736 pieces of news with 6 emotions, including *touched, shocked, amused, sad, surprised,* and *angry*. *Sina-6* was rated with the total number of 20, thus the votes are more valid. Both datasets are split into training, development and testing sets with 70/10/20. To evaluate the proposed approach, we compare the predicted probabilities with the actual distribution over emotions. We use accuracy to evaluate the performance of all models, here "accuracy" is actually the same as micro-averaged F_1 [27] (Table 1).

Table 1. Statistics of datasets.

Data	Emotion	# of articles	# of rating	Data	Emotion	# of articles	# of rating
Sina-6	touched	4,565	1,283,931	*Sina-8*	touched	5,672	254,488
	amused	9,294	2,024,269		amused	6,812	320,824
	sad	4,442	1,565,738		sad	2,888	212,316
	angry	15,429	4,006,858		angry	18,287	830,927
	surprised	569	476,249		surprised	1,571	105,585
	shocked	1,437	1,188,267		bored	2,603	151,599
					sympathetic	2,676	149,883
					heartwarming	388	58,196

4.2 Baselines

The focus on this paper is to train a model that can be generalized on two similar data sets, therefore the main baseline models we chose are CNN that trained separately. Here are some basic baseline models.

- **CNN** a same CNN-based structure as in the proposed model trained only on one single data set.
- **CNN+1** same as CNN but with one more fully connect layer, which is designed for extracting more features that may potentially neglected by the proposed model.

- **W2V** a CNN structure trained with the pre-trained Word2Vectors embedding [28], which can bring more context information for better classification results.
- **Dropout** a CNN structure with an additional Dropout layer, which can effectively prevent the overfitting problem.

In addition to these basic models that are modified from the proposed model, we also test the strong baselines as follows.

- **WE**, called word-emotion method, is a lexicon based method proposed by [5]. The method constructs emotion dictionary and generates a distribution of word w over document d as well as a distribution of emotion e conditioned on word w, which can be looked up from the emotion dictionary. Then it can predict emotion e by $\hat{P}(e|d) = \sum_{w \in W} p(w|d)p(e|w)$. We use the *minimum* pruning strategy to remove common words as it performs the best in their paper.
- **RPWM**, reader perspective weighted model, is a topic-based method designed by [12], in which emotion entropy is proposed to estimate the weight of different training documents and thus to relieve the influence of noise training data. Latent topics have been used to associate words with documents to avoid word emotional ambiguity. The number of topic is set as 14, which performs the best.
- **SVM**, we use bag-of-unigrams as features and train a SVM classifier with LibLinear [29]. The features are selected after removing the very rare words with document frequency lower than 0.001. We also test with bigrams but the performance became worse.
  We still use unigrams as features and remove the rare words with the least document frequency 0.001.

4.3   Experiment Results

We implemented all the baselines and compare them to our proposed model, the accuracy is shown in Figs. 3 and 4. Note that the proposed model is trained jointly on two data sets but the accuracy measurement is calculated separately in different test data sets, therefore we derive two accuracy values for two data sets.

As we can see from Figs. 3 and 4, the proposed method achieved the best performance or obtained competitive results. On *Sina-6*, the jointly trained CNN structure outperformed the other baselines, especially the lexicon-based model **WE** and the machine learning based model **PRWM**, with large margins. This lies in the fact that words in different context may convey different meanings and trigger different emotions. Such models mainly based on word-level features so they cannot beat models that can consider more complicated features, such as context-level features, document-level features, especially the relationships between features.

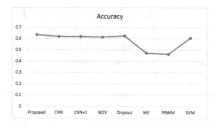

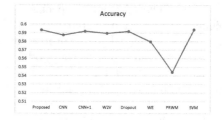

Fig. 3. Accuracy on Sina-6.   **Fig. 4.** Accuracy on Sina-8.

SVM is a strong baseline in many text classification tasks, we can observe this from both figures where *SVM* generated very competitive results. In *Sina-8* it actually achieved the best accuracy among all the models, our model is keeping up with *SVM* and still works better than the other baselines. We then analyze two data sets and found that the data set created by [5] contains large proportion of news articles with only one vote, whereas *Sina-6* contains news only with number of votes larger than 20. *Sina-8* contains many less valid data, which means some news may not belong to the top rated emotion category. Surprisingly, *SVM* can handle such noise with only bag-of-unigrams. This may be because *SVM* only focuses on words: it is trained on every word in training data, which makes *SVM* has more training resource so the bias caused by noise data can be alleviated; while deep learning models focus not only on words but the structures of text. When many input texts have incorrect labels, the neural networks will be more likely misled by such noise during training phase.

PRWM behaves poor on both data sets. In fact, it is very difficult to tune a topic model. The model itself is quite complicated and the training is rather time consuming. The model setting in our result is already the best situation we can obtain, yet its result points in both figures are still lower than any other models with big gaps.

We can also observe from figures that neural networks generally work better than other methods. In general, among all the neural networks, the proposed jointly trained networks outperform other models. Although two data sets share different emotion categories, our strategy still solves the classification problem and achieves the best or competitive result.

5 Summary

In this paper, we deal with an online news emotion analysis task, which is a text classification problem. The online news always share similar text content and have common basic emotion categories, to the best of our knowledge, no other works have taken such special features of online news services into consideration.

We come up with a jointly trained convolutional neural network based model. First, the model has shared embedding layers for lower-level features extraction. Then CNN part will extract both local and global context features for better document-level representations. Due to the overlap of emotion categories, a partially shared emotion-specific layer structure is introduced for extracting higher-level emotion-related features. The final output is generated from softmax layer, which calculates probabilities over emotion categories. Given a piece of news, we can input to the model and then get the predicted emotion that readers would most likely have after reading news. The experiments are conducted on real-world data and the results prove the effectiveness of the proposed approach.

In the future, we will consider to integrate more information sources such as news comments, and we will continue to explore neural networks for text classification tasks, in order to solve online news emotion classification more accurately.

Acknowledgment. This research is supported by Natural Science Foundation of Tianjin (No. 16JCQNJC00500) and Fundamental Research Funds for the Central Universities.

References

1. LeCun, Y., et al.: Gradient-based learning applied to document recognition. Proc. IEEE **86**(11), 2278–2324 (1998)
2. Zhang, Y., Zhang, N., Si, L., Lu, Y., Wang, Q., Yuan, X.: Cross-domain and cross-category emotion tagging for comments of online news. In: Proceedings of the 37th International ACM SIGIR Conference on Research and Development in Information Retrieval, pp. 627–636. ACM, July 2014
3. Mohammad, S.M., Turney, P.D.: Emotions evoked by common words and phrases: using mechanical turk to create an emotion lexicon. In: Proceedings of the NAACL HLT 2010 Workshop on Computational Approaches to Analysis and Generation of Emotion in Text, pp. 26–34. Association for Computational Linguistics (2010)
4. Mohammad, S.M., Turney, P.D.: Crowdsourcing a word-emotion association lexicon. Comput. Intell. **29**(3), 436–465 (2013)
5. Rao, Y., Lei, J., Wenyin, L., Li, Q., Chen, M.: Building emotional dictionary for sentiment analysis of online news. World Wide Web **17**(4), 723–742 (2014)
6. Staiano, J., Guerini, M.: DepecheMood: a lexicon for emotion analysis from crowd-annotated news. arXiv preprint arXiv:1405.1605 (2014)
7. Tang, Y., Chen, H.-H.: Mining sentiment words from microblogs for predicting writer-reader emotion transition. In: LREC 2012
8. Yang, C., Lin, K.H.-Y., Chen, H.-H.: Building emotion lexicon from weblog corpora. In: Proceedings of the 45th Annual Meeting of the ACL on Interactive Poster and Demonstration Sessions. Association for Computational Linguistics (2007)
9. Yang, C., Lin, K.H.-Y., Chen, H.-H.: Writer meets reader: emotion analysis of social media from both the writer's and reader's perspectives. In: IEEE/WIC/ACM International Joint Conferences on Web Intelligence and Intelligent Agent Technologies, WI-IAT 2009, vol. 1. IEEE (2009)
10. Rao, Y., et al.: Affective topic model for social emotion detection. Neural Netw. **58**, 29–37 (2014)
11. Rao, Y., et al.: Sentiment topic models for social emotion mining. Inf. Sci. **266**, 90–100 (2014)

12. Li, X., et al.: Social emotion classification via reader perspective weighted model. In: AAAI (2016)
13. Bao, S., et al.: Joint emotion-topic modeling for social affective text mining. In: Ninth IEEE International Conference on Data Mining, ICDM 2009. IEEE (2009)
14. Bao, S., et al.: Mining social emotions from affective text. IEEE Trans. Knowl. Data Eng. **24**(9), 1658–1670 (2012)
15. Lin, K.H.-Y., Yang, C., Chen, H.-H.: What emotions do news articles trigger in their readers? In: Proceedings of the 30th Annual International ACM SIGIR Conference on Research and Development in Information Retrieval. ACM (2007)
16. Lin, K.H.-Y., Yang, C., Chen, H.-H.: Emotion classification of online news articles from the reader's perspective. In: Proceedings of the 2008 IEEE/WIC/ACM International Conference on Web Intelligence and Intelligent Agent Technology, vol. 1. IEEE Computer Society (2008)
17. Strapparava, C., Mihalcea, R.: SemEval-2007 task 14: affective text. In: Proceedings of the 4th International Workshop on Semantic Evaluations. Association for Computational Linguistics (2007)
18. Kim, Y.: Convolutional neural networks for sentence classification. arXiv preprint arXiv:1408.5882 (2014)
19. Lai, S., et al.: Recurrent convolutional neural networks for text classification. In: AAAI, vol. 333 (2015)
20. Zhang, X., Zhao, J., LeCun, Y.: Character-level convolutional networks for text classification. In: Advances in Neural Information Processing Systems (2015)
21. Ciresan, D.C., et al.: Flexible, high performance convolutional neural networks for image classification. In: IJCAI Proceedings-International Joint Conference on Artificial Intelligence, vol. 22, no. 1 (2011)
22. Johnson, R., Zhang, T.: Semi-supervised convolutional neural networks for text categorization via region embedding. In: Advances in Neural Information Processing Systems (2015)
23. dos Santos, C., Gatti, M.: Deep convolutional neural networks for sentiment analysis of short texts. In: Proceedings of COLING 2014, the 25th International Conference on Computational Linguistics: Technical Papers (2014)
24. Collobert, R., Weston, J.: A unified architecture for natural language processing: deep neural networks with multitask learning. In: Proceedings of the 25th International Conference on Machine Learning. ACM (2008)
25. Ruder, S.: An overview of multi-task learning in deep neural networks. arXiv preprint arXiv:1706.05098 (2017)
26. Zhang, Y., Yang, Q.: A survey on multi-task learning. arXiv preprint arXiv:1707.08114 (2017)
27. Christopher, D.M., Prabhakar, R., Hinrich, S.: An Introduction to Information Retrieval. Cambridge University Press, Cambridge (2008)
28. Collobert, R., et al.: Natural language processing (almost) from scratch. J. Mach. Learn. Res. **12**(Aug), 2493–2537 (2011)
29. Chang, C.C., Lin, C.J.: LIBSVM: a library for support vector machines. ACM Trans. Intell. Syst. Technol. (TIST) **2**(3), 27 (2011)

Evaluation of Score Standardization Methods for Web Search in Support of Results Diversification

Zhongmin Zhang[1], Chunlin Xu[2], and Shengli Wu[1($\boxtimes$)]

[1] School of Computer Science, Jiangsu University, Zhenjiang 212013, China
swu@ujs.edu.cn
[2] School of Computing, Ulster University, Newtownabbey BT370QB, UK

Abstract. Score standardization is a necessary step for many different types of Web search tasks in which results from multiple components need to be combined or re-ranked. Some recent studies suggest that score standardization may have impact on the performance of some typical explicit search result diversification methods such as XQuAD. In this paper, we evaluate the performance of six score standardization methods. Experiments with TREC data are carried out with two typical explicit result diversification methods XQuAD and PM2. We find that the reciprocal standardization method performs better than other score standardization methods in all the cases. Furthermore, we improve the reciprocal standardization method by scaling those scores up so as to better satisfy the requirement of probability scores and obtain better results with XQuAD. We confirm that score standardization has significant impact on the performance of explicit search result diversification methods and such a fact can be used to obtain more profitable score standardization methods and result diversification methods.

Keywords: Score standardization · Explicit search result diversification
Web search · Information retrieval

1 Introduction

Search result diversification has been widely used to deal with ambiguous queries of web users in recent years. The main idea of search result diversification is to provide a resultant list that may cover all possible aspects of the query in top k results so as to promote the user's satisfaction [3, 8].

In the last couple of years, a number of approaches have been proposed to deal with search result diversification. Many of these approaches can be divided into two categories: implicit and explicit. An implicit method assumes no prior knowledge of the query aspects and it only considers the contents of the documents and some statistics of the document collection. MMR [1] and KL divergence-based model [10] are two typical methods in this category. An explicit method assumes that possible query aspects and some related information (e.g., relative importance) of the aspects can be explicitly identified. Thus an explicit result diversification algorithm tries to cover all aspects of the query in the top-k documents. XQuAD [8] and PM-2 [3] are two

© Springer Nature Switzerland AG 2018
X. Meng et al. (Eds.): WISA 2018, LNCS 11242, pp. 182–190, 2018.
https://doi.org/10.1007/978-3-030-02934-0_17

well-known approaches in this category. Previous studies show that the implicit methods are not as effective as the explicit methods [3, 8].

For the documents in the resultant list, almost all explicit diversification methods need to know the probabilities of them being relevant to a given query and its aspects, but the raw scores produced by many information search systems are not always in the range of [0, 1], thus these scores cannot be employed in an explicit diversification method directly. Sometimes information search systems do not provide scores for the documents retrieved at all. In such situations, score standardization is necessary to transform the relevance-related scores or the rankings of a list of ranked documents into reasonable probabilities. According to [6], it is important to have proper probability scores for those documents involved when using an explicit result diversification method. However, when presenting those methods such as XQuAD, the authors do not indicate how to obtain proper probability scores [3, 8]. For the commonly used score standardization method MinMax [4], Ozdemiray and Altingovde pointed out the potential weakness of it when used in XQuAD [6]: once a highly-scored document for a certain aspect is selected, then the impact of covering this aspect will fade away or be nullified for the following iterations.

Score standardization (or score normalization) has been investigated on some other tasks in web search such as data fusion [9] and federated search [11]. It is often served as a preliminary step to results merging or fusion. MinMax [4], the fitting method [9], and the reciprocal standardization method [2] are typical score standardization methods that have been used in those tasks. To the best of our knowledge, none of the previous researches explore the influence of these score standardization methods on explicit search result diversification thoroughly.

In this paper, we evaluate the performance of six score standardization methods for search result diversification. Using TREC data, we carry out an experiment to evaluate the performance of two typical explicit result diversification methods with various score standardization methods. The results show that the reciprocal standardization method performs better than the other score standardization methods in all the cases.

In the remainder of this paper, we first introduce the 2 typical explicit methods for search results diversification (Sect. 2), and then 6 score standardization methods (Sect. 3). In Sect. 4 we present experimental setting and results for the evaluation of these score standardization methods. Section 5 is the conclusions.

2 Two Explicit Methods for Results Diversification

Typically, search engines take a two-step process to generate a diversified resultant list: first it does an initial search that only concerns relevance. Then the initial ranked list is re-ranked for diversification. XQuAD and PM2 are two well-known re-ranking methods. Let q denotes a user query, $T = \{q_1, q_2, ..., q_t\}$ denotes all the aspects, or subtopics for q. $L = \{l_1, l_2, ..., l_n\}$ is the initial list of documents retrieved. XQuAD [8] and PM2 [3] are greedy algorithms. XQuAD uses the following formula

$$d^* \leftarrow \arg \max_{d \in L} \{(1 - \lambda)P(d|q) +$$
$$\lambda \sum_{q_i \in Q} [P(q_i|q)P(d|q_i) \prod_{d_j \in S} (1 - P(d_j|q_i))]\} \tag{1}$$

to compute scores for all the candidate documents d_i and the document with the highest score is picked up one by one. This process is repeated until all required positions are occupied. In Eq. 1, $P(d|q)$ is the likelihood of document d being observed given the query q, $P(q_i|q)$ is the relative importance of subtopic q_i, and $P(d|q_i)$ is the likelihood of document d being observed given the subtopic q_i. Parameter λ controls the tradeoff of the two parts: relevance and diversity.

PM-2 [3] is a probabilistic adaptation of the Sainte-Laguë method for assigning seats to members of competing political parties. PM-2 starts with a ranked list L with k empty positions. For each of these positions, it computes the quotient for each subtopic q_i following the Sainte-Laguë method and the position is assigned to the subtopic q_{i*} with the largest quotient. The next step is to find a suitable document d^* from the set of candidate documents of the selected subtopic q_{i*} according to the following formula

$$d^* \leftarrow \arg \max_{d_j \in L} \left\{ (1 - \lambda)qt[i^*]P(d_j|q_{i*}) + \lambda \sum_{i \neq i^*} qt[i] * P(d_j|q_i) \right\} \tag{2}$$

where $P(d_j|q_i)$ is the likelihood of document d_j being observed given the subtopic q_i, qt[i] is the abbreviation of quotient[i].

3   Six Score Standardization Methods and Improvement

In this section, we shall brief six score standardization methods. Among them, **Min-Max, the fitting method, Sum**, and **Z-scores** are score-based methods, while **the reciprocal function** and **the logarithmic function** are ranking-based methods. Then we compare the equations used for data fusion and results diversification. Based on that, possible improvement of score standardization for results diversification is discussed.

First let us introduce four score-based methods. MinMax is a commonly used method in various search tasks including search result diversification [6]. For a ranked list of documents, MinMax maps the highest score to 1, the lowest score to 0, and any other scores to a value between 0 and 1 by using the following formula

$$P(d|q) = \frac{s(d,q) - \min_{d_j \in L} s(d_i, q)}{\max_{d_j \in L} s(d_j, q) - \min_{d_j \in L} s(d_i, q)} \tag{3}$$

where $s(d, q)$ denotes the raw score to be normalized, $\max_{d_j \in L} s(d_j, q)$ and $\min_{d_j \in L} s(d_i, q)$ denote the highest and the lowest scores of the resultant list in question, respectively. [0,1] may not be the best interval for normalized scores because it is not guaranteed that the top-ranked documents are always useful while bottom-ranked documents are always useless. The fitting method [9] tries to improve MinMax by normalizing scores into a range [a, b] (0 < a < b < 1). Therefore we have

$$P(d|q) = a + \frac{s(d,q) - \min_{d_i \in L} s(d_i, q)}{\max_{d_j \in L} s(d_j, q) - \min_{d_i \in L} s(d_i, q)} * (b - a) \qquad (4)$$

Z-score [7] is a kind of standard score in statistics which indicates how many standard deviations a datum is above or below the mean. Z-score standardization is defined as

$$P(d|q) = \frac{s(d,q) - \mu}{(\sigma)} \qquad (5)$$

where μ is the simple mean defined as $\mu = \frac{1}{n} \sum_{i=1}^{n} s(d_i, q)$, and σ is the standard deviation defined as $\sigma = \sqrt{\frac{1}{n} \sum_{1}^{n} (s(d_i, q) - \mu)^2}$. Note that the Z-scores do not necessarily located in the range of [0, 1]. Therefore, Z-scores cannot be used directly by XQuAD. Sum is another method that maps the lowest score to 0 and the sum of the all scores to 1 [5]. It can be expressed as

$$P(d|q) = \frac{s(d,q)}{\sum_{i=1}^{n} s(d_i, q)} \qquad (6)$$

Another type of method is ranking-based. The following two methods assign a fixed score to documents at a specific ranking position. The first one is a reciprocal function-based method [2]. For any resultant list, a score of $P(d_i|q) = \frac{1}{i+60}$ is assigned to document d_i at rank i. The other is a logarithmic function-based method. It uses the formula $P(d_i|q) = \max\{1 - 0.2 * ln(i+1), 0\}$ to normalize score of the document at rank i. This method is taken from our ongoing work on fusing diversified search results. Our experiments show that it is likely more effective than the reciprocal function-based method for that purpose.

However, when using these score standardization methods for data fusion and results diversification, there are some subtle differences. Let us take CombSum, a typical data fusion method, to illustrate this. CombSum uses the following formula

$$F(d, q) = \sum_{j=1}^{m} s_j(d|q) \qquad (7)$$

where $F(d|q)$ is the combined score of document d, and $s_j(d|q)$ is the score of document d obtained in component result j ($1 \leq j \leq m$). If all $s_j(d|q)$ are scaled up or down linearly, then $F(d|q)$ will be scaled up or down correspondingly. However, such a linear transformation will not affect the ranking of all the documents involved at the fusion stage. Thus, we observe that the exact probability score may not be necessary for a large group of data fusion methods like CombSum. However, if we look at Eq. 2, $P(d|q)$ and $P(d|q_i)$ are probabilities that document d is relevant to the main query q or to one of its sub-topics q_i. When $P(d|q)$ and $P(d|q_i)$ are scaled up or down linearly, it will affect document selecting in the re-ranking stage for XQuAD. Therefore, accurate probabilities should be assigned to both $P(d|q)$ and $P(d|q_i)$ so as to make them comparable in XQuAD. On the other hand, only $P(d|q_i)$ but not $P(d|q)$ is required in PM2, its score standardization of $P(d|q_i)$ is similar to some data fusion methods. But the latter normalizes $P(d|q)$ rather than $P(d|q_i)$.

4   Experimental Settings and Results

The experiment is carried out with the "ClueWeb09" collection[1] and 4 groups of topics used in the TREC 2009, 2010, 2011 and 2012 Web tracks. The "ClueWeb09" collection comprises over 1 billion documents crawled from the Web. Each group consists of 50 topics and we have a total number of 200 topics. For each of the queries, the organizer of the event provides the official relevance judgment file that indicates which document is relevant and which is not (served as ground truth). Thus for any retrieved results they can be evaluated by some performance metrics such as ERR-IA@20 and Alpha-nDCG@20. However, two of the topics (#95 and #100) are dropped because they do not have any relevant documents. Each topic contains a query field and 3 to 8 sub-topic fields. We use the query field as the main query of each topic and the description of subtopics as queries for aspects. Figure 1 is an example of an ambiguous query (Query 25) used in TREC. This query has 4 different explanations or subtopics.

Our experiment comprises three steps. First we use an open-source information search engine Indri (Version 5.5)[2] to search "ClueWeb09" to obtain all the initial document lists related to a query q and subtopics of q. Top 100 documents are used in re-ranking. Note that all the relevance-related scores generated by Indri are negative and sore standardization is a necessity. The second step is scoring standardization and six above-mentioned score standardization methods are used to transform the scores or rankings into probabilities. For the fitting method, the values of a and b are set to 0.06 and 0.6, respectively. This is the same as in [9] (page 25). In order to let Z-scores to fit in the range of 0 and 1, we treat Z-scores by a method similar to MinMax. However, the maximum and the minimum scores used are not at query-level but over all 198 queries. The same score standardization method is used for both main query results and sub-topic results for any re-ranking method. The last step is search result diversification, in which XQuAD and PM-2 are used. Both of them have a control parameter λ and we let λ take values from {0.01, 0.02, 0.03, ..., 1.0}. Two typical metrics, ERR-IA@20 and α-nDCG@20, are used for retrieval evaluation. The best performances of XQuAD and PM-2 with different score standardization methods are presented in Table 1.

From Table 1, we can see that the performances of XQuAD and PM-2 with all six score standardization methods are better than the initial result without diversification,

---

[1] Its webpage is located at http://lemurproject.org/clueweb09/.

[2] Its webpage is located at http://www.lemurproject.org/indri/. Its retrieval model is based on a combination of the language modeling and inference network retrieval frameworks.

```
<topic number="75" type="faceted">
  <query>tornadoes</query>
  <description>
  Find information about tornadoes, what causes them, and where they occur.
  </description>
  <subtopic number="1" type="inf">
  Find information about tornadoes, what causes them, and where they occur.
  </subtopic>
  <subtopic number="2" type="nav">
  Find videos and pictures of tornadoes.
  </subtopic>
  <subtopic number="3" type="inf">
  What were the deadliest tornadoes in history?
  </subtopic>
  <subtopic number="4" type="inf">
  Find information about forecasting tornadoes.
  </subtopic>
</topic>
```

Fig. 1. An ambiguous query (Query 25)

Table 1. The best performances of XQuAD and PM-2 with different score standardization methods. The highest scores are shown in boldface. The superscripts of a result denote a statistically significant difference from the initial result (I), MinMax (M), Log (L), Sum (S), Z-score (Z), Fitting (F) at the 0.05 level)

	Norm score	λ	ERR-IA@20	Alpha-nDCG@20
Initial result	–	–	0.2120	0.3209
PM2	MinMax	0.92	0.2598^I	0.3667^I
	Logarithmic	0.91	0.2637^I	0.3701^I
	Sum	0.91	0.2618^I	0.3681^I
	Z-scores	0.73	$0.2690^{I,M}$	$0.3748^{I,M}$
	Fitting	0.75	0.2659^I	0.3728^I
	R60	**0.67**	$\mathbf{0.2755}^{I,M,L,F,S}$	$\mathbf{0.3794}^{I,M,L,S,F}$
XQuAD	MinMax	0.99	0.2639^I	0.3728^I
	Logarithmic	0.91	0.2637^I	0.3701^I
	Sum	0.85	0.2660^I	0.3698^I
	Z-scores	0.98	$0.2728^{I,M}$	$0.3816^{I,M,L}$
	Fitting	0.76	0.2676^I	0.3704^I
	R60	**0.74**	$\mathbf{0.2821}^{I,M,L,S,F}$	$\mathbf{0.3821}^{I,L,S}$

and XQuAD performs better than PM-2 in almost all the cases except using the logarithmic function-based method. The reciprocal function-based method is the best for both of XQuAD and PM-2 among all six score methods. We run a two-tailed T test to compare any two groups of methods. The results are also shown in Table 1. We can see that R60 performs better than the other methods for both XQuAD and PM-2. The reason behind this is because R60 is able to estimate rank-relevance curve more

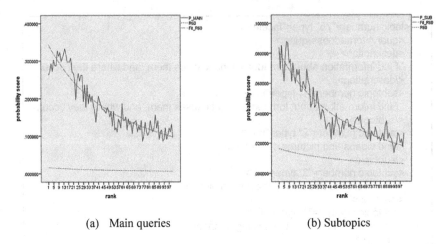

(a)   Main queries   (b) Subtopics

Fig. 2. Curves of the observed posterior probability scores, R60, and the fitted linear model of R60

accurately than the others in a linear style. Figure 2 shows the standardized coefficients of the regression using observed rank-relevance curve as the objective variable and each of the functions as the independent variable. For both main queries and sub-topics, R60 is the best with the highest coefficients (Table 2).

Table 2. Standardized coefficients of all six methods fitting the observed curve linearly

Method	Main query	Sub-topics
MinMax	0.882	0.918
Logarithmic	0.876	0.922
Sum	0.864	0.910
Z-scores	0.872	0.913
Fitting	0.882	0.918
R60	**0.930**	**0.936**

Even it performs the best among all six score standardization methods, R60 has its potential weak point and can be improved, as analyzed in Sect. 3. The scores it generates are $\{1/61 = 0.0164, 1/62 = 0.0161, 1/63 = 0.0159,...\}$ for documents ranked at $\{1, 2, 3,...\}$. It is very likely these scores are far too small to serve as proper probability scores (refer to Fig. 2). For all 198 queries and the results we obtain by running Indri, we first observe the posterior probability of those documents being relevant to the main query at each rank. We also do the same thing for all the sub-queries together. Figure 2 shows the curves of them and the scores assigned by R60. We can see that the curve of R60 is always far below the posterior probability curve. We try to estimate the posterior probability by a linear function of R60. The optimum fitting function of R60 is also shown in Fig. 2 as fit_R60. The linear functions we obtain for main queries and sub-topics are 24.261*R60-0.054 and 6.224*R60-0.021, respectively. 24.261 is almost 4 times as large as 6.244. Therefore, using exactly the same reciprocal function for both main queries and sub-topics is not a good solution. We can generate a new version of R60, which is referred to as Fit_R60, by using the scalar coefficients but ignoring the constant. We have The performance of Fit_R60 with XQuAD is presented in Table 3.

Table 3. Performances of R60 and Fit_R60; XQuAD is the re-ranking method used. The superscripts of a result denote a statistically significant difference from the initial result (▲), R60 (■) at the 0.05 level

Method	λ	ERR-IA@20	Alpha-nDCG@20
initial result	–	0.2120	0.3209
R60	0.74	0.2821$^{▲}$	0.3821$^{▲}$
Fit_R60	0.92	**0.2833$^{▲■}$**	**0.3848$^{▲}$**

From Table 3, we can see that Fit_R60 performs slightly better than R60. This means XQuAD indeed performs better if more accurate probability scores for $P(d|q)$ and $P(d|q_i)$ are used. Two-tailed T test is carried out to compare different groups of results. We find that Fit_R60 is better than R60 at a significance level of 0.05 in the case of ERR-IA@20.

5   Conclusions

In this paper we have investigated the impact of score standardization on the explicit search result diversification methods. Six typical score standardization methods are tested with two typical explicit search result diversification methods. Through extensive experiments with TREC data, we find that the reciprocal function-based method is the best choice for both main query score normalization and sub-topic scores normalization among all six methods. Both results diversification methods PM-2 and XQuAD can benefit from this. Furthermore, we find that XQuAD demands accurate probability scores for both main query results and sub-topics results. This finding leads to improved reciprocal function and better results diversification results using XQuAD. We conclude that score standardization has significant impact on the performance of explicit search result diversification methods. More profitable score standardization methods are achievable by using more accurate probability scores of all retrieved documents for both main queries and any of the sub-topics.

References

1. Carbonell, J., Goldstein, J.: The use of MMR, diversity-based reranking for reordering documents and producing summaries. In: SIGIR, pp. 335–336 (1998)
2. Cormack, G.V., Clarke, C.L.A., Büttcher, S.: Reciprocal rank fusion outperforms condorcet and individual rank learning methods. In: SIGIR, pp. 758–759 (2009)

3. Dang, V., Croft, W.B.: Diversity by proportionality: an election-based approach to search result diversification. In: SIGIR, pp. 65–74 (2012)
4. Lee, J.H.: Analyses of multiple evidence combination. In: SIGIR, pp. 267–276 (1997)
5. Montague, M., Aslam, J.A.: Relevance score standardization for meta-search. In: CIKM, pp. 427–433 (2001)
6. Ozdemiray, A.M., Altingovde, I.S.: Explicit search result diversification using score and rank aggregation methods. JASIST **66**(6), 1212–1228 (2015)
7. Renda, M.E., Straccia, U.: Web meta-search: rank vs. score based rank aggregation methods. In: SAC, pp. 841–846 (2003)
8. Santos, R.L.T., Macdonald, C., Ounis, C.I.: Exploiting query reformulations for web search result diversification. In: WWW, pp. 881–890 (2010)
9. Wu, S.: Data Fusion in Information Retrieval. Springer, Berlin (2012). https://doi.org/10.1007/978-3-642-28866-1
10. Zhai, C.X., Cohen, W.W., Lafferty, J.: Beyond independent relevance: methods and evaluation metrics for subtopic retrieval. In: SIGIR, pp. 10–17 (2003)
11. Shokouhi, M., Si, L.: Federated search. Found. Trends Inf. Retr. **5**(1), 1–102 (2011)

An Integrated Semantic-Syntactic SBLSTM Model for Aspect Specific Opinion Extraction

Zhongming Han[1,2(✉)], Xin Jiang[1], Mengqi Li[1], Mengmei Zhang[1], and Dagao Duan[1]

[1] School of Computer and Information Engineering,
Beijing Technology and Business University, Beijing 100048, China
webir@163.com
[2] Beijing Key Laboratory of Big Data Technology for Food Safety,
Beijing, China

Abstract. Opinion Mining (OM) of Internet reviews is one of the key issues in Natural Language Processing (NLP) field. This paper proposes a stacked Bi-LSTM aspect opinion extraction model in which semantic and syntactic features are both integrated. The model takes embedded vector which is composed by word embedding, POS tags and dependency relations as its input while taking label sequence as its output. The experimental results show the effectiveness of this structural features embedded stacked Bi-LSTM model on cross-domain and cross-language datasets, and indicate that this model outperforms the state-of-the-art methods.

Keywords: Aspect · Opinion extraction · Dependency tree
Stacked Bi-LSTM

1 Introduction

With the evolution of the Internet, OM has become one of the most vigorous research areas in NLP field. An aspect is a concept in which the opinion is expressed in the given text [1]. Aspect specific OM task can be divided into four main subtasks: aspect extraction, opinion extraction, sentiment analysis and opinion summarization [2]. This paper focuses on the second subtasks: opinion extraction. In this paper, we propose a hierarchical model based on stacked Bi-LSTM using both semantic information and syntactic information as input to extract aspect specific opinions.

Internet reviews OM can be carried out from three directions: document-level OM [3], sentence-level OM and aspect-level OM. Aspect-level OM is to extract both the aspects and the corresponding opinion expressions in sentences [4]. The extraction of opinion towards its corresponding aspect is a core task in Aspect-level OM. In recent years, the neural network has reached remarkable effect in NLP. Pang et al. [5] committed a survey of the current deep models used to handle text sequence issues. Socher et al. [6] proposed the recursive neural tensor network and represent phrases by distributed vectors. RNN [7] and its variants such as LSTM [8] and GRU [9] stood out from various deep learning methods. Huang et al. [10] proposed a bidirectional LSTM-CRF model for sequence labeling, and on this basis, Ma et al. [11] joined the CNNs in

© Springer Nature Switzerland AG 2018
X. Meng et al. (Eds.): WISA 2018, LNCS 11242, pp. 191–199, 2018.
https://doi.org/10.1007/978-3-030-02934-0_18

the model to encode character-level information of a word into its character-level representation. Du et al. [12] proposed an attention mechanism based RNN model which contains two bidirectional LSTM layers to label sequences so that to extract opinion phrases. Nevertheless, the neural networks' performance drops rapidly when the models solely depend on neural embedding as input [11].

2 SBLSTM Model

2.1 SBLSTM Model Structure

We model aspect opinion extraction as a sequence labeling. The input of the model includes embedded vector, POS tags and dependency relations. The output is the corresponding label sequence of the input text sequence. We use a stacked Bi-LSTM between the input layer and output layer. Opinion expressions extraction has often been treated as a sequence labeling task. This kind of method usually uses the conventional B-I-O tagging scheme.

The basic idea of LSTM is to present each sequence forwards and backwards to two separate hidden states to capture past and future information. Then the two hidden states are concatenated to form the final output. The bidirectional variant of one unit's hidden state's update at time step t is as following.

$$\overrightarrow{h_t} = \vec{g}\left(\overrightarrow{h_{t-1}}, x_t\right)\left(\overrightarrow{h_0} = 0\right) \tag{1}$$

$$\overleftarrow{h_t} = \overleftarrow{g}\left(\overleftarrow{h_{t+1}}, x_t\right)\left(\overleftarrow{h_T} = 0\right) \tag{2}$$

$h_t = \left[\overrightarrow{h_t}, \overleftarrow{h_t}\right]$ can be regarded as an intermediate representation containing the information from both directions to predict the label of the current input x_t.

Stacked RNNs is stacked by k(k $\geq$ 2) RNN networks. The first RNN receives the word embedding sequences as its input and the last RNN forms the abstract vector representation of the input sequence which is used to predict the final labels. Suppose the output of jth RNN on time-step t is h_t^j, the stacked RNNs can be formulated as following.

$$h_t^j = \begin{cases} g\left(h_{t-1}^j, x_t\right) & j = 0 \\ g\left(h_{t-1}^j, h_t^{j-1}\right) & otherwise \end{cases} \tag{3}$$

The function g in (3) can be replaced by any RNN transition functions. We expect to capture the important opinion elements. Therefore, we choose the 2 layer Stacked-BiLSTM network as the basic model, and adds attention mechanism to it. In this attention model, the second BLSTM's input i_t^2 at time t can be expressed as:

$$\mathrm{i}_t^2 = \sum\nolimits_{s=1}^{T} \alpha_{ts} h_s^1 \tag{4}$$

where h_s^1 is the output vector of the first BLSTM at time s, α_{ts} is the weight of the output vector sequence $[h_1^1, h_2^1, h_3^1, \ldots, h_T^1]$, the product of which is the input of the second BLSTM at the time t. The weight α_{ts} is calculated as follows:

$$e_{ts} = tanh\left(W^1\,{}_s^1 + W^2\,{}_{t-1}^2 + b\right) \tag{5}$$

$$\alpha_{ts} = \frac{exp\left(e_{ts}^T e\right)}{\sum_{k=1}^{T} exp\left(e_{tk}^T e\right)} \tag{6}$$

where W^1 and W^2 are the parameter matrixs that update in the model training process. b is the bias vector. e and e_{ts} has the same dimension and also update with the above adjustable parameters in the model training process.

Figure 1 demonstrates a stacked Bi-LSTM model consisting two Bi-LSTMs with an attention layer. The input is distributed word vectors of texts while the output is a series of B-I-O tags predicted from the network. In order to make the stacked RNNs to be extended easily, we use stacked bidirectional LSTMs with depth of 2 as our basic model in this paper.

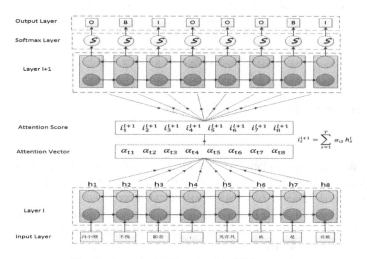

Fig. 1. A stacked bidirectional LSTM network

2.2   Features in SBLSTM Model

In SBLSTM model, the features used is as following

- *Word embeddings.* The word embedding is a kind of distributed vector which contains the semantic information.

- *POS tags.* We use Stanford Tagger to obtain the POS tags.
- *Syntactic tree.* Here we particularly apply the syntactic information, dependency tree in the model. Figure 2 displays the dependency tree for a movie review.

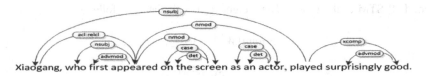

Fig. 2. Dependency tree of an example context

The syntactic representation of one word is defined as its m ($m \geq 0$) children in a dependency tree, where m denotes the window size to limit the amount of the dependency relations of one word for the learning models. Introducing the window size could prevent excessive usage on VRAM.

Finally, the three type's features will be concatenated as the input vector and fed to the SBLSTM model. Figure 3 shows the final features composition of one word.

Fig. 3. Input composition of one word

3   Experiments Design and Analysis

3.1   Datasets

For now, there are no available benchmark datasets that mark phrase boundaries of aspect specific expressions. Therefore, two manually constructed datasets are used in our experiment. Mukherjee constructed an annotated corpus which considers 1, 2, and 3-star product reviews from Amazon in English. The other dataset consists of online reviews of three Chinese movies Mr. Six, The Witness and Chongqing Hotpot collected from Douban, Mtime and Sina microblog. The movie reviews dataset is manually annotated.

The statistics information of these two datasets are showed in Table 1. Figures 4 and 5 display the sentences length distributions of the two datasets.

Table 1. Statistics of dataset

Domain	Review	Aspect	Min Length	MaxLength	Language
Product	12102	Earphone, gps, keyboard, mouse, mp3player, router domains	4	292	English
Movie	50000	13 Actors of 3 movies: *Mr. Six, The Witness* and *Chongqing Hotpot*	3	209	Chinese

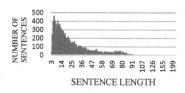

Fig. 4. Distribution of movie dataset sentences length

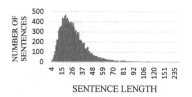

Fig. 5. Distribution of product dataset sentences length

3.2 Experimental Setting

In experiments, we use Stanford parser to obtain the syntactic information. The SBLSM models is implemented in python 2.7 and we use the Keras framework to construct the deep neural networks. The input length is limited to be 60 in LSTMs' models and the amount of LSTM input units is 60 while the number of the output units is 64. The word embeddings dimension is set to be 100. The window size in extracting dependency features is set to be 4 initially. In training process, the ratio between training set and validation set is 4:1. The activation function chosen is softmax function and the batch size to train the model is 256.

3.3 Quantitative Analysis

Evaluation Metrics. Precision, recall and F1 score are commonly used to evaluate the performance of OM models. In OM task, the boundaries of opinion expressions are hard to define. Therefore, we use *proportional overlap* as a soft measure to evaluate the performance.

Model Comparative Analysis. To illustrate the performance boost of our SBLSTM model, we firstly compare our model with some baseline methods on both two datasets. Since we use stacked bidirectional LSTM with depth of 2 as the core model, we choose the LSTM network and the bidirectional LSTM as the baselines. Furthermore, we also compare our model with the CRF model and rules based method which depends on the dependency tree.

As shown in Table 2, we reported the accuracy, the precision, the recall and the F1-score across all single runs for each approach. We could found that the proposed SBLSTM model outperforms the baseline methods in terms of accuracy, recall and F1. Bi-LSTM outperforms all of the others in terms of precision in the movie dataset, and compared with the CRF model, which achieves the highest precision in the product dataset, our proposed model also provides a comparable precision. Another observation is for both datasets, Bi-LSTM outperforms the normal LSTM model with absolute gains of 4.73% and 4.87% in terms of F1 score.

Table 2. Results of our proposed model against baseline methods

		Accuracy (%)	Precision (%)	Recall (%)	F1-score
Product	Rules-based	78.35	50.78	51.32	0.5105
	CRF	80.66	**74.82**	47.38	0.5802
	LSTM	81.09	72.41	53.26	0.6138
	Bi-LSTM	83.22	73.02	60.39	0.6611
	SBLSTM	**85.78**	73.62	**65.21**	**0.6916**
Movie	Rules-based	83.36	61.13	71.03	0.6571
	CRF	85.64	76.97	54.43	0.6377
	LSTM	86.55	75.41	58.31	0.6576
	Bi-LSTM	90.01	**83.21**	61.35	0.7063
	SBLSTM	**90.73**	77.54	**75.11**	**0.7631**

Feature Comparative Analysis. In training process, the batch size is set to be 256 and the epoch number is set to be 30. Table 3 shows the comparison of experimental results using different feature sets.

Table 3. Comparison of the models performance using different features

	Accuracy (%)	Precision (%)	Recall (%)	F1-score
Word embedding	85.76	75.30	52.23	0.6165
Word embedding + POS	88.13	81.09	55.32	0.6577
POS + dependency relation	89.58	71.29	74.17	0.7270
ALL(W + P + D)	90.73	77.54	75.11	0.7631

Our proposed methods which introduces all of the three types feature performs best in term of accuracy, recall and F1 score. Refer to the third line and the fourth line of Table 3, adding word embeddings into the feature set makes the performance of the model improved in a similar way. This indicates that both word embedding and POS tags have some help in extracting the aspect specific opinion expressions. Particularly, we can observe that the recall measure and the F1-score are improved by 20% and 10%

respectively when the dependency relations have been added into features, providing the evidence that syntactic information does play an important role in extracting the opinions.

Window Size Analysis. We conduct a series of experiments with different window sizes to compare and analyze the impact of the children amount in dependency tree on model performance on movie dataset. The batch size in training process is set to be 256 and each trial was carried out in 300 epochs. Table 4 shows the comparison of the predictive performance of the proposed stacked Bi-LSTM models with both the semantic features and the syntactic features.

From Table 4, we found that F1-score increase with the growth of the window size in general, tending to be stable when the window length is greater than 4.

Table 4. Performance of different window sizes

Window Size	Accuracy (%)	Precision (%)	Recall (%)	F1-score
1	88.85	81.09	55.32	0.6577
2	90.23	69.15	71.32	0.7022
3	90.29	75.17	72.06	0.7358
4	90.73	77.54	75.11	**0.7631**
5	89.73	76.49	74.35	0.7581
6	90.38	75.61	73.56	0.7526
7	88.39	76.47	72.26	0.7431
8	88.83	69.17	78.23	0.7465
9	89.98	69.42	76.47	0.7486
10	89.43	74.42	71.02	0.7387

3.4 Qualitative Analysis

To explore the contribution of this paper, we conducted a qualitative analysis experiment on five chinese movie comments below and the aim aspect is *FengXiaogang*.

The experiment uses rules-based model, CRF model, LSTM network and Bi-LSTM network as baseline methods. The aspect specific opinion extraction results of different methods are shown in Table 5. The green words are annotated opinion expression which we want models to extract, and the red words refer to those words haven't been extracted by the model while these blue ones are those words not in annotation.

The dependency rule based method is more effective when the sentence is short and simple. When here comes a complex sentence, it is impossible to obtain more information when the comment contains a demonstrative pronoun. Most importantly, no matter the length of the sentence, our model can extract the opinion information well.

Table 5. Aspect specific opinion extraction results of different methods

Comments	Director Feng's acting is just right, and the nature of the characters of the men in Beijing is in place.	Feng is really worthy of winning the Golden Horse Award, acting just the right place.	A group of old drama kings the old Beijing charm was showed out, making people unforgettable, the last tears is still in my mind.	The best actor title of Feng is well deserved. Tolerance, responsibility and quality which men should have incisively and vividly by him.	The movie's plot is good, but I do not understand the Feng's logic. The lines are always annoying with dirty words
Rules-based	Director Feng's acting is just right, and the nature of the characters of the men in Beijing is in place.	Feng is really worthy of winning the Golden Horse Award, acting just the right place.	A group of old drama kings really acted well, the old Beijing charm was showed out, making people unforgettable, and the last tears is still in my mind.	The best actor title of Feng is well deserved. Tolerance, responsibility and quality which men should have were acted incisively and vividly by him.	The movie's plot is good, but I do not understand the Feng's logic. The lines are always annoying with dirty words
CRF	Director Feng's acting is just right, and the nature of the characters of the men in Beijing is in place.	Feng is really worthy of winning the Golden Horse Award, acting just the right place.	Agroup of old drama kings really acted well, the old Beijing charm was showed out, making people unforgettable, and the last tears is still in my mind.	The best actor title of Feng is well deserved. Tolerance, responsibility and quality which men should have were acted incisively and vividly by him.	The movie's plot is good, but I do not understand the Feng's logic. The lines are always annoying with dirty words
LSTM	Director Feng's acting is just right, and the nature of the characters of the men in Beijing is in place.	Feng is really worthy of winning the Golden Horse Award, acting just the right place.	A group of old drama kings really acted well, the old Beijing charm was showed out, making people unforgettable, and the last tears is still in my mind.	The best actor title of Feng is well deserved. Tolerance, responsibility and quality which men should have were acted incisively and vividly by him.	The movie's plot is good, but I do not understand the Feng's logic. The lines are always annoying with dirty words
Bi-LSTM	Director Feng's acting is just right, and the nature of the characters of the men in Beijing is in place.	Feng is really worthy of winning the Golden Horse Award, acting just the right place.	A group of old drama kings really acted well, the old Beijing charm was showed out, making people unforgettable, and the last tears is still in my mind.	The best actor title of Feng is well deserved. Tolerance, responsibility and quality which men should have were acted incisively and vividly by him.	The movie's plot is good, but I do not understand the Feng's logic. The lines are always annoying with dirty words
OurModel	Director Feng's acting is just right, and the nature of the characters of the men in Beijing is in place.	Feng is really worthy of winning the Golden Horse Award, acting just the right place.	A group of old drama kings really acted well, the old Beijing charm was showed out, making people unforgettable, and the last tears is still in my mind.	The best actor title of Feng is well deserved. Tolerance, responsibility and quality which men should have were acted incisively and vividly by him.	The movie's plot is good, but I do not understand the Feng's logic. The lines are always annoying with dirty words

4   Conclusions

In this paper, we proposed a method to embed syntactic information into the deep neural models. Experimental results on two domains and different languages data sets showed that the proposed stacked bidirectional LSTM model outperform all of the baseline methods, proofing that the syntactic information did play a significant role in correctly locating the aspect-specific opinion expressions.

References

1. Poria, S., Cambria, E., Gelbukh, A.: Aspect extraction for opinion mining with a deep convolutional neural network. Knowl.-Based Syst. **108**, 42–49 (2016)
2. Hu, M., Liu, B.: Mining and summarizing customer reviews. In: Proceedings of the 10th ACM SIGKDD International Conference on Knowledge Discovery and Data Mining (KDD 2004), pp. 168–177. ACM, New York (2004)
3. Valakunde, N.D., Patwardhan, M.S.: Multi-aspect and multi-class based document sentiment analysis of educational data catering accreditation process. In: International Conference on Cloud & Ubiquitous Computing & Emerging Technologies (CUBE 2013), pp. 188–192. IEEE Computer Society, Washington (2013)
4. Singh, V.K., Piryani, R., Uddin, A., et al.: Sentiment analysis of movie reviews: a new feature-based heuristic for aspect-level sentiment classification. In: International Multi-Conference on Automation, Computing, Communication, Control and Compressed Sensing (iMac4 s 2013), pp. 712–717. IEEE Computer Society, Washington (2013)
5. Pang, L., Lan, Y.Y., Xu, J., et al.: A survey on deep text matching. Chin. J. Comput. **40**(04), 985–1003 (2017). (in Chinese with English abstract)
6. Socher, R., Perelygin, A., Wu, J.Y., et al.: Recursive deep models for semantic compositionality over a sentiment treebank. In: Proceedings of the 2013 Conference on Empirical Methods in Natural Language Processing (EMNLP 2013), pp. 1631–1642. ACL, Stroudsburg, PA (2013)
7. Goller, C., Kuchler, A.: Learning task-dependent distributed representations by backpropagation through structure. In: IEEE International Conference on Neural Networks, vol. 1, pp. 347–352. IEEE (2002)
8. Hochreiter, S., Jurgen, J.: Long short-term memory. Neural Comput. **9**, 1735–1780 (1997)
9. Cho, K., Merrienboer, B.V., Bahdana, D., Bengio, Y.: On the properties of neural machine translation: encoder-decoder approaches. In: Proceedings of SSST-8, Eighth Workshop on Syntax, Semantics and Structure in Statistical Translation, Doha, Qatar, 25 October 2014, pp. 103–111 (2014)
10. Huang, Z., Xu, W., Yu, K.: Bidirectional LSTM-CRF models for sequence tagging. Computing Research Repository, abs/1508.01991 (2015)
11. Ma, X., Hovy, E.: End-to-end sequence labeling via bi-directional LSTM-CNNs-CRF (2016)
12. Du, J., Gui, L., Xu, R.: Extracting opinion expression with neural attention. In: Li, Y., Xiang, G., Lin, H., Wang, M. (eds.) SMP 2016. CCIS, vol. 669, pp. 151–161. Springer, Singapore (2016). https://doi.org/10.1007/978-981-10-2993-6_13

Natural Language Processing

Deep Learning Based Temporal Information Extraction Framework on Chinese Electronic Health Records

Bing Tian and Chunxiao Xing[(⊠)]

RIIT, Beijing National Research Center for Information Science and Technology,
Department of Computer Science and Technology, Institute of Internet Industry,
Tsinghua University, Beijing, China
tb17@mails.tsinghua.edu.cn, xingcx@tsinghua.edu.cn

Abstract. Electronic Health Records (EHRs) are generated in the clinical treatment process and contain a large number of medical knowledge, which is closely related to the health status of patients. Thus information extraction on unstructured clinical notes in EHRs is important which could contribute to huge improvement in patient health management. Besides, temporal related information extraction seems to be more essential since clinical notes are designed to capture states of patients over time. Previous studies mainly focused on English corpus. However, there are very limited research work on Chinese EHRs. Due to the challenges brought by the characteristics of Chinese, it is difficult to apply existing techniques for English on Chinese corpus directly. Considering this situation, we proposed a deep learning based temporal information extraction framework in this paper. Our framework contains three components: data preprocessing, temporal entity extraction and temporal relation extraction. For temporal entity extraction, we proposed a recurrent neural network based model, using bidirectional long short-term memory (LSTM) with Conditional Random Fields decoding (LSTM-CRF). For temporal relation extraction, we utilize Convolutional Neural Network (CNN) to classify temporal relations between clinical entities and temporal related expressions. To the best of our knowledge, this is the first framework to apply deep learning to temporal information extraction from clinical notes in Chinese EHRs. We conduct extensive sets of experiments on real-world datasets from hospital. The experimental results show the effectiveness of our framework, indicating its practical application value.

Keywords: Deep learning · Temporal information extraction
Electronic health records · Chinese

1 Introduction

EHRs are rich in a variety of data that can facilitate timely and efficient surveillance on the prevalence of, health care utilization for, treatment patterns for, and outcomes of a host of diseases, including obesity, diabetes, hypertension,

© Springer Nature Switzerland AG 2018
X. Meng et al. (Eds.): WISA 2018, LNCS 11242, pp. 203–214, 2018.
https://doi.org/10.1007/978-3-030-02934-0_19

and kidney disease etc. [2]. And since clinical notes are designed to capture states of patients over time, temporal information extraction on these unstructured data is of great importance which can be beneficial to many fields such as clinical decision support systems, question answering and text summarization etc. [1,10,16,17,26].

In the English domain, there are already lots of research work and besides, some open evaluation tasks conducted on English EHRs such as I2B2 (Informatics for Integrating Biology & the Bedside) and SemEval (The semantic evaluati) have also greatly promoted the development of temporal information extraction research on English documents. For example, I2B2 2012 aimed at extracting temporal expressions which refers to the dates, times, durations, or frequencies phrases in the clinical text and temporal relations between the clinical events and temporal expressions [19] on English clinical notes. The semantic evaluations (SemEval) workshop organized by Association for Computational Linguistics (ACL) has also referred to the temporal evaluation (TempEval) task several times [27].

However, when it comes to Chinese domain, very limit work has been done. Comparing to English language, temporal information extraction on Chinese clinical notes have many new challenges due to the different characteristic of Chinese language. On one hand, annotated Chinese clinical corpora are not only expensive but also often unavailable for research due to patient privacy and confidentiality requirements. On the other hand, most commonly used medical domain dictionaries and knowledge base for helping clinical temporal information extraction task such as UMLS and SNOMED CT, do not have the Chinese version. What's more, due to the characteristics of Chinese, there seems to be more clinicians' frequent use of interchangeable terms, acronyms and abbreviations, as well as negation and hedge phrases in EHRs which may cause ambiguity. Additionally, some Chinese function words which are important for semantic understanding, such as "的", "了" are often omitte and complex and diverse logical expression of Chinese also further increased the difficulty of temporal information extraction.

To address these challenges, we proposed a deep learning based temporal information extraction framework on Chinese EHRs. Our framework contains three components: data preprocessing, temporal entity extraction and temporal relation extraction. In data preprocessing component, our mainly job is to annotate the corpus. All of our corpus are come from real world hospital. To obtain high quality datasets, we invited two experts from hospital to help annotate the corpus. And we even implemented an annotation tool for more convenient data annotation. In temporal entity extraction component, we aimed at extracting the clinical entity such as disease, treatment etc. and temporal related entities such as time expressions. And we proposed a model using a bi-directional LSTMs with a random sequential conditional layer (LSTM-CRF) beyond it. Bi-directional LSTM-CRF is an end-to-end solution. It can learn features from a dataset automatically during the training process without human labor. The temporal relation extraction component focuses on classifying a given event-

event pair or event-time pair in a clinical note as one of a set of pre-defined temporal relations. Temporal relation classification is one of the most important temporal information extraction (IE) tasks. And we creatively utilized CNN to do the classification. Compared with state-of-the-art methods on English documents which heavily depend on manual feature engineering [24], our CNN-based model can achieve better performance. The extensive experimental results show the effectiveness of our framework, indicating its practical application value.

The rest of paper is organized as follows. Section 2 provides an overview of the existing temporal information extraction approaches. Section 3 and Sect. 4 respectively describe our temporal clinical entity extraction and temporal relation extraction methods in detail. Next, Sect. 5 reports the experiments and discusses the results. Finally, we draw our conclusions in Sect. 6.

2   Related Work

In this section, we first introduce the recent work of temporal information extraction on English clinical domain. And then we focus on the Chinese domain and introduce some information extraction work based on non-clinical corpus.

Recently, as the automatic detection of temporal relations between events in electronic medical records has the potential to greatly augment the value of such records for understanding disease progression and patients' responses to treatments [4], a large amount of work has focused on temporal information extraction on English clinical notes. Due to the unstructured nature, most work utilize the statistical machine learning methods. For example, Cheng et al. [4] presented a three-step methodology for labeling temporal relations using machine learning and deterministic rules over an annotated corpus provided by the 2012 i2b2 Shared Challenge. D'Souza et al. [9] examined the task of temporal relation classification for the clinical domain. And they employed sophisticated knowledge derived from semantic and discourse relations by combining the strengths of rule-based and learning-based approaches. Differently, Mirza et al. [15] found that for temporal link labelling, using a simple feature set results in a better performance than using more sophisticated features based on semantic role labelling and deep semantic parsing.

Besides traditional machine learning based methods, recent years have seen an increasing interest for the use of neural approaches. For example, Dligach et al. [8] experimented with neural architectures for temporal relation extraction and established a new state-of-the-art for several scenarios. And they found that neural models with only tokens as input outperform state-of-the-art hand-engineered feature-based models. Tourille et al. [22] relied on Long Short-Term Memory Networks (LSTMs) for narrative container identification between medical events (EVENT) and/or temporal expressions (TIMEX3). And Lin et al. [13] described a method for representing time expressions with single pseudo-tokens for Convolutional Neural Networks (CNNs).

Despite the challenges brought by Chinese language, there are already lots of work focusing on information extraction on Chinese corpus [3]. For example,

Zheng et al. [27] annotated a document-level Chinese news corpus to be used for the recognition of temporal relations between Chinese events. And they introduced several effective features according to the characteristics of Chinese, such as trigger semantics, special words, event arguments, event co-reference relation, etc. to improve the performance. Differently, Li et al. [12] proposed and implemented a temporal relation resolution (TRR) engine for Chinese text based on Chinese-English bilingual parallel corpus.

Different from the existing work, we focus on extracting temporal related entities and temporal relation on Chinese clinical notes using deep learning based framework. And we will introduce our model in detail in the following section.

3   Temporal Entity Extraction Model

Temporal entity extraction aims at extracting clinically significant events, including both clinical concepts such as problems, tests, treatments, and clinical departments, and events relevant to the patients' clinical timeline such as time, date etc. In this section, we first introduce the LSTM network and CRF algorithm and then propose our LSTM-CRF based model.

3.1   LSTM Networks

Recent years, Recurrent Neural Networks (RNN) have been employed in many Natural Language Processing (NLP) problems and obtained promising results such as machine translation [6], sentiment classification [21], and information retrieval [18] etc. The most widely used RNN variants is LSTM. Figure 1 shows the architecture of basic LSTM. It is a special kind of RNN, capable of learning long-term dependencies. In this paper, we use LSTM to sequence tagging. As

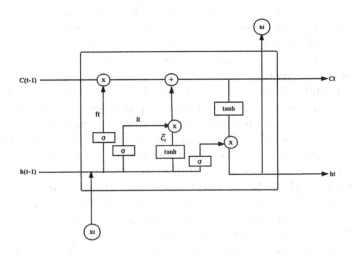

Fig. 1. LSTM Architecture

shown in Fig. 1, A common LSTM unit is composed of four neural network layers which called cells, input gates, output gates and forget gates. And LSTM memory cell is implemented as the following:

$$i_2 = \sigma(W_{xi}x_t + W hih_{t-1} + W_{ci}ct - 1 + b_i)$$
$$f_t = \sigma(W_{xf}x_t + W_{hf}ht - 1 + W_{cf}ct - 1 + b_f)$$
$$c_t = f_t c_{t-1} + i_t \tanh(W_{xc}x_t + W_{hc}h_{t-1} + b_c)$$
$$o_t = \sigma(W_{xo}x_t + W_{ho}h_{t-1} + W_{co}c_t + b_o)$$
$$h_t = o_t \tanh(c_t)$$

σ is the logistic sigmoid function, and i, f, o and c are the input gate, forget gate, output gate and cell vectors, all of which are the same size as the hidden vector h. And weight matrix W is the parameters we need to train during the training process. Figure 2 shows a LSTM sequence tagging model.

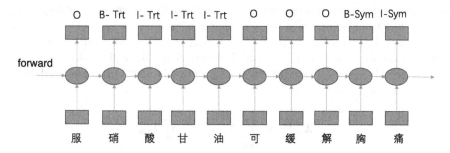

Fig. 2. A LSTM sequence tagging model

3.2   CRF Based Model

CRF was proposed by Lafferety et al. [11]. In natural language processing domain, CRF is mainly used to solve sequence annotation problems. It combines the strengths of Maximum Entropy Model (ME) and Hidden Markov Model(HMM). On one hand, it can capture a large amount of human observational experience and on the other hand it enable capture Markov chain dependencies between different tags.

Figure 3 describes the CRF based sequence tagging model. $\boldsymbol{X} = (x_1, x_2, \ldots, x_n)$ is the observation sequence and the corresponding labelling sequence $\boldsymbol{Y} = (y_1, y_2, \ldots, y_n)$ is the labelling sequence. What a CRF based model really do is construct the conditional probability model $P(\boldsymbol{Y}|\boldsymbol{X})$. And training the parameters by maximum the gold conditional probability $P(\boldsymbol{Y}|\boldsymbol{X})$ from training data.

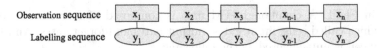

Fig. 3. A CRF based sequence tagging model

3.3   LSTM-CRF Based Model

Figure 4 displays our LSTM-CRF based temporal entity extraction model. The bottom is the word embedding layer. It has been shown in [7] that word embedding plays a vital role to improve sequence tagging performance. In this paper, we conducted the word embedding according to the methods proposed by Mikolov et al. [14]. After obtaining the word embedding, we feed it into a Bi-LSTM layer. Compared with the single directional LSTM, the Bi-LSTM can capture the information from both side so that become more expressive. Next, we combine the information from both LSTM and feed them into a CRF layer which can hold the strong dependencies across output labels. Finally, we get our outputs using softmax function. Specifically, for an input sentence: $\boldsymbol{X} = (x_1, x_2, \ldots, x_n)$ and its corresponding labelling sequence $\boldsymbol{Y} = (y_1, y_2, \ldots, y_n)$. We consider $\boldsymbol{P}$ is the matrix of scores output by the Bi-LSTM layer. And P is of size $n * k$ where k is the number of tags. So $P_{i,j}$ represents the jth word in a sentence. For a labelling sequence $\boldsymbol{Y}$, its score is as Follows:

$$s(\boldsymbol{X}, \boldsymbol{Y}) = \sum_{i=0}^{n} A_{y_i, y_{i+1}} + \sum_{i=1}^{n} P_{i, y_i} \qquad (1)$$

Where A is the matrix of transition scores produced in the CRF layer. The element $A_{i,j}$ corresponds the score of a transition from the tag i to tag j. y_0 and

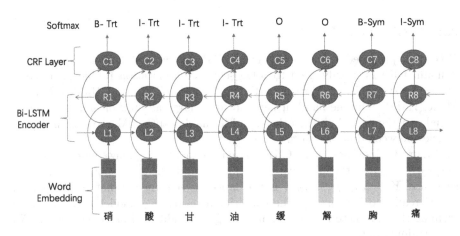

Fig. 4. LSTM-CRF based temporal entity extraction model

y_n are the start and end tags of a sentence. Then, A softmax function yields a probability for the sequence $\boldsymbol{Y}$:

$$p(\boldsymbol{Y}|\boldsymbol{X}) = \frac{e^{s(\boldsymbol{X},\boldsymbol{Y})}}{\sum_{hat(y)\in Y_x} e^{s(\boldsymbol{X},hat(y))}} \tag{2}$$

So, during training, we maximize the log-probability of the correct tag sequence:

$$log(p(\boldsymbol{Y}|\boldsymbol{X})) = s(X,y) - log(\sum_{hat(y)\in Y_x} e^{s(X,hat(y))}) \tag{3}$$

4   Temporal Relation Extraction

Temporal relation classification, one of the most important temporal information extraction (IE) tasks, involves classifying temporal relations among all the clinically relevant events such as patient's symptoms (pre- and post-admission), medical test results, doctor's diagnoses, administrated drugs and patient's responses etc. And the research on temporal relations between events plays an important role in studying development of diseases and effectiveness of potential treatments. Recent years, there are already some work using CNN in natural language processing domain [23]. In this section, we propose a CNN based model to do the classification.

4.1   CNN Based Model

As shown in Fig. 5, in the training process, there are seven layers in our CNN based model. The outermost layer of the model is initial input. It is the sentence in clinical notes. The last layer refers the output which is a vector and each value of the vector corresponds the possibility of a temporal relation. Besides these, there are 5 more layers in the model including feature layer, embedding layer, convolution layer, pooling layer and fully connected layer.

Feature Layer. In the feature layer, we introduce five features to represent each word which are word itself (W), distance to clinical entity one (D1), distance to clinical entity two (D2), POS tag (POS) and entity type of the word (T). After obtaining these features, we construct a feature dictionary and all the features are ultimately represented by a numerical matrix.

Embedding Layer. In the embedding layer, each feature corresponds to a vector of the embedding feature matrix. Supposing $M^i = \in R^{n*N}, i = 1, 2, 3, 4, 5$ is the embedding feature matrix of the ith feature(here n represents the dimension of the feature vector, N represents the number of possible values of the feature or the size of the feature dictionary), then each column in the matrix M^i is the representation of the value of ith feature. We can extract the feature vector by taking product of embedding feature matrix with the one hot vector of the feature value. Assuming that one hot representation of the jth value of the ith

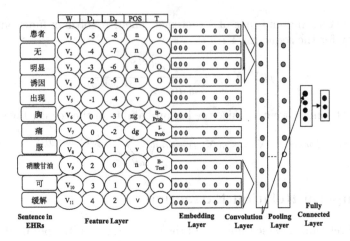

Fig. 5. CNN based temporal relation extraction model

feature is a^i_j then when the value of the ith feature is j, its vector representation f^i_j is expressed as follows:

$$f^i_j = M^i a^i_j \tag{4}$$

Convolution Layer. In convolution layer, we obtain the local features of the sentence by convolution operations. Supposing $x^1 x^2 x^3 \ldots x^m$ is a feature vector sequence of a sentence with length m, where x^i is the feature vector of the ith word and the length of the filter is c, then the output sequence of the convolution layer is computed as given below:

$$h^i = f(w \cdot x^{i:i+c-1} + b) \tag{5}$$

$f(x)$ is the ReLU function: $f(x) = max(0, x)$. w and b are the parameters we need to train.

Pooling Layer. In the pooling layer, we choose the max-pooling to obtain the global feature of each sentence. Not only does this reduce the dimensions of the output, but it still retains the most salient features. For example, in Fig. 5, we filter sentences of length 11 with four filters of length 3, and get four vectors of dimension 9. After max-pooling, we finally extract four features.

Fully Connected Layer. In the fully connected layer, we use the forward propagation to determine the predicted output. Supposing the output of the pooling layer is a vector z whose values come from different filters, then the output o of connecting layer computed as given below:

$$o = W \cdot z + b, (W \in R^{[r] \times l}, b \in R^{[r]}) \tag{6}$$

$[r]$ indicates the number of relationship types.

5   Evaluation

We conducted extensive experiments to evaluate the effectiveness of our model. In this section, we first introduce our experimental settings and then report our experimental results.

5.1   Experimental Settings

All the experiments are done on the real-world clinical notes collected from a famous medical institute. To obtain well-labeled corpus, we first implemented a tool for annotating entities and relations conveniently. And then we invited two medical experts to help annotate the corpus. Specifically, the corpus contains 2200 clinical notes and more than 2039000 words. For temporal related clinical entities, we extracted 7 entities including "疾病" (disease), "疾病诊断分类" (disease type), "自诉症状" (symptom), "异常检查结果" (test result), "检查" (test), "治疗" (treatment) and time. And for temporal relation, we define 4 types : "before", "after", "overlap" and "unknown".

5.2   Experimental Results of Temporal Entities Extraction

Figure 6 shows the experimental results of temporal entities extraction. Different templates refer to different hyper-parameters. When all the hyper-parameters are fine tuned, the model achieved best performance with the precision 91.031%, recall 87.781% and F1 score 89.376%.

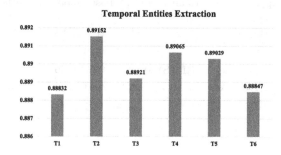

Fig. 6. Experimental results of temporal entities extraction

5.3   Experimental Results of Temporal Relations Extraction

Implementation. While implementing our model, we set the word embedding dimension to be 50 and the other 4 feature dimensions to be 5. In other words, the dimension of each word is 70. In convolution layer, we set filters of different lengths and for each length, there are 100 filters. Moreover, we use dropout with a probability of 0.50 to prevent overfitting.

Influence of Filter Lengths. The length and number of filters greatly affect the performance of the CNN-based model. In order to choose the optimal filter length, we conducted a series of experiments. And Table 1 shows the results. We can see that the best performance with F1 score as 0.814 is achieved by using filter lengths of 3, 4 and 5 together.

Table 1. Performance with different length and number of filters

Filter	[2]	[3]	[4]	[5]
F1 score	0.747	0.783	0.793	0.804
Filter	[6]	[3,4]	[3,5]	[4,5]
F1 score	0.807	0.794	0.780	0.797
Filter	[4,6]	[5,6]	[3–5]	[4–6]
F1 score	0.786	0.800	0.814	0.802

Comparison with Featured Based Models. Table 2 shows the temporal relations extraction results compared with the best 3 models in i2b2 evaluation task [5,20,25]. And it can be seen that our model outperform all the other methods. Different with previous methods, our methods added the relation type defined as Sect. 5 as our feature which made a great contribution to such good results.

Table 2. Comparative performance of CNN based model and other models

Model	P	R	F1 score
CNN	0.87	0.76	0.81
Rule based+CRF+SVM	0.71	0.67	0.69
ME+SVM+rule based	0.75	0.64	0.69
SVM	0.66	0.71	0.68

6   Conclusions

In this paper, we proposed a deep learning based temporal information extraction framework in this paper. Our framework contains three components: data preprocessing, temporal entity extraction and temporal relation extraction. For temporal entity extraction, we proposed a recurrent neural network based model, using bidirectional long short-term memory (LSTM) with Conditional Random Fields decoding (LSTM-CRF). For temporal relation extraction, we utilize Convolutional Neural Network (CNN) to classify temporal relations between clinical entities and temporal related expressions. To the best of our knowledge, this is the first framework to apply deep learning to temporal information extraction from clinical notes in Chinese EHRs. The extensive experimental results show the effectiveness of our framework, indicating its practical application value.

Acknowledgement. Our work is supported by NSFC (91646202), the National Hightech R&D Program of China (SS2015AA020102), Research/Project 2017YB142 supported by Ministry of Education of The People's Republic of China, the 1000-Talent program, Tsinghua University Initiative Scientific Research Program.

References

1. Ao, X., Luo, P., Wang, J., Zhuang, F., He, Q.: Mining precise-positioning episode rules from event sequences. IEEE Trans. Knowl. Data Eng. **30**(3), 530–543 (2018)
2. Birkhead, G.S., Klompas, M., Shah, N.R.: Uses of electronic health records for public health surveillance to advance public health. Annu. Rev. Public Health **36**, 345–359 (2015)
3. Chen, X., Zhang, Y., Xu, J., Xing, C., Chen, H.: Deep learning based topic identification and categorization: mining diabetes-related topics on Chinese health websites. In: Navathe, S.B., Wu, W., Shekhar, S., Du, X., Wang, X.S., Xiong, H. (eds.) DASFAA 2016. LNCS, vol. 9642, pp. 481–500. Springer, Cham (2016). https://doi.org/10.1007/978-3-319-32025-0_30
4. Cheng, Y., Anick, P., Hong, P., Xue, N.: Temporal relation discovery between events and temporal expressions identified in clinical narrative. J. Biomed. Inform. **46**, S48–S53 (2013)
5. Cherry, C., Zhu, X., Martin, J.D., de Bruijn, B.: À la recherche du temps perdu: extracting temporal relations from medical text in the 2012 i2b2 NLP challenge. JAMIA **20**(5), 843–848 (2013)
6. Cho, K., et al.: Learning phrase representations using RNN encoder-decoder for statistical machine translation. arXiv preprint arXiv:1406.1078 (2014)
7. Collobert, R., Weston, J., Bottou, L., Karlen, M., Kavukcuoglu, K., Kuksa, P.: Natural language processing (almost) from scratch. J. Mach. Learn. Res. **12**(Aug), 2493–2537 (2011)
8. Dligach, D., Miller, T., Lin, C., Bethard, S., Savova, G.: Neural temporal relation extraction. In: Proceedings of the 15th Conference of the European Chapter of the Association for Computational Linguistics, Short Papers, vol. 2, pp. 746–751 (2017)
9. D'Souza, J., Ng, V.: Temporal relation identification and classification in clinical notes. In: ACM Conference on Bioinformatics, Computational Biology and Biomedical Informatics, ACM-BCB 2013, Washington, DC, USA, 22–25 September 2013, p. 392 (2013). https://doi.org/10.1145/2506583.2506654
10. Hristovski, D., Dinevski, D., Kastrin, A., Rindflesch, T.C.: Biomedical question answering using semantic relations. BMC Bioinform. **16**(1), 6 (2015)
11. Lafferty, J., McCallum, A., Pereira, F.C.: Conditional random fields: probabilistic models for segmenting and labeling sequence data (2001)
12. Li, L., Zhang, J., He, Y., Wang, H.: Chinese temporal relation resolution based on Chinese-English parallel corpus. Int. J. Embed. Syst. **9**(2), 101–111 (2017)
13. Lin, C., Miller, T., Dligach, D., Bethard, S., Savova, G.: Representations of time expressions for temporal relation extraction with convolutional neural networks. BioNLP **2017**, 322–327 (2017)

14. Mikolov, T., Sutskever, I., Chen, K., Corrado, G.S., Dean, J.: Distributed representations of words and phrases and their compositionality. In: Advances in Neural Information Processing Systems, pp. 3111–3119 (2013)
15. Mirza, P., Tonelli, S.: Classifying temporal relations with simple features. In: Proceedings of the 14th Conference of the European Chapter of the Association for Computational Linguistics, pp. 308–317 (2014)
16. Mishra, R., et al.: Text summarization in the biomedical domain: a systematic review of recent research. J. Biomed. Inform. **52**, 457–467 (2014)
17. Musen, M.A., Middleton, B., Greenes, R.A.: Clinical decision-support systems. In: Shortliffe, E.H., Cimino, J.J. (eds.) Biomedical Informatics: Computer Applications in Health Care and Biomedicine, pp. 643–674. Springer, London (2014). https://doi.org/10.1007/978-1-4471-4474-8_22
18. Palangi, H., et al.: Deep sentence embedding using long short-term memory networks: analysis and application to information retrieval. IEEE/ACM Trans. Audio, Speech Lang. Process. (TASLP) **24**(4), 694–707 (2016)
19. Sun, W., Rumshisky, A., Uzuner, O.: Evaluating temporal relations in clinical text: 2012 i2b2 challenge. J. Am. Med. Inform. Assoc. **20**(5), 806–813 (2013)
20. Tang, B., Wu, Y., Jiang, M., Chen, Y., Denny, J.C., Xu, H.: A hybrid system for temporal information extraction from clinical text. JAMIA **20**(5), 828–835 (2013)
21. Tang, D., Qin, B., Liu, T.: Document modeling with gated recurrent neural network for sentiment classification. In: Proceedings of the 2015 Conference on Empirical Methods in Natural Language Processing, pp. 1422–1432 (2015)
22. Tourille, J., Ferret, O., Neveol, A., Tannier, X.: Neural architecture for temporal relation extraction: a Bi-LSTM approach for detecting narrative containers. In: Proceedings of the 55th Annual Meeting of the Association for Computational Linguistics, Short Papers, vol. 2, pp. 224–230 (2017)
23. Wang, J., Wang, Z., Zhang, D., Yan, J.: Combining knowledge with deep convolutional neural networks for short text classification. In: Proceedings of the 26th International Joint Conference on Artificial Intelligence, pp. 2915–2921. AAAI Press (2017)
24. Xu, Y., Wang, Y., Liu, T., Tsujii, J., Chang, E.I.C.: An end-to-end system to identify temporal relation in discharge summaries: 2012 i2b2 challenge. J. Am. Med. Inform. Assoc. **20**(5), 849–858 (2013)
25. Xu, Y., Wang, Y., Liu, T., Tsujii, J., Chang, E.I.: An end-to-end system to identify temporal relation in discharge summaries: 2012 i2b2 challenge. JAMIA **20**(5), 849–858 (2013)
26. Zhang, Y., Li, X., Wang, J., Zhang, Y., Xing, C., Yuan, X.: An efficient framework for exact set similarity search using tree structure indexes. In: 2017 IEEE 33rd International Conference on Data Engineering (ICDE), pp. 759–770. IEEE (2017)
27. Zheng, X., Li, P., Huang, Y., Zhu, Q.: An approach to recognize temporal relations between Chinese events. In: Lu, Q., Gao, H. (eds.) Chinese Lexical Semantics. LNCS (LNAI), vol. 9332, pp. 543–553. Springer, Cham (2015). https://doi.org/10.1007/978-3-319-27194-1_55

Improving Word Embeddings
by Emphasizing Co-hyponyms

Xiangrui Cai, Yonghong Luo, Ying Zhang$^{(\boxtimes)}$, and Xiaojie Yuan

College of Computer Science, Nankai University, Tianjin, China
{caixiangrui,luoyonghong,zhangying,yuanxiaojie}@dbis.nankai.edu.cn

Abstract. Word embeddings are powerful for capturing semantic similarity between words in a vocabulary. They have been demonstrated beneficial to various natural language processing tasks such as language modeling, part-of-speech tagging and machine translation, etc. Existing embedding methods derive word vectors from the co-occurrence statistics of target-context word pairs. They treat the context words of a target word equally, while not all contexts are created equal for a target. Some recent work learns non-uniform weights of the contexts for predicting the target, while none of them take the semantic relation types of target-context pairs into consideration. This paper observes co-hyponyms usually have similar contexts and can be substitutes of one another. To this end, this paper proposes a simple but effective method to improve word embeddings. It automatically identifies possible co-hyponyms within the context window and optimizes the embeddings of co-hyponyms to be close directly. Compared to 3 state-of-the-art neural embedding models, the proposed model performs better on several datasets of different languages in terms of the human similarity judgement and the language modeling tasks.

1 Introduction

Word representations are central to various Natural Language Processing (NLP) tasks. The conventional approach represents words with one-hot vectors which have the dimensionality of the vocabulary size and only one non-zero entry. They suffer from not only high dimensionality and sparsity, but also lack of semantic relationships. For example, the one-hot representations of "walk" and "run" are not related (the distance between them is 0), which are not useful for inferring "run to school" when we know "walk to school". To tackle these problems, word representations that are able to express word similarity are necessary and have become a mainstream research topic recently.

Based on the distributional hypothesis [16] that words share similar contexts are similar, words are represented by high dimensional but sparse vectors where each entry measures the correlation between the word and the corresponding context [2,37]. Some works employ SVD [7] or LDA [9,31,35] to reduce the dimensionality. Recently, there has been a surge of work representing words as

X. Meng et al. (Eds.): WISA 2018, LNCS 11242, pp. 215–227, 2018.
https://doi.org/10.1007/978-3-030-02934-0_20

dense, real valued vectors. Their main idea is to express the generative probability of sentences with the vectors of the words in the sentences and to learn the vectors and the parameters of the probability function simultaneously. These representations, referred to as word embeddings, come to prominence through a sequence of works of Mikolov et al. [26,27,30] which depart from Bengio et al. [4] and follows the suggestions of Minh et al. [28]. Word embeddings have been shown to perform well in various NLP tasks [1,5,10–12,20,34].

Previous work on neural word embeddings takes the contexts of a target word equally, where the contexts refer to the words that precede and follow the target, typically within a k-token window. Nevertheless, not all contexts are created related to the target. Ling et al. [21] extend the Continuous Bag-of-Words (CBOW) model by assigning non-uniform weights to the contexts according to the relative positions. Liu et al. [22] propose an extension to the exponential family embeddings (CS-EFE) model to model the probability of the target conditioned on a subset of the contexts. However, both of them are based on the same conjecture that context words at **different relative positions** (to the target word) contribute differently for predicting the target word. Neither of them are aware that **the semantic relationships between each context and the target could be different**. For example, in the sentence "Shenzhen and Shanghai are tier-1 cities in China", "Shenzhen" and "Shanghai" are co-hyponyms, and "cities" is a hypernym of "Shenzhen" and "Shanghai" when given context window to be 5 tokens. Existing methods considers these two relationships as the same type, which indicates that "Shenzhen" and "cities" contribute equally for learning the embedding of "Shanghai". However, the co-hyponym "Shenzhen" should have higher influence on learning the embedding of "Shanghai" since it is more similar to "Shanghai" compared to "cities".

As we observed, it is not enough to treat all context-target pairs as the same relations. Especially, co-hyponyms are closely related words and usually can be substitutes of one another. This paper proposes a new objective which is able to detect the co-hyponyms automatically and differentiate the influence of co-hyponyms and other words for learning the embedding vector of the target word. Our model, referred to as CHSG, is built based on the Subword Information Skip-gram (SISG) model [30]. After identifying the possible co-hyponyms of the target within the contexts, the CHSG model optimizes the co-hyponyms to be close to the target word directly, which is different from the indirect optimization in SISG model. The method is simple but has been proven effective. The experiments on several Wikipedia datasets of different languages demonstrate the effectiveness of the CHSG model in terms of the human similarity judgement and the language modeling tasks compared to 3 state-of-the-art neural embedding models.

Generally, the contributions of this paper are summarized as follows:

- This paper, for the first time, observes that the relation types of target-context pairs should be distinguished during training of word embeddings. Specifically, the co-hyponyms have particular contributions for learning the embeddings of the target due to their close correlations.

- This paper proposes the CHSG model that treats the co-hyponyms within the contexts discriminately and optimizes the vectors of the co-hyponym contexts to be close to the target word directly.
- We conduct experiments on several Wikipedia datasets in terms of the human similarity judgement and the language modeling tasks. The results demonstrate the effectiveness of the CHSG model compared to 3 state-of-the-art embedding models.

The rest of this paper is organized as follows: Sect. 2 introduces the Skip-gram and the SISG models, and describe the details of the proposed model; Sect. 3 reports the experimental results on the two tasks; we review the related literature in Sect. 4 and conclude this paper in Sect. 5.

2   Model

In this section, we first briefly review the Skip-gram and the SISG models and introduce the definition of co-hyponyms. Then we describe the CHSG model in detail.

2.1   Skip-Gram and SISG

Given a vocabulary V, where each word is identified by its index $w \in \{1, 2, \ldots, |V|\}$, the goal of word embeddings is to learn a vector representation of each word. The Skip-gram model is shown in Fig. 1. The model consists of 3 layers: the input layer that fetches a central word, the hidden layer that projects the input word index to a vector, and the output layer that predicts the context words.

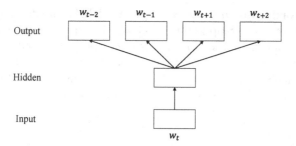

Fig. 1. The Skip-gram model.

Generally, given a training corpus, represented as a word sequence $w_1, w_2, \ldots, w_T$, the Skip-gram model [27] learns word embeddings by using the central word to predict each word within a predefined context window. This is achieved by maximizing the average log probability of the contexts given the central word:

$$\frac{1}{T}\sum_{t=1}^{T}\sum_{t-L\leq c\leq t+L,c\neq t}\log p(w_c|w_t), \tag{1}$$

where L is the size of the context window. The probability of observing the context w_c given w_t is parameterized using the word vectors. Specifically, given a score function s, mapping each pair of word-context to a real value score, the probability of a context word is defined by the following softmax function:

$$\log p(w_c|w_t) = \frac{e^{s(w_t,w_c)}}{\sum_{w\in V}e^{s(w_t,w)}}. \tag{2}$$

To reduce the computational complexity of this formulation, Mikolov et al. [27] proposed a negative sampling method to approximate the softmax:

$$-\log(1 + e^{-s(w_t,w_c)}) - \sum_{k=1}^{K}\log(1 + e^{s(w_t,w_k)}), \tag{3}$$

where K is the number of negative samples. The negative sample w_k follows the unigram distribution raised to 3/4th power, i.e., $P(w_k) = U(w_k)^{\frac{3}{4}}/Z$ (Z is a constant). This objective is to maximize the probability of observed co-occurrences in the corpus and to minimize the probability of the negative sampled word pairs.

In the **Skip-gram** model, each word is represented with two vectors. One input vector $\boldsymbol{v}_{w_t}$ represents w_t as an input central word, and the other one $\boldsymbol{v}'_{w_t}$ represents w_t as an output context word. Given a target w_t and a context w_c, the score function in the Skip-gram model is defined as the dot product of their corresponding vectors:

$$s(w_t, w_c) = \boldsymbol{v}_{w_t}^{\top}\boldsymbol{v}'_{w_c}. \tag{4}$$

The **Subword Information Skip-gram (SISG)** model [30] follows the same architecture of the Skip-gram model, but incorporates the subword information. The main contribution of SISG is to compose the input word vector by the sum of the character n-gram vectors, i.e., $\boldsymbol{v}_{w_t} = \sum_{g\in\mathcal{G}_{w_t}}\boldsymbol{z}_g$, where $\mathcal{G}_{w_t}$ is the set of character n-gram of w_t and $\boldsymbol{z}_g$ represents the vector of the character n-gram g. The rest parts of SISG are the same as Skip-gram.

2.2   Co-hyponyms

The concept of co-hyponyms is derived from hypernyms and hyponyms. In linguistics, a hyponym is a word whose semantic field is included within that of another word, which is its hypernym. For example, as shown in Fig. 2, on the one hand, "Food" is a hypernym of "Fruit", "Meat" and "Vegetable". On the other hand, "Fruit", "Meat" and "Vegetable" are hyponyms of "Food".

If two hyponyms X and Y belong to a hypernym Z, X and Y are identified as co-hyponyms. In Fig. 2, "Banana" and "Mango" are co-hyponyms and they share the same hypernym "Fruit".

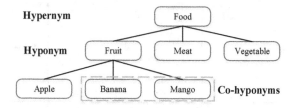

Fig. 2. Illustration of hypernyms, hyponyms and co-hyponyms.

2.3 CHSG Model

We observe word pairs that are co-hyponyms are usually share the similar context words. For example, the contexts of "Shanghai" are {city, China, finance, trade, stock, ... }, which are very close to the contexts of "Shenzhen". Based on the distributional hypothesis [16] that words share similar contexts are similar. The two co-hyponyms "Shanghai" and "Shenzhen" should be embedded to close positions.

However, by reviewing the training process of the Skip-gram and the SISG models, we find that the vectors of such similar words are optimized to be close in a indirectly way. Figure 3 demonstrates the typical training process for two co-hyponyms. First, supposing "city" is a shared context of both "Shanghai" and "Shenzhen", the two models initialize the context (output) vector of "Shanghai", and the embeddings (input vectors) of "Shanghai" and "Shenzhen" randomly as shown in Fig. 3(a). Second, the models encounter "city" is a context of "Shenzhen", and optimize the embedding of "Shenzhen" to be close to the context vector of "city" according to Eqs. (3) and (4), which is to maximize the cosine similarity of the two vectors, as shown in Fig. 3(b). Third, the models encounter another case that "city" is a context of "Shanghai" and then they optimize the embedding of "Shanghai" to be close to "city", as shown in Fig. 3(c). Finally, we obtain that "Shenzhen" and "Shanghai" are close to each other. This indirect optimization leads to substandard results and slow convergence. Furthermore, even "Shanghai" appears to be a context of "Shenzhen", the Skip-gram and SISG models only optimize the context vector of "Shanghai" and the embedding of "Shenzhen" to be together. In this case, the models cannot embed the embeddings of "Shanghai" and "Shenzhen" to be close either.

We call this optimization process the indirect way, because the co-hyponyms are embedded through the output vectors of their shared contexts. This indirect optimization is sensitive to noisy context words and some similar words cannot be embedded closely as a result. One straightforward method to address this problem is to treat the contexts of a word differently, where the co-hyponyms are optimized directly. Therefore, this paper proposes a modified score function based on Eq. (4):

$$s(w_t, w_c) = \max\{\boldsymbol{v}_{w_t}^\top \boldsymbol{v}'_{w_c}, \boldsymbol{v}_{w_t}^\top \boldsymbol{v}_{w_c}\} \tag{5}$$

The framework is the same as the Skip-gram model (Eq. (3)), According to Eq. (5), our model can detect the co-hyponyms automatically when the dot

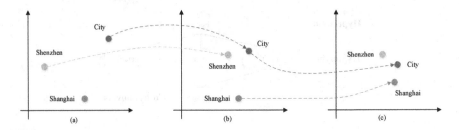

Fig. 3. Three steps of optimization in the Skip-gram and SISG models. Given "city" to be a shared context of the co-hyponyms "Shanghai" and "Shenzhen". The co-hyponyms are optimized close indirectly. However, the two words could be closer if we directly optimize the word vectors.

product of embeddings (input vectors) is large. Then our model optimizes their input vectors directly. Compare to the score function in Eq. (4), this score function can lead the model to be aware of possible co-hyponym word pairs and then pull them to close positions in the embedding space directly.

2.4 Optimization Algorithm

Similar to previous work, the model is trained by the Stochastic Gradient Descent (SGD) algorithm. Given a objective function $\mathcal{L}(\theta)$ parameterized by a model parameter set θ, the SGD minimizes the objective by updating the parameters in the opposite direction of the gradient of the objective function with regard to the parameters for each training sample, i.e.,

$$\theta^{(new)} = \theta^{(old)} - \eta \nabla_\theta \mathcal{L}(\theta) \tag{6}$$

We derive the updates for the parameters layer by layer from top to bottom. The objective of the negative sampling method for each word-context pair is to minimize the following loss function:

$$\mathcal{L} = \log(1 + e^{-s(w_t, w_c)}) + \sum_{k=1}^{K} \log(1 + e^{s(w_t, w_k)}), \tag{7}$$

For the output vector, the derivative of $\mathcal{L}$ with regard to v'_{w_j} is

$$\frac{\partial \mathcal{L}}{\partial v'_{w_j}} = \left(\sigma(s(w_t, w_j)) - t_j\right) \cdot \frac{\partial s(w_t, w_j)}{\partial v'_{w_j}}, \tag{8}$$

where $t_j = 1$ only when w_j is the actual observed context in the training corpus ($j = c$), otherwise $t_j = 0$. The $\sigma(\cdot)$ refers to the sigmoid function. Similarly, we can obtain the derivative of $\mathcal{L}$ with regard to v_{w_t}:

$$\frac{\partial \mathcal{L}}{\partial v_{w_t}} = \sum_{w_j \in \{w_c\} \cup W_{neg}} \left(\sigma(s(w_t, w_j)) - t_j\right) \cdot \frac{\partial s(w_t, w_j)}{\partial v_{w_t}}, \tag{9}$$

where $\mathcal{W}_{neg}$ refers to the set of negative contexts for each pair of word and context. We optimize this model by updating the input and output vectors alternatively with Eqs. (8) and (9).

3   Experiments

In this section, we present the experiments to evaluate the proposed model. We first introduce the datasets and the implementation details. Then we present the experimental results of the human similarity judgement and the language modeling tasks.

3.1   Datasets

To conduct the human similarity judgement experiment, we train the models on Wikipedia datasets of 6 languages,[1] namely, Enligsh (EN), Arabic (AR), German (DE), Spanish (ES), French (FR), Russian (RU). The datasets are preprocessed with the Matt Mahoney's script[2] to remove all non-Roman characters and to map digits to English words. The statistics (the data sizes, the numbers of total tokens and the vocabulary sizes) of these datasets are shown in Table 1.

Table 1. Statistics of Wikipedia datasets.

	EN	AR	DE	ES	FR	RU
Size	16 GB	1.6 GB	7.6 GB	4.4 GB	3.2 GB	8.4 GB
#Tokens	3.16B	181M	1.39B	866M	654M	883M
$|V|$	2.17M	667K	2.39M	1.04M	778K	2.19M

For the language modeling task, we make use of the datasets (Data-1M) introduced by Botha and Blunsom [6].[3] The datasets contain data of 5 languages: Czech (CS), German (DE), Spanish (ES), French (FR) and Russian (RU). Each training set has about 1 million tokens, and we perform the same preprocessing and data splits as [6]. The statistics (the data sizes, the numbers of total tokens and the vocabulary sizes) of the datasets are shown in Table 2.

3.2   Implementation Details

We implement the CHSG model based on the FastText toolkit[4] that implements the SISG model. We compare the CHSG model against the Skip-gram,

---

[1] https://dumps.wikimedia.org.
[2] http://mattmahoney.net/dc/textdata.html.
[3] https://bothameister.github.io/.
[4] https://github.com/facebookresearch/fastText.

Table 2. Statistics of Data-1M datasets.

	CS	DE	ES	FR	RU
Size	6.4 MB	6.3 MB	5.5 MB	5.6 MB	12 MB
#Tokens	1M	1M	1M	1M	1M
$|V|$	46K	36K	27K	25K	62K

the CBOW and the SISG models. All models are optimized by the stochastic gradient descent algorithm and the learning rate is decayed linearly. That is, the learning rate of the stochastic gradient descent algorithm is $\gamma_0(1 - \frac{t}{TN})$, where γ_0 is the initial learning rate, T the total number of words, N the total training epoch and t the current time. The model is trained in parallel. All threads share the same parameters and the vectors of the model are updated asynchronously.

The dimensionality of word embeddings is set to be 300 for all the models. The number of negative samples is set to be 5. Words within 5 tokens before and after the target words are treated as contexts. To avoid highly frequent words, the rejection threshold is set to be $1e-4$. The starting learning rate for the CBOW model is 0.05, and it is 0.025 for the Skip-gram, the SISG and the CHSG models. The out-of-vocabulary words in the evaluation sets are represented using the sum of their subwords in the SISG and the CHSG models. All these settings are the default values for the baselines and work well for our model as well.

3.3 Human Similarity Judgement

We evaluate the quality of the word embeddings on the task of word similarity. This task is to compute the Spearman's rank correlation coefficient (SRCC) [36] between the human labeled similarity scores and the cosine similarity computed from word embeddings. We benchmark the models on several evaluation sets: for English, we compare the models on RW [23] and WS353 [14]; for Arabic and Spanish, we use the datasets introduced in [17]; for German, the models are compared on GUR65, GUR350 [15] and ZG222 [39]; the French word vectors are evaluated on RG65, which is translated from English by [18]; Russian word vectors are evaluated on HJ [29].

The results of human similarity judgement task is reported in Table 3. The best results on the corresponding evaluation sets are in **boldface**. We can observe that our model, CHSG, performs the best and achieve a good margin on most of the evaluation sets compared the three baselines. This demonstrates the advantages of mapping co-hyponyms to close positions directly. Besides, we also observe that the effective of the proposed objective is more important for English, Arabic, Spanish, French and Russian than for German. Many German words are compound words, such as "Lebenslangerschicksalsschatz" (lifelong treasure of destiny), which comes from "lebenslanger" + "Schicksals" + "Schatz". Such compound words consist of wealthy information and the surrounding contexts are most likely not their co-hyponyms. Therefore, our method does not achieve any improvements.

Table 3. SRCC (%) on the 6 Wikipedia datasets and several evaluation sets.

	EN		AR	DE			ES	FR	RU
	RW	WS353	WS353	GUR65	ZG222	GUR350	WS353	RG65	HJ
Skip-gram	43	72	51	78	35	61	57	70	59
CBOW	43	73	52	78	38	62	58	69	60
SISG	47	71	55	81	**44**	**70**	59	75	66
CHSG	**49**	**74**	**56**	**83**	43	**70**	**60**	**77**	**68**

3.4 Language Modeling

In this section, we describe the evaluations of the word embeddings on a language modeling task. We conduct this experiment on the Data-1M datasets with 5 languages (CS, DE, ES, FR, RU). To compare against the baselines, we build the same Recurrent Neural Network (RNN) model as that in [30]. The model has 650 LSTM units, regularized by a dropout layer (with probability of 0.5). The model is trained by Adagrad [13]. The starting learning rate is set to be 0.1 and the weight decay 0.00001.

We train word vectors with the CHSG, the Skip-gram and the SISG models respectively. Then the embedding layer of the language model is initialized by these pre-trained word embeddings respectively. We report the text perplexity of the RNN model without using pre-trained word embeddings (referred to as RNN) and with embeddings learned by the Skip-gram model (Skip-gram), the SISG model (SISG) and our model (CHSG). The results on the five datasets are presented in Table 4.

Table 4. Test perplexity on the 5 datasets of different languages.

	CS	DE	ES	FR	RU
RNN	366	222	157	173	262
Skip-gram	339	216	150	162	237
SISG	312	206	145	159	206
CHSG	**308**	**203**	**141**	**154**	**201**

We highlight the best perplexity results in **boldface**. As we can see, it benefits the plain RNN model by initializing the embedding layer with the pre-trained word vectors in terms of the text perplexity. The most important observation is that the word embeddings learned by the CHSG model performs the best compared to the baselines on all of the 5 datasets. This shows the significance of identifying possible co-hyponyms for word embeddings on the language modeling task.

4   Related Work

The earliest idea of word embeddings dated back to 1988 [33]. The Neural Probabilistic Language Model (NPLM) [4] was the first attempt to learn word embeddings with a neural network model based on the distributional hypothesis [16]. The model shared word embeddings across different sentences and documents. It used a softmax function to predict the contexts of given words, which leaded to highly computational cost. Collobert et al. [10] proposed a unified framework that utilized Convolutional Neural Network (CNN) to learn word embeddings and demonstrated the use of word embeddings in multiple NLP tasks, including part-of-speech tagging, chunking, named entity resolution, semantic role labeling and language modeling. The model was trained jointly on these tasks using weight sharing. To address the high complexity of the NPLM model, Minh et al. [28] proposed a fast hierarchical language model. After a series of work by Mikolov et al. [26,27], i.e., the Skip-gram and the CBOW model, neural word embeddings came to prominence and were widely employed as the preprocessing step for various NLP applications. Recently, Bojanowski et al. [30] enrich word vectors further by incorporating subword information. All of these models focused on the scalability of neural-network models. None of them were aware that differences between contexts.

Some recent work has explored the influence of contexts on word embeddings. Melamud et al. [25] investigated the impact of different context types (window-based, dependency-based and substitute based) on learning word embeddings by the Skip-gram model [26,27]. Their results suggested that the intrinsic tasks tended to have a preference to particular types of contexts, while the extrinsic tasks needed more careful tuning to find the optimal settings. Considering the context words contributed differently for predicting the central word, Ling et al. [21] proposed to learned non-uniform weights for contexts based on the CBOW model. They incorporated an attention model into the CBOW model to compute the hidden representation. Similarly, Liu et al. [22] built a probabilistic generative model based on the exponential family embedding [32] to select related contexts within the predefined context window. The relatedness of context and word was modeled by a Bernoulli variable. However, word-context pairs in these models are still learned in an indirect way, which results in substandard embeddings as we explained in Sect. 2.

Another line of word embeddings was along paradigmatic representations, which represented a word position with a probable substitute vector. For example, the blank in "I like eating ____" could be filled in with several words (e.g., "apples", "chickens"), which are substitutes. Each substitute in the vector is associated with a probability. Currently, the probable substitute vectors were derived from co-occurrence statistics [38], which extended the SCODE framework [24]. The SCODE framework were also used in Word Sense Induction [3,19], and other NLP tasks [8]. The main downside of these models is that the possible substitutes were obtained from a pre-trained language model.

5   Conclusion

This paper proposes a simple but effective objective function for learning word embeddings based on the SISG model. The objective encourages to embed co-hyponyms to close positions directly. The intrinsic and extrinsic evaluation on several datasets of different languages demonstrate the effectiveness of our method compared to the three state-of-the-art word embedding models. One future work is to take more complex semantic correlations between words (e.g., hypernymy and hyponymy) into account. Another interesting direction is to learn word embeddings and types of word correlations together in a single model.

Acknowledgements. This research is supported by National Natural Science Foundation of China (No. 61772289), Natural Science Foundation of Tianjin (No. 16JCQNJC00500) and Fundamental Research Funds for the Central Universities.

References

1. Ballesteros, M., Dyer, C., Smith, N.A.: Improved transition-based parsing by modeling characters instead of words with LSTMs. In: EMNLP, pp. 349–359 (2015)
2. Baroni, M., Lenci, A.: Distributional memory: a general framework for corpus-based semantics. Comput. Linguist. **36**(4), 673–721 (2010)
3. Baskaya, O., Sert, E., Cirik, V., Yuret, D.: AI-KU: using substitute vectors and co-occurrence modeling for word sense induction and disambiguation. In: Second Joint Conference on Lexical and Computational Semantics (*SEM), Proceedings of the Seventh International Workshop on Semantic Evaluation (SemEval 2013), vol. 2, pp. 300–306 (2013)
4. Bengio, Y., Ducharme, R., Vincent, P., Jauvin, C.: A neural probabilistic language model. J. Mach. Learn. Res. **3**(2), 1137–1155 (2003)
5. Bojanowski, P., Joulin, A., Mikolov, T.: Alternative structures for character-level RNNs. In: Workshop on International Conference of Learning Representation (2016)
6. Botha, J., Blunsom, P.: Compositional morphology for word representations and language modelling. In: International Conference on Machine Learning, pp. 1899–1907 (2014)
7. Bullinaria, J.A., Levy, J.P.: Extracting semantic representations from word co-occurrence statistics: stop-lists, stemming, and SVD. Behav. Res. Methods **44**(3), 890–907 (2012)
8. Cirik, V., Yuret, D.: Substitute based SCODE word embeddings in supervised NLP tasks. arXiv preprint arXiv:1407.6853 (2014)
9. Cohen, R., Goldberg, Y., Elhadad, M.: Domain adaptation of a dependency parser with a class-class selectional preference model. In: Proceedings of ACL 2012 Student Research Workshop, pp. 43–48. Association for Computational Linguistics (2012)
10. Collobert, R., Weston, J.: A unified architecture for natural language processing: deep neural networks with multitask learning. In: ICML, pp. 160–167. ACM (2008)
11. Collobert, R., Weston, J., Bottou, L., Karlen, M., Kavukcuoglu, K., Kuksa, P.: Natural language processing (almost) from scratch. J. Mach. Learn. Res. **12**(Aug), 2493–2537 (2011)

12. Dos Santos, C.N., Gatti, M.: Deep convolutional neural networks for sentiment analysis of short texts. In: COLING, pp. 69–78 (2014)
13. Duchi, J., Hazan, E., Singer, Y.: Adaptive subgradient methods for online learning and stochastic optimization. J. Mach. Learn. Res. **12**(Jul), 2121–2159 (2011)
14. Finkelstein, L., et al.: Placing search in context: the concept revisited. In: Proceedings of the 10th International Conference on World Wide Web, pp. 406–414. ACM (2001)
15. Gurevych, I.: Using the structure of a conceptual network in computing semantic relatedness. In: Dale, R., Wong, K.-F., Su, J., Kwong, O.Y. (eds.) IJCNLP 2005. LNCS (LNAI), vol. 3651, pp. 767–778. Springer, Heidelberg (2005). https://doi.org/10.1007/11562214_67
16. Harris, Z.S.: Distributional structure. Word **10**(2–3), 146–162 (1954)
17. Hassan, S., Mihalcea, R.: Cross-lingual semantic relatedness using encyclopedic knowledge. In: Conference on Empirical Methods in Natural Language Processing: Volume, pp. 1192–1201 (2009)
18. Joubarne, C., Inkpen, D.: Comparison of semantic similarity for different languages using the Google n-gram corpus and second-order co-occurrence measures. In: Butz, C., Lingras, P. (eds.) AI 2011. LNCS (LNAI), vol. 6657, pp. 216–221. Springer, Heidelberg (2011). https://doi.org/10.1007/978-3-642-21043-3_26
19. Jurgens, D., Klapaftis, I.: Semeval-2013 task 13: word sense induction for graded and non-graded senses. In: Second Joint Conference on Lexical and Computational Semantics (*SEM), Proceedings of the Seventh International Workshop on Semantic Evaluation (SemEval 2013), vol. 2, pp. 290–299 (2013)
20. Kusner, M., Sun, Y., Kolkin, N., Weinberger, K.: From word embeddings to document distances. In: International Conference on Machine Learning, pp. 957–966 (2015)
21. Ling, W., et al.: Not all contexts are created equal: better word representations with variable attention. In: Proceedings of the 2015 Conference on Empirical Methods in Natural Language Processing, pp. 1367–1372 (2015)
22. Liu, L., Ruiz, F., Athey, S., Blei, D.: Context selection for embedding models. In: NIPS, pp. 4819–4828 (2017)
23. Luong, T., Socher, R., Manning, C.D.: Better word representations with recursive neural networks for morphology. In: Proceedings of the Seventeenth Conference on Computational Natural Language Learning, pp. 104–113 (2013)
24. Maron, Y., Lamar, M., Bienenstock, E.: Sphere embedding: an application to part-of-speech induction. In: Advances in Neural Information Processing Systems, pp. 1567–1575 (2010)
25. Melamud, O., McClosky, D., Patwardhan, S., Bansal, M.: The role of context types and dimensionality in learning word embeddings. In: NAACL, pp. 1030–1040 (2016)
26. Mikolov, T., Chen, K., Corrado, G., Dean, J.: Efficient estimation of word representations in vector space (2013)
27. Mikolov, T., Sutskever, I., Chen, K., Corrado, G.S., Dean, J.: Distributed representations of words and phrases and their compositionality. In: Advances in Neural Information Processing Systems, pp. 3111–3119 (2013)
28. Mnih, A., Hinton, G.E.: A scalable hierarchical distributed language model. In: Advances in Neural Information Processing Systems, pp. 1081–1088 (2009)
29. Panchenko, A., et al.: Human and machine judgements for Russian semantic relatedness. AIST 2016. CCIS, vol. 661, pp. 221–235. Springer, Cham (2017). https://doi.org/10.1007/978-3-319-52920-2_21

30. Bojanowski, P., Grave, E., Joulin, A., Mikolov, T.: Enriching word vectors with subword information. In: Annual Meeting of the Association for Computational Linguistics, pp. 135–146 (2017)
31. Ritter, A., Mausam, Etzioni, O.: A latent Dirichlet allocation method for selectional preferences. In: Proceedings of the 48th Annual Meeting of the Association for Computational Linguistics, pp. 424–434. Association for Computational Linguistics (2010)
32. Rudolph, M., Ruiz, F., Mandt, S., Blei, D.: Exponential family embeddings. In: Advances in Neural Information Processing Systems, pp. 478–486 (2016)
33. Rumelhart, D.E., Hinton, G.E., Williams, R.J.: Learning representations by back-propagating errors. Cogn. Model. **5**(3), 1 (1988)
34. Santos, C.D., Zadrozny, B.: Learning character-level representations for part-of-speech tagging. In: ICML, pp. 1818–1826 (2014)
35. Séaghdha, D.O.: Latent variable models of selectional preference. In: Proceedings of the 48th Annual Meeting of the Association for Computational Linguistics, pp. 435–444. Association for Computational Linguistics (2010)
36. Spearman, C.: The proof and measurement of association between two things. Am. J. Psychol. **15**(1), 72–101 (1904)
37. Turney, P.D., Pantel, P.: From frequency to meaning: vector space models of semantics. J. Artif. Intell. Res. **37**, 141–188 (2010)
38. Yatbaz, M.A., Sert, E., Yuret, D.: Learning syntactic categories using paradigmatic representations of word context. In: Proceedings of the 2012 Joint Conference on Empirical Methods in Natural Language Processing and Computational Natural Language Learning, pp. 940–951. Association for Computational Linguistics (2012)
39. Zesch, T., Gurevych, I.: Automatically creating datasets for measures of semantic relatedness. In: The Workshop on Linguistic Distances, pp. 16–24 (2006)

Double Attention Mechanism for Sentence Embedding

Miguel Kakanakou, Hongwei Xie[(⊠)], and Yan Qiang

College of Computer Science and Technology,
Taiyuan University of Technology, Taiyuan 030024, China
xiehongwei@tyut.edu.cn

Abstract. This paper proposes a new model for sentence embedding, a very important topic in natural language processing, using a double attention mechanism to combine of a recurrent neural network (RNN) and a convolutional neural network (CNN). First, the proposed model uses a bidirectional Long Short Term Memory Recurrent Neural Network (RNN-LSTM) with a self-attention mechanism to compute a first representation of the sentence called primitive representation. Then the primitive representation of the sentence is used along with a convolutional neural network with a pooling based attention mechanism to compute a set of attention weights used during the pooling step. The final sentence representation is obtained after concatenation of the output of the CNN neural network with the primitive sentence representation. The double attention mechanism helps the proposed model to retain more information contained in the sentence and then to be able to generate a more representative feature vector for the sentence. The model can be trained end-to-end with limited hyper-parameters. We evaluate our model on three different benchmarks dataset for the sentence classification task and compare that with the state-of-art method. Experimental results show that the proposed model yields a significant performance gain compared to other sentence embedding methods in all the three dataset.

Keywords: Bidirectional LSTM · Convolutional neural network
Sentence embedding · Pooling-based attention mechanism
Self-attention mechanism

1 Introduction

Sentence embedding has become increasingly useful for a variety of Natural Language Processing (NLP) tasks such as sentiment analysis, document summarization, machine translation, discourse analysis, etc. The goal is to train a model that can automatically transform a sentence to a vector that encodes the semantic meaning of the sentence which are inputs of various Natural Language Processing (NLP) tasks.

Sentence embedding methods generally fall into two categories. The first consists of using unsupervised learning to produce universal sentence embedding (Hill et al. 2016). This includes methods like SkipThought vectors (Kiros et al. 2015), Paragraph Vector (Le and Mikolov 2014), recursive auto-encoders (Socher et al. 2011, 2013), Sequential Denoising Autoencoders (SDAE), FastSent (Hill et al. 2016) etc.

© Springer Nature Switzerland AG 2018
X. Meng et al. (Eds.): WISA 2018, LNCS 11242, pp. 228–239, 2018.
https://doi.org/10.1007/978-3-030-02934-0_21

The other category consists of models trained specifically for a certain task. They are usually combined with downstream applications and trained by supervised learning. One generally finds that specifically trained sentence embedding performs better than generics ones, although generic ones can be used in a semi-supervised setting, exploiting large unlabeled corpora. Along this category, the Recurrent Neural Networks (RNN) with Long Short-Term Memory (LSTM) cells and the Convolutional Neural Network (CNN) are the most popular artificial neural network models used.

Originally invented for computer vision (LeCun et al. 1995), CNN models have subsequently been shown to be effective for NLP and have achieved excellent results on problems such as sequence labeling (Collobert et al. 2011), semantic parsing (Yih et al. 2014), search query retrieval (Shen et al. 2014), and sentence modeling (Kalchbrenner et al. 2014).

The LSTM neural network model analyzes a text word by word and stores the semantics of all the previous text in a fixed-sized hidden layer (Elman 1990). The advantage of this model is its ability to better capture long term memory. It accumulates increasingly richer information as it goes through the sentence, and when it reaches the last word, the hidden layer of the network provides a semantic representation of the whole sentence. A common approach in the LSTM based model consists in creating a simple vector representation by using the final hidden state of the model. This approach is biased because later words are more dominant than earlier words. Also Carrying the semantics along all time steps of a recurrent model is relatively hard and not necessary (Zhouhan and Minwei 2017). To tackle this bias problem of the RNN model, people propose to use attention mechanism on the hidden state of the LSTM model.

The concept of attention has gained popularity recently in training neural networks. In sentence modelling task, it allows the model to learn to selectively focus on the important parts of the sentence.

In this work, we design a new model that uses a double attention mechanism to combine the bidirectional LSTM model with the CNN model. The proposed model consists of two parts. The first part is to compute a primitive representation of the sentence by using a self-attentive mechanism, which provides a set of summation weight for the LSTM hidden states. These set of summation weight vectors are dotted with the LSTM hidden states, and the resulting weighted LTSM hidden states are considered as the primitive representation of the sentence. The second part of the proposed model is a convolutional neural network on top of which we use and attention pooling mechanism. The attention weights are obtained by comparing the local representation generated by the convolutional operation with the primitive sentence representation. At last, the primitive representation is combined with the convolutional structure to form the final representation of the sentence.

We conduct experiments on several sentence classification tasks. Experiment results demonstrate that the new model outperforms state-of-the-art approaches on three benchmark datasets.

This paper is organized as follow: Sect. 2 gives a brief review of related work. Section 3 details on our proposed model. The performance of the proposed method is compared with state-of-the-art methods in Sect. 4. Section 5 summarizes the contributions of this work.

2 Related Work

2.1 Unsupervised Models for Sentence Embedding

The ParagraphVector DBOW model proposed by Le and Mikolov (2014) is an unsupervised model that learns sentence or paragraph representation in the same way as learning word vectors using CBOW and skip-grams. The ParagraphVector can then use softmax distribution to predict words contained in the sentence given the sentence vector representation.

Kiros et al. (2015) propose a SkipThought model that train an encoder-decoder model with recurrent neural networks where the encoder maps the input sentence to a sentence vector and the decoder generates the sentences surrounding the original sentence. The model is similar to Skip-gram model in the sense that surrounding sentences are used to learn sentence vectors.

FastSent (Hill et al. 2016) is a sentence-level log-linear bag-of-words model. It can be seen as a Skip-thought where the encoder is a sum over word embedding, which tries to predict independently the words from the previous and following sentences.

Sequential (Denoising) Autoencoders (SDAE) employs an encoder-decoder framework, similar to neural machine translation (NMT), to denoise an original sentence (target) from its corrupted version (source). The model composes sentence vectors sequentially, but it disregards context of the sentence.

2.2 Supervised Models Without Attention Mechanism for Sentence Embedding

Several supervised models have been proposed along this line, by using recurrent networks (Hochreiter and Schmidhuber 1997; Chung et al. 2014), recursive networks (Socher et al. 2013) and Convolutional networks (Kalchbrenner et al. 2014; dos Santos and Gatti 2014) as an intermediate step in creating sentence representations to solve a wide variety of tasks including classification and ranking (Yin and Schütze 2015; Palangi et al. 2016; Tan et al. 2016; Feng et al. 2015).

2.3 Supervised Models with Attention Mechanism for Sentence Embedding

Liu et al. (2016) proposed a sentence encoding-based model that use self-attention for recognizing text entailment. They utilize the mean pooling over LSTM states as the attention source, and use that to re-weight the pooled vector representation of the sentence.

Joo et al. (2016) proposed an attention pooling-based convolutional neural network for sentence modelling. The model uses a bidirectional LSTM to generate an intermediate sentence representation and then use the generated intermediate sentence representation as a reference for local representations produced by the convolutional layer to obtain attention weights. Different from that model, our model proposes a double attention mechanism by a new level of attention along the LSTM hidden states.

Lin et al. (2017) proposed a structured self-attentive sentence embedding model that use a two layer multilayers perceptron without bias to compute a set of summation weight vectors for the LSTM hidden states. These set of summation weight vectors are dotted with the LSTM hidden states and the resulting weighted LSTM hidden states are considered as an embedding for the sentence. Different from their model, our proposed model use one perceptron without bias over the LSTM hidden states to compute the attention weight. Also our model combines the LSTM with a CNN with attention pooling.

3 The Proposed Model

The proposed model consists of two part. The first part is a bidirectional LSTM with self-attention mechanism, and the second part is a convolutional neural network (CNN) with an attention pooling based mechanism. The embedding vector of the sentence is obtained by combining the output of the attention pooling based CNN with the primitive sentence representation obtained from the bidirectional LSTM.

In the following sections, we will first describe the word embedding which is the input of our model (Sect. 3.1), then will present the bidirectional LSTM model (Sect. 3.2) and finally we will describe the attention pooling based CNN model and how we compute the final representation of the sentence.

3.1 Word Embedding

The input of our model is N variable-length sentences. Each sentence denoted S is constituted by words which are represented by vectors. Word representation methods generally fall into two categories. The first consists of methods such as one-hot vectors. This method is problematic due to homonymy and polysemy words. The other category consists of using unsupervised learning method to obtain continuous word vector representations. Recent research results have demonstrated that continuous word representations are more powerful.

In this paper, we use word embedding based on word2vec (Mikolov et al. 2013). Word2vec word-embedding procedure learns a vector representation for each word using (shallow) neural network language model. While a task specific word2vec may perform better on some specifics tasks, we choose to use the pre-trained google word2vec embedding. The model is trained on 100 billion words from Google News by using the Skip-gram method and maximizing the average log probability of all the words using a softmax function.

3.2 The Bidirectional LSTM with Self-attention Mechanism

For this section, we will use the Fig. 1 to describe the bidirectional LSTM with self-attention.

Suppose we have sentence, which has n words. We represent each word with its word2vec embedding vectors. Let's denote by d the word vector dimension and by i the embedding vector of the i^{eme} word in the sentence. The sentence S is thus a sequence

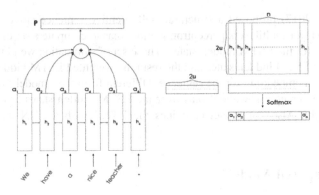

Fig. 1. A sample model structure showing the architecture of the self-attentive bidirectional LSTM.

represented as two-dimension matrix, which concatenates all the word embedding vector together. The shape of S is n-by-d.

$$S = (\omega_1, \omega_2, \ldots, \omega_n) \tag{1}$$

Now we apply the LSTM structure over the sentence in the left-to-right and right-to-left directions, respectively, resulting in a forward state sequence $h_1', h_2', \ldots, h_n'$ and a backward state sequence $h_n'', h_{n-1}'', \ldots, h_1''$, respectively. For each word embedding vector at a position i, a concatenation of h_i' and h_i'' is used as its bidirectional state vector.

Let's denote by h_i the bidirectional state vector of the i^{eme} word ω_i.

$$h_i = \{h_i', h_i''\}i \tag{2}$$

We note the output of the bidirectional LSTM as H.

$$H = (h_1, h_2, \ldots, h_n) \tag{3}$$

Let's denote the hidden unit number for each unidirectional LSTM be u. The shape of H is n-by-2u.

Following the traditional approach of sentence embedding, we will have chosen h_n as the embedding vector of the whole sentence. But this approach is biased because doing so we ignore all the states before h_n. Instead of doing so, we will use a self-attention mechanism to compute a set of summation weight for the bidirectional hidden states vector.

The attention mechanism takes the whole LSTM hidden states H as input and outputs a vector of weights A. We achieve that by using the following formula:

$$A = \text{softmax}\left(W_S \cdot H^T\right) \tag{4}$$

Here W_S is a weight vector with dimension 2u and H^T is the transpose of H. Since H shape is n-by-2u, the output weight vector A will have a size n. The softmax function ensure all the computed weight sum up to 1.

We then sum up H according to the weight provided by A to get a vector representation P of the input sentence. We call P the primitive representation of the sentence.

$$P = AH \tag{5}$$

3.3   The Convolution Neural Network Based on Attention Pooling

The convolution neural network is a state-of-the-art method to model semantic representations of sentences. The convolution action has been commonly used to synthesize lexical n-gram information. In our model, we use three different convolutional filters with varying convolution window size to form parallel CNNs so that they can learn multiple type of embedding of local regions so as to complement each other to improve model accuracy. The final output is the concatenation of the output of each convolution layer. We will use the Fig. 2 to describe how attention pooling based perform on each convolution layer.

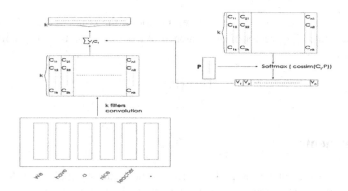

Fig. 2. A sample model structure showing the architecture of the attention pooling based convolutional neural network of our model

The input of the convolution layer is a sentence represented by a two-dimension matrix. We perform one-dimension convolution with k filters over the input sentence. We use zero-paddings to guarantee that the feature map obtained after the convolution has the same length with the input sentence. The shape of the feature map is thus n-by-k. We denote by C the feature map obtained after the convolution.

$$C = [c_1, c_2, \ldots, c_n] \tag{6}$$

where each c_i is a vector of dimension k.

We will now compute the attention weight that we will use for the pooling. For that, we will compare each feature map vector ci with the primitive sentence representation P that we get from the self-attentive bidirectional LSTM. We use cosine similarity to measure the similarity. After that we apply the softmax function to the similarity vector in order to ensure all the computed weight sum up to 1. We denote the computed weights by V.

$$V = \text{softmax}(\text{cossim}(c_i, P)) \tag{7}$$

Where cossim denote the cosine similarity function.

These set of weights are dotted with the feature map and the resulting weighted feature map are considered as the output of the convolution layer. We denote by O_1 the output of the convolution layer.

$$O_1 = \Sigma v_i \cdot c_i \tag{8}$$

Similarly, we obtain the O_2 and O_3 for the other two convolutional layer with different window size. To capture global semantics of the sentence, we average the outputs of three filters to generate the output M.

$$M = (1/3)(O_1 + O_2 + O_3) \tag{9}$$

Finally, we concatenate the primitive representation of the sentence with the output M to obtain the embedding vector of our sentence. We denote by G the embedding vector of the sentence.

$$G = \{P, M\} \tag{10}$$

4   Experiments

In this section, we evaluate the performance of our model by applying it to sentence classification task with three benchmark datasets and compare it with sate-of-the-art approaches.

4.1   Datasets

We test our model on three benchmark datasets for sentence classification task. The datasets are described as follows:

MR: This is the dataset for movie review with one sentence per review. The objective is to classify each review into either positive or negative by its overall sentiment polarity. The class distribution of this dataset is 5331/5331. The average sentence length is 20 words and the maximum sentence length is 56 words.

SUBJ: This is the subjectivity dataset where the goal is to classify a sentence as being subjective or objective. The class distribution is 5000/5000. The average sentence length is 23 words and the maximum sentence length is 120 words.

MPQA: This is the opinion polarity detection subtask of the MPQA dataset. The class distribution is 3310/7295. The average sentence length is 3 words and the maximum sentence length is 36 words.

4.2   Comparison Systems

The performance of the proposed model is compared with different state-of-art neural sentence models. We use the same comparison system with Meng et al. (2016). The comparison systems are described as follows:

NB-SVM and MNB (Naïve Bayes SVM and Multinomial Naïve Bayes): They are developed by taking Naives Bayes log-count ratios of uni and bi-gram features as input to the SVM classifier and Naïve Bayes classifier respectively. The word embedding technique is not used.

cBOW (Continuous Bag-of-words): This model use average or max pooling to compose a set of word vectors into a sentence representation.

RAE, MV-RNN and RNTN (Recursive Auto-encoder, Matrix-vector Recursive Neural Network and Recursive Neural Tensor Network): These three models belong to recursive neural networks and recursively compose word vectors into sentence vector along a parse tree. Every word in the parse tree is represented by a vector, a vector and a matrix and a tensor-based feature function in RAE, MV-RNN and RNTN respectively.

RNN and BRNN (Recurrent Neural Network and Bidirectional Recurrent Neural Network): The RNN composes words in a sequence from the beginning to the end into a final sentence vector while the BRNN does the composition from both the beginning to the end and the end to the beginning.

CNN, one-hot CNN DCNN (Standard Convolutional Neural Network, one-hot vector Convolutional Neural Network and Dynamic Convolutional Neural Network): The CNN and the DCNN use pre-trained word vectors while the one-hot CNN employs high dimensional one-hot vector representation of words as input. The CNN and the one-hot CNN employs max pooling and DCNN uses k-max pooling at the pooling stage.

P.V. (Paragraph Vector): It learns representations for sentences or paragraphs in the same way as learning word vectors using CBO W and skip-grams.

APCNN. (Attention Pooling-based Convolutional Neural Network): It use a bidirectional LSTM without attention mechanism to generate an intermediate representation of the sentence that is used during the attention pooling mechanism.

4.3   Experimental Settings

For all our experiments, we use the Stochastic Gradient Descent optimization algorithm with a learning rate of 0.1 and a weight decay of 0.95. We conduct the experiments with 50 epochs and we use mini-batches of size 64. We evaluate the model every 100 steps. We use google pre-trained word2vec thus the dimension of each word vector is 300.

We study the sensitivity of the proposed model to the convolutional region size, the number of convolutional feature and the dropout rate.

We found that we achieve the best performance when we use the settings values listed in the Table 1.

Table 1. Parameter setting of different datasets

Dataset	Parameters		
	Region size	Feature maps	Dropout rate
MR	(4, 5, 6)	150	0.4
SUBJ	(6, 7, 8)	200	0.5
MPQA	(2, 3, 4)	150	0.5

Our model is developed in Python with Tensorflow and Numpy libraries. The experiments are conducted on a MAC OS PC with 2.9 GHz Intel Core i7 processor and 8 GB RAM.

4.4   Experimental Results

The Fig. 3 shows the effect of the variation of the region size window on the accuracy of our model. We can see that the optimal window size depends on the dataset. Also when we choose a very big window size (e.g.: 9, 10, 11), the accuracy of the model tends to decrease. We believe that it is because we are taking in consideration some words that does not belong to the context of the current word.

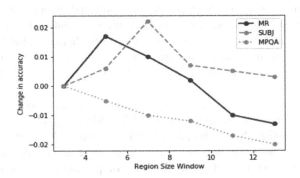

Fig. 3. Effect of the variation of the region size window on the accuracy of our model.

The Fig. 4 shows the effect of the variation of the number of feature maps on the accuracy of our model. We can see that when the number of feature maps is too small (less than 50) or is very high (more than 400), the accuracy of our model tends to decrease. This is because we have performed our experience on short sentences dataset. This behavior may be different if we were using a long sentences dataset.

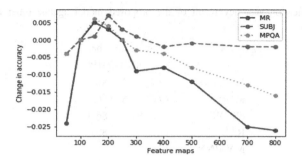

Fig. 4. Effect of the variation of the number of feature maps on the accuracy of our model.

Figure 5 shows the effect of the variation of the dropout rate on the accuracy of our model. We can see that we achieve our best accuracy when the dropout rate is around 0.4 or 0.5. In order words, our model achieve its best performance when we randomly turn off half of the learned features.

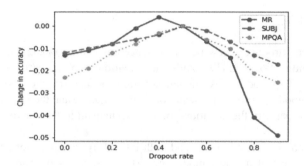

Fig. 5. Effect of the variation of the dropout rate on the accuracy of our model

The performance of the proposed model compared with other approaches is given in the Table 2.

We can conclude from the table that our model consistently outperforms the other systems in all the three datasets. We believe that it is the attention mechanism we have applied over the bidirectional LSTM that helps the new model outperform the APCNN model who is using also the attention pooling mechanism. We think that the attention mechanism over the bidirectional LSTM helps to get a better primitive representation of the sentence comparing to the simple bidirectional LSTM used in the APCNN model.

Table 2. Classification accuracy result of the proposed model against other approaches on benchmark datasets

Model	Table column head		
	MR	SUBJ	MPQA
NB-SVM	79.4	93.2	86.3
MNB	79.0	93.6	86.3
cBOW	77.2	91.3	86.4
RAE	77.7	–	86.4
MV-RNN	79.0	–	–
RNN	77.2	92.7	90.1
BRNN	81.6	93.2	90.3
CNN	81.5	93.4	89.5
One-hot CNN	77.8	91.1	83.9
P.V.	74.8	90.5	74.2
APCNN	82.5	94.3	90.7
Our Model	83.2	94.7	91.2

5 Conclusion

In this work, we introduce a new neural sentence embedding model that combine the self-attentive bidirectional LSTM with the attention pooling based Convolutional neural network. The model uses the self-attention mechanism over the bidirectional LSTM to get a good primitive representation of the sentence and then use that primitive representation to perform the attention pooling mechanism in the convolutional neural network.

Our model can be trained end-to-end with limited hyper-parameters. Experimental results demonstrate that the new model outperforms state-of-art approaches on three benchmark datasets for text classification.

For future work, we could focus on investigating the use of other recurrent neural network such as the gated recurrent neural network with the self-attentive mechanism.

References

Er, M.J., Zhang, Y., Wang, N., Pratama, M.: Attention pooling-based convolutional neural network for sentence modelling. J. Inf. Sci. **373**, 388–403 (2016)

Lin, Z., et al.: A structured self-attentive sentence embedding. In: Proceedings of International Conference on Learning Representations Conference, pp. 34–49 (2017)

Bengio, Y., Ducharme, R., Vincent, P.: A neural probabilistic language mode. Mach. Learn. Res. **3**, 932–938 (2003)

Bowman, S.R., Angeli, G., Potts, C., Manning, C.D.: A large annotated corpus for learning natural language inference. https://nlp.stanford.edu/pubs/snli_paper.pdf. Accessed 17 Sept 2017

Bowman, S.R., Gauthier, J., Rastogi, A., Gupta, R., Manning, C.D., Potts, C.: A fast unified model for parsing and sentence understanding. https://arxiv.org/abs/1603.06021. Accessed 10 Sept 2017

Cheng, J., Dong, L., Lapata, M.: Long short-term memory-networks for machine reading. https://arxiv.org/abs/1601.06733. Accessed 21 Oct 2017

Chung, J., Gulcehre, C., Cho, K.H., Bengio, Y.: Empirical evaluation of gated recurrent neural networks on sequence modeling. https://arxiv.org/abs/1412.3555 (2014). Accessed 12 Jan 2018

dos Santos, C., Gatti, M.: Deep convolutional neural networks for sentiment analysis of short texts. In: Proceedings of COLING 2014, the 25th International Conference on Computational Linguistics: Technical Papers, pp. 69–78 (2014)

dos Santos, C., Tan, M., Xiang, B., Zhou, B.: Attentive pooling networks. https://arxiv.org/abs/1602.03609. Accessed 23 Oct 2017

Feng, M., Xiang, B., Glass, M.R., Wang, L., Zhou, B.: Applying deep learning to answer selection: a study and an open task. In: Proceedings of IEEE Workshop on Automatic Speech Recognition and Understanding, pp. 813–820 (2015)

Hill, F., Cho, K., Korhonen, A.: Learning distributed representations of sentences from unlabeled data. http://www.aclweb.org/anthology/N16-1162 (2016). Accessed 12 Jan 2018

Hochreiter, S., Schmidhuber, J.: Long short-term memory. Neural Comput. 9(8), 1735–1780 (1997)

Kalchbrenner, N., Grefenstette, E., Blunsom, P.: A convolutional neural network for modelling sentences (2014). http://www.aclweb.org/anthology/P14-1062. Accessed 25 Oct 2017

Kim, Y.: Convolutional neural networks for sentence classification. http://www.aclweb.org/anthology/D14-1181. Accessed 10 Dec 2017

Kiros, R., et al.: Skip-thought vectors. In: Advances in Neural Information Processing Systems, pp. 3294–3302 (2015)

Le, Q.V., Mikolov, T.: Distributed representations of sentences and documents. In: Proceedings of ICML Conference, vol. 14, pp. 1188–1196 (2014)

Lee, J.Y., Dernoncourt, F.: Sequential short-text classification with recurrent and convolutional neural networks. https://arxiv.org/abs/1603.03827. Accessed 26 Jan 2018

Li, P., et al.: Dataset and neural recurrent sequence labeling model for open-domain factoid question answering. https://arxiv.org/abs/1607.06275. Accessed 10 Mar 2018

Ling, W., Lin, C.-C., Tsvetkov, Y., Amir, S.: Not all contexts are created equal: better word representations with variable attention. In: Proceedings of Natural Language Processing Conference, pp. 1367–1372 (2015)

Liu, Y., Sun, C., Lin, L., Wang, X.: Learning natural language inference using bidirectional LSTM model and inner-attention (2016). https://arxiv.org/abs/1605.09090. Accessed 10 Mar 2018

Ma, M., Huang, L., Xiang, B., Zhou, B.: Dependency-based convolutional neural networks for sentence embedding. http://www.aclweb.org/anthology/P15-2029. Accessed 25 Mar 2018

Margarit, H., Subramaniam, R.: A batch-normalized recurrent network for sentiment classification. In: Proceedings of Neural Information Processing Systems (2016)

Memisevic, R.: Learning to relate images. IEEE Trans. Pattern Anal. Mach. Intell. 35(8), 1829–1846 (2013)

Word Embedding Based Document Similarity for the Inferring of Penalty

Tieke He[(⊠)], Hao Lian, Zemin Qin, Zhipeng Zou, and Bin Luo

State Key Laboratory for Novel Software Technology, Nanjing University,
Nanjing 210093, China
hetieke@gmail.com

Abstract. In this paper, we present a novel framework for the inferring of fine amount of judicial cases, which is based on word embedding when calculating the distances between documents. Our work is based on recent studies in word embeddings that learn semantically meaningful representations for words from local occurrences in sentences. This framework considers the context information of words by adopting the *word2vec* embedding, compared to traditional processing methods such as hierarchical clustering, kNN, k-means and traditional collaborative filtering that rely on vectors. In the area of judicial research, there exists the problem of deciding the amount of fine or penalty of legal cases, in this work we deal with it as a recommendation task, specifically, we divide all the legal cases into 7 classes by the amount of fine, and then for a target legal case, we try to infer which class this case belongs to. We conduct extensive experiments on a legal case dataset, and the results show that our proposed method outperforms all the comparative methods in metrics *Precision*, *Recall* and *F1-Score*.

Keywords: Collaborative Filtering · Word embedding
Penalty inferring

1 Introduction

Today, we live in the information age and are confronted with much information from all scopes of our lives. With the rapid development of the internet, we have entered an era of information explosion [16]. Due to this enormous amount of information, it is necessary to rely on techniques which are capable of filtering available data and allow the search for suitable data. In this sort of circumstance, recommender systems appear as a practical methodology [10]. Recommender systems adopts information filtering to recommend information of interest to a user and are defined as the system which suggests an appropriate product or service after learning the users' preferences and requirements [11,28].

This work is supported in part by the National Key Research and Development Program of China (2016YFC0800805), and the Fundamental Research Funds for the Central Universities (021714380013).

X. Meng et al. (Eds.): WISA 2018, LNCS 11242, pp. 240–251, 2018.
https://doi.org/10.1007/978-3-030-02934-0_22

As recommender systems are playing an increasingly critical role in many fields, the purpose of this paper is to seek opportunity of adopting this kind of technology in the judicial study, as well to introduce a novel framework in tackling the specific task as inferring the amount of fine or say penalty in legal cases.

Judicial study has witnessed progress in an amount of directions, such as speech recognition and computer vision in remote trial, as well as data compression, storage and transmission. However, there is little effort devoted to the knowledge discovery of law cases, which constitutes the most crucial part of the judicial big data. A legal case mainly consists of the description of the fact of the case, the rules or conditions that are applied, the basic information of the accuser and accused, as well as the lawyer and the court information. Getting to the final decision of a law case has always been a complex task for the court, it often involves with many rounds of negotiation and bargaining, along with referring to applicable codes and conditions, as well as historical similar cases. Each step of this process calls for intensive human effort and expertise. Thus, offering aid in some of these steps can greatly benefit the judicial cause, so as to improve the efficiency of the whole judicial system. In an attempt to mimic the process of court judgement, we seek to provide quantitative analysis of certain steps from the view of computer science. It is widely acknowledged that, for most parties the threat of being fined or punished provides incentives to take care not to harm others [1]. For instance, industrial firms may resist fouling the air, motorists may obey traffic regulations, and manufactures try to produce safe toys – all to avoid fines for violation of standards. Motivated by this, we start out by inferring the penalty or the amount of a fine, which functions as a practical utility in guiding judges to the final decision of a legal case.

Previous studies such as incorporating the court trial in their models [12,18], is not often explicitly analyzed. The outcome of the court is somehow predetermined and the decision moment in their models falls before the actual verdict. In their study, Polinsky and shavell [19] made an extensive economic analysis of the optimal level of punitive damages in a variety of circumstances. Following, Daughety and Reinganum [6] model both the settlement and the litigation process for decisions made in court, allowing for incomplete information about the damages incurred by the plaintiff on the part of both the defendant and the court. According to the analysis of Earnhar [7], driving factors behind penalty decisions can be divided into five categories: causes of accidents, measured damages, environmental factors, regional factors and political influence. And their study shows significant differences between the different regions and the between different political systems. Also, the size and the cause of the violation influence the level of the fine.

From a traditional point of view, we start off the task of inferring the penalty in the way of classic recommendation. That is, we hire methods such as hierarchical clustering, kNN, k-means and traditional collaborative filtering, which are based on the results of TF-IDF [20]. Specifically, TF-IDF is adopted to extract the words in the textual description of legal cases, and then a vector of words is generated for each legal case, which constitutes as the foundation of following processing in clustering algorithms and similarity calculation. The drawback of this approach is that, it overlooks the fact that words possess context information in a document, except the weight they carry according to the document or the whole corpus. And this leads to the sacrifice of accuracy of recommendation when generating the recommendation list.

In seeing this, we bring in the technique of word embedding, it deals with documents as a matrix of words, i.e., the embedding methods learns a vector for each word following the way of *word2vec* by using a shallow neural network language model, specifically, a neural network architecture (the skip-gram model) that consists of an input layer, a projection layer, and an output layer to predict nearby words is adopted, and then a matrix is constructed for each document through these vectors. There are some plausible embeddings that are available, such as [14,15], [4], [17] and [24].

We are dealing with the task of inferring the penalty in two ways, the first is by clustering the legal cases, and then the penalty of the targeted case is obtained by the majority voting strategy, the other is through collaborative filtering, that is, finding the neighbors of the targeted case, and then again adopting the voting strategy. Technique details will be introduced in the following section that presenting our framework.

In summary, our work makes the following contributions:

- We are among the beginners in the judicial research of introducing the recommendation techniques to tackle the problem of penalty inferring, and a recommendation framework is presented to deal with this task, which consists of two sorts of methodologies, one is clustering methods, and the other is collaborative filtering.
- Embedding methods are adopted to improve the accuracy of the recommendation, which considers the context information of words in the document, rather than traditional methods such as TF-IDF that treats word as independent to document.
- Extensive experiments are conducted on a judicial dataset, and the results show that the recommendation framework is effective, and the embedding method outperforms all the traditional methods.

The rest of this paper is organized as follows. Section 2 reviews the related work. We present our framework in Sect. 3, as well as the embedding method in aiding the similarity calculation. We report the experimental results in Sect. 4. We conclude our paper in Sect. 5.

2 Related Work

In our work, we mainly compare the effects of four different methods on the penalty recommendations. These four methods are hierarchical clustering, kNN, k-mans and collaborative filtering methods. Following this, we introduce the related work in this section.

Hierarchical Clustering is a simple while practical clustering algorithm. In detail, there are two main types of Hierarchical Clustering which are top-down approach and bottom-up approach. When performing Hierarchical Clustering on tags, using the co-occurrence as the distance, Hierarchical Clustering can get the relative sets of tags with chosen distance, which can then be used as the 'topics'. In Hierarchical Clustering, each word or tag can only be allocated in only one set of tags or 'topics'.

There are two relatively novel algorithms of Hierarchical clustering method. BIRCH [25] (Balanced Iterative Reducing and Clustering Using Hierarchies) is mainly used when the amount of data is large, and the data type is numerical. In this method, the tree structure is first used to divide the object set, and then using other clustering methods to optimize these clusters. ROCK [8] (A Hierarchical Clustering Algorithm for Categorical Attributes) are mainly used on categorical data type.

The advantages of hierarchical clustering are mainly in the following aspects: (1) In this algorithm, the similarity between the distance and the rules is easy to define and less limited. (2) The algorithm does not need to pre-set the number of clusters. (3) The hierarchical relationship of the classes can be found. However, hierarchical clustering has the disadvantages of high computational complexity and sensitivity to singular values.

The K-means [27] algorithm is a single iterative clustering algorithm that partitions a given dataset into a user-specified number of clusters, i.e., K. It is simple to implement and run, relatively fast, easy to adapt, and common in practice. Via a clustering algorithm, data are grouped by some notion of "closeness" or "similarity". In K-means, the default measure of closeness is Euclidean distance. The K-means algorithm has some drawbacks: (1) it is very sensitive to its initial value as different ones may lead to different solutions; and (2) it is based on an objective function simply and usually solves the extreme value problem by a gradient method.

K-means is simple, which is easy to understand and implement with low time complexity. However, there are also the following defects: (1) K-means needs manually set the number of classes and it is sensitive to the initial value setting, so K-means++, intelligent K-means, and genetic K-means are used to make up for this deficiency; (2) K-means is very sensitive to noise and outliers, so K-medoids and K-medians are available to compensate for this defect; (3) K-means is only used for numerical data, not for categorical data.

KNN [9] is an instance-based learning algorithm. Although it is most often used for classification, it also can be used in estimation and prediction. Given a set of training data, a new data may be classified simply by comparing it to the most similar data in the training dataset. The process of building KNN classifier involves identifying k value, the number of the most similar classes to be considered in the training dataset. The process involves also measuring the similarity based on defining the distance function. The most commonly used distance function is Euclidean distance.

The KNN algorithm has high accuracy and is insensitive to outliers. It can be used both for classification and regression, including nonlinear classification. While the training time complexity is only $O(n)$. However, this algorithm requires a large amount of storage memory, which is one of its disadvantages. There is also the problem of sample imbalance (i.e., there are a large number of samples in some categories and a small number of other samples).

Collaborative filtering (CF) is a technique mostly used by recommender systems. There are two categories of Collaborative Filtering, user-based collaborative and item-based collaborative. User-based method first find out several similar users to the active user, and then aggregates items preferred by these similar users to generate a recommendation list. Item-based method has been proposed to address the scalability problem of user-based method. Many adjustments to original collaborative filtering have been proposed. Resnick et al. [21] computed user similarity based on ratings on items which have been co-rated by both users. Chee et al. [2] used the average ratings of a small group of users as default ratings. Inverse user frequency was proposed in [3], and the concept of inverse user frequency is that popular items preferred by most users should not be recommended.

3   Framework

We introduce the proposed framework in this section. As mentioned above, our solution of tackling the problem of penalty inferring consists of two sorts of methodologies, one is by clustering, and the other is through collaborative filtering. The framework is illustrated in Fig. 1.

From Fig. 1 we can see the overall workflow of penalty inferring process. We first perform some data preprocessing on the original documents. We perform the data preprocessing as segmentation, filtering the stop words and property labelling. We left the option of stemming because previous research [5] shows that it has less impact.

Segmentation is first performed on the textual description. For the Chinese document, the results of segmentation greatly affect the effect of the whole task, we compared a series of Chinese lexical analyzer and finally choose the *NLPIR* [29] to do the segmentation. Once we get the result of segmentation, we need to remove the stop words. Stop words are the set of common words such as *the*, *is*, *an* etc., these words are not important for our task of bug management and should be removed. Mainly, we adopt a commonly used Chinese stop word list in this step. Afterwards, we adopt the *NLPIR* to label the words. And an original feature space is obtained for each legal case in this step.

The left part of Fig. 1 depicts the methods that are based on feature vector, while the right part describes the method based on feature matrix.

For the left part, to best illustrate our idea in this step but without loss of generality, we choose TF-IDF to construct the vector space, *Term Frequency-Inverse Document Frequency* [22] is a numerical statistic model that is intended to show how important a word is to a document in a collection or corpus. We first

 **Workflow**

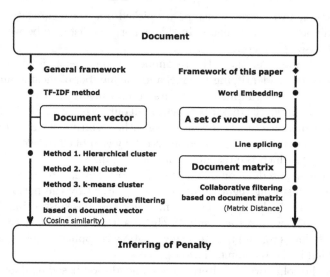

Fig. 1. The proposed framework for penalty inferring

compute the TF-IDF for all the words. That is, after choosing the feature words in the previous step, we should identify the weight of them, as different words vary in importance to legal cases, a numerical value is required to indicate the difference. Normally, words that are important to legal cases shall have a large value, and those less important will have a small value. In this paper, we adopt the TF-IDF to compute the numerical value for the feature words, the TF-IDF not only considers the frequency of words, it also takes into account the influence of the frequency of documents, the formula is defined as follows:

$$w_{ik} = tf_{ik} * \log \frac{N}{df_i}$$

in which, tf_{ik} is the frequency of feature word in legal case k, N is the number of legal cases, and df_i is the frequency of feature word in all the legal cases. In this way, we fulfill the process of constructing vector space for all the legal cases.

Based on the results of TF-IDF, clustering methods such as hierarchical clustering, kNN and k-means are conducted to cluster all the legal cases, and then the clusters are used to infer the penalty of target legal case, through a voting strategy.

The technique details are as follows. Hierarchical Clustering is a simple while useful clustering algorithm [23,26]. In detail, there are two main types of Hierarchical Clustering which are top-down approach and bottom-up approach. We adopt the bottom-up approach in this paper. Given a set of vectors $T = \{t_1, t_2, \ldots, t_n\}$, t_i denotes a certain vector. At first, each word is placed in a single cluster, so the initial set of clusters is

$$C = \{c_1 = \{t_1\}, c_2 = \{t_2\}, \ldots, c_n = \{t_n\}\}$$

in each iteration, two nearest clusters are picked out and aggregated together, using some distance measurements. In this paper, the distance between clusters is computed based on vectors.

For the kNN based method, the recommendation goes as follows. We build the KNN classifier by identifying k value, and the number of most similar classes to be considered in the training dataset, its process involves measuring the similarity based on defining the distance function, in our case, we adopt the Euclidean distance as our distance function. Once we get the resulting clusters, for the test case, we find its cluster, and then a voting strategy is applied to select an amount category, through which the recommendation is realized.

The K-means based recommendation is most similar to the kNN based method. Via a clustering algorithm, data are grouped by the notion of "closeness" or "similarity". In K-means, we measure the closeness using Euclidean distance. Similarly, we get the resulting clusters, then for the test case, we find the cluster it belongs to, and then a voting strategy is applied to select an amount category, which is the recommendation process.

While for the collaborative filtering recommendation based on document vector, the technique details are as follows. Collaborative filtering (CF) has been widely used in business situations. CF methods consist of User-Based CF, Item-Based CF and other variations. The main idea of CF is that similar items may share similar preferences. A higher similarity means they are much more similar. Given legal case list $U = \{u_1, u_2, \ldots, u_n\}$ and feature list $\{i_1, i_2, \ldots, i_m\}$, a case u can be represented by its feature vector $r_u = (r_{u,1}, r_{u,2}, \ldots, r_{u,m})$. The similarity between case u and v can be measured by the distance between r_u and r_v, using Cosine Similarity. After all the similarities are calculated, for the target test case, the most of k similar cases are used to predict the final amount of penalty of it, then just as other methods, a voting strategy is applied to generate the final category of penalty.

Word embedding is a method of using neural networks to calculate the degree of association of adjacent words. Compared with other methods, it not only considers word frequency, but also considers the context of the article. Therefore, when using this method, the relative distance of the article will be closer to reality.

For the right part of Fig. 1. The recommendation goes like this, the first stage is the same as the left part, i.e., segmentation, remove the stop words, and then the labelling. Afterwards, we adopt the iterative process of word embedding, through which a set of word vector is generated. Then by the line splicing, a matrix is composed for each document (legal case). In the next step, the collaborative filtering based on document matrix is used to generate the final recommendation list, the main difference is that it is based on the matrix distance, in this paper we use the word travel cost approach [13] to compute the distances between matrixes. And afterwards, for the target test case, the most of k similar cases are used to predict the final amount of penalty of it, then just as other methods, a voting strategy is applied to generate the final category of penalty.

4   Experiments

4.1   Experimental Settings

Datasets. We obtained the legal case dataset by crawler, which is encrypted. Each legal case is originally a textual description, which means it is not directly usable as input for our evaluation. Also, the explicit penalty amount is hidden from the law case, we can only get the category that amount falls into. Specifically, there are 8 categories of penalty amount, which are (0–1,000], (1,000–2,000], (2,000–3,000], (3,000–4,000], (4,000–5,000], (5,000–10,000], (10,000–500,000] and (500,000–max]. Then, through a filtering process, that is, we filter out those legal cases. A threshold of 30 feature words is adopted to remove law cases with no sufficient topic words. In our measurement, we only adopt the first 7 categories, and the number of each category is 500, which means we obtained 3,500 legal cases in total. In this way, we get a legal case dataset that is applicable for our comparison, as well as other comparative methods, i.e., hierarchical clustering, kNN, k-means and traditional collaborative filtering that is based on the feature vector.

Comparative Approaches. In this part, we present a set of comparative approaches for the evaluation of our proposed framework, in detail, we want to exploit the ability of the word embedding based method in inferring the penalty or the amount of a fine. We mainly compare its effectiveness with the following methods:

- *Hierarchical Clustering.* The clustering method based on hierarchical clustering.
- *kNN.* The clustering method based on kNN.
- *K-means.* The clustering method based on K-means.
- *Vector based CF.* This is the traditional feature vector based collaborative filtering method.
- *Martrix based CF.* This is the word embedding based collaborative filtering method proposed in this paper.

Evaluation Methods. To make an overall evaluation of the performance of our proposed PTM model, we first design the following setting, i.e., penalty inferring with PTM.

In the primary setting, we adopt 5/6 of the ground-truth dataset as the training set, and the rest 1/6 as the testing set. Furthermore, we apply different dividing strategies as to see the influence of ratios. For the testing, we just mark-off the class of penalty amount of testing law cases, and then the inferred class of amount is calculated through all the comparative methods.

For the ease of description, we define some notations as follows. Al is the number of test cases that are correctly clustered into the specific original cluster, Bl is the number of test cases that are incorrectly assigned to this certain cluster, and Cl is the number of test cases that are not assigned to their original cluster of certain cluster, while k is the number of clusters. And in this way, we define the *Precision* and *Recall* in the following way.

$$Precision = \frac{\sum \frac{Al}{Al+Bl}}{k} \qquad (1)$$

$$Recall = \frac{\sum \frac{Al}{Al+Cl}}{k} \qquad (2)$$

We expect both *Precision* and *Recall* to be good. However, they usually conflict with each other, improving one is usually at the expense of the other. Thus, F_1 measure is introduced to combine *Precision* and *Recall*. F_1 measure is calculated as follows:

$$F_1 = \frac{2 \times Precision \times Recall}{Precison + Recall} \qquad (3)$$

4.2   Experimental Results

Table 1 shows the result of matrix based collaborative filtering method, we mainly compared between different ratios, i.e., training set V.S testing set as the $2:1, 3:1, 4:1, 5:1$, and choose K nearest legal cases for the voting process, which K we set as 5, 10, 15, 20, 25.

From the result, we can infer that, with the increasing of training set, the effect of recommendation increases. Also, with more neighbors voting for the final penalty, the result tends to be more accurate.

We also compared with other methods mentioned in the previous section, the result is illustrated in Fig. 2.

Table 1. Matrix based collaborative filtering results

		5	10	15	20	25
2:1	*Precision*	0.2857	0.3091	0.3564	0.2219	0.3061
	Recall	0.2087	0.2027	0.2415	0.2743	0.3039
	F1	0.2412	0.2449	0.2879	0.2453	**0.3049**
3:1	*Precision*	0.3615	0.3768	0.3688	0.3355	0.3365
	Recall	0.1977	0.2506	0.2928	0.2608	0.2720
	F1	0.2556	0.3010	0.3264	0.2934	**0.3008**
4:1	*Precision*	0.4158	0.2441	0.3648	0.3254	0.4680
	Recall	0.2057	0.3000	0.2940	0.3175	0.2800
	F1	**0.2752**	**0.2619**	**0.3256**	**0.3214**	**0.3504**
5:1	*Precision*	0.4089	0.2793	0.4150	0.2784	0.4654
	Recall	0.2057	0.3047	0.3000	0.3422	0.3928
	F1	0.2738	0.2915	0.3482	0.3070	**0.4260**

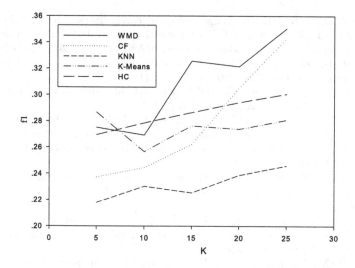

Fig. 2. Results of comparative methods in F1

From Fig. 2, we can see the proposed word embedding method outperforms all the comparative methods, in which it adopts the WMD distance measure, i.e., the Word Mover Distance. And, of all the methods, KNN works the worst.

5   Conclusions

In this paper, we introduced the penalty recommendation techniques for the judicial study, which provides efficient utility as to help judges on decision of the final penalty or amount of the fine. Specifically, we developed a word embedding based collaborative filtering method to generate recommendations for penalty. We conduct extensive experiments to evaluate the performance of the proposed framework on a real legal case dataset. The experimental results demonstrated the superiority of the proposed framework.

References

1. Boyer, M., Lewis, T.R., Liu, W.L.: Setting standards for credible compliance and law enforcement. Can. J. Econ./Rev. Can. D'économique **33**(2), 319–340 (2000)
2. Chee, S.H.S., Han, J., Wang, K.: RecTree: an efficient collaborative filtering method. In: Kambayashi, Y., Winiwarter, W., Arikawa, M. (eds.) DaWaK 2001. LNCS, vol. 2114, pp. 141–151. Springer, Heidelberg (2001). https://doi.org/10.1007/3-540-44801-2_15
3. Chowdhury, G.G.: Introduction to Modern Information Retrieval. Facet Publishing, London (2010)
4. Collobert, R., Weston, J.: A unified architecture for natural language processing: deep neural networks with multitask learning. In: Proceedings of the 25th International Conference on Machine Learning, pp. 160–167. ACM (2008)

5. Čubranić, D.: Automatic bug triage using text categorization. In: SEKE (2004)
6. Daughety, A.F., Reinganum, J.F.: Keeping society in the dark: on the admissibility of pretrial negotiations as evidence in court. RAND J. Econ. 203–221 (1995)
7. Earnhart, D.: Enforcement of environmental protection laws under communism and democracy. J. Law Econ. **40**(2), 377–402 (1997)
8. Guha, S., Rastogi, R., Shim, K.: ROCK: a robust clustering algorithm for categorical attributes. Inf. Syst. **25**(5), 345–366 (2000)
9. Guo, G., Wang, H., Bell, D., Bi, Y., Greer, K.: KNN model-based approach in classification. In: Meersman, R., Tari, Z., Schmidt, D.C. (eds.) OTM 2003. LNCS, vol. 2888, pp. 986–996. Springer, Heidelberg (2003). https://doi.org/10.1007/978-3-540-39964-3_62
10. He, T., Chen, Z., Liu, J., Zhou, X., Du, X., Wang, W.: An empirical study on user-topic rating based collaborative filtering methods. World Wide Web **20**(4), 815–829 (2017)
11. He, T., Yin, H., Chen, Z., Zhou, X., Sadiq, S., Luo, B.: A spatial-temporal topic model for the semantic annotation of POIs in LBSNs. ACM Trans. Intell. Syst. Technol. (TIST) **8**(1), 12 (2016)
12. Kilgour, D.M., Fang, L., Hipel, K.W.: Game-theoretic analyses of enforcement of environmental laws and regulations. JAWRA J. Am. Water Resour. Assoc. **28**(1), 141–153 (1992)
13. Kusner, M., Sun, Y., Kolkin, N., Weinberger, K.: From word embeddings to document distances. In: International Conference on Machine Learning, pp. 957–966 (2015)
14. Mikolov, T., Chen, K., Corrado, G., Dean, J.: Efficient estimation of word representations in vector space. arXiv preprint arXiv:1301.3781 (2013)
15. Mikolov, T., Sutskever, I., Chen, K., Corrado, G.S., Dean, J.: Distributed representations of words and phrases and their compositionality. In: Advances in Neural Information Processing Systems, pp. 3111–3119 (2013)
16. Milani, B.A., Navimipour, N.J.: A systematic literature review of the data replication techniques in the cloud environments. Big Data Research (2017)
17. Mnih, A., Hinton, G.E.: A scalable hierarchical distributed language model. In: Advances in Neural Information Processing Systems, pp. 1081–1088 (2009)
18. P'ng, I.P.: Strategic behavior in suit, settlement, and trial. Bell J. Econ. 539–550 (1983)
19. Polinsky, A.M., Shavell, S.: Punitive damages: an economic analysis. Harv. Law Rev. **111**, 869–962 (1998)
20. Ramos, J., et al.: Using TF-IDF to determine word relevance in document queries. In: Proceedings of the First Instructional Conference on Machine Learning, vol. 242, pp. 133–142 (2003)
21. Resnick, P., Iacovou, N., Suchak, M., Bergstrom, P., Riedl, J.: GroupLens: an open architecture for collaborative filtering of netnews. In: Proceedings of the 1994 ACM Conference on Computer Supported Cooperative Work, pp. 175–186. ACM (1994)
22. Salton, G., Wong, A., Yang, C.S.: A vector space model for automatic indexing. Commun. ACM **18**(11), 613–620 (1975)
23. Shepitsen, A., Gemmell, J., Mobasher, B., Burke, R.: Personalized recommendation in social tagging systems using hierarchical clustering. In: Proceedings of the 2008 ACM Conference on Recommender Systems, pp. 259–266. ACM (2008)
24. Turian, J., Ratinov, L., Bengio, Y.: Word representations: a simple and general method for semi-supervised learning. In: Proceedings of the 48th Annual Meeting of the Association for Computational Linguistics, pp. 384–394. Association for Computational Linguistics (2010)

25. Virpioja, S.: BIRCH: balanced iterative reducing and clustering using hierarchies (2008)
26. Wang, W., Chen, Z., Liu, J., Qi, Q., Zhao, Z.: User-based collaborative filtering on cross domain by tag transfer learning. In: Proceedings of the 1st International Workshop on Cross Domain Knowledge Discovery in Web and Social Network Mining, pp. 10–17. ACM (2012)
27. Wilkin, G.A., Huang, X.: K-means clustering algorithms: implementation and comparison. In: 2007 Second International Multi-Symposiums on Computer and Computational Sciences. IMSCCS 2007, pp. 133–136. IEEE (2007)
28. Yin, H., Wang, W., Wang, H., Chen, L., Zhou, X.: Spatial-aware hierarchical collaborative deep learning for POI recommendation. IEEE Trans. Knowl. Data Eng. 29(11), 2537–2551 (2017)
29. Zhou, L., Zhang, D.: NLPIR: a theoretical framework for applying natural language processing to information retrieval. J. Assoc. Inf. Sci. Technol. 54(2), 115–123 (2003)

Syntactic and Semantic Features Based Relation Extraction in Agriculture Domain

Zhanghui Liu[1,2], Yiyan Chen[1,2], Yuanfei Dai[1,2], Chenhao Guo[1,2], Zuwen Zhang[1,2], and Xing Chen[1,2(✉)]

[1] College of Mathematics and Computer Science, Fuzhou University,
Fuzhou 350116, China
`chenxing@fzu.edu.cn`
[2] Key Laboratory of Network Computing and Intelligent Information Processing,
Fuzhou University, Fuzhou 350116, China

Abstract. Relation extraction plays an important role in many natural language processing tasks, such as knowledge graph and question answering system. Up to the present, many of the former relation methods work directly on the raw word sequence, so it is often subject to a major limitation: the lack of semantic information, which leads to the problem of the wrong category. This paper presents a novel method to extract relation from Chinese agriculture text by incorporating syntactic parsing feature and word embedding feature. This paper uses word embedding to capture the semantic information of the word. On the basis of the traditional method, this paper integrates the dependency parsing, the core predicate and word embedding features, using the naive Bayes model, support vector machine (SVM) and decision tree to build experiment. We use the websites knowledge base to construct a dataset and evaluate our approach. Experimental results show that our proposed method achieves good performance on the agriculture dataset. The dataset and the word vectors trained by Word2Vec are available at Github (https://github.com/A-MuMu/agriculture.git).

Keywords: Relation extraction · Word embedding
Syntactic feature · Semantic feature · Agriculture field

1 Introduction

Relation extraction plays a key role in the field of information extraction, and can be widely used in various Natural Language Processing applications, such as information extraction, knowledge base completion, and question answering. Relation extraction refers to automatically identify which semantic relation between two marked entities. For instance, given the input:

This work is supported by the Guiding Project of Fujian Province under Grant NO. 2018H0017 and the National Key R&D Program of China under Grant No. 2017YFB1002000.

[鳄梨$_{e_1}$]原产于[中美洲$_{e_2}$]，为人所知已有好几个世纪了。

the target is to identify the relation between "鳄梨" and "中美洲" as "原产地".

In the field of Chinese agriculture, Wang et al. [1] put forward the method for recognition of chinese agriculture named entity. Hu et al. [2] proposed a question answering system in agriculture. However, those work do not give specific method of agricultural relation extraction and systematic agricultural dataset. Therefore, in this paper the construction of dataset in the field of agriculture was carried out before relation extraction. Previous work has shown that the relation extraction method basing on statistical learning has become the mainstream, and the features based method has attracted much attention due to its superior performance. In traditional methods, lexicon, entity types and dependency relations are used. Words represented by dictionary indexes do not contain semantic information and cannot express the relationship between two semantically similar words, which leads to inaccurate classification. To address the problem, we use word embedding to capture the semantic information of words. Word embedding [3] denotes the word as a dense low-dimensional real value vector and serializes discrete word semantics. In addition, A large number of natural language processing tasks have proved that the POS (Part of Speech) tagging, context, NER (Named Entity Recognition) tagging and dependency syntactic features contain rich semantic information, which is conducive to improving the relation extraction model. Therefore, these traditional features and word embedding are adopted for the extraction of relation in this paper.

The contribution can be sum up as follows:

- We propose a Chinese relation extraction dataset in the field of agriculture.
- We propose a relation extraction method in the field of agriculture, and the experimental results on the relation extraction of the dataset in the agricultural field shows that the method obtains a good discriminant performance.

2   Related Work

The choice of features has a great influence on the performance of relation extraction, which leads to the focus of research on how to acquire linguistic features such as lexical, syntactic and semantic accurately.

Kambhatla et al. [4] comprehensively considered the characteristics of entity words, instance types, entity reference, overlap, dependency tree and parsing tree. The relation classifier of the maximum entropy model was raised. Jiang et al. [5] systematically studied the feature space in relation extraction, which showed that the basic characteristics of feature subspaces can effectively improve the performance of relation extraction. Li et al. [6] presented a method of relation extraction combining location and semantic features. In the way of extracting syntactic and semantic features, Guo et al. [7] integrated features such as dependency syntactic relations, core predicates and semantic role labeling. Chen et al. [8] used Omni-word features and soft constraints to achieve Chinese relation extraction. He et al. [9] put forward a method of relation extraction based on syntactic analysis and relational dictionary. Zhang et al. [10] proposed an entity relation extraction method based on embedding features of fusion words, adding features with semantic information into vectors. Ma et al. [11] presented a feature combination method of Subject-Action-Object (SAO) structure and dependency analysis to extract semantic relations between entities. On the basis of the above work, this paper adds the word embedding feature to obtain the semantic information of the word, and using the syntactic feature to express the grammatical information of the sentence.

3   The Proposed Method

Figure 1 depicts the framework of our work. The content in dashed box is the core of this method. Firstly, web crawler is constructed to collect agricultural data from the web. The original corpus is manually annotated to construct the dataset. Then, basing on the results of lexical and syntactic analysis by Stanford NLP, the triple groups are generated. A series of features such as word embedding, POS tagging and dependency parsing are extracted. Finally, using feature vectors as input for training the model.

In this paper, we use feature vector based machine learning method to extract relations. Feature vector is a numerical representation of the triple instance. In other words, an instance (e_1, e_2) is transformed into a feature vector x, where x_i is the $i\text{-}th$ element of the N dimensions feature vector x. For multi-classification problems $y_i \epsilon \{y_1, y_2, \cdots, y_n\}$, the purpose of machine learning algorithm basing on feature vectors is to learn a classification function f by using a given set of data $(x_1, y_1), (x_2, y_2), \ldots, (x_n, y_n)$, so that for a given new feature vector x', f can correctly classify it, namely $f(x') = y'$. In this paper, we use a wealth of lexical, syntactic and semantic features.

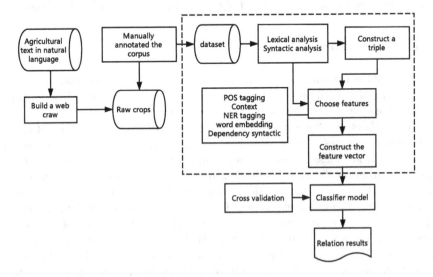

Fig. 1. The framework of relation extraction method in agricultural field.

Features of Lexical Level. In order to describe the characteristics of lexical level more clearly, it is subdivided into POS tags and Named Entity Recognition (NER) types. The part of speech features mainly consider the POS tags of entities (e_1, e_2), the context (w_{pre1}, w_{post1}) of entity e_1, and the context (w_{pre2}, w_{post2}) of entity e_2. NER refer to the types of entity, such as *PERSON*, *ORGANIZATION*, *LOCATION* etc.

Features of Semantic Level. We use word embedding to represent entities and entity's semantic. A sentence S can be regarded as a collection of multiple words, expressed as $S = \{w_1, w_2, \ldots, w_j, \ldots w_m\}$, where w_j is the *j-th* word in the sentence. The relation R can be described as $(w_{pre1}, e_1, w_{post1}, w_{pre2}, e_2, w_{post2})$. The semantic information of entity (e_1, e_2) and entities context (w_{prei}, w_{posti}) are distributed into dense, low-dimensional real value vectors by means of the word embedding matrix.

Feature of the Syntactic Analysis. The syntax level feature mainly refers to the information contained in the dependency parsing tree. Those kind of features include dependency object of the entity and the distance between the entity and the core predicate. Generally speaking, the dependency parsing tree of a sentence is derived from its parsing tree, including which word the entity dependency and their dependency relation. Specifically, the dependency words are defined as de_1 and de_2 respectively. Word embedding is used to map de_1 and de_2 into vectors to get their semantic information. In addition, a large number of experiments on the results of syntactic analysis show that core predicates play a key role in acquiring entity boundaries and undertaking entity relations. Therefore, the shortest distance between the entity and the core predicate in the dependency parsing tree are recorded as a feature.

4 Experiments

4.1 Dataset and Evaluating Metric

We evaluate our method on a new dataset, which we constructed based on the agriculture data from the Interactive Encyclopedia[1], the China Agricultural Information Network[2], the China Agricultural Network[3] and other websites. These websites contain rich and comprehensive chinese agricultural information, which are relatively authoritative. Firstly, we construct the web crawler to collect agricultural data. Then, the original corpus is manually annotated to construct the dataset. Finally, basing on the results of lexical and syntactic analysis by Stanford NLP, the triple groups $(e_1, e_2, relation)$ are generated. These sentences and triples form the dataset.

According to the Chinese agricultural text, we defined 7 types of relation: source area relation between crop and location, the inclusion relation between

---

[1] http://www.baike.com/.
[2] http://www.agri.cn/.
[3] http://www.zgny.com.cn.

crops and nutrients, the alias relation between the agricultural entities, ancestral relations between agricultural entities, the honorary title relation of the semantically identical entities, relation between crop entities and medical value and no-Relation. Because the relational extraction process is regarded as a classification process, the evaluation method adopts the conventional accuracy rate, recall rate and F_1-score.

4.2 Experimental Settings and Results

The dataset used in this paper is composed of 24764 tuples. Feature selection was applied to construct the text feature vectors. In order to prove the effectiveness of the selected features in the dataset of agriculture domain, 3 sets of comparative experiments were carried out as shown in Table 1. The first and third groups of experiments were used to demonstrate the role of syntactic features in relation extraction. The second and third groups of experiments were set up to emphasize the influence of semantical features on relation extraction.

Table 1. Feature sets

Group	Feature sets
1	POS, NER, dependency parsing tree
2	POS, NER, word embedding
3	POS, NER, dependency parsing tree, word embedding

Table 2 shows the results of training which uses traditional and syntactic features. The precision of SVM is 95%, and Decision tree is 96%. Both are of the same performance. However, precision of multiple relation categories (for example, value) is relatively low, indicating that the model trained by this group of features can not be well categorized for these types of relations. It can be seen from Table 2 that although the precision of Bayesian algorithm is 96%, the recall rate is only 30%. This is due to the imbalance in the number of relational instances.

Table 2. The training result of adding dependency parsing features.

Relation type	SVM			naive Bayes			Decision tree		
	Precision	Recall	F_1-score	Precision	Recall	F_1-score	Precision	Recall	F_1-score
Alias	0.61	0.93	0.74	0.79	0.08	0.14	0.73	0.93	0.82
Source area	0.88	0.86	0.87	0.90	0.83	0.86	0.88	0.91	0.89
Ancestor	0.64	0.53	0.58	0.85	0.08	0.14	0.75	0.61	0.67
Value	0.00	0.00	0.00	0.03	0.35	0.06	0.30	0.12	0.18
Honnorary title	0.00	0.00	0.00	0.00	1.00	0.00	0.00	0.00	0.00
Include	0.38	0.20	0.26	0.00	0.00	0.00	0.79	0.57	0.66
no-Relation	1.00	0.98	0.99	1.00	0.31	0.47	0.99	0.99	0.99
Avg/total	0.95	0.95	0.95	0.96	0.30	0.45	0.96	0.96	0.96

Table 3. The training result of adding semantic features.

Relation type	SVM			naive Bayes			Decision tree		
	Precision	Recall	F_1-score	Precision	Recall	F_1-score	Precision	Recall	F_1-score
Alias	0.61	0.91	0.73	0.60	0.11	0.18	0.83	0.89	0.86
Source area	0.85	0.89	0.87	0.21	0.98	0.34	0.93	0.86	0.89
Ancestor	0.58	0.25	0.35	0.00	0.00	0.00	0.72	0.68	0.70
Value	0.33	0.03	0.06	1.00	0.04	0.07	0.62	0.42	0.50
Honnorary title	0.00	0.00	0.00	0.00	0.00	0.00	0.67	0.22	0.33
Include	0.70	0.61	0.65	0.81	0.29	0.43	0.83	0.71	0.77
no-Relation	0.99	0.99	0.99	1.00	0.98	0.99	0.99	1.00	1.00
Avg/total	0.95	0.95	0.95	0.92	0.89	0.89	0.97	0.97	0.97

Table 4. The training result of adding all features.

Relation Type	SVM			naive Bayes			Decision tree		
	Precision	Recall	F_1-score	Precision	Recall	F_1-score	Precision	Recall	F_1-score
Alias	0.95	0.99	0.97	0.80	0.87	0.83	0.90	0.90	0.90
Source area	0.94	0.97	0.96	0.79	0.92	0.85	0.91	0.91	0.91
Ancestor	0.93	0.93	0.93	0.75	0.79	0.77	0.83	0.86	0.86
Value	1.00	0.87	0.93	0.83	0.56	0.67	0.76	0.79	0.79
Honorary title	1.00	0.67	0.80	0.19	1.00	0.32	0.00	0.00	0.00
Include	0.96	0.89	0.92	0.68	0.79	0.73	0.81	0.80	0.80
no-Relation	1.00	1.00	1.00	1.00	0.98	0.99	0.99	0.99	0.99
Avg/total	0.99	0.99	0.99	0.97	0.96	0.97	0.98	0.98	0.98

Table 3 shows the result of using traditional features and word embedding feature for training. We can find that in the three models, the decision tree algorithm has the best performance and the average precision is 97%. The precision and F_1-score of SVM model is the same as that of Table 2, which shows that the influence of dependency parsing and word embedding on SVM model is similar. Comparing with Tables 2 and 3, it can be found that the Bayes model of word embedding feature increased by 50% in recall rate, indicating that word embedding is beneficial to improve the performance of Bayes model. According to Tables 2 and 3, word embedding has greater effect on improving the efficiency of relation classification.

Table 4 shows the results of training after adding dependency parsing and word embedding to traditional features. From the table, we can clearly be aware of that the results of the three models have been improved, the precision and F_1 reaching above 97%. The precision of the SVM model is the highest, reaching 99%. We can conclude that SVM model can obtain more accurate classification results for the relation with fewer training instances. By contrast, decision trees and biases are less effective.

By comparing Tables 2, 3 and 4, it can be concluded that when word embedding and dependency parsing are added in the experiment, the model achieves the best classification effect. The results show that the proposed method has certain significance.

5   Conclusion

In this paper, we present a word embedding and dependency parsing based method relation extraction method for Chinese agriculture text. We use the website knowledge base to construct a dataset of multiple relations per sentence and to evaluate our approach. Experimental results prove the effectiveness of our method.

References

1. Wang, C.Y., Wang, F.: Study on recognition of Chinese agricultural named entity with conditional random fields. J. Agric. Univ. Hebei **37**(1), 132–135 (2014)
2. Hu, D.P.: The research of question analysis based on ontology and architecture design for question answering system in agriculture. Ph.D. thesis, Chinese Academy of Agricultural Sciences (2013)
3. Chen, E., Qiu, S., Chang, X., Fei, T., Liu, T.: Word embedding:continuous space representation for natural language. J. Data Acquis. Process. **29**(1), 19–29 (2014)
4. Kambhatla, N.: Combining lexical, syntactic, and semantic features with maximum entropy models for extracting relations. In: ACL 2004 on Interactive Poster and Demonstration Sessions, p. 22 (2013)
5. Jiang, J., Zhai, C.X.: A systematic exploration of the feature space for relation extraction. In: Human Language Technology Conference of the North American Chapter of the Association of Computational Linguistics, Proceedings, 22–27 April 2007, Rochester, New York, USA, pp. 113120 (2007)
6. Li, H.G., Wu, X.D., Li, Z., Wu, G.G.: A relation extraction method of Chinese named entities based on location and semantic features. Appl. Intell. **38**(1), 115 (2013)
7. Guo, X., He, T., Hu, X.: Chinese named entity relation extraction based on syntactic and semantic features. J. Chin. Inf. Process **28**(6), 183–189 (2014)
8. Chen, Y., Zheng, Q., Zhang, W.: Omni-word feature and soft constraint for Chinese relation extraction. In: Meeting of the Association for Computational Linguistics, pp. 572–581 (2014)
9. He, Y., Lyu, X., Xu, L.: Extraction of non-taxonomic relations between ontological concepts from Chinese patent documents. Comput. Eng. Des. **38**(1), 97–102 (2017)
10. Zhang, Q., Guo, H., Zhang, Z.: Extracting entity relationship with word embedding representation features. Data Anal. Knowl. Discov. **1**(9), 8–15 (2017)
11. Ma, X., Zhou, C., Lv, X.: Extraction of non-taxonomic relations based on SAO structure. Comput. Eng. Appl. **54**(8), 220–225 (2018)

Classification for Social Media Short Text Based on Word Distributed Representation

Dexin Zhao, Zhi Chang[✉], Nana Du, and Shutao Guo

Tianjin University of Technology, Tianjin 300384, China
qiqiharxin@163.com, changzhi1123@outlook.com, cloud-dn@163.com,
alsonsmileshine@hotmail.com

Abstract. According to the brief and meaningless features of the social media content, we propose a text classification algorithm based on word vectors, which can quickly and effectively realize the automatic classification of the short text in social media. In view of the lack of word order and position considerations in the Word2vec model, we combine the Word2vec trained word vector with the convolutional neural network (CNN) model to propose SW-CNN and WW-CNN classification algorithms. The methods are evaluated on the three different datasets. Compared with existing text classification methods based on convolutional neural network (CNN) or recurrent neural network (RNN), the experimental results show that our approach has superior performance in short text classification.

Keywords: Text classification · Word2vec · Word vector
Social media

1 Introduction

With the rapid development of mobile terminals, a large number of social software has become an indispensable part of people's lives. Although people's lives become more convenient, the social media data generated by the users have made it difficult to extract useful information to fully satisfy the needs of people. Most of the social media produces short texts that look like spoken texts and more casual than traditional text. Social media short text data stream contains a lot of noise and unstructured information, slang and lack of words. These have brought great difficulty to the classification of the short text in social media.

In the traditional text classification problems, there have been a lot of related research, such as classification method based on rule characteristics [1,2], a classification method combined with SVMs and Naive Bayesian [3], classification method of building dependency tree based on the conditional random field [4]. In recent years, deep learning has shown good results in image processing and Natural Language Processing tasks, including convolutional neural network (CNN)

© Springer Nature Switzerland AG 2018
X. Meng et al. (Eds.): WISA 2018, LNCS 11242, pp. 259–266, 2018.
https://doi.org/10.1007/978-3-030-02934-0_24

[5,6] and recurrent neural network (RNN) [7]. In general, CNN is used to deal with the problem of image classification, convolution and pooling structure can extract various textures and features in images and finally integrate and output information with the fully connected network. In short text analysis tasks, due to the compact structure and independent expression of meaning, CNN becomes possible in dealing with this kind of problem. The author Yoon Kim [5] based on distributed word vector to represent text by using a simple CNN network to classify text from two aspects of specific tasks and static vectors and achieved excellent results. Rie Johnson and Tong Zhang [8] take into account the problem of word order, they improved the word bag model and classified the text by the convolution layer of CNN and proposed the seq-CNN and bow-CNN models. [11] propose a new neural language model incorporating both word order and character order in its embedding. Chunting Zhou *et al.* [9] and Ji Young Lee *et al.* [10] studied a combination of CNN and RNN respectively. The former uses the combination of CNN and RNN as text classifier; the latter uses CNN and RNN to train the vectors and uses the common artificial neural network (ANN) as a classifier. These classification methods using deep learning have also achieved excellent results.

However, since social media short text has the characteristics of short length, huge interference, irregular, features sparseness and other characteristics, these traditional text feature representation methods based on vector space model are difficult to obtain satisfying text representation results. Although word vector representation technology based on deep learning can well express the grammatical and semantic relations of words, it loses the word order information in training process and does not take into account the influence of word order on text meaning.

Based on the above problems, this paper begins with words feature representation, studies on social media short text classification based on Word2vec model and Convolution Neural Network model (CNN). Given the lack of word order and position considerations in the Word2vec and the CNN model, we propose w-Word2vec(WW) and seq-Word2vec(SW) algorithm. Finally, it has input the word vector that related to the word order and the position to the CNN model for training.

2   Model

2.1   Word Vector Model Based on Word Order

In previous research, Rie Johnson and Tong Zhang [8] have improved the one-hot expression according to the word order, but because the one-hot expression does not consider the semantic information between words like word2vec, it's too redundant and sparse. As a result, we use their idea to improve the word2vec model based on the problem of word order information.

We add the word vectors of the two words that are connected before and after about word order problem, as shown in Fig. 1, the sentence "Please do not tell me that", we have word vectors of the adjacent words "please" and "do", "do" and "not", "not" and "tell", and so on, adding two word vectors together, and then implementing a consideration of word order by this way. We assume that a sentence is composed of, and the formula is as follows:

$$x_w = e(w_i) + e(w_{i+1}), \begin{cases} w_i \in c \\ i \neq k \end{cases} \qquad (1)$$

Among them, the c represents all the context words of the word w, and $e(w_i)$ represents the word vector of the word w_i. This method is called seq-word2vec.

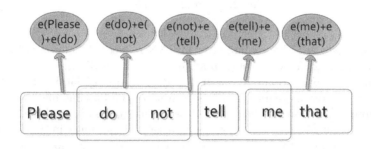

Fig. 1. Seq-word2vec model.

In addition to the above improvement, we also consider another situation: in two adjacent words, the word of the front is the vector of its own word, and the latter is not the first word because of its position. According to the previous example, the sentence "Please do not tell me that", the original word vector of the word "please" plus the original word vector of the word "do" that represents the word vector of the word "please", the original word vector of the word "do" plus the original word vector of the word "not" that represents the word vector of the word "do". Now we consider the position factor, adding weight to the word vector, the update formula is as follows:

$$x_w = e(w_i) + ae(w_{i+1}), \begin{cases} w_i \in c \\ i \neq k \end{cases} \qquad (2)$$

Among them, a is the weight of the second word in the two adjacent words, it represents the important relationship between two adjacent words. The method structure shown in the following figure, which defines the method named w-word2vec.

2.2 CNN for Text Classification

Now we consider the application of CNN in text classification. CNN can be widely applied to text classification, CNN model is similar to the N-gram model, and filter window in CNN can be seen as a N-gram method. Besides, the use of

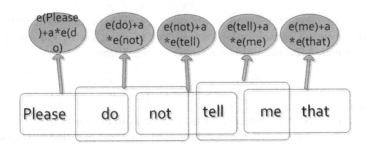

Fig. 2. w-word2vec model.

convolution layer and the pooling layer in CNN reduces the number of training parameters and extracts information at the higher level (Fig. 2).

In 2014, the application of CNN proposed by Kim *et al.* [5]. For text classification tasks. In recent years, CNN has been used as a baseline method, we will make improvements on this basis and classify short texts of social media.

If a document $D = (s_1, s_2, \ldots, s_k)$, s represents the sentence in the document, the number of different words is N, and the word vector dimension of every word is d. Since each sentence has a different length, therefore, we choose a unified sequence of words with a sentence length K, document D has N words without repetition, the length of each sentence is K, and the word vector dimension of each word is d, we set the word vector of the i-th word as e_i. We use the word vector of each sentence as the input of the CNN model, that is, the matrix of the $k \times d$ dimension.

2.3   Text Classification Model Based on Word Vector

Previous work shows that the word2vec model and the CNN model has a good effect on Natural Language Processing, but the word2vec model does not take into account the word order. Therefore, for the characteristics of social media short text own short length, huge interference, irregular, features sparseness, we combine CNN model and improved word vector to classify social media short texts. In next section, we will test the methods proposed in this paper. This section mainly introduces the model of the combination of word vectors considering word order with CNN. The model structure is shown in Fig. 3.

First, considering the input text data in CNN model are the word vector of each sentence, a sentence is expressed as matrix of dimension $k \times d$, K is the number of words in this sentence, D is the word vector dimension of every word, then the matrix input to CNN. We change the word vector to the two word vectors considered in word order in the previous chapter, respectively.

We use the improved word vector matrix as the input matrix of CNN, then we can see the convolution layer consisting of multiple filters. Here, we refer to the setting of the filter by Kim *et al.* [5], the size of the three filters are 3, 4, 5 respectively. They represent 3 words, 4 words, and 5 words, respectively, and

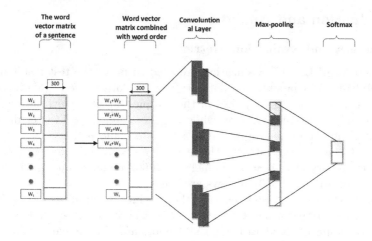

Fig. 3. SWCNN model.

according to the size of different filters, convolution operations are carried out with a variety of filters, and the corresponding local features of different sizes are obtained. In Natural Language Processing, some crucial keywords can reflect the category of this sentence. In this paper, we also choose the best feature of each feature dictionary, the features have been extracted into the pool layer, through the use of the pool layer, the model further selects the most representative features of each feature information of the dictionary. We select the max-pooling, which is to record only the maximum value of each feature dictionary. The pool layer can reduce not only the number of model parameters but also filter some noise. In this case, generating a string of single variable feature vector from each dictionary, and then splicing these characteristics obtained by pooling and form a long feature vector. Then, the feature vector enters the full connection layer and add dropout regularization during the training process of the fully connected layer to prevent overfitting. Finally, using the regularized feature vector as input and sentences classification by softmax classification function, where $softmax(z_i) = e^{z_i}/\sum_j e^{z_j}$. Finally, categorizing the sentence to the category with the highest probability to finish the text classification.

2.4 Loss Function

Using the model outputs we can define the loss function. The loss is a measurement of the error our network makes, and our goal is to minimize it. The standard loss function for categorization problems it the cross-entropy loss.

$$L(p, q) = -\sum_x p(x) \log q(x) \tag{3}$$

where $p(x)$ represents the true probability distribution and $q(x)$ represents the probability distribution of the prediction.

3   Evaluation and Analysis

3.1   Dataset and Evaluation Metric

Binary and Multi-Label Classification are used to prove the test result in this paper. Multi-Label Classification dataset: Twitter data is used as data sources in this test, a total of 10539 pieces in three topics, among which 90% of each classification was used for training while the remaining 10% Twitter data was used for testing. The three topics are disaster, auto, and politics.

Binary Classification dataset: Movie Review data from Rotten Tomatoes are used for binary classification. It includes 10662 reviews, among which 5331 reviews are positive and 5331 reviews are negative.

Accuracy is used to evaluate effect of the classifier. Accuracy is for evaluation of all data. No matter which class it is, as long as it is correct, it is molecule with total amount of the data being the denominator. Or we can say, accuracy is an evaluation of the correct rate on the classifier as a whole.

$$Accuracy = (TP + TN)/(P + N) \qquad (4)$$

TP is the number of cases being correctly classified as the positive case, TN is the number of cases being correctly classified as the negative case; P and N represent the amount of all positive and negative cases.

3.2   Experimental Design

(1) Filter all punctuation marks and special characters.
(2) Managed all non-standard text in original source, such as messy code or English character of full width. The non-standard text in the original source should be converted. There is no need to separate word because the text we chose is in English which uses space to separate word.
(3) Word embedding training was carried out with word2vec model in gensim in python to input continuous sense for training to generate word embedding with the size of 300.
(4) Expand the processed sentences in Step 2 to the longest to construct glossary index and then reflect each word to an integer with the size of the glossary to make each sentence an integer embedding.
(5) Correspond integer of each sentence to the trained word embedding in Step 3 to make a word embedding matrix and then process word embedding according to seq-word2vec and w-word2vec to generate new word embedding matrix.
(6) Finally, input the word embedding matrix generated in Step5 into CNN for training to get the final classification.

3.3   Experimental Results and Analysis of the Result

The comparison methods used in this paper are all deep learning related methods. The experiment test six different classification methods, including the seq-word2vec+CNN and w-word2vec+CNN methods proposed in this paper, which we referred to as SW-CNN and WW-CNN, respectively. Moreover, the performance of these six methods results are shown below:

Table 1. Accuracy.

Model	Binary classification	Multi classification
Char-CNN	72.7%	94.7%
FastText	79.9%	96.2%
LSTM	75.3%	97.3%
CNN	72.8%	95.4%
SW-CNN	75.2%	98.1%
WW-CNN	75%	98.7%

The second column in Table 1 is the accuracy of movie review classification with the six classification approaching. We can see that the accuracy for Char-CNN is 72.7%, the lowest among the six, followed by 72.8% of CNN. The best is the FastText with accuracy of 79.9% while that of SW-CNN proposed in this paper is 75.2%. Accuracy of WW-CNN, 75.0%, is slightly lower than that of SW-CNN. The thrid column in Table 1 is the comparison of the six classification approaches in three classes of Twitter data sources, from which we can see that the two approaches proposed in this paper enjoy the highest accuracy among the rest, 2.7% and 3.3% higher than that of CNN text, followed by LSTM, with FastText being the worst. Accuracies of the six approaches are all higher than 90%.

Compared with traditional classification approaches, our two approaches proposed in this paper had good improvement on classification of short text. Both improved word embedding via word order relationship. Moreover, the WW-CNN model takes into account the importance of location, and then tests on two datasets show that the method proposed in this paper is better than the previous CNN method. In terms of Twitter data sources, their performance compared with the rest four approaches is exhibited, with the accuracy of WW-CNN being slightly higher than that of the SW-CNN. For other classification method, we need to make specific judgments based on different datasets.

The reason for the above is summarized as follows: First, embeddings learned from the very beginning were not used, which was replaced by that trained in Word2vec. The processed embeddings are of higher group characters than those random ones. Second, the original word embedding didn't consider the relationship between words while the text considered it and the major factors of different position are considered in WW-CNN. Finally, text processed by CNN can express text characteristics better.

4   Conclusion

In this paper, we use a word vector considering word order relation as input, and then use CNN to classify text. Different from the traditional word vector model, although the word2vec model contains the semantic features between words, it can better express the text features, but it does not take into account the relationship between word order. We propose SW-CNN and WW-CNN model to improve the expression of word vectors. Experiments show that the proposed method (SW-CNN) algorithm and WW-CNN algorithm) have increased the accuracy of 2.7% and 3.3% compared with that of traditional CNN algorithm on Multi-label classification of short texts in social media.

References

1. Silva, J., Coheur, L., Mendes, A.C.: From symbolic to sub-symbolic information in question classification. Artif. Intell. Rev. **35**(2), 137–154 (2011)
2. Huang, Z., Thint, M., Qin, Z.: Question classification using head words and their hypernyms. In: EMNLP, pp. 927–936 (2008)
3. Wang, S., Manning, C.D.: Baselines and bigrams: simple, good sentiment and topic classification. In: Association for Computational Linguistics, pp. 90–94 (2012)
4. Delaye, A., Liu, C.L.: Text/Non-text classification in online handwritten documents with conditional random fields. Commun. Comput. Inf. Sci. **321**(53), 514–521 (2012)
5. Kim, Y.: Convolutional neural networks for sentence classification. In: Conference on Empirical Methods on Natural Language Processing (2014)
6. Kalchbrenner, N., Grefenstette, E., Blunsom, P.: A Convolutional neural network for modelling sentences. Eprint Arxiv. (2014)
7. Socher, R., Huval, B., Manning, C.D., et al.: Semantic compositionality through recursive matrix-vector spaces. In: Joint Conference on Empirical Methods in Natural Language Processing and Computational Natural Language Learning, pp. 1201–1211 (2012)
8. Johnson, R., Zhang, T.: Effective use of word order for text categorization with convolutional neural networks. Eprint Arxiv (2014)
9. Zhou, C., Sun, C., Liu, Z., et al.: A C-LSTM neural network for text classification. Comput. Sci. **1**(4), 39–44 (2015)
10. Lee, J.Y., Dernoncourt, F.: Sequential short-text classification with recurrent and convolutional neural networks. 515–520 (2016)
11. Trask, A., Gilmore, D., Russell, M.: Modeling order in neural word embeddings at scale. Comput. Sci., 2266–2275 (2015)

CNN-BiLSTM-CRF Model for Term Extraction in Chinese Corpus

Xiaowei Han[✉], Lizhen Xu, and Feng Qiao

School of Computer Science and Engineering, Southeast University,
Nanjing, China
220161557@seu.edu.cn

Abstract. Neural networks based term extraction methods regard term extraction task as sequence labeling task. They make better modeling of natural language and eliminate the dependence of traditional term extraction methods on artificial features. CNN-BiLSTM-CRF model is proposed in this paper to minimize the influence of different word segmentation results on term extraction. Experiment results show that CNN-BiLSTM-CRF has higher stability than the baseline model for different word segmentation results.

Keywords: Term extraction · Recurrent neural networks · Word embedding
Convolutional neural networks

1 Introduction

Automatic Term Extraction (ATE) is one of fundamental tasks for data and knowledge acquisition. Term Extraction aims to extract domain terminology which contains words and collocations from domain corpus, usually followed by several complex Nature Language Processing (NLP) tasks such as information retrieval (LingPeng et al. 2005) and ontology construction (Biemann and Mehler 2014). Typical ATE task consists of several sub tasks: word segmentation, word embedding, feature learning and term recognition.

In recent few years, it is a common method to transform term extraction task into sequence labeling task. Taking a given sentence as a sequence of words, the sequence labeling task identifies the semantic role of each word in the source sentence. Sequence labeling is considered as supervised machine learning problems; actually, researchers usually treat it as a multi-step classification task. Hidden Markov model and conditional random field model are commonly used machine learning models to implement sequence labeling. End-to-end system using deep neural networks solves the problem of artificial feature extraction in traditional machine learning methods, makes feature extraction is no longer a bottleneck in machine learning tasks. In particular, deep bi-directional long short-term memory model can well analyze the contextual features of sentences as the basis of conditional random field tasks (Zhou and Xu 2015).

In this paper, CNN-BiLSTM-CRF is proposed to identify terminology that contains more than one word but has been segmented into continuous parts by word segmentation. Two word embedding models, GloVe and word2vec, are compared in this paper.

X. Meng et al. (Eds.): WISA 2018, LNCS 11242, pp. 267–274, 2018.
https://doi.org/10.1007/978-3-030-02934-0_25

2   Related Work

2.1   Word Embedding

Word2vec (Mikolov et al. 2013) is the most commonly used word embedding model; it uses sliding window method to learn the local context features of words. That means there is a shorter distance between vectors trained form words that have similar context in the corpus. Matrix decomposition methods decompose the co-occurrence matrix of words and documents in corpus to get vector representation of words and documents respectively. Vectors obtained by this method contain the overall statistics information of words. GloVe (Pennington et al. 2014) considers the combination of both; it uses sliding windows to extract co-occurrence words in word context.

According to (Levy et al. 2015), GloVe is not necessarily better than word2vec in practical application; word2vec is state-of-the-art and a simple word2vec model is trained faster. But in sequence labeling task, co-occurrence information is considered to be a very important overall feature, so the performance of two word embedding models is compared in this paper.

2.2   Sequence Labeling and Term Extraction in Chinese Corpus

Instead of traditional methods (Collobert et al. 2011) introduced a neural network model for sequence labeling which consists of word embedding layer, convolutional layers and CRF layer. The convolutional layers learn features of words in the fixed context window; it has limitations because interrelated words in a certain sentence may not be continuous. Graves and Wayne (2014) implemented Recurrent Neural Networks (RNN) to Sequence generation tasks, RNN can effectively model sequence problems, but too long sequences lead to difficulties in training. Long short-term memory model (Hochreiter and Schmidhuber 1997) uses memory cell and three gating units to enrich RNN's ability in processing long sequences. Bi-directional Long Short-term Memory model enhance LSTM by combines forward and backward LSTM layers to get context information.

Term Extraction in Chinese Corpus is very different from other languages, because Chinese has distinctive syntactic grammar and word formation. Besides, large parts of Chinese terminology come from the translation of English terminology, which also has a great impact on the extraction of terms. RNN and CRF solve the first part of the problem, the second part is supposed to be considered.

The result of word segmentation has a greater impact on subsequent tasks. The same corpus can get different results through different word segmentation systems. As a example, terminology "循环神经网络" (Recurrent Neural Networks) could be divided into three independent words: "循环", "神经", "网络" or two independent words: "循环", "神经网络", or as a complete word "循环神经网络". Term extraction system should be compatible with different word segmentation results and carry out analysis and correction to obtain a unified domain concept, and extract correct terms. This paper proposes to add convolutional layer in BiLSTM-CRF model and try to solve these problems.

3 Model

3.1 Convolutional Neural Networks

Image classification tasks use image pixels as input. There are usually too many pixels in images; even in the simplest handwritten numeral recognition task, the input vector is as high as 784 dimensions. Pooling layer is very necessary to minimizes the number of parameters in the model, at the same time, helps to suppress the over fitting. When the convolutional neural networks are applied to NLP, convolutional layer is still used to extract the local features of a sentence, but Pooling is an operation on the whole sentence scale. That means pooling layer extracts the most significant features of sentences, actually, no matter how big the granularity of the pooling operation is, some of the local features in sentences are ignored.

Therefore, convolutional neural networks model used in this paper is shown in Fig. 1, pooling layer is not used in this model.

Convolutional layer accepts sequence of words (word embedding in this paper) as input. By adding padding zero vector at the front and back ends of the sentence, the dimension of feature vectors extracted by convolution would equal to the length of source sentence. Each item in the vector represents the local feature of the sentence extracted from the convolution filter. Different convolution filters, including convolution filters of different sizes, could be used to extract different local features.

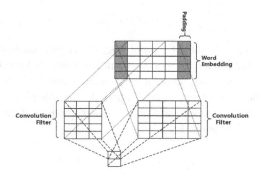

Fig. 1. Convolutional layer for features extraction

Denoting the length of the input sentence as W, the size of the convolution filter as F, padding width as P and stride as S, the formula for calculating W', the number of feature vectors, shown as follows:

$$W' = \frac{W - F + 2 * P}{S} + 1 \tag{1}$$

For convolution filters of different sizes, it is necessary to make sure that the dimension of the feature vector is always equal to the length of the sentence, so that the corresponding elements of different feature vectors could be concatenated, and it could be regarded as vector representation of each word in source sentences.

3.2 Bi-Directional Long Short-Term Memory (BiLSTM)

LSTM is an RNN architecture specifically designed to address the vanishing gradient and exploding gradient problems. LSTM uses memory unit and three multiplicative gates instead of hidden neural units in ordinary RNN.

Denoting the LSTM as function $F()$, the t_th word (vector representation of t_th word obtained from convolutional layer) from input as x_t and the latent state from the last time-step as h_{t-1}, it can be formulated as follows:

$$h_t = F(x_t, h_{t-1}) \tag{2}$$

Function F contains following formulations:

$$i_t = \sigma(W_{xi}x_t + W_{hi}h_{t-1} + W_{ci}c_{t-1} + b_i) \tag{3}$$

$$f_t = \sigma\left(W_{xf}x_t + W_{hf}h_{t-1} + W_{cf}c_{t-1} + b_f\right) \tag{4}$$

$$c_t = f_t \odot c_{t-1} + i_t \odot tanh(W_{xc}x_t + W_{hc}h_{t-1} + b_c) \tag{5}$$

$$o_t = \sigma(W_{xo}x_t + W_{ho}h_{t-1} + W_{co}c_{t-1} + b_o) \tag{6}$$

$$h_t = o_t \odot tanh(c_t) \tag{7}$$

In these equations, i_t, f_t, c_t, o_t stand for input gate, forget gate, memory cell and output gate respectively, W and b stand for model parameters, $tanh()$ is hyperbolic tangent and $\odot$ denotes an element-wise product operation. The input gate controls the magnitude of the new input into the memory cell c; the forget gate controls the memory propagated from the last time step; the output gate controls the magnitude of the output. A single LSTM unit structure is shown in Fig. 2.

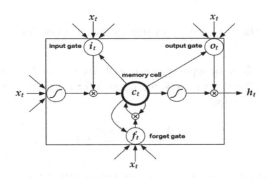

Fig. 2. LSTM cell

h_t contains input information from previous time-steps, an RNN with reverse-direction is applied to collect information from follow-up time-steps. The structure of the whole RNN unfolded by time steps is shown in Fig. 3.

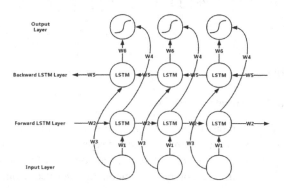

Fig. 3. RNN structure

3.3   Conditional Random Field (CRF)

In this paper, CRF is built on top of the network in order to perform the final prediction to tag sequences; it takes representations provided by the last BiLSTM layer as input. CRF layer learns the conditional probability $P(Y|X)$, where $X = (x_1, x_2, \ldots, x_n)$ are sequences of input and $Y = (y_1, y_2, \ldots, y_n)$ are label sequences. By the fundamental theorem of random fields, the joint distribution over the label sequence Y for the given X is as follow:

$$P(Y|X) = \frac{1}{Z(X)} exp\left(\sum_{i=1}^{n} \left(\sum_{j} \lambda_j t_j(y_{i-1}, y_i, X, i) + \sum_{k} \mu_k s_k(y_i, X, i) \right) \right) \quad (8)$$

$Z(X)$ is normalization constant, t_j represents the feature functions defined on edges called the transition feature, which denotes the transition probabilities from y_{i-1} to y_i for the given input sequence X. s_k represents the feature function defined on nodes, called the state feature, denoting the probability of y_i for the given input sequence X. λ_j and μ_k are weights for t_j and s_k. Alternatively, t and s could be written in the same form, $P(Y|X)$ can be written as:

$$P(Y|X, W) = \frac{1}{Z(X)} exp\left(\sum_{k} \omega_k f_k(Y, X) \right) \quad (9)$$

ω stands for weights to the feature function that the CRF learns, and $f()$ defines the feature function. For the given input sequences and label sequences

$D = [(X_1, Y_1), (X_2, Y_2), \ldots, (X_n, Y_n)]$, constructing the following objective function by maximum likelihood estimation (MLE):

$$L(\lambda, D) = -log\left(\prod_{m=1}^{N} p(Y_m|X_m, W)\right) + C\frac{1}{2} \parallel W \parallel^2 \qquad (10)$$

4 Experiments

4.1 Data Set and Parameter Settings

1200 short papers form computer field in Chinese are extracted from CNKI and used as corpus. Segment original corpus with IKAnalyzer segmentation tool, the general lexicon and the computer domain lexicon were used as the basis for word segmentation respectively. The terms in the computer lexicon are derived from Dictionary of Computer and Internet. The processed corpus contains two different word segmentation results. This paper uses the BIEOS annotation method to mark the word segmentation results manually (Table 1).

Table 1 Results of word segmentation

Raw text	Test Set1	Test Set2
命名实体识别是典型的序列标注问题，而循环神经网络是一种很有效地解决序列标注问题的神经网络模型，能够有效地利用数据的序列信息，具有一定的记忆功能	命名/b实体/i识别/e是/o典型/o的/o序列/b标注/i问题/e而/o循环/b神经/i网络/e是/o一/o种/o很/o有效/o地/o解决/o序列/b标注/i问题/e的/o神经/b网络/i模型/e能够/o有效/o地/o利用/o数据/o的/o序列/b信息/e具有/o一/o定/o的/o记忆/o功能/o	命名实体识别/s是/o典型/o的/o序列标注问题/s而/o循环神经网络/s是/o一/o种/o很/o有效/o地/o解决/o序列标注问题/s的/o神经网络模型/s能够/o有效/o地/o利用/o数据/o的/o序列/b信息/e具有/o一/o定/o的/o记忆功能/o

Split the corpus into training set and test set in 4:1 ratio. The corpus is shuffled with the granularity of the sentence; make sure there is no big difference between training set and test set in distribution of terms. The test set consists of two parts, Test Set 1 is obtained by segmentation using general lexicon and Test Set 2 is segmented by computer domain lexicon.

Word embedding tool Google word2vec and Stanford GloVe are used in this paper for word2vec and GloVe respectively. And both results of two word segmentation in corpus are used in the training of word embedding.

In this paper, the dimension of the word vector represented by word embedding is 32. In convolutional layer, convolution filter size is set to 3, padding size is 1, and stride is 1, a total of 128 convolution filters are used. In BiLSTM layer, set state activation as *tanh()* and gate activation as *sigmoid()*. The number of hidden layers is 512.

The objective function of formula (10) is optimized by using the basic batch gradient descent method and batch is set to 8. Learning rate is set to $1e-3$, and the $L2$ regularization weight is $8e-4$. The construction and training of the model are implemented by using paddlepaddle framework.

4.2   Evaluation Methodology

Model proposed in this paper aims to reduce the influence of different word segmentation results on term extraction. The impact of word2vec and GloVe on experiment results also should be measured.

BiLSTM-CRF model proposed in (Huang et al. 2015) is used as the baseline for this experiment. Train set after word embedding using word2vec and GloVe is used as input for both baseline model and CNN-BiLSTM-CRF model proposed in this paper. Then the training models are applied to two different word segmentation test sets respectively. Experiment results are shown in Tables 2 and 3, precision, recall rate and F1-value were used as the criteria.

Table 2 Precision, recall and F1-value in test set 1

Model	Word embedding	P	R	F1
Baseline model	GloVe	0.7692	0.7337	0.7510
	Word2vec	0.7523	0.7289	0.7404
CNN-BiLSTM-CRF	GloVe	0.7983	0.7813	0.7897
	Word2vec	0.7725	0.7687	0.7706

Table 3 Precision, recall and F1-value in test set 2

Model	Word embedding	P	R	F1
Baseline model	GloVe	0.7992	0.7865	0.7927
	Word2vec	0.8053	0.7932	0.7992
CNN-BiLSTM-CRF	GloVe	0.7812	0.7973	0.7892
	Word2vec	0.7762	0.7734	0.7749

5   Analysis

According to the results of experiment, baseline model performs a little bit better than CNN-BiLSTM-CRF model on Test Set2 which is segmented by domain lexicon. It is considered that the complexity of the model and difficulty of model training increased due to the addition of convolutional layers. However, in Test Set 1, the performance of the baseline method has been greatly affected, but the CNN-BiLSTM-CRF model has hardly been affected. This is considered the addition of the convolutional layer helps to extract the contextual features of the word; and to a certain extent, eliminates the influence of different word segmentation results.

The result of Test Set 2 shows model has not be improved by using GloVe compared with word2vec. But when it comes to Test Set 1, GloVe brings a marked improvement. Compared with the results obtained from domain lexicon, long term terminology might be segmented into multiple words in segmentation using general lexicon, such as "循环神经网络". Word embedding by GloVe contains the concurrence information of the words is considered to be helpful in such problems. In particular, word embedding by GloVe is more conducive for convolutional layer learning overall features.

6 Conclusion

In this paper, CNN-BiLSTM-CRF model is proposed to alleviate the effect of word segmentation quality and identify terminology that contains more than one word but has been segmented into continuous parts by word segmentation. Experiments show that this model has higher stability than the baseline model in different word segmentation results. At the same time, the results of experiment show that using word embedding by GloVe in CNN-BiLSTM-CRF model can achieve better performance.

References

Lingpeng, Y., Donghong, J., Guodong, Z., Yu, N.: Improving retrieval effectiveness by using key terms in top retrieved documents. In: Losada, David E., Fernández-Luna, Juan M. (eds.) ECIR 2005. LNCS, vol. 3408, pp. 169–184. Springer, Heidelberg (2005). https://doi.org/10.1007/978-3-540-31865-1_13

Biemann, C, Mehler, A.: Text Mining: From Ontology Learning to Automated Text Processing Applications. Festschrift in Honor of Gerhard Heyer (2014)

Zhou, J., Xu, W.: End-to-end learning of semantic role labeling using recurrent neural networks. In: Proceedings of the Annual Meeting of the Association for Computational Linguistics (2015)

Koomen, P., Punyakanok, V., Dan, R., et al.: Generalized inference with multiple semantic role labeling systems. In: Conference on Computational Natural Language Learning, pp. 181–184 (2005)

Mikolov, T., Chen, K., Corrado, G., et al.: Efficient estimation of word representations in vector space. Comput. Sci. (2013)

Pennington, J., Socher, R., Manning, C.: Glove: global vectors for word representation. In: Conference on Empirical Methods in Natural Language Processing, pp. 1532–1543 (2014)

Levy, O., Goldberg, Y., Dagan, I.: Improving distributional similarity with lessons learned from word embeddings. Bull. De La Soc. Bot. France **75**(3), 552–555 (2015)

Collobert, R., Weston, J., Karlen, M., et al.: Natural language processing (almost) from scratch. J. Mach. Learn. Res. **12**(1), 2493–2537 (2011)

Hochreiter, S., Schmidhuber, J.: Long short-term memory. Neural Comput. **9**(8), 1735–1780 (1997)

Graves, A., Wayne, G., Danihelka, I.: Neural turing machines. arXiv:1410.5401 (2014)

Huang, Z., Xu, W., Yu, K.: Bidirectional LSTM-CRF models for sequence tagging. Comput. Sci. (2015)

Medical Data Acquisition Platform Based on Synthetic Intelligence

Zhao Gu[1], Yongjun Liu[2,3(✉)], and Mingxin Zhang[3]

[1] College of Software Engineering, Northeastern University, Shenyang, China
kv.ch40@gmail.com
[2] College of Computer Science and Engineering, Northeastern University,
Shenyang, China
yongjun1981@126.com
[3] Department of Computer Science and Engineering,
Changshu Institute of Technology, Changshu, China
mxzhang163@163.com

Abstract. Nowadays, most medical device manufacturers are still using on-device data integrating and transmitting. However the real hospital situation is complex. Medical devices have different interfaces. Some of them are even outdated. This situation makes medical data can't be automatically exported. Data can only be copied by hospital staff manually. In order to solve this data extraction problem caused by interface incompatibility and device version incompatibility, we implemented this medical data acquisition platform base on synthetic information. It uses OCR (Optical Character Recognition) technology to collect intuitive data directly from the screen interface. This platform also includes an embedded voice recognition module implemented on Raspberry Pi. The voice recognition system is used to solve the time consuming and inconvenience problem caused by manually data recording through transforming the voice signal into texts and instructions. Finally, we upload medical data to the server through the socket for effective data integration. The system hardware structure is simple; cost is under control. It has good stability and can be used in a wide range of applications.

Keywords: Embedded systems · Intelligent medical · Character recognition
Speech blind separation · Speaker recognition

1 Introduction

At present, clinical information systems become increasingly in common use during the construction of medical informatization. The same hospital may use various kinds of devices, some of them are out of date and don't support data output. The development of hospital information construction has accumulated massive data. However, due to the complexity and incompatibility between systems, these data are separated and accords with different standards. Data exchange and sharing became very difficult. Nowadays, data has become valuable asset, especially clinical medical data. During the process of medical research, the first step is to achieve the medical data from real life. However, the isolated data resources in hospital become a great barrier to cross.

© Springer Nature Switzerland AG 2018
X. Meng et al. (Eds.): WISA 2018, LNCS 11242, pp. 275–283, 2018.
https://doi.org/10.1007/978-3-030-02934-0_26

Through constructing the unified data acquisition platform, we could solve all these problems for good. Furthermore, with the rapid growth of big data and data mining method, new technologies call for useable data resources with large scale.

This unified data acquisition platform meets the requirement of interconnection between information systems required by the National Medical Commission. It accords with HL7, DICOM communication standard and provides data acquisition and data storage function. As to the devices which don't have the output function, we use the technology of Artificial Intelligence, combining both software and hardware [4–9]. In order to extract and store these medical data, we use smart cameras to capture the image showed on the device screen and extract the data information from the image.

The advantage of using OCR technology is saving the time and money to unify complex device interfaces [1–3]. We could acquire the real-time clinical data of patients in different information systems of the hospital. The system could record the real-time data as log entities associated with a medical visit. Through data processing, we could provide effective data resources on Internet Data Analysis Platform. These data resources have wide application prospectsin many fields like science research, punctual medicine and pharmaceutical industry.

Doctors in emergency department usually don't have time to write down medical order. Through the voice process and recognition algorithm, the system automatically performs speaker identification and speaker verification, and then signs them with voice digital signature algorithm [5–9]. Voice information can be transformed into text and then be typed into the expected text fields in systems like HIS, PACS and CIS. This system's cost is under control and has good portability.

2   Implementation Principle and Process

The data acquisition platform is divided into video acquisition module, voice recognition module and synthesis data integration module

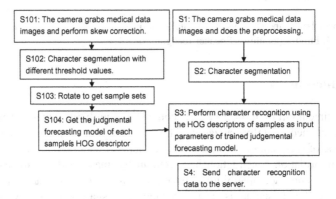

Fig. 1. Flowchart of image recognition process

As is shown in Fig. 1, the technical plan includes the following steps

S1. The camera collects medical data images and does the preprocessing.

S2. Character segmentation: Segment the text string images to be recognized in foreground into character images.

S3. Perform character recognition using the HOG descriptors of the character images as input parameters of trained judgmental forecasting model.

S4. Send the character recognition data to the server.

The trained judgmental forecasting modelis trained through the following steps

S101. Perform preprocessing on acquired medical data image.

S102. Character Segmentation: Segment the images into fore and background with threshold value i, then segment the text string images into sample character images. The sample characters include 0-9. Repeat the process with a different i for m times; we could get 10 m samples.

S103. Rotate image with the step length of 1 degree in both clockwise and anticlockwise direction for n times, we could get $20 * m * n$ sample character images.

S104. Extract the HOG descriptors of each sample image; We could finally get a descriptor subset as a training sample to solve the SVM discriminant model.

2.1   Voice Recognition Module

This module is responsible for collecting and recognizing the voice data of doctors and nurses, input the result text in expected text fields in system. It adopts ICA [5–7] along with PCA [8] as preprocessing to realize the bind signal separation. The voice recognition uses recognition algorithm based on MFCC and GMM [9] to realize speaker recognition and voice digital signature. The bind signal separation and speaker recognition process are shown in Fig. 2.

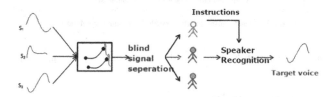

Fig. 2. Flowchart of voice separation and recognition

Detailed steps are as the following:

S1. The microphone array acquires multichannel audio signal, note it as observation matrix $x(t) = [x_1(t), x_2(t), \ldots, x_M(t)]^T$

S2. Perform blind signal separation on acquired observation signal $x(t)$.

S3. Get audio source matrix after seperation

$$s(t) = [s_1(t), s_2(t), \ldots, s_N(t)]^T$$

S4. Perform speaker recognition on separated signal $s(t)$.
S5. Finally, we could get the target audio signal.

We perform identical feature extraction processing on test voice signals and compare them with the models produced by system training. If the similarity is large enough, we judge that the speaker is valid.

2.2 Synthesis Data Integration Model

We use individual information system with network gateway to store Views, Web Service, etc. The system accesses HIS, PACS, RIS, LIS, EMR server on demand and selectively synchronizes hospitalization information to the databas. Effective data can be output and returned as references. The structure graph is shown in Fig. 3.

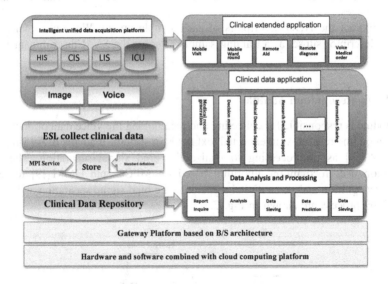

Fig. 3. Structure chart of the Whole Intelligent Medical Data Acquisition platform

3 System Tests and Analysis

We train the classifier using the digital character image set collected from medical device screen. Pi camera is adjusted until tilt angles of characters are less than 20 degrees. Then, we perform character recognition. Test devices include: Raspberry Pi, Pi Camera, resolution 8-megapixel, etc.

3.1   Image Recognition Test

The test data in the table below is part of data images extracted from a video took in hospital. We use data images with different tilt angles on the same machine as test data. The step length is 5°; we use 0° as reference. Test data are clockwise and anticlockwise, plus and minus 20°, 15°, 10°, 5°. The following is test case of 0° and minus 20°.

The result image shows statistical character recognition system didn't have error. However, during the video shooting, some jitters would make "0.26" be mistaken as "2...2.6...", "395" be mistaken as ".39.5", "and 712" be mistaken as "712..." This is because after the first frame, the system would locate the parameter values in the block. To promote the efficiency of OCR, the later frames directly perform character recognition according to the first frame's location. Through observing we find that the jitter would change the location of sign parameters in later frames and eventually cause wrong recognition result. Thus, this system needs to fix the location of camera (Table 1).

3.2   Voice Recognition Test

Collect mixed voice signal of three people speaking at the same time, the length of mixed audio signal is 3 s. Select one of them as the target speaker. First, we perform blind signal separation to get individual recovered voice, and then perform characteristic parameter extraction and match. Then, choose the closest one as the same person and play the amplified output of recovered voice signals.

As we could see in the graph below, recognition rate could be 98.33% with the SNR equals 30 dB, close to 100%. However, recognition rate goes low as the SNR reduces.

The reason why recognition rate goes low is because the power of noise gets higher as SNR reduces. As is shown in the comparison graph below, graph (a) is the original audio signal without noise; graph (b) is the audio signal with high power noise. We could see clearly the influence noise has on audio signal. High power noise would cause error in short-time energy estimation. The model would mistakenly recognize the noise as audio, which would greatly reduce the recognition performance (Figs. 4, 5).

Thus, in order to improve the recognition rate under low SNR, we apply Multi-taper Spectral Subtraction Estimation as preprocessing, which helps us filter out the high power noise effectively and improve the robustness of GMM model.

And we could see from both Figs. 6 and 7 that compared with normal fragmentation method, using Multi-taper Spectral Subtraction Estimation as preprocessing could achieve better performance in low SNR environment.

The average execution time of the module is shown in the Table 2.

As we can see from the table, the total time cost of voice separation and recognition is less than the length of voice signal itself. We basically satisfy the real-time requirement of medical informatization.

Table 1. Table of test cases

0degree

Ppeak	Pmean	PEEP	respiratory rate	oxygen concentration	Last respiratory	TI/Ttot	MVI	MVe	VTI	VTe	Mesp
16	7	4	15	33	0	0.26	9.0	9.2	395	712	9.2
15	6		14		1	0.33	9.1	9.1	567	223	9.1
14			16			0.27	8.5	9.0	352	517	9.0
			18			0.31	8.8	8.7	615	264	8.7
						0.21	8.7	8.9	738	843	8.9
						0.32		8.6	661	637	8.6

Minus 20 degrees

Ppeak	Pmean	PEEP	respiratory rate	oxygen concentration	Last respiratory	TI/Ttot	MVI	MVe	VTI	VTe	Mesp
16	7	4	14	33	0	0.27	9.1	9.1	567	517	9.2
15	6		16		1	0.29	8.7	9.2	615	712	8.9
14			18			0.21	8.5	9.0	661	223	9.1
			15			0.37	8.8	8.9	738	843	9.0
						0.32	9.0	8.7	395	691	8.7
						0.33		8.6	352	637	8.6
						0.26		8.5		264	8.5

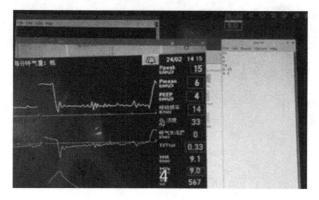

Fig. 4. Screenshot of image recognition system

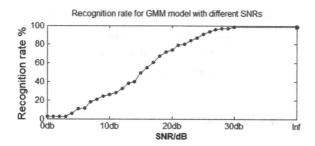

Fig. 5. Speaker recognition performance in the background of different SNR

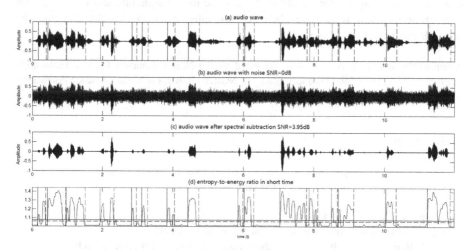

Fig. 6. The comparison graph between raw audio signal and signal with low SNR

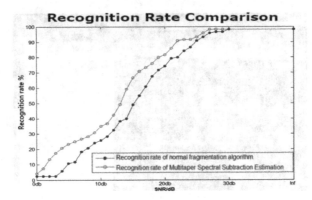

Fig. 7. Comparison graph between former and latter fragmentation algorithm

Table 2. Average execution time of the module

Operation	Time
Blind signal separation	2.33 s
Speaker recognition	0.32 s

4   Conclusion and Prospect

This system performs intelligent recognition and store process based on application of OCR technology in the field of WIT-120. The voice recognition can recognize a concrete person thus doctors could perform voice digital signature and transform oral instructions into text. It improves the efficiency, integrates data, thus the data could be used for medical data mining.

Acknowledgements. This work was supported by the Next Generation Internet Technology Innovation Project of CERNTE under grant No. NGII20170709, the Natural Science Foundation of Jiangsu Province under grant No. 15KJB520001.

References

1. Abadpour, A., Kasaei, S.: A new parametric linear adaptive color space and its implementation. In: Proceedings of Annual Computer Society of Iran Computer Conference (2004)
2. Antani, S., Crandall, D., Kasturi, R.: Robust extraction of text in video. In: Proceedings of IEEE International Conference on Pattern Recognition (2000)
3. Ai, C., Hou, H., Li, Y., Beyah, R.: Authentic delay bounded event detection in heterogeneous wireless sensor networks. Ad Hoc Netw. **7**(3), 599–613 (2009)
4. Gao, W., Zhang, X., Yang, L., Liu, H.: An improved Sobel edge detection. In: 2010 3rd International Conference on Computer Science and Information Technology, Chengdu, pp. 67–71 (2010)

5. Reynolds, D.A.: Speaker identification and verification using Gaussian mixture speaker models. Speech Commun. **17**(1), 91–108 (1995)
6. Tan, R., Liu, J., Li, Z.: Speaker recognition technology and its applications. Inf. Technol. 23–25 (2008)
7. Hyvärinen, A., Oja, E.: Independent component analysis: algorithms and applications. Neural Netw. **13**(4), 411–430 (2000)
8. Liang, S., Zhang, Z., Cui, L., Zhong, Q.: Dimensionality reduction method based on PCA and KICA, Syst. Eng. Electron. 2144–2148 (2012)
9. Reynolds, D.A., Rose, R.C.: Robust text-independent speaker identification using Gaussian mixture speaker models. IEEE Trans. Speech Audio Process. **3**(1), 72–83 (1995)

Data Privacy and Security

Generalization Based Privacy-Preserving Provenance Publishing

Jian Wu[1,2($\boxtimes$)], Weiwei Ni[1,2], and Sen Zhang[1,2]

[1] Department of Computer Science and Engineering, Southeast University,
Nanjing 211189, China
wujahu@163.com
[2] Key Laboratory of Computer Network and Information Integration
in Southeast University, Ministry of Education, Nanjing 211189, China

Abstract. With thriving of data sharing, demands of data provenance publishing become increasingly urgent. Data provenance describes about how data is generated and evolves with time. Data provenance has many applications, including evaluation of data quality, audit trail, replication recipes, data citation, etc. Some in-out mapping relations and related intermediate parameters in data provenance may be private. How to protect the privacy in the data provenance publishing attracts increasing attention from researchers in recent years. Existing solutions rely primarily on Γ-privacy model, hiding certain properties to solve the module's privacy-preserving problem. However, the Γ-privacy model has the following disadvantages: (1) The attribute domains are limited. (2) It's difficult to set consistent Γ value for the workflow. (3) The attribute selection strategy is unreasonable. Concerning these problems, a novel privacy-preserving provenance model is devised to balance the tradeoff between privacy-preserving and utility of data provenance. The devised model applies the generalization and introduces the generalized level. Furthermore, an effective privacy-preserving provenance publishing method based on generalization is proposed to achieve the privacy security in the data provenance publishing. Finally, theoretical analysis and experimental results testifies the effectiveness of our solution.

Keywords: Data provenance · Privacy-preserving
Generalization · Generalized level

1 Introduction

With the thriving of data sharing, demands for data provenance publishing are increasingly urgent. Data provenance describes how data is generated and evolves with time. The most common form of data provenance is workflow, which is usually represented by a directed acyclic graph, or DAG. Nodes in the DAG correspond to modules, and edges connecting these nodes represent relations between modules. Data provenance has many applications, including evaluation of data quality, audit trail, replication recipes, data citation, etc. Some in-out mapping relations and related intermediate parameters may be privacy in data provenance. The res, output of M2, is a sensitive attribute for patients with heart disease in the Fig. 1. An attacker can acquire the res to leak the user privacy. The M2 is the predictive module about heart disease for

X. Meng et al. (Eds.): WISA 2018, LNCS 11242, pp. 287–299, 2018.
https://doi.org/10.1007/978-3-030-02934-0_27

patients, and the provenance owner doesn't want users to speculate on the implementation mechanism of M2. However, the attacker may analyze the in-out mapping relations of M2 to reverse the function of M2 with a certain probability thereby resulting in the leakages of the privacy of M2. Thus the issue of the privacy security in the data provenance publishing has attracted increasing attention from researchers in recent years.

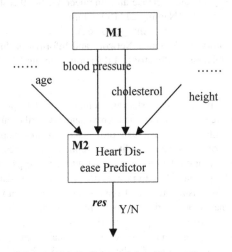

Fig. 1. Prediction module of heart disease

The existing methods mainly adopt in-out parameters hiding and data provenance abstract view to achieve the privacy security in the data provenance publishing. Davidson in 2010 proposed privacy-preserving provenance model -Γ-privacy. Modules in a workflow take a set I of input attribute and produce a set O of output attribute. Modules are modelled as a relation R over a set of attributes A = I $\cup$ O that satisfies the functional dependency I $\rightarrow$ O, as shown in Fig. 2. Each tuple in R shows an execution of the workflow, providing only visible view RV by hiding some of the attributes in R to guarantee that given any input to any module, the attacker can guess the correct output with probability at most 1/Γ. However, the model still has the following disadvantages:

(1) Γ-privacy model works only for modules in which all input and output attributes have finite domains. For the attributes with infinite domains, the Γ value cannot be set, resulting in the failure of the Γ-privacy model.

(2) The Γ-privacy model needs to set consistent Γ value for the workflow. The difference of the number of attributes for all modules and the variation of the domains of attributes make it difficult to set consistent Γ value.

(3) Attributes selection strategy of Γ-privacy model is not reasonable enough. Once the Γ value is set, the provenance owner cannot determine the selection of attributes. The privacy of intermediate parameters and in-out mapping relations are not taken into account, which leaks data provenance privacy.

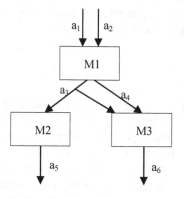

(a). Example of a workflow execution

(b). Module relation for M1

I		O	
a_1	a_2	a_3	a_4
0	1	0	1
1	1	1	1
0	0	1	0
1	0	0	1

a_1	a_2	a_3	a_4	a_5	a_6
0	1	0	1	1	1
1	1	1	1	0	1
0	0	1	0	0	0
1	0	0	1	0	1

(c). Workflow relation

Fig. 2. Workflow and relation mapping

Concerning these problems, a novel privacy-preserving provenance model, α-GLprivacy, is devised to balance the tradeoff between privacy-preserving and utility of data provenance. Furthermore, an effective privacy-preserving provenance publishing method based on generalization, GPPub, is proposed to achieve the privacy security in the data provenance publishing. Our contributions in this paper are summarized as follows:

(1) A novel privacy-preserving provenance model is devised and an effective privacy-preserving provenance publishing method based on generalization is proposed to achieve the privacy security in the data provenance publishing and solve the problem of limited attribute domains in the workflow.
(2) The concept of generalization level is introduced. The publisher is allowed to set different generalization levels for attributes, which solves the problem that the Γ value is difficult to set.
(3) Considering the sensitive attributes and in-out mapping relations, a reasonable and effective attributes selection strategy is presented.

The remaining parts of this paper are organized as follows. A brief survey of related work regarding provenance privacy preserving is given in Sect. 2. The problem statement and related definitions are described in Sect. 3. A privacy-preserving provenance Publishing method and algorithms are presented in Sect. 4. Experimental results and a conclusion are discussed in Sects. 5 and 6, respectively.

2 Related Work

With the ever-increasing demands for data provenance sharing, the issue of privacy-preserving provenance publishing has been paid increasing attention by researchers.

A method based on node grouping was proposed in [2]. Several modules were merged and abstracted into one to ensure the privacy of sensitive modules by making grouping rules. Based on the method of [2], a similar method was presented to automatically identify which modules could be merged, and delete the dependencies between the modules in [3].

Grouping approaches provide an abstract graph view that preserves the privacy of sensitive graph components, the major drawback is that the graphical structure of the workflow and the dependencies among the modules are destroyed, and it is even possible to add dummy dependencies.

A privacy-preserving provenance model, Γ-privacy was proposed in [8–10]. Modules are modelled as a relation R over a set of attributes $A = I \cup O$ that satisfies the functional dependency $I \rightarrow O$. Each tuple in R shows an execution of the workflow, providing only visible view RV by hiding some of the attributes in R to guarantee that given any input to any module, the attacker can guess the correct output with probability at most $1/\Gamma$. However, the model mainly has the following problems:

(1) The Γ-privacy model works only for modules in which all input and output attributes have finite domains. For the attributes with infinite domains, the Γ value cannot be set. Figure 3 shows a function $f(x) = |x| + 1$, then y is a positive integer. Therefore, if y is hidden directly, according to the Γ-privacy model, the Γ value cannot be set.

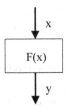

Fig. 3. Functional processing module

(2) The Γ-privacy model needs to set consistent Γ value for the workflow, resulting in difficulty in setting Γ. The difference of the number of attributes for all modules and the variation of the domains of attributes make it difficult to set consistent Γ value. Besides, the privacy protection strength of modules is also different. The consistent Γ value is difficult to meet the privacy protection requirements of all modules. As shown in Fig. 2(a), assuming $|a_1| = 3$, $|a_2| = 3$, $|a_3| = 2$, $|a_4| = 3$,

$|a_5| = 2$, $|a_6| = 3$. If $\Gamma = 3$, hiding a_4 makes M1 and M3 satisfy Γ, but no matter which attribute of M2 is hidden, M2 cannot set Γ value. If $\Gamma = 4$, M1 and M3 cannot set Γ value no matter which attribute of M1 and M3 is hidden. Therefore, it is more difficult to set a consistent Γ for the workflow.

(3) Attributes selection strategy of Γ-privacy model is not reasonable enough. The privacy of intermediate parameters and in-out mapping relations are not taken into account, which may leak data provenance privacy. As shown in Fig. 2(a), assuming $|a_1| = 3$, $|a_2| = 3$, $|a_3| = 4$, $|a_4| = 3$, $|a_5| = 2$, $|a_6| = 3$, If $\Gamma = 4$, all modules can satisfy the value of Γ if and only if a_3 is hidden. However, if a_6 is a sensitive attribute, the publisher cannot hide a_6, resulting in privacy leakages. Once the Γ value is set, the publisher cannot decide on the selection of attributes, which increases the risk of privacy leakages. The Γ-privacy model assumes that the contributions of all inputs to the output are equal. However, that is not reasonable enough. As shown in Fig. 1, age, blood pressure, cholesterol, and height about user are in-attributes, and the output is two-valued variable, indicating whether the user has a heart disease or not. From a medical point of view, height almost has no effect on heart disease, and blood pressure is an important predictor of heart disease. Even if the height attribute is hidden, the probability that the attacker can guess the correct result will be higher than 50%.

3 Problem Statement and Definitions

3.1 Problem Statement

As mentioned above, the existing privacy-preserving provenance model relies primarily on Γ-privacy model. However, the Γ-privacy model also has the disadvantages of limited attribute domains, difficulty in setting Γ value and unreasonable attributes selection strategy.

The new model needs to address the following issues:

(1) Attribute domains are no longer limited.
(2) The privacy of all modules can be protected well, in other words, the privacy protection requirements of all modules can be achieved.
(3) The owner can determine the selection of attributes.

Applying the generalization idea, the in-out parameters of the modules are generalized to solve the problem of the limited domains of attributes. The generalization hierarchy trees for attributes are constructed and the generalization level is introduced to meet the privacy protection requirements of all modules. A reasonable and effective attributes selection strategy is presented to make the owner determine the selection of attributes.

3.2 Definitions

Concerning these problems of Γ-privacy model, a novel privacy-preserving provenance model is devised to balance the tradeoff between privacy-preserving and utility of data provenance. The devised model applies the generalization and introduces the generalized level. Furthermore, an effective privacy-preserving provenance publishing method based on generalization is proposed to achieve the privacy security in the data provenance publishing. Relevant definitions are as follows:

Definition 1. Generalization hierarchy tree (GH-Tree): Given an attribute domain D, the set of nodes $S = \{R, v_1, v_2, \ldots, v_n, s_1, s_2, \ldots, s_n\}$, R represents the root of the tree, v_i represents the intermediate node, s_i represents the leaf node, g represents the mapping function of S to D, and two nodes a and b are of parent-child relationship in S, which satisfy g (a) $\subseteq$ g (b). Besides $g(si) = 1(n \geq i \geq 1)$, $g(s_1) \cup g(s_2) \cup \ldots \cup g(s_n) = g(R)$.

GH-Tree is a hierarchical structure composed of attribute values and concepts based on different levels of abstraction and it is a generalization semantic description of attributes. An example of the generalization hierarchy tree is shown in Fig. 4.

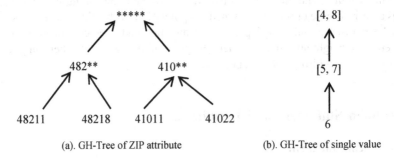

(a). GH-Tree of ZIP attribute   (b). GH-Tree of single value

Fig. 4. Examples of GH-Tree

Definition 2. Generalized attribute set P: Given a workflow W, Modules of W are modelled as a relation R over a set of attributes. A restricted subset of R is selected to allow users to access accurately. The remaining attributes of R are called generalized attribute set P.

Definition 3. Generalized level (GL): Given attribute A, the corresponding generalization hierarchy tree is GH-Tree (A). A path passing from a root node to a leaf node represents a generalized sequence of attribute A, such as A1 $\xrightarrow{g1}$ A2 $\xrightarrow{g2}$ ... $\xrightarrow{gn-1}$ An. The subscript of attribute A is denoted as the generalization level of attribute A ($1 \leq n \leq H_{GH-Tree(A)}$, and $H_{GH-Tree(A)}$ represents the height of the generalization hierarchy tree of attribute A).

According to the Fig. 4(a), generalization process of ZIP is 41011 $\xrightarrow{g1}$ 410 ** $\xrightarrow{g2}$ *****, therefore the generalized level of ZIP is 3.

GL describes the privacy protection strength of the data provenance, the higher generalized level is, and the higher privacy protection strength is.

Definition 4. α-GLprivacy model: Given a workflow W, Modules are modelled as a relation R over a set of attributes $A = I \cup O$ that satisfies the functional dependency

$I \rightarrow O$, then selecting the appropriate attributes to generate generalized attribute set P, as shown in Table 1, P = {a_2, a_3, a_n}, generalizing all the attributes of P, building GH - Trees and setting the generalized levels. If the following condition is satisfied: given threshold α, if $\forall$ a $\in$ P, $\frac{GL(a)}{H_{GH-Tree(a)}} \geq \alpha$, then the data provenance publishing satisfies α-GLprivacy.

Table 1. Mapping relation of workflow W and relation R

a_1	a_2	a_3	...	a_{n-1}	a_n
0	[2,3]	[2,3]	...	0	[-3,0]
0	[11,12]	[0,1]	...	0	[-3,0]
1	[31,32]	[2,3]	...	1	[-1,2]

4 Method

4.1 Overview

According to the privacy of intermediate parameters and strength of in-out mapping relation, the in-attributes that contribute most to the output and sensitive attributes are selected, then the set P and GH-Trees for the attributes are got and constructed, and the generalized level is set. Finally, the attributes in P are generalized and the data provenance publishing satisfies α-GLprivacy.

The process of GPPub method is shown in Fig. 5. The steps of the method are as follows:

Step 1: Formulating the attributes selection strategy, and generating the generalized attribute set P.
Step 2: GH-Trees for attributes are constructed in P.

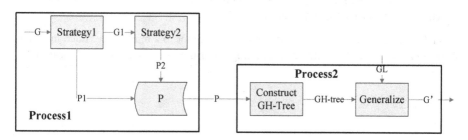

Fig. 5. The process of GPPub method

Step 3: Given the generalized level threshold, the attributes in P are generalized to make the data provenance publishing satisfy α-GLprivacy.

It is easy to see that the core of GPPub method lies in the formulation of attributes selection strategy and the construction of GH-Tree.

4.2   Attributes Selection and Construction of GH-Tree

Some in-out mapping relations and related intermediate parameters in data provenance may be private. Attackers can acquire this information to leak provenance privacy. In-out parameters may contain sensitive information, which may leak user privacy.

Concerning these problems, a generalized attributes selection strategy is proposed. The strategy is as follows:

First, the in-out parameters marked as sensitive need to be selected. For example, as shown Fig. 1, the *res*, output of M2, is a sensitive attribute for patients with heart disease. The patients don't want the sensitive attribute to be published. If the attacker obtains accurate *res*, patient privacy will be leaked. Therefore before provenance publishing, the sensitive attributes should be generalized. From the perspective of the owner, since it is easier to decide which attributes are sensitive, sensitive attributes is generally specified by the owner, corresponding to Strategy 1 in Fig. 5.

Second, the input items that contribute most to the output item in the module are added to the P. The specific selection strategy is as follows:

Given a data provenance graph, DAG, G, $\forall M \in G$, if $\forall a_{out}$, $\exists a_{in} \in M \bigwedge (\forall w \in W_M, w_{a_{in}} \geq w \Rightarrow a_{in} \in P$.

Symbol a_{out} and a_{in} denote the output and input attribute of M. W_M represents the contributions of all input items of M to the output result. The inputs of a module have different contributions on the output. In the heart attack prediction module shown in Fig. 1, compared with height, blood pressure is a more important factor affecting heart disease. The probability that the attacker guesses the correct output based on the blood pressure may be higher than 50%, while the risk of module privacy leakages may be increased. Therefore, the input items that contribute most to the output item in the module should be generalized before publishing, corresponding to Strategy 2 in Fig. 5.

Then GH-Trees are constructed for all attributes in P. According to the definition of the generalization hierarchy tree, GH-Tree should have the following features:

(1) Any intermediate node is generalization of its children.
(2) The root node is a generalization of all nodes in the tree.
(3) Generalization tree describes the relation between attribute original values and their corresponding generalized values.

GH-Tree is constructed from the bottom up, similar to a binary tree. Each node has at most two child nodes. The corresponding algorithm is presented by Algorithm 1.

Algorithm 1 GH-Tree construction algorithm

Input: domains of a in the P: $U\{u_0,u_1,...,u_{n-1}\}(n>=2)$

Output: **GH-Tree**

---

1. create an empty **GH-Tree, T**=null;
2. While(true)
3.   nns=n/2+1;
4.   **NNodes**=$\{v_0,v_1,...v_{nns-1}\}$;
5.   for i < nns
6. v_i.**addchild(u_{2i} and u_{2i+1})** ;
7.   end for
8. v_{nns-1}.**addchild(u_{2i-1} and u_{2i})** ;
9.   if(nns==1)
10.   **T= v_{nns-1}**;
11.   break;
12.   end if
13.   if(nns>1)
14.   n=nns;
15.   **U**=$\{ u_0,u_1,...,u_{nns-1}\}$;
16.   end if
17.   end while
18.   return **GH-Tree**;

4.3   Algorithm Analysis

The algorithm mainly has three steps: firstly, it invokes the attributes selection strategy to generate P; secondly, it calls the algorithm 1 to construct the GH-Trees; finally, it generalizes the attributes according to the generalized level.

Algorithm 2 GPPub algorithm

Input: the original graph *G(V,E)*.

Output: the Generalized graph *G'*.

---

1. P = ∅;
2. for each *e* in *E*
3. if(*e.value.isPrivacy()*)
4.   P.put(*e*);.
5. end if
6. end for
7. for each *m* in **M**
8. for each e_{out}
9.   P.put(*m.getBigestProperty()*);
10.   end for
11. end for
12. for each **p** in **P**
13. **p**.createGH-Tree();
14. **p**.value(p.generlizatedValue(level));
15. end for
16 return *G'* ;

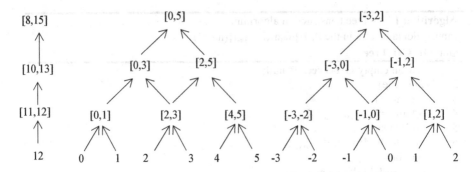

Fig. 6. GH-Trees of a_2, a_3, a_6

It is assumed $a_1 = \{0, 1\}$, a_2 is a positive integer, and $a_5 = a_3 = a_2 \bmod 6$, $a_4 = a_1$, $a_6 = a_3 - 3$. $P = \{a_2, a_3, a_6\}$ obtained by Process1 in Fig. 2(a). Since a_2 is a positive integer, the Γ value cannot be set in Γ-privacy model. However, the generalization can be applied to solve the problem that the attribute domains are limited. The value of a_2 can directly be generalized. For example, $a_2 = 12$, can be generalized to [10, 13]. The generalization trees of attributes a_2, a_3, a_6 are shown in Fig. 6. Table 2 shows an example of execution of $\alpha = 0.5$, corresponding to GL = {2, 2, 3}. Table 3 shows an example of execution of $\alpha = 0.75$ with corresponding GL = {3, 3, 4}.

Table 2. Example of a workflow execution ($\alpha = 0.5$)

a_1	a_2	a_3	a_4	a_5	a_6
0	[2,3]	[2,3]	0	0	[-3,0]
0	[11,12]	[0,1]	0	0	[-3,0]
1	[31,32]	[2,3]	1	1	[-1,2]

Table 3. Example of a workflow execution ($\alpha = 0.75$)

a_1	a_2	a_3	a_4	a_5	a_6
0	[2,5]	[0,3]	0	0	[-3,2]
0	[10,13]	[0,3]	0	0	[-3,2]
1	[31,34]	[2,5]	1	1	[-3,2]

From the above analysis, we can see that the GPPub method can solve the problem that the domains of attributes are limited to a certain extent. Attributes selection strategy takes full account of the privacy of the intermediate parameters and in-out mapping relations. The introduction of generalized level instead of Γ value, fundamentally solves the problem in set consistent Γ value.

5   Experimental Evaluation

This section analyzes the effectiveness of the proposed α-GLprivacy model and the GPPub method. The data is simulated by the data generator. The algorithm is implemented in Java. The hardware and software environment of the computer is configured As follows: AMDPhenomIIN6603.0GHZ, 8G memory, Win7.

Figures 7 and 8 show the effectiveness on achieving the privacy security in the data provenance publishing with the α-GLprivacy model and the Γ-privacy model in the keyword search and the structured queries [11]. α and Γ values have no effect on the accuracy of keyword search. For structured queries, the accuracy of α-GLprivacy query is obviously better than the accuracy of Γ-privacy model. It can be seen that if α is set properly, it can guarantee the effectiveness of the queries.

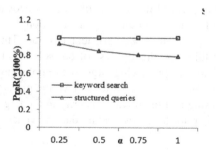

Fig. 7. Queries accuracy of α-GLprivacy

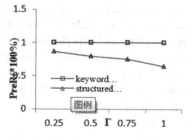

Fig. 8. Queries accuracy of Γ-privacy

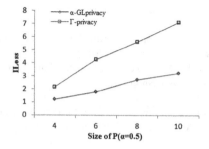

Fig. 9. Information loss

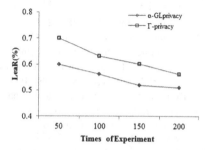

Fig. 10. Rate of privacy leakages

On this basis, a comparative analysis between the α-GLprivacy model and the Γ-privacy model is made considering the information loss and the rate of privacy leakages. The measurement of information loss adopts the method based on generalization in [12]. Figure 9 shows that the information loss of Γ-privacy is significantly larger than the information loss of α-GLprivacy with the same attribute set P.

When the output domain of a module is 2, the attacker can correctly guess the statistical probability of the result as the rate of privacy leakages. As shown in Fig. 10, α-GLprivacy decreases the rate of the privacy leakages compared to Γ-privacy due to the α-GLprivacy fully considers the privacy of intermediate parameters and in-out mapping relations, which makes attributes selection strategy more reasonable.

6   Conclusion

In the privacy-preserving provenance publishing, Γ-privacy model has the disadvantages of limited attribute domains, difficulty in set Γ value and unreasonable attributes selection strategy. Concerning these problems, a novel privacy-preserving provenance model, α-GLprivacy, is devised to balance the tradeoff between privacy-preserving and utility of data provenance. Furthermore, an effective privacy-preserving provenance publishing method, GPPub, is proposed to achieve the privacy security in the data provenance publishing. The GPPub applies generalization to solve the problem that the domains of attributes are limited. A reasonable and effective attributes selection strategy is presented. The strategy takes full account of the privacy of the intermediate parameters and in-out mapping relations. The introduction of generalization level instead of Γ value, fundamentally solves the problem in set consistent Γ value. Theoretical analysis and experimental results testifies the effectiveness of our solution. In the future work, considering the strength of privacy-preserving target data Publishing, researching privacy-preserving provenance publishing methods to achieve the privacy security in the provenance publishing and ensure that the privacy security in the target data publishing when target data publishing is prior to corresponding data provenance publishing.

References

1. Ming, G.A.O., Che-Qing, J.I.N., et al.: A survey on management of data provenance. Chin. J. Comput. **33**(3), 373–389 (2010)
2. Missier, P., Bryans, J., Gamble, C., et al.: Provenance Graph Abstraction by Node Grouping. Computing Science, Newcastle University, Newcastle upon Tyne (2013)
3. Mohy, N.N., Mokhtar, H.M.O., El-Sharkawi, M.E.: A comprehensive sanitization approach for workflow provenance graphs. In: EDBT/ICDT Workshops (2016)
4. Komadu. http://d2i.indiana.edu/provenancekomadu/
5. Davidson, S.B., et al.: Privacy issues in scientific workflow provenance. In: Proceedings of the 1st International Workshop on Workflow Approaches to New Data-centric Science. ACM (2010)
6. Chebotko, A., Chang, S., Lu, S., Fotouhi, F., Yang, P.: Scientific workflow provenance querying with security views. In: WAIM, pp. 349–356 (2008)

7. Davidson, S.B., Khanna, S., Milo, T., et al.: Provenance views for module privacy. In: Proceedings of the Thirtieth ACM SIGMOD-SIGACT-SIGART Symposium on Principles of Database Systems, pp. 175–186. ACM (2011)

8. Davidson, S.B., Khanna, S., Panigrahi, D., et al.: Preserving module privacy in workflow provenance (2010)

9. Davidson, S.B., Khanna, S., Roy, S., et al.: Privacy issues in scientific workflow provenance. In: Proceedings of the 1st International Workshop on Workflow Approaches to New Data-centric Science, p. 3. ACM (2010)

10. Davidson, S.B., Khanna, S., Roy, S., et al.: On provenance and privacy. In: Proceedings of the 14th International Conference on Database Theory, pp. 3–10. ACM (2011)

11. Fung, B., Wang, K., Chen, R., et al.: Privacy-preserving data publishing: a survey of recent developments. ACM Comput. Surv. (CSUR) **42**(4), 14 (2010)

12. Oinn, T., et al.: Taverna: a tool for the composition and enactment of bioinformatics workflows. Bioinformatics **20**, 2004 (2004)

13. Simmhan, Y.L., Plale, B., Gannon, D.: A framework for collecting provenance in data-centric scientific workflows. In: Proceedings of the IEEE International Conference on Web Services, ICWS 2006, pp. 427–436. IEEE Computer Society, Washington, D.C. (2006). http://dx.doi.org/10.1109/ICWS

14. Shui-Geng, Z.H.O.U., Feng, L.I., et al.: Privacy preservation in data applications: a survey. Chin. J. Comput. **32**(5), 847–861 (2009)

15. Ludäscher, B., et al.: Scientific workflow management and the kepler system: research articles. Concurr. Comput. Pract. Exper. **18**(10), 1039–1065 (2006). https://doi.org/10.1002/cpe.v18:10

A Kind of Decision Model Research Based on Big Data and Blockchain in eHealth

Xiaohuan Wang[1], Qingcheng Hu[2], Yong Zhang[3], Guigang Zhang[4], Wan Juan[6], and Chunxiao Xing[5(✉)]

[1] 31008 Troops of PLA, Beijing 100091, China
[2] Research Institute of Information Technology, Tsinghua University, Beijing 100084, China
[3] Beijing National Research Center for Information Science and Technology, Tsinghua University, Beijing 100084, China
[4] Department of Computer Science and Technology, Tsinghua University, Beijing 100084, China
[5] Institute of Internet Industry, Tsinghua University, Beijing 100084, China
xingcx@tsinghua.edu.cn
[6] 69036 Troops of PLA, Kuerle 841000, China

Abstract. Big Data is known as the ability of solving data sources, improving productions, and analyzing problems more faster and effectively. Blockchain is good at solving the problems of data security, sharing and reconciliation. In this paper, we introduce how to use the Big Data and Blockchain to change the strategies of decision model. We take the application of medical and health industry for example, recommend the precision analysis technology based on the Big Data and Blockchain, adopt the closed loop analysis and feedback mechanism, establish an intelligent contract system, use the weighted decision model to implement the decision-making process effectively.

Keywords: Big data · Blockchain · Decision model

1 Introduction

Before the era of Big Data, the dimension of data information is relatively simple, and data information interaction is limited, Internet is anonymous. But in this age, with the connection of the Mobile Internet, Social Networking, E-commerce Payment and Paperless Office, everyone has more and more dimensions to interact with each other, everynode is unique, and anonymity is impossible. Everyone's personalized resources and unique values will be effectively explored in the era of Big Data. In large multi-dimensional data, we can not only get the description of things more accurately and more abundantly, but also get a series of information about behavior, integrity and so on. Google Alfa dog, with the combination of the Deep Learning and Big Data, can really achieve an equal dialogue between artificial intelligence and human brain. Modern science and technology, which creates the artificial intelligence, is connected to the optimized Blockchain of the personalized Internet, and participate in social and economic activities as an independent Internet user, achieve equal dialogue between the

X. Meng et al. (Eds.): WISA 2018, LNCS 11242, pp. 300–306, 2018.
https://doi.org/10.1007/978-3-030-02934-0_28

two sides in the form of intelligent life. Big Data and Artificial Intelligence are a kind of productivity, which improve the efficiency of production. Blockchain is a kind of production relationship, which changes the distribution of production data and makes the society more rational and intelligent. Big Data [1, 4] solves the problem of data sources. Blockchain [2, 3, 5] solves the problems of data anti-counterfeiting, sharing, checking accounts and so on. Big Data and Blockchain greatly improve the overall integration level of various industries and enhance the ability of information acquisition, resource investigation and problem solving. Of course, the application of Big Data at this stage is still limited, but with the combination of Big Data and Blockchain, more and more comprehensive and individualized application becomes a trend. Intelligent production, service and creation will pass through the whole human society, a real realization of self worth age is coming [2].

2 Related Work

With the symbiotic development of Big Data and Blockchain, Blockchain provides a possible solution for large data breakthrough bottlenecks; on the other hand, the increasingly mature Blockchain also needs the platform provided by Big Data [6, 7]. The strategies of decision model based on big data and blockchain generally have five stages as follows:

2.1 Blockchain Aims to Integrate Big Data Collection and Sharing

Breaking the data islands and forming an open data sharing ecosystem is the key to use Big Data in the future. Blockchain, as a distributed database storage technology which is untampered, can effectively solve the problems transparently and safely, form a complete, traceable, unauthorized and multi-party trusted data history of key information. Using the Blockchain as a unified data structure and interface for distributed storage, can realize the interconnection and sharing of key important data with low cost, break the data island and form a multiparty trust data chain [2, 3].

2.2 Blockchain as Access to Big Data Analysis Platform

The traceability of the Blockchain retains the data from each step of data collection, sorting, transaction, circulation and calculation, makes the quality of the data obtain an unprecedented strong trust endorsement, guarantees the correctness of data analysis. Blockchain enables authorized access to data through multi-signature private keys, encryption technology, secure multi-party computing technology [2]. The data is stored in the Blockchain or the related platform which is based on the Blockchain. It can not only protect the privacy of the data, but also provide social sharing in a safe way without accessing the original data [3]. Therefore, the data saved by the Blockchain is used as the data source of large data analysis to supplement the accurate key data for Big Data analysis. At the same time, the anonymity of Blockchain is used to guarantee data privacy [3].

2.3   Data as Asset Trade in Blockchain Networks

Blockchain is known as the underlying technology of bitcoin. In the third generation of Blockchain network, everything can be treated as asset, which can be digitalized to register, validate and trade. The ownership of intelligent assets is mastered by the person who holds the private key. The owner can carry out the sale of assets by transferring private key or asset to the other party [3]. Because the Blockchain platform can support the interconnection and exchange of various assets, and Big Data assets can participate in the Blockchain platform, which use the intelligent matching mechanism of Blockchain platform to support the applications such as large data exchange. Big Data as a kind of asset, which is combined with Blockchain, uses the mechanism of market and distribution, will be the solution to break the data island. These methods become another powerful driving force for government to promote industry self-discipline, substantive flow and industrial application of Big Data [3].

2.4   Blockchain Supports Whole Lifecycle of Big Data

As a central network platform, Blockchain can include all kinds of assets in the whole society, so that different transaction subjects and different types of resources have the possibility of transboundary trading. Blockchain can ensure the security of capital and information, and through mutual trust and value transfer system to achieve previously impossible transactions and cooperation [2]. The future Blockchain will become the infrastructure for value sharing, and most of the human economy can run on the Blockchain in a way which is undetectable by the end users [3]. The upper and lower reaches of the industry make intelligent production with the share information. All types of assets are traded and exchanged on the basis of Blockchain as the underlying technology exchange. Unfamiliar parties can cooperate based on trusted records on Blockchain. The government's public welfare and social charity increase credibility and transparency through the Blockchain. Blockchain is not only the infrastructure of all kinds of economic activities, but also the source of all kinds of data [3]. In technical level, Blockchain can not only provide data which is not easily tampered, but also provide data from different sources, angles and dimensions. Big Data analysis which is based on distributed storage of structured and unstructured data on a whole network, and increases capacity through new storage techniques. Blockchain can provide high quality and audited data for the common value of the Internet, and the Blockchain itself is promoted from the supplementary data source of big data analysis to the major data source of the Big Data lifecycle [7].

2.5   Intelligent Contract [6] Promots Social Governance

With the development of digital economy, Big Data can handle more and more realistic prediction tasks. While Blockchain implements intelligent contracts with DAO, DAC, DAS, automatically runs a large number of tasks which help these predictions into action. In the future social governance, the resource providers can use the Blockchain as the intermediary, and Big Data as the precision analysis tool of the public product demand can automaticly process the intelligent contract with the standardized public products. It's transparent, fair and auditable. It reduces the human intervention and redundancy through the intelligent contract. Through the scanning and accurate analysis of the full amount of data, the combination of the Blockchain network and the automatic implementation of the intelligent contract, which is trusted with unfamiliar multiparty, push the governance of human society to a new level.

3   A Closed Loop Analytical Architecture in Medical and Health Industry Based on Big Data and Blockchain

In order to ensure the quality of final decision, we design a closed loop analytical architecture which is based on Big Data and Blockchain in Medical and Health Industry. The system's architecture, show in Fig. 1, consists of three primary components. First, the preprocessing module [7] takes use of big data and blockchain technology to process data and extracts the key features of the medical and health data. Second, the intelligent contract system module [8, 9] includes sorts of intelligent contract system. Finally, the user interaction module [6] supports user manipulations of inspection and evaluateion.

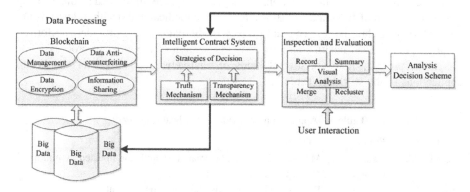

Fig. 1. A closed loop analytical architecture based on big data and blockchain

The architecture of closed loop analysis system [6, 8] which is set up according to the precision analysis technology, uses the reliability weighted decision mechanism with the principles of extreme transparency and truth, shows all various of the views and ideas, and desides the final optimization decision with demonstration and analysis. At the same time, the architecture will record and sum up the experience rationally. The complete and rational reflection on each mistake which is the data of the feedback to the intellient contract system, will be corrected [10]. At this time, a closed loop of decision analysis is formed through forward decision and reverse feedback. When the users, who have all levels of the problem or enter the key words, they will find the relevant principles and technical solutions in a set of tools. All the problems encountered will be documented and summarized into principles, and when you encounter them, you can find a solution and execute it accordingly.

Intelligent contract system [6, 8] is the core of decision-making mechanism. The system, which is based on the different types of events, establishes the principles of different situations in multiple perspectives, weights the different views, and embeds the decision criteria in the form of algorithms into the intelligent contract system. The ability to handle large scale information at a faster pace without emotion can make better decisions faster. Big Data and Blockchain can use the technology to guarantee mutual trust of data, realize transaction endorsement, provide technical guarantee for the final construction of asset exchange and value transaction in the digital economic world. The combination of Blockchain and Big Data will promote social reform and self-restraint, reestablish the trust of society without completely discarding human trust to develop a completely untrustworthy technology system.

4 The Applications Based on Big Data and Blockchain in Medical and Health Industry

4.1 The Applications in Medical and Health Industry

The applications of Big Data in the field of medical and health is very promising, and the biggest challenge is data collection, privacy, security, credibility and so on. The key to the application of Big Data and Blockchain in the medical and health industry is not the performance, but the security and privacy of the data, which provides the formulation of personalized health solutions and optimization of the efficient process.

At the present, there are several kinds of relatively mature medical and health applications in the world as Table 1.

Table 1. Applications in medical and health industry

Company	Problems	Solutions
Gem health [11]	Data leakage	Medical records can be created, shared, and updated by multiple organizations
Factom [12]	Data quality	Protect medical records and track the accounts
Block verify [13]	Data dispersion	Develop the application of anti-counterfeiting and verification in pharmaceutical enterprises

4.2 The Solutions in Medical and Health Industry

Data Security and Privacy. Data privacy can be better solved in Blockchain. Because of high redundancy, no tampering, low cost, multi-signature and complex authority management, Blockchain may be the best data storage scheme that human can find at present. Through the multi signature private key and encryption technology, when the data is processed and placed on the Blockchain through the hash algorithm, only those who are authorized can be allowed to access the data. Doctors, nurses and patients who need permission to use blockchain technology, and access data according to certain rules [2].

Data Untampered. Using Blockchain, you can safely and accurately store files permanently and protect customer records. The patient information is connected to the Blockchain in an encrypted way, which can not only ensure that the data are not tampered, but also the privacy of the patient can be protected more safely by setting up a number of private keys. Blockchain no longer controls medical data by a single group, but allows all participants to share responsibility for data security and authenticity [2].

Personalized Health Solution. In the past, the medical system charged more according to the amount of treatment rather than the results of treatment, and hospitals and doctors did not have a big driving force to improve the results of treatment. The introduction of a new medical risk sharing model can curb costs and encourage the use of resources wisely. Under the new mechanism, the remuneration of a doctor is linked to the effect of the patient's treatment or the total cost control, and the doctor's medical plan for each patient is based on the best scientific evidence available [8].

Optimizing Process and Improving Efficiency. Due to the complexity of the medical charging system, the government and medical institutions spend a lot of manpower and material resources to maintain the system every year. If the insurance company, the hospital charge department, the loan party and the patient all use the same blockchain for management, it can protect the patient's privacy and improve the efficiency of the medical charge process. The stability of the Blockchain allows all stakeholders to quickly access, view, and obtain distributed total accounts that are not dependent on the existence of third parties, and promote the safe flow of medical data between institutions [9].

5   Conclusions and Future Work

Big Data and Blockchain are the two independent technological development directions. Big data has the need for system version upgrades, whether the Blockchain can withstand the test of large capacity and high frequency or not, and truly provides a competitive scheme. These are all challenges and opportunities. However, at present, there is still no uniform data access platform in all applications. Breaking the barriers between internal, external, public and commercial has become a major problem. The task of the future work is to integrate the information content of different channels together, to build a shared database with Blockchain as a unified standard, to provide information pool for the calculation and analysis of large data. It is necessary to cooperate with the experts of Blockchain, domain experts and IT experts so as to effectively promote the development and application of Blockchain in various industries.

Acknowledgment. This work was supported by NSFC (91646202), the 1000-Talent program, Tsinghua University Initiative Scientific Research Program.

References

1. Mcafee, A., Brynjolfsson, E.: Big data: the management revolution. Harvard Bus. Rev. **90** (10), 60–66 (2012)
2. Czepluch, J.S., Lollike, N.Z., Malone, S.O.: The use of blockchain technology in different application domains
3. Peters, G.W., Panayi, E.: Understanding modern banking ledgers through blockchain technologies: future of transaction processing and smart contracts on the internet of money. In: Tasca, P., Aste, T., Pelizzon, L., Perony, N. (eds.) Banking Beyond Banks and Money. NEW, pp. 239–278. Springer, Cham (2016). https://doi.org/10.1007/978-3-319-42448-4_13
4. https://en.wikipedia.org/wiki/Big_data
5. https://en.wikipedia.org/wiki/Blockchain
6. Baker, P., Gourley, B.: Data Divination: Big Data Strategies. Cengage Learning PTR, Boston (2014)
7. Tapscott, D., Tapscott, A.: Blockchain Revolution: How the Technology Behind Bitcoin is Changing Money, Business, and the World (2016)
8. Tian, J., Wu, Y., Zhao, G., Liu, W.: Blockchain and Big Data: Building an Intelligent Economy. Post & Telecom Press, Beijing (2017)
9. Narayanan, A., Bonneau, J., Felten, E., Miller, A., Goldfeder, S.: Bitcoin and Cryptocurrency Technologies. Princeton University Press, Princeton (2016)
10. Dalio, R.: Principles. Simon & Schuster, New York (2018)
11. https://www.gemhealth.com.au/
12. https://www.factom.com/
13. http://www.blockverify.io/

Comparative Analysis of Medical P2P
for Credit Scores

Ranran Li[1], Chongchong Zhao[1], Xin Li[6], Guigang Zhang[2,3,4,5],
Yong Zhang[2,3,4,5], and Chunxiao Xing[2,3,4,5(✉)]

[1] School of Computer and Communication Engineering,
University of Science and Technology Beijing, Beijing 100083, China
G20168669@xs.ustb.edu.cn
[2] Research Institute of Information Technology, Tsinghua University,
Beijing 100084, China
[3] Beijing National Research Center for Information Science and Technology,
Tsinghua University, Beijing 100084, China
[4] Department of Computer Science and Technology, Tsinghua University,
Beijing 100084, China
[5] Institute of Internet Industry, Tsinghua University, Beijing 100084, China
xingcx@tsinghua.edu.cn
[6] Department of Rehabilitation, Beijing Tsinghua Changgung Hospital,
Beijing 100084, China

Abstract. Due to the convenience of the Peer to Peer (P2P) platform loan, the P2P platform is becoming more and more popular. Medical care has always been the most concerned issue. The emergence of medical P2P solves the insufficient funds problems of many people. In order to predict default customers, reduce the credit risk of credit institutions, and shorten credit approval time, this article uses historical data to establish a credit scoring model.

Keywords: P2P · XGBoost · Credit scoring

1 Introduction

Despite the rapid development of technology and economy, medical care has always been one of the most concerned issues. For families with disease patient, sometimes they have no enough money for medical treatment. Moreover, Statistics show that the average annual growth rate of the medical beauty industry in the past few years is 15%. Although there are not many health care loan platforms, due to the "Internet + Finance + Medical" model, medical peer platforms have emerged.

Loan applications for traditional P2P platforms have many uses, such as repaying credit cards, buying cars, and so on. But some personal medical P2Ps have gradually emerged in the market. However, due to the sensitivity of personal medical loan data to the medical P2P platform, it is difficult to obtain loan

© Springer Nature Switzerland AG 2018
X. Meng et al. (Eds.): WISA 2018, LNCS 11242, pp. 307–313, 2018.
https://doi.org/10.1007/978-3-030-02934-0_29

transaction records. We can only choose loan data that is publicly available from Lending Club. Many traditional P2P platforms accept or reject decisions by evaluating credit scores [1]. This paper uses the medical loan data of the Lending Club platform to establish a loan default prediction model to predict whether a new loan applicant will default. Therefore, we can decide whether to approve the loan based on the evaluation results.

This paper is organized as follows. In the next section, related work is given. Section 3 describe the original medical loan data of Lending Club, presents our process of data preprocessing and describes how to implement the work of Feature Engineering [2]. Section 4 Model comparison and evaluation. Section 5 make conclusions of this article.

2   Related Work

A few decades ago, the evaluation of personal credit was based on the experience of the staff [3] and there were significant subjective factors. With the development of Internet technology, statistical methods, artificial intelligence, etc. have gradually become mainstream methods. C.R. Durga devi analyzes Bagging, Boosting and Random Forest three Ensemble Learning of relative Performance in Credit Scoring [4]. Xiao and Fei used grid search technology to perform 5 cross-validation to find the optimal parameter values of various kernel functions of SVM [5]. G. E. Road make a classification comparison of individual and integrated methods in credit evaluation [6]. The main idea of the ensemble method is to combine several sets of models that solve a similar problem to obtain a more accurate model. In this study, we compared the results of XGBoost and Adaboost with the SVM and LR methods of K-fold cross-validation.

3   Data Pre-processing and Feature Engineering

3.1   Data Description

When the loan applicant applies for loan from the Lending Club platform, the Lending Club platform need the customer to fill out the loan application form online or offline. The model was studied based on Lending Club's personal medical loan data which included 18594 records with 145 attributes of clients' personal information (demographic, financial, employment and behavioral attributes), and some additional information related to their credit history. Traditionally, credit scoring systems make use of data relating to the 5Cs of credit: Capacity, Character, Capital and Collateral, and Conditions [7]. According to the study of the previous personal credit literature, we summed up the three basic principles: Condition, Credit History and Current Loan. Table 1 shows the three principles.

Table 1. The features are divided according to three principles.

Principles	Features
Demographic	id,member_id, emp_title, emp_length, url,zip_code,addr_state
Condition	home_ownership, annual_inc, verification_status, dti, mths_since_last_delinq, open_acc, revol_bal, revol_util,
Credit History	grade, sub_grade, delinq_2yrs, earliest_cr_line, inq_last_6mths, mths_since_last_record, pub_rec, collections_12_mths_ex_med
Current Loan	loan_amnt, funded_amnt, funded_amnt_inv,term, int_rate installment, issue_d, loan_status, pymnt_plan

3.2 Data Pre-processing

For the same data based on different data mining purposes, many times we do not need to train all the data. Whether the user information is complete affects the credit loan rejection or acceptance. Complete user information is easier to borrow than a portion of user information borrowers. The almost unique characteristics and constant variables have little effect on the credit scoring model. In the original datasets, there are also redundant and irrelevant features, which will degenerate the performance of clustering and decrease the accuracy of classifiers. Therefore, we apply feature selection methods to remove these features. We chose to delete the corresponding user data due to the low missing rate of the variable. The missing values of the remaining categorical variables are filled with "Unknown". Loan status includes 'Current', 'Charged Off', 'Full Paid', 'Late (16–30 days)', 'Later (31–120 days)', 'Grace Period', 'Default'. Loan status divides the attributes of the data into seven categories, each of which fills in missing values with corresponding mean values. At the same time, the variables are mapped to 1 (negative sample) and 0 (positive sample). Other ordered categorical variables do the same data preprocessing. We convert other categorical variable into dummy variables.

The "addr_state" feature refers to the state in which the user is located. The preprocessing of the"addr_state" feature is relatively specific. The default rate is calculated as follows:

$$\frac{number\ of\ default}{total\ users\ number} = rate\ of\ default(about\ every\ state) \tag{1}$$

According to the level of default rate, 50 states is divided into four categories. In order to ensure the usefulness of the information contained in the outliers and make the algorithm less affected by outliers, we standardized the data.

$$\frac{v - min}{std} = v{'} \tag{2}$$

v min and std respectively refer to the value, minimum and variance of a variable.

Pearson Correlation of Features

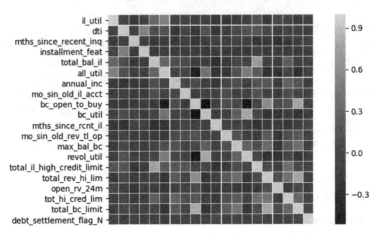

Fig. 1. Pearson correlation map.

3.3 Feature Engineering

Feature engineering is a repeated iterative process. Most of the time it is analyzing the business, analyzing the case, and constantly looking for features. Better features mean that only simple models are needed. Better features also mean that better results can be obtained to predict outcomes. Usually using XGBoost can filter out more effective features. The learning process and feature selection of the model are performed simultaneously [8]. The importance of the feature is output through the model. But we find out redundant features from the Pearson correlation map in Fig. 1.

In this study, we use the filter and XGBoost method based on the training data as the criterion to obtain the optimal feature Subset. The most important top 20 features are showed in the following Fig. 2 [9].

4   Experiment Analysis

The dataset is from medical of Lending Club to evaluate credit assessment. Through data preprocessing, there are 18,487 data samples. Of these, 16,068 were in normal loan status, and 2,419 were in default. Based on the above feature extraction, we use these data as input to the final model (Table 2).

The implementation of the various machine learning algorithms is categorized into two categories i.e. individual and ensemble [6]. In the experiment, we compared XGBoost with other algorithms, including ensemble learning Adaboost algorithm and two individual algorithms (Logistic Regression and Support Vector Machine). We also demonstrate the superiority of XGBoost algorithm [10]. The XGBoost model is a model proposed by Chen Tianqi [11] in recent years

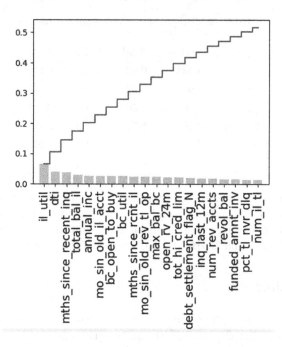

Fig. 2. Feature importance by XGBoost.

Table 2. Description of the data sets.

Data sets	Samples	Classes	Percent
Training set	12941	2	0.7
Testing set	5546	2	0.3

and has been widely used in various data mining areas, especially Kaggle competition. The regularization goal optimization formula as follows:

$$Obj^{(t)} \simeq \sum_{i=1}^{n}[l(y_i, y_i^{(t-1)}) + g_i f_t(x_i) + \frac{1}{2}h_i f_t^2(x_i)] + \Omega(f_t) + Constant \quad (3)$$

where g_i is the first order gradient statistics of the loss function and h_i is the second order gradient statistics of the loss function [12]. The CART tree in the experiment was 160 times and the training results were the best. In the experiment, we compared XGBoost with other algorithms, including ensemble learning Adaboost algorithm and two individual algorithms (Logistic Regression [13] and Support Vector Machine). In order to reduce the error rate, we performed a 10-fold cross-validation on LR and SVM. To show the evaluation of these algorithm, we Adopt performance evaluation metrics including Precision, Recall, F1-Score, Accuracy, and AUC (Area under the ROC Curve). The results are shown in the table:

Table 3. Comparison of classifier results.

	XGBoost	Adaboost	LR	SVM
Accuracy	0.941	0.942	0.812	0.898
Precision	0.943	0.925	0.725	0.851
Recall	0.601	0.567	0.603	0.578
F1-Score	0.763	0.717	0.615	0.701
AUC	0.924	0.919	0.719	0.828

The results show that XGBoost and Adaboost outperform all baselines significantly in terms of F1-Score, AUC, and Accuracy, demonstrating its superior performance for credit scoring as the final classifier. However, For the Adaboost and XGBoost algorithms, the XGBoost algorithm performs even better (Table 3).

5   Conclusion

Created Credit Scoring Model showed unexpectedly good accuracy. This will not only help our Microfinance institution to accelerate their business processes but also to manage their risks better and reduce costs as well. In this paper, we conducted data analysis, data preprocessing, feature engineering, and model building. We used a combination of XGBoost and Fileter for feature selection. In the xgboost model, the positive and negative sample data ratio is 1:13. Although the recall is not very good, the evaluation (Accuracy, Precision, AUC) is still very good. Relatively speaking, we have achieved good results. Internet finance people said that in general, beauty care is a very good consumer scenario, and relevant institutions are also very willing to cooperate. The serious illnesses may face the problem of repayment sources. Therefore, medical P2P is currently mainly concerned with dentistry, obesity, Plastic and reproductive four types of medical credit business. Hope that through the future efforts, medical P2P shortens the approval time of medical loans, reduces costs, and serves more people.

Acknowledgement. This work was supported by NSFC(91646202), National Social Science Foundation of China No. 15CTQ028, Research/Project 2017YB142 supported by Ministry of Education of The People's Republic of China, the 1000-Talent program.

References

1. Jasmina, N., Amar, S.: Using data mining approaches to build credit scoring model. In: 17th International Symposium INFOTEH-JAHORINA, East Sarajevo (2018)
2. Luis Eduardo Boiko, F., Heitor Murilo, G.: Improving credit risk prediction in online peer-to-peer (P2P) lending using imbalanced learning techniques. In: 2017 International Conference on Tools with Artificial Intelligence, Boston, pp. 175–181 (2017)

3. Wang, Y.: Applied analysis of social network data in personal credit evaluation. In: Aiello, M., Yang, Y., Zou, Y., Zhang, L.-J. (eds.) AIMS 2018. LNCS, vol. 10970, pp. 221–232. Springer, Cham (2018). https://doi.org/10.1007/978-3-319-94361-9_17

4. Durga, C.R., Manicka, R.: A relative evaluation of the performance of ensemble learning in credit scoring. In: 2016 IEEE International Conference on Advances in Computer Applications (ICACA), Coimbatore, pp. 161–165 (2017)

5. Xiao, W.B.: A study of personal credit scoring models on support vector machine with optimal choice of kernel function parameters. Syst. Eng.-Theor. Pract. **26**(10), 73–79 (2006)

6. Road, G.E.: Comparative study of individual and ensemble methods of classification for credit scoring. In: 2017 International Conference on Inventive Computing and Informatics (ICICI), 23–24 November 2017 (2017)

7. Guo, G., Zhu, F., Chen, E., Liu, Q., Wu, L., Guan, C.: From footprint to evidence: an exploratory study of mining social data for credit scoring. Assoc. Comput. Mach. (ACM) **10**(4), Article No. 22 (2016)

8. Zhang, X., Zhou, Z., Yang, Y.: A novel credit scoring model based on optimized random forest. In: IEEE 8th Annual Computing and Communication Workshop and Conference (CCWC). Las Vegas, pp. 60–65 (2018)

9. Shi, X., Li, Q.: An accident prediction approach based on XGBoost. In: 2017 12th International Conference on Intelligent Systems and Knowledge Engineering (ISKE), Nanjing, pp. 1–7 (2017)

10. Zhong, J., Sun, Y., et al.: XGBFEMF: an XGBoost-based framework for essential protein prediction. IEEE Trans. NanoBiosci. (Early Access) 1–8 (2018)

11. Chen, T., Guestrin, C.: XGBoost: reliable large-scale tree boosting system. In: Proceedings of the 22nd ACM SIGKDD International Conference on Knowledge Discovery and Data Mining, KDD 2016, USA, pp. 785–794 (2016)

12. Armin, L., Firman, A.: Ensemble GradientBoost for increasing classification accuracy of credit scoring. In: 2017 4th International Conference on Computer Applications and Information Processing Technology (CAIPT), Kuta Bali, pp. 1–4 (2017)

13. Cox, D.R.: The regression analysis of binary sequences. J. Royal Stat. Soc. Ser. B (Methodol.), 215–242 (1958)

Knowledge Graphs and Social Networks

An Evolutionary Analysis of DBpedia Datasets

Weixi Li[1], Lele Chai[1], Chaozhou Yang[1], and Xin Wang[1,2(✉)]

[1] School of Computer Science and Technology, Tianjin University, Tianjin, China
{xabbfuc,lelechai,yangchaozhou,wangx}@tju.edu.cn
[2] Tianjin Key Laboratory of Cognitive Computing and Application, Tianjin, China

Abstract. Linked Data, a method to publish interrelated data on the Semantic Web, has rapidly developed in recent years due to new techniques which enhance the availability of knowledge. As one of the most important central hubs of Linked Data, DBpedia is a large crowd-sourcing encyclopedia that contains diverse and multilingual knowledge from various domains in terms of RDF. Existing research has mostly focused on the basic characteristics of a specific version of the DBpedia datasets. Currently, we are not aware of any evolutionary analysis to understand the changes of DBpedia versions comprehensively. In this paper, we first present an overall evolutionary analysis in graph perspective. The evolution of DBpedia has been clarified based on the comparison of 6 versions of the datasets. Then we select two specific domains as subgraphs and calculate a series of metrics to illustrate the changes. Additionally, we carry out an evolutionary analysis of the interlinks between DBpedia and other Linked Data resources. According to our analysis, we find that although the growth of knowledge in DBpedia is an overall trend in recent five years, there does exist quite a few counter-intuitive results.

Keywords: DBpedia · Evolutionary analysis · RDF graph

1 Introduction

With the Linked Data [1] initiative, an increasingly large number of RDF data [6] have been published on the Semantic Web. DBpedia [5], as one of the most important central hub of Linked Data, has been evolving rapidly over the past decade. As DBpedia is a crowd-sourced community effort to extract structured information from Wikipedia, which contains comprehensive knowledge from various domains in over one hundred languages, DBpedia is undoubtedly a high-quality multilingual encyclopedia of Linked Data. Thus far, the latest version of DBpedia datasets has covered 127 languages and described 38.3 millions things. Several previous research works have conducted basic statistical analyses of a specific DBpedia version. For example, the experiments in [3] counted the number of triples and connected components in DBpedia. In addition, the official website of DBpedia has listed statistical data about the number of entities, instances, statements, etc. Although these data reflect some DBpedia characteristics, they merely focus on a specific version of DBpedia. Since the first version of DBpedia in 2007, 15 more versions have been published during the decade. Hence, at present, it is necessary to analyze the evolution of DBpedia. However, to the best of our knowledge, we are not aware of any existing evolutionary analysis of DBpedia. Thus, in this paper, we select 6 distinct high-quality datasets of DBpedia, i.e., the *mapping-based objects* (MBO) that are extracted from Wikipedia infoboxes manually, to conduct a series of evolutionary analyses in graph and social-network perspective [9]. Our evolutionary analyses of MBO datasets can be classified into three parts:

Basic Analysis. This analysis measures basic characteristics of MBO datasets including degree distribution, clustering metrics, and connectivity, which illustrate the over-view of the datasets and their evolution.

Domain-Specific Analysis. We select two specific domains, i.e., music genre genealogy and plant classification, to study the characteristics and evolution from the complex-network viewpoint. We measure the average path length and clustering coefficient of those domains and visualize the links between entities.

External Link Analysis. With the increasingly growth of Linked Data, the quantity of links between DBpedia and other Linked Data datasets should grow as well. However, the comparison among the 6 versions of more than 30 datasets counters our intuition.

Our Observations. Based on the above analyses, the following observations are obtained: (1) The entity number of the datasets tends to increase except the version 14 and 15-04. (2) The MBO datasets have a low clustering coefficient and a good connectivity. (3) There exists the small-world phenomenon in specific domains of the MBO datasets. (4) The type of evolution in a domain is not only the increase of entities, but the changes of identifiers and categories of entities. (5) The quantity of most datasets remains relatively stable in the first four versions, and the decrease in quantity occurs in many datasets in the last two versions. (6) Only the external links to the YAGO dataset increase in the latest 16-04 version of DBpedia, while links to other external datasets basically keep unchanged.

2   Our Approach

This section first introduces necessary background knowledge to explain our experiments, metrics, and the experimental settings we used in this paper. There are also some definitions given in this section to help better illustrate our approach to analyzing DBpedia.

Mapping-Based Objects (MBO) Datasets. Most of datasets in DBpedia only contain links from Wikipedia to other data sources, which means the relationships between entities are simple and only overall analysis like quantity changes can be conducted. Thus, those datasets are not suitable for an evolutionary analysis of DBpedia. Although mapping-based objects (MBO) datasets are subsets of DBpedia, almost all entities of DBpedia and relationships between the entities are included in MBO datasets. MBO datasets are representative datasets of entity relationships in DBpedia, which are suitable for analyses from several different perspectives. Thus, we select MBO datasets to carry out our analysis of DBpedia. The information of MBO datasets is listed in Table 1.

Table 1. Mapping-based objects datasets

Version	Dump date	Triples
3.8	2012.06.01	18,295,012
3.9	2013.04.03	17,887,489
14	2014.05.02	15,457,626
15-04	2015.02.05	16,090,021
15-10	2015.10.02	13,842,298
16-04	2016.04	11,324,345

RDF Triple. Let U and L denote the disjoint set of URIs and literals, respectively. A triple $\langle s, p, o \rangle \in U \times U \times (U \times L)$ is called a RDF triple, where s is the *subject*, p is the *predicate*, and o is the *object*. A finite set of RDF triples is called an *RDF graph*.

Degree Distribution. The degree of a node in a network is the number of connections it has to other nodes. If a network is directed, then the nodes have two kinds of degrees, i.e., the in-degree, which is the number of incoming edges, and the out-degree, which is the number of outgoing edges. By counting the numbers of nodes that have each given number of degrees, degree distributions can be generated. If a network is found to have a degree distribution that approximately follows a power law [12], which means the number of nodes with degree k is proportional to $k^{-\gamma}$ for $\gamma > 1$. This type of networks is referred to as scale-free [13] networks.

Global Clustering Coefficient (GCC). This metric measures how close the neighbors of a node are [14]. A triplet is three nodes that are connected by either two (open triplet) or three (closed triplet) undirected edges. $GCC = \frac{number\ of\ closed\ triplets}{number\ of\ triplets}$. In a random network consisting of N nodes, GCC is around $1/N$ [13], which is very small as compared to most real networks. Most large-scale real networks have a trend toward clustering, in the sense that their clustering coefficients are much greater than $O(1/N)$.

Connectivity. In this paper, we measure the connectivity of a graph or network by the largest weakly connected component (WCC). For a directed graph, WCC is a subgraph in which any two nodes can reach each other through some undirected path and to which no more nodes or edges can be removed while still preserving its reachability. The largest WCC means a WCC that contains the most number of nodes in a network.

Average Path Length (APL). The average path length (APL) [2] of the network is the average number of edges along the shortest paths for all possible pairs of network nodes. In fact, the APL of most real complex networks is relatively small, much smaller than that of random networks with a equal number of nodes. This phenomenon is called the small-world effect. The APL is one of the most important parameters to measure the efficiency of communication networks.

On Apache Spark, we first load the RDF triples of MBO datasets and generate nodes and edges to form a graph. Then we conducted a series of experiments on 6 versions of MBO datasets, including entity number, degree distribution, global clustering coefficient, connectivity, and average path length. These experiments can illustrate many characteristics of DBpedia, which help to explore DBpedia in more detail. After an overall analysis, we filter the data and focus on two representative domains where we can obtain more knowledge about DBpedia. Finally, We use Gephi[1], an open-source network analysis and visualization software package written in Java, to draw the relationships between entities, and explore differences by comparing experimental results.

Experimental Setup. The prototype program, which is implemented in Scala using Spark, is deployed on an 8-site cluster connected by a gigabit Ethernet. Each site has a Intel(R) Core(TM) i7-7700 CPU with 4 cores of 3.60 GHz, 16 GB memory, and 500 GB disk. All the experiments were carried out on Linux (64-bit CentOS) operating systems.

3   Entity Analysis

To better describe DBpedia and find its characteristics, we conducted several experiments. An overall analysis can illustrate some basic features of DBpedia, which is the first step to know about the evolution of DBpedia and important to select appropriate metrics to precisely measure the evolution. After that, domain specific analysis concretely depicts the evolution and makes it easier to focus on graph and complex-network perspectives.

3.1   Basic Analysis

We start to analyze DBpedia in terms of entity number, degree distribution, global clustering coefficient and connectivity. The experiments were conducted on 6 versions of MBO datasets to find out the changes among different versions of datasets.

---

[1] https://gephi.org/

Entity Number. In Fig. 1a, the number of entities of version has an increasing trend except version 15-04 and 14. According to the statistics, the entity number of version 14 is 4,547,757, while the number of version 15-04 is 4,338,493. The decrease of 209,264 entities counters our intuition that the number of entities should always grow during the five years, since the number of Wikipedia articles and Linked Data keep increasing. Thus, a more in-depth comparison has been done to find out the reason. After the comparison, a typical example is found to help explain the decrease. In version 14, there are 531,929 entities linking to a class containing the knowledge of geographic information, while in version 15-04, 450,060 entities link to the same class about geographic information. According to our observation, on the one hand, most entities of the two classes are identical, for instance, 429,824 out of 450,060 entities in version 15-04 are identical to the entities in version 14. On the other hand, the difference between the datasets is also evident. In version 15-04, there are 20,236 new entities collected and 81,869 old entities removed. According to the fact that Wikipedia articles are growing, we infer that the unusual decrease between version 15-04 and 14 is caused by the change of the methods of the data extraction and modification of entities and classes in DBpedia.

Degree Distribution. Figure 1bcd shows the degree distribution of the MBO datasets of the version 16-04. Obviously, the degree distribution exhibits the power-law pattern characteristic of scale-free networks that allows for a few

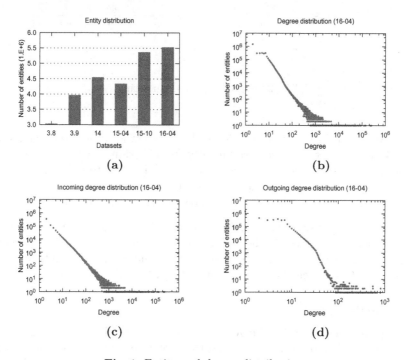

Fig. 1. Entity and degree distribution

nodes of very large degree to exist, which is quite in accord with the structure of knowledge graph. Since the degree distributions of the other five datasets are almost the same as version 16-04, we omit remaining figures of degree distributions here. In Fig. 2a, the average degrees of these datasets fluctuate around 7.0 which tends to be stable.

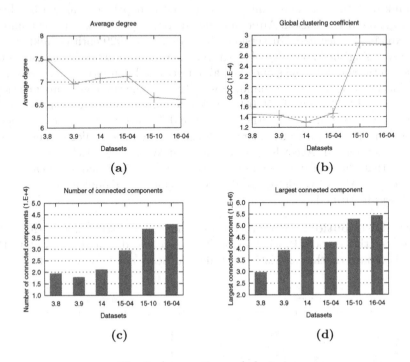

Fig. 2. Aggregation and clustering

GCC. As shown in Fig. 2b the GCC of the datasets is stable from version 3.8 to 15-04, however, it increases prominently from version 15-04 to 15-10, which means the links between the entities strengthened. Then the global clustering coefficient remains stable. Although there is an increase in global clustering coefficient, this value is still too low to show the clustering result.

Connectivity. Figure 2c shows the number of WCC. Figure 2d shows the entity number of the largest WCC. The entity number of the largest WCC depicts the connectivity of the graph. Among these versions, more than 98% entities are in the largest WCC, which shows good connectivity.

3.2   Domain-Specific Analysis

The basic analysis illustrates the overall evolution of the MBO knowledge graph. Furthermore, it is also necessary to find the evolutionary trends at a finer granularity like [4]. Thus, from MBO datasets, we select two representative domains to conduct a further analysis: one domain conforms to a complex-network structure, and the other is of a hierarchical shape [7], both of which can reflect the evolution of other domains as well due to their representation of various domains in the DBpedia knowledge graph. The analysis indicates that the evolution of these two specific domains is relatively stable, which accords with the overall DBpedia evolution.

Music Genre Genealogy. In past few years, a variety of new music genres have emerged which originated from existing genres. In addition, we found a predicate in the MBO datasets that can link the old and new music genres, which is dbo:stylisticOrigin. This predicate means exactly a music genre that originates from another music genre. In Fig. 3, we visualize the music genre genealogy of version 16-04. In order to make the graph more concise, genres with PageRank [8] lower than 0.005 are filtered out. According to Fig. 3, Folk, Blues, and Jazz are more influential just as the intuition that these genres play an essential role in music. In Table 2, the growth trend of the music genres is similar to the growth of the MBO datasets. Besides, the music genre genealogy is a complex network. Thus, we utilize the methods of complex-network analysis to analyze the domain. As shown in Table 2, the APL of the network fluctuates

Table 2. Specific domain statistics

Version	Music genre			Plant classification		
	Entity	APL	GCC	Entity	APL	GCC (10E-4)
3.8	714	4.347	0.101	43,921	1.053	7.818
3.9	747	4.230	0.114	45,232	1.054	7.592
14	1,142	4.257	0.115	48,083	1.199	7.306
15-04	1,096	4.323	0.114	50,087	1.219	7.041
15-10	1,218	4.234	0.108	52,632	1.221	6.710
16-04	1,229	4.205	0.109	53,642	1.221	6.583

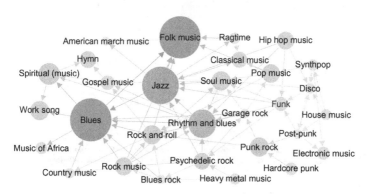

Fig. 3. Music genres 16-04

around 4.3 that is significantly shorter than the random graph, whose APL is 6.4, with the same numbers of nodes and edges. The clustering coefficient is higher than 0.1 among all the versions, and the clustering coefficient of the same random graph is only 0.002. Thus, the statistical data indicate a good connectivity among the genres and reveal the small-world phenomenon.

We also extract all the music genres in DBpedia from the datasets Instance Types which contains the type information of all the entities. The entities belonging to the type dbo:MusicGenre are all music genres. We find that some music genres in earlier datasets disappeared from newer ones, which indicates that the change is not just the increase of genres. After comparing all entities in different versions, we find the reasons. First, the types of some music genres were changed to something instead of dbo:MusicGenre, so they are removed from the music genres datasets. Second, the identifiers of some entities were modified in newer version whereas they are identical to the older ones in semantic meanings, since they refer to the same Wikipedia articles.

Plant Classification. Plant classification is aimed to place the known plants into groups that have similar characteristics. In MBO datasets, there are also RDF triples about plant classification which have the predicates: dbo:kingdom, dbo:division, dbo:class, dbo:order, dbo:family, and dbo:genus. Thus, we extract these triples and visualize the plant classification. Figure 4 depicts 39 entities with higher PageRank than the others. As shown in Fig. 5b, the relatively stable average degree of 6 versions reflects a higher density of the plant classification compared to the whole MBO datasets.

In Fig. 5a, the entity numbers of different hierarchies among different versions are shown, which keep increasing from the version 3.8 to 16-04. According to our observation, all entities definitely link to the kingdom due to its highest hierarchy, hence, the entity number of the kingdom type should be the largest. However, it is an abnormality that the entity number of class and family is larger than the one of kingdom. After a thorough comparison among the entities of family category, we surprisingly find that some entities belong to more than one family,

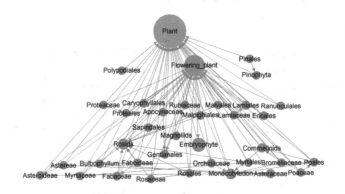

Fig. 4. Plant classification 16-04

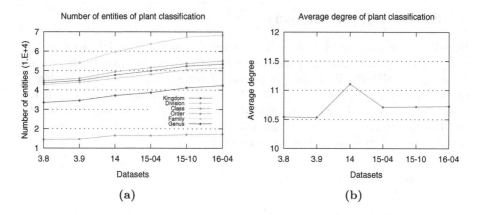

Fig. 5. Plant classification

which is contradictory to the principle of plant classification. This phenomenon is associated with the imprecise classification of plant in MBO datasets. In fact, there exist other taxonomic categories like subfamily, which ranks below family and above genus. However, in MBO datasets, there is no subfamily category, so the subfamily category of the plant is represented by the family category, which causes the above phenomenon.

We also analyze the plant classification by using the methods of the complex-network analysis. In Table 2 the APL of the network keeps increasing, but it is still significantly small, which is lower than 1.3 among all the versions. The clustering coefficient is very small, and it keeps decreasing among the versions, which is strange but reasonable. This phenomenon is caused by the hierarchical structure of plant classification. Most of the newly added plant entities appear at the bottom of the structure, and they have no connection with each other due to the hierarchical structure of plant classification, i.e., the bottom entities should not link to other bottom entities, which greatly reduces the number of new closed triplets and increases the number of new open triplets. Besides, this network shows the small-world effect, but does not show clustering [14] due to its small APL and small clustering coefficient.

The Differences. In Table 2, there are two major differences between Music Genre and Plant Classification. The APL and the GCC of the latter are much smaller than the former. The reason for the smaller APL is that every plant entity which is at the bottom of the hierarchical structure has not only a dbo:genus predicate to another entity but also all the other predicates listed in Plant Classification, which means that the maximum path length of any two entities in Plant Classification is 2. The reason why Plant Classification has a smaller GCC is due to the different structure of two domains. Plant Classification follows a hierarchical structure while Music Genre follows a social network structure [10].

4   External Link Analysis

Based on our analysis, we are able to have an insight into DBpedia in two perspectives, the overall entity distribution and specific domains, however, only focusing on DBpedia itself cannot provide a comprehensive understanding of the evolution of DBpedia. Thus we select several external datasets with the same publication cycle, which are linked to DBpedia. Then we assume that the number of external datasets should also increase, since the external datasets contain entities that are as same as ones of DBpedia.

4.1   Overall Analysis

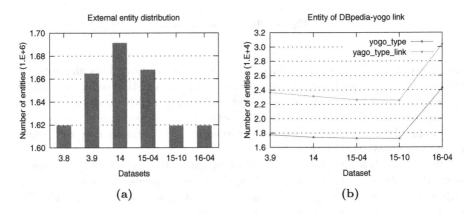

(a)   (b)

Fig. 6. External entity

Number of Datasets. According to our statistics, the number of external datasets fluctuates around 35 while it is in contrast to our intuition that the number of external datasets should increase year by year. These datasets (16-04 version) are listed in Table 3. Then we select 30 datasets (top30 in Table 3) that appeared in all the versions to evaluate the variation of external links of DBpedia.

Number of Entities. By comparing the datasets, we can know that the number of entities of 27 datasets does not change from 3.8 to 15-04, which means the number of external links is unchanged although DBpedia has a significant growth. In addition, the entity number of 17 out of 27 datasets reduces in 15-10, and based on previous statistics, the entities of DBpedia as well reduce in 2015-04. Considering the possible latency of update of external datasets, the reason for the decrease is related to the reduction of the entity number of the DBpedia. Another 10 datasets are still stable in those versions, while compared with 15-10, all the 16-04 external datasets are changed. Then we sum up all links of 30 datasets and draw Fig. 6a. The figure shows the declines in version 15-10 and 16-04, and the reduction in 15-10 is significantly great.

Table 3. External datasets

Sequence number	Name	Links
1	amsterdammuseum_links	617
2	bbcwildlife_links	443
3	bookmashup_links	8,735
4	bricklink_links	10,090
5	cordis_links	295
6	dblp_links	192
7	dbtune_links	805
8	diseasome_links	2,301
9	eunis_links	11,029
10	eurostat_linkedstatistics_links	253
11	eurostat_wbsg_links	137
12	factbook_links	545
13	gadm_links	42,330
14	geospecies_links	15,972
15	gho_links	196
16	gutenberg_links	2,505
17	italian_public_schools_links	5,822
18	linkedgeodata_links	99,073
19	linkedmdb_links	13,750
20	musicbrainz_links	22,457
21	nytimes_links	9,672
22	opencyc_links	26,940
23	openei_links	674
24	revyu_links	6
25	sider_links	1,967
26	tcm_links	904
27	umbel_links	861,797
28	uscensus_links	12,592
29	wikicompany_links	7,993
30	wordnet_links	459,139
31	yago_links	3,896,562
32	yago_types	57,879,000
33	yago_type_links	568,897
34	yago_taxonomy	570,276
35	lmdb_links	13,750

Entity Change. As the inconsistency of the change of external datasets and DBpedia, it is necessary to do a further analysis. First, we compare 27 datasets that remain stable, which shows that there is no difference among the entities in datasets from 3.8 to 15-04. Then the comparison of 15-04 datasets and 15-10 datasets shows that the entities of 24 datasets have changed. For example, some changes are swaps of subjects and objects, and others are the changes of entity URIs. Additionally, by comparing 15-10 and 16-04 datasets, it is shown that those modifications also occur in 16-04 datasets.

4.2   YAGO Dataset

YAGO is another typical external dataset of knowledge graph which includes the Is-a hierarchy as well as non-taxonomic relations between entities [11]. It is linked to DBpedia entities with the predicate `yago:sameAs` and it has the same publication cycle with the DBpedia so that it is possible to do a comparative analysis with DBpedia. We first list the link evolution in Table 4 to evaluate the change of links of YAGO to DBpedia, which shows that the quantity almost remains stable from 3.8 to 15-10, however, there is a notable growth in 16-04. To better describe the trend, we select all the plants that appear in both DBpedia datasets and YAGO datasets (`YAGO_links` and `YAGO_types`) then draw a statistical figure. In Fig. 6b, the number of links between YAGO and DBpedia increases substantially in 16-04.

Table 4. YAGO triples

Dataset Version	Links	Types	Type Links	Taxonomy
3.8	18,060,238	N/A	N/A	N/A
3.9	2,886,306	41,190,198	450,155	455,028
14	2,886,308	41,190,140	450,157	455,030
15-04	2,886,308	41,190,140	450,157	455,030
15-10	2,886,308	41,190,140	450,157	455,030
16-04	3,896,562	57,879,000	568,897	570,276

5   Conclusion and Future Work

In this paper, we have conducted some evolutionary analyses of 6 versions of typical datasets in DBpedia. In the basic analysis, we find the increasing trend of the entities in MBO datasets then analyze the abnormal decrease between version 14 and 15-04. The reasons for the decrease are the change of the methods of the data extraction and modification of entities and classes. In domain analysis, the entities of two domains both grow in quantity. Then a small-world network is found when analyze the domain of music genre genealogy, and we also indicate an imprecise classification in the domain of plant classification. In external link analysis, the fluctuation of the number of external datasets along with the stability of the entity number of the datasets are not in accord with our intuition that the number of datasets and entities should both increase. This phenomenon indicates that the external datasets may not update timely. Our next work is to analyze the evolution of DBpedia from different perspectives. Furthermore, except for the MBO datasets and the external datasets, there are still many valuable datasets that are worth for in-depth analyses.

Acknowledgement. This work is supported by the National Natural Science Foundation of China (61572353), the Natural Science Foundation of Tianjin (17JCY-BJC15400), and the National Training Programs of Innovation and Entrepreneurship for Undergraduates (201710056091).

References

1. Bizer, C., Heath, T., Berners-Lee, T.: Linked data-the story so far. Int. J. Semant. Web Inf. Syst. **5**(3), 1–22 (2009)
2. Ceccarello, M., Pietracaprina, A., Pucci, G., et al.: Space and time efficient parallel graph decomposition, clustering, and diameter approximation. In: SPAA 2015, pp. 182–191 (2015)
3. Ge, W., Chen, J., Hu, W., Qu, Y.: Object link structure in the semantic web. In: Aroyo, L., et al. (eds.) ESWC 2010. LNCS, vol. 6089, pp. 257–271. Springer, Heidelberg (2010). https://doi.org/10.1007/978-3-642-13489-0_18
4. Hu, W., Qiu, H., Dumontier, M.: Link analysis of life science linked data. In: Arenas, M., et al. (eds.) ISWC 2015. LNCS, vol. 9367, pp. 446–462. Springer, Cham (2015). https://doi.org/10.1007/978-3-319-25010-6_29
5. Jens, L., Robert, I., Max, J., et al.: DBpedia - a large-scale, multilingual knowledge base extracted from wikipedia. Semant. Web J. **6**(2), 167–195 (2015)
6. Klyne, G., Carroll, J.J.: Resource description framework (RDF): concepts and abstract syntax (2006)
7. Newman, M.: Networks: An Introduction. Oxford University Press, Oxford (2010)
8. Page, L., Brin, S., Motwani, R., Winograd, T.: The pagerank citation ranking: bringing order to the web. Technical report, Stanford InfoLab (1999)
9. Rodriguez, M.A.: A graph analysis of the linked data cloud. arXiv preprint arXiv:0903.0194 (2009)
10. Scott, J.: Social Network Analysis. Sage, Newcastle upon Tyne (2017)
11. Suchanek, F.M., Kasneci, G., Weikum, G.: Yago: a core of semantic knowledge. In: WWW 2007, pp. 697–706 (2007)
12. Virkar, Y., Clauset, A.: Power-law distributions in binned empirical data. Ann. Appl. Stat. **8**(1), 89–119 (2014)
13. Wang, X.F., Chen, G.: Complex networks: small-world, scale-free and beyond. IEEE Circuits Syst. Mag. **3**(1), 6–20 (2003)
14. Watts, D., Strogatz, S.: Collective dynamics of 'small-world' networks. Nature **393**(6684), 440–442 (1998)

Spatio-Temporal Features Based Sensitive Relationship Protection in Social Networks

Xiangyu Liu, Mandi Li, Xiufeng Xia$^{(\boxtimes)}$, Jiajia Li, Chuanyu Zong, and Rui Zhu

School of Computer Science,
Shenyang Aerospace University, Shenyang 110136, Liaoning, China
xiaxiufeng@163.com

Abstract. Without considering the spatio-temporal features of check-in data, there exists privacy leakage risk for the sensitive relationships in social networks. In this paper, a Spatio-Temporal features based Graph Anonymization algorithm (denoted as STGA) is proposed, in order to protect sensitive relationships in social networks. We firstly devise a relationship inference algorithm based on users' neighborhoods and spatio-temporal features of check-in data. Then, in STGA, we propose an anonymizing method through suppressing the edges or check-in data to prevent the inference of sensitive relationships. We adopt a heuristic to obtain the inference secure graph with high data utility. A series of optimizing strategies are designed for the process of edge addition and graph updating, which reduce the information loss meanwhile improving the efficiency of the algorithm. Extensive experiments on real datasets demonstrate the practicality, high data utility and efficiency of our methods.

Keywords: Social network · Sensitive relationship · Privacy protection
Spatio-temporal feature · Data utility

1 Introduction

A sensitive relationship mainly refers to an edge (or link) between two people who are unwilling to post on social network. Before publishing data, sensitive relationships are always removed, but it can still be predicted with high probability based on background knowledge and related inference technologies. In recent years, sensitive relationships have mostly been predicated through social network topology, i.e., we consider there existing an edge between two unconnected vertices when the user similarity is larger than a threshold δ. For instance, Fig. 1(a) shows that user u_1 and u_2 respectively have two neighbors, and each of them are their common neighbors. Obviously, we can predict that these two users are more likely to know each other. With the rapid development of mobile intelligence services, many social network applications support

The work is partially supported by the National Natural Science Foundation of China (Nos. 61502316, 61502317, 61702344), Key Projects of Natural Science Foundation of Liaoning Province (No. 20170520321).

X. Meng et al. (Eds.): WISA 2018, LNCS 11242, pp. 330–343, 2018.
https://doi.org/10.1007/978-3-030-02934-0_31

check-in service. Since users with sensitive relationship usually have similar behaviors, there is a spatio-temporal association between their check-in data. Therefore, user's check-in data provides new background knowledge for predicting sensitive relationships. As shown in Fig. 1(a), there is a common neighbor both for vertex pairs u_1, u_5 and u_2, u_5. According to exiting methods, sensitive edges (u_1, u_5) and (u_2, u_5) could be inferred with equal probabilities. However, in Fig. 1(b), we clearly observe that users u_1, u_2 and u_5 have checked in at location l_a at time t_1, and u_1, u_5 have also checked in at location l_b at time t_2. Intuitively, it can be inferred that the probability of there existing a link between u_1 and u_5 is larger than u_2 and u_5. Previous work on inferring and protecting privacy is relatively simple in terms of inference mechanism and data types. The spatio-temporal correlation provides a new feature for inferring sensitive relationships while the existing privacy protection methods cannot satisfy the privacy protection requirement.

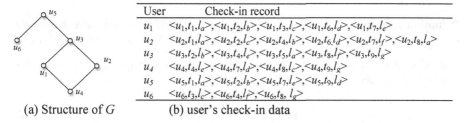

User	Check-in record
u_1	$<u_1,t_1,l_a>,<u_1,t_2,l_b>,<u_1,t_3,l_c>,<u_1,t_6,l_d>,<u_1,t_7,l_e>$
u_2	$<u_2,t_1,l_a>,<u_2,t_2,l_c>,<u_2,t_4,l_b>,<u_2,t_6,l_d>,<u_2,t_7,l_f>,<u_2,t_8,l_a>$
u_3	$<u_3,t_2,l_b>,<u_3,t_4,l_e>,<u_3,t_5,l_a>,<u_3,t_8,l_f>,<u_3,t_9,l_g>$
u_4	$<u_4,t_4,l_e>,<u_4,t_7,l_d>,<u_4,t_8,l_c>,<u_4,t_9,l_g>$
u_5	$<u_5,t_1,l_a>,<u_5,t_2,l_b>,<u_5,t_7,l_e>,<u_5,t_9,l_d>$
u_6	$<u_6,t_3,l_c>,<u_6,t_4,l_f>,<u_6,t_8, l_g>$

(a) Structure of G (b) user's check-in data

Fig. 1. A social network graph $G(V, E, C)$ with check-in data

2   Related Work

Recent work on protecting sensitive relationships are mainly classified into the following categories: k-anonymous, cluster-based anonymization, deductive control and differential privacy. Cheng et al. [1] proposed k-isomorphism to divide and modify the graph into k isomorphic subgraphs with the same number of vertices. Jiang et al. [2] proposed to clustering vertices into super vertices, where all the vertices with attribute information in a super vertex are indistinguishable from each other. In work [3–6], the authors studied how to prevent link inference attacks in social networks. Liu et al. [3] adopted inference control to make targeted modifications to social network, so that the prediction model cannot accurately infer the sensitive relationships. Fu et al. [7] protected user privacy by splitting the attribute connections and social connections. Different from only considering the graph structure, [4, 8] focused on check-in data to protect the location privacy. Wang et al. [9–12] used differential privacy to protect privacy but ignore the check-in data.

Without considering the combination of graph structure and users' check-in data, the existing research has imperfect privacy inference and protection methods. In this work, we try to protect sensitive relationships oriented to the spatio-temporal features.

3 Preliminaries and Problem Definition

In this paper, we model a social network as a graph $G(V, E, C)$, where V is the set of vertices, E is the set of edges and C is the set of check-in data of vertices. We define each vertex in set V represents a user in social network. The edge (u, v) in set E indicates that vertices u and v have a relationship. The check-in data in set C represents as $<u, t, l>$, where u is the user ID, t is the check-in time, and l is the check-in location.

Definition 1 *(common neighbor similarity). The common neighbor similarity of vertices u and v is defined as Eq. 1,*

$$Sim_{Ngr}(u, v) = \frac{|N(u) \cap N(v)|}{|N(u) \cup N(v)|} \tag{1}$$

where the more common neighbors u and v have, the higher their common neighbor similarity is.

Definition 2 *(common check-in). Let l be a place, if vertices u and v check-in in the same place l within a given time range, we say u and v have once common check-in.*

Definition 3 *(spatio-temporal relevance similarity). The spatio-temporal relevance similarity of vertices u and v is defined as Eq. 2,*

$$Sim_{ST}(u, v) = \frac{n}{\sqrt{|C_u| \cdot |C_v|}} \tag{2}$$

where n refers to the common check-in number of vertices u and v, C_u refers to the check-in set of vertex u.

For two individuals with sensitive relationship, the greater the number of their common check-in data, the higher the similarity of their spatio-temporal correlation.

Definition 4 *(user similarity). Given the similarity parameter α $(0 \leq \alpha \leq 1)$, the user similarity of the vertices u and v based on the users' neighbor and check-in data, which is calculated as Eq. 3,*

$$Sim(u, v) = \alpha Sim_{Ngr}(u, v) + (1 - \alpha) Sim_{ST}(u, v) \tag{3}$$

Definition 5 *(Sensitive relationship). Given graph $G(V, E, C)$, if vertices u and v have a link and their relationship is unwilling to expose, we define that edge(u, v) is sensitive, and the vertex pair $<u, v>$ is denoted as a sensitive pair.*

Assuming that some (not all) edges in $E(G)$ are *sensitive*. Then, such edges should be removed from $G(V, E, C)$ to protect sensitive relationships. However, adversaries can still infer these edges based on users' common neighbors and check-ins.

Definition 6 *(relational inference). Given the similarity parameter α and the similarity threshold δ, for the unconnected vertex pair $<u, v>$, if their user similarity satisfies $Sim(u, v) > \delta$, it is inferred that vertex u and v have an edge.*

The minimum threshold δ can be provided by the data owner and these two sensitive individuals, indicating that the tolerable limits of revealing their relationship.

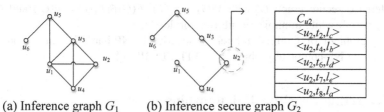

(a) Inference graph G_1 (b) Inference secure graph G_2

Fig. 2. Social network graph

Inferring all unconnected vertex pairs' relationships in G in Fig. 1 with $\alpha = \ = 1/2$, we obtain G_1 in Fig. 2(a), where $\{(u_1, u_2), (u_1, u_5), (u_3, u_4)\}$ are newly inferred edges.

Definition 7 *(inference secure). Given graph G(V, E, C), sensitive edge set S, similarity parameter α (0 $\leq$ α $\leq$ 1) and similarity threshold δ (0 $<$ δ < 1), if each edge (u, v) in S is not in E(G) and satisfies Sim(u, v) $\leq$ δ, then G is inference secure.*

For graph $G(V, E, C)$ in Fig. 1, let $S = \{(u_1, u_2), (u_1, u_5)\}$ and $\alpha = \delta = 1/2$. To prevent the attack, an inference secure graph G_2 of G in Fig. 2(b) is obtained by removing the edge (u_1, u_3) and check-in $<u_2, t_6, l_d>$, where $Sim(u_1, u_2) = 0.45 \leq 1/2$ and $Sim(u_1, u_5) = 0.335 \leq 1/2$.

In order to obtain the inference secure graph, we mainly deal with the social network graph from the following two aspects: First, we remove edges to reduce the common neighbor similarity. Second, we remove check-in data to reduce the spatio-temporal correlation similarity. The user similarity of sensitive pairs can be reduced through the above methods. In addition, it is also necessary to consider the information loss because of removing edges and check-in data.

Definition 8 *(information loss). Given graph G(V, E, C) and the information loss parameter (0 < β < 1). Let the removed data set R consists of the removed edge set R_e and the removed check-in set R_c. We define the information loss of vertex u caused by inference secure is IL_p, and the total information loss is IL.*

$$IL_p(u) = \beta \cdot \frac{countif_{v \in N(u)}((u, v) \in R_e)}{|N(u)|} + (1 - \beta) \cdot \frac{|C_u \cap R_c|}{|C_u|} \tag{4}$$

$$IL = \frac{1}{|V|} \cdot \sum_{i=1}^{|V|} IL_p(u_i) \tag{5}$$

In this work, IL_p considers the impact of removing vertex u's neighbors and check-ins. Where $countif_{v \in N(u)}((u, v) \in R_e)$ denotes the neighbors that vertex u reduces. $|C_u \cap R_c|$ indicates the number of check-in data that u removed. For all vertices with information loss, the sum of their IL_p is the total information loss IL.

Problem 1 (Optimal Inference Security). Given graph $G(V, E, C)$, sensitive edge set S, similarity threshold δ, similarity parameter α and the information loss parameter β, find an inference secure graph G' with $V(G) = V(G')$, $E(G) \cap E(G') = E(G')$ and $C(G) \cap C(G') = C(G')$ while minimizing information loss IL.

The problem of Optimal Inference Security is NP-hard. It can be proved by reducing the NP-complete problem of SATISFIABILITY.

4   Protecting Sensitive Relationships

In this section, we propose a Spatio-Temporal features based Graph Anonymization algorithm (denoted as STGA) to prevent sensitive relationship inference attacks based on graph structure and check-in data.

The main idea of Algorithm 1 is to obtain an inference secure graph by removing the common neighbors and the co-check-in data of sensitive pairs, the similarity of which is decreased. We design a heuristic to evaluate the impact of each edge or check-in data on inferring the sensitive edges and incurred information loss. Then, we select the object, which has more inference contribution and less information loss, to remove. Algorithm 1 iteratively performs the above operations until the inference secure graph is obtained.

Algorithm 1: Spatio-Temporal features based Graph Anonymization (STGA)

Input: Graph $G(V,E,C)$, sensitive edge set S, similarity threshold δ, similarity parameter α and the information loss parameter β

Output: inference secure graph G'

1. $E(G) \leftarrow E(G) \backslash S$;
2. $S_0 \leftarrow \{(u, v) | \forall (u, v) \in S$ & $Sim(u, v) > \delta\}$;
3. for $\forall (u, v) \in S_0$ do
4.   for $\forall w \in N(u) \cap N(v)$ do
5.   Calculate $IC_{(u, v)}(u, w)$ and $IC_{(u, v)}(v, w)$;
6.   Add edges (u, w) and (v, w) to E_0 ;
7.   for each common check-in record ci of u and v do
8.   Calculate $IC_{(u, v)}(ci)$;
9.   Add ci to C_0 ;
10. repeat
11. $data \leftarrow$ **FindRemovalData**$(G, S_0, E_0, C_0, \alpha, \beta)$;
12.   if $data$ is an edge then
13.   Remove $data$ from $E(G)$;
14.   Add the nodes of edge $data$ to $CNode$;
15. $e' \leftarrow$ **FindAddEdge**$(G, S, data, \alpha, CNode)$;
16.   if $e' \neq$ null then
17. $E(G) = E(G) \cup \{e'\}$;
18.   Remove the node of edge e' from $CNode$
19.   else if $data$ is a check-in record then
20.   Remove $data$ from $C(G)$;
21.   Add the node of check-in record $data$ to $CNode$;
22.   Update($data, S_0, E_0, C_0$) ;
23. until $S_0 == \emptyset$;
24. return G' ;

Algorithm 1 first removes the sensitive edges from $E(G)$ (Line 1). And set S_0 is initialized with sensitive edges with $Sim > \delta$ (Line2). Then, we obtain the edges in set E_0 by finding the sensitive pairs' common neighbors and save the influence contribution $IC_{(u,v)}$ of the edges (discussed in detail in Algorithm 2). Similarly, the set C_0 save the sensitive pairs' co-check-in data with their influence contribution IC (Lines 7–9). Algorithm 1 iteratively removes to obtain an inference secure graph (Lines 10–23). In the process of iteration, we use procedure **FindRemovalData** to select an object to remove in E_0 or C_0 through a heuristic (Line 11). If the object to removed is an edge, removing it from $E(G)$. Then, the procedure **FindAddEdge** finds a new edge to reduce information loss due to the removed edge (Lines 15–17). Else if this object is a check-in data, removing it from $C(G)$ (Lines 19–21). After that, we partly update S_0, E_0 and C_0 through the procedure **Update**. Finally, when S_0 is empty, Algorithm 1 stops iterating and outputs the inference secure graph G'.

4.1 Selecting an Object to Delete

In the process of obtaining the inference secure graph, randomly selecting an object to remove would result in high information loss. In this section, we propose a heuristic to evaluate the impact of removing each edge or check-in data on inferring sensitive relationships and information loss.

In Algorithm 2, we evaluate each removable object with heuristic function *Score*, and select the one with largest value to remove. The *Score* value is calculated based on the inference contribution IC and the information loss IL.

The function IC measures the inference contribution of the candidate removing object, which is calculated as follows:

$$IC(u,w) = \sum_{(u,v)\in S_0, v\in N(w)} \left[Sim(u,v) - Sim_{(u,w)\notin E(G)}(u,v) \right] \\ + \sum_{(w,n)\in S_0, n\in N(u)} \left[Sim(w,n) - Sim_{(u,w)\notin E(G)}(w,n) \right] \tag{6}$$

$$IC(ci_u) = \sum_{(u,v)\in S_0} \left[Sim(u,v) - Sim_{ci_u\notin C(G)}(u,v) \right] \\ + \sum_{(n,u)\in S_0} \left[Sim(n,u) - Sim_{ci_u\notin C(G)}(n,u) \right] \tag{7}$$

$IC(u, w)$ represents the inference contribution of edge (u, w), which is calculated by summing all the sensitive pairs' decrement of user similarity after removing the edge (u, w) (Line 8 in STGA). Similarly, $IC(ci_u)$ represents the inference contribution of check-in data ci_u to the set S_0. The more the inference contribution of candidate deleted object, the more the user similarity reduced by removing this object.

The information loss IL needs to be calculated by Eqs. 4 and 5. For edge (u, w), its information loss can be simplified as $IL(u, w) = (1/|N(u)| + 1/|N(w)|) \times \beta/|V|$. For the check-in data ci_u, its information loss is $IL(ci_u) = (1-\beta)/(|V| \times |C_u|)$.

The $Score$ of a removable edge (u, w) is calculated as $Score(u,w) = IC(u,w)/IL(u, w)$. The $Score$ of the removable check-in data ci_u is calculated as $Score(ci_u) = IC(ci_u)/IL(ci_u)$.

Algorithm 2: FindRemovalData

Input: Graph G(V,E,C), sensitive edge set S_0, deletable edge set E_0, deletable check-in set C_0, similarity parameter α and the information loss parameter β

Output: the object *data* to be deleted

1. $Score_{max} = 0$;
2.   for %$(u, w) \in E_0$ do
3.   Calculate the $Score(u, w)$ of edge (u, w) ;
4.   if $Score(u, w) > Score_{max}$ then
5. $data \leftarrow (u, w)$;
6. $Score_{max} = Score(u, w)$;
7.   for %$ci \in C_0$ do
8.   Calculate the $Score(ci)$ of check-in record ci ;
9.   if $Score(ci) > Score_{max}$ then
10. $data \leftarrow ci$;
11. $Score_{max} = Score(ci)$;
12.   return $data$;

4.2   Searching an Edge to Add

Although the edge deletion protects the sensitive relationships, it will also affect the graph structure of the social network. In this section, we design an edge addition strategy to reduce the graph information loss, which makes the new edge have similar properties to the removed on in terms of graph structure and check-in data. In addition, the newly added edge should not contribute for inferring the sensitive edges.

After removing edge (u,w), Algorithm 3 finds and adds the new edge by selecting a non-sensitive (not sensitive to others) vertex w in edge(u,w). Connecting with the vertex n, which should meet the following conditions: (1) $(w, n) \notin E(G)$. (2) n is a non-sensitive vertex. (3) Vertex n is affected by the delete operation. (4) Vertices n and u have high similarities. Where, condition (1) is necessary for constructing a new edge, condition (2) ensures that the new edge (w, n) doesn't contribute to inferring sensitive edges, and condition (3) enables the newly added edge have no effect on graph attributes while reducing the total information loss. In Algorithm 3, we use a greedy strategy to find and add the most similar edge to removed. Since only one vertex is changed between the old and new edge, the similarity between them is equal to user similarity $Sim(n, u)$, where n is the candidate vertex that has been found and u is the sensitive vertex in the removed edge (u, w). At this time, the condition.

$E(G) \cap E(G') = E(G')$ in Question 1 can be relaxed to $E(G) \cap E(G') \approx E(G')$.

Algorithm 3: FingAddEdge

Input: Graph $G(V,E,C)$, sensitive edge set S, deleted object $data = (u, w)$, similarity parameter α and the vertex set $CNode$ with information loss

Output: the newly edge e'

1. $e' =$ null, $Sim_{max} = 0$;
2.   for $\%(u, v) \in S$ do
3.   Add u and v to $SNode$;
4.   if $w \in \{u, w\}$ & $w \notin SNode$ then
5.   for $\%n \in SNode$ & $n \notin SNode$ do
6.   if $n \neq m$ & $n \notin N(m)$ then
7.   if $Sim(w, n) > Sim_{max}$ then
8. $e' \leftarrow (w, n)$;
9. $Sim_{max} = Sim(w, n)$;
10.   return e' ;

In Algorithm 3, $SNode$ is initialized to a vertex set that the vertex in it is sensitive to others (Lines 2, 3). Given the removed edge (u, w) and the vertex set $CNode$ (vertices in it have information loss), Algorithm 3 looks for the candidate vertex according to the condition (1–3) (Lines 5, 6), and chosses the new edge with the highest similarity to the removable one (Lines 7–9).

4.3   Data Updating

Algorithm 1 (STGA) needs to update the sensitive edge set S_0, the removable edge set E_0 and the removable check-in set C_0 after the deletion operation (Lines 11–21 in STGA). One simple method is to recalculate S_0 and obtain E_0 and C_0 after each deletion, which have repeated calculations and low execution efficiency. Therefore, the following updating strategy is proposed in Algorithm 4 to improve the execution efficiency. The mainly idea of Algorithm 4 is to evaluate the impact of the sensitive edges by the removed object first, and then judging and partial updating the data in E_0 and C_0.

Algorithm 4: Update

Input: Deleted object data, sensitive edge set S0, deletable edge set E0, deletable check-in set C0 and similarity threshold δ

Output: Updated S0, E0 and C0

1. $S_0\_del = \emptyset$;
2. for $\%(u,v) \in S_0 \, \& IC_{(u,v) \in S_0}(data) \neq 0$ do //Update S_0
3. $\quad Sim(u, v) = Sim(u, v) - IC_{(u, v)}(data)$;
4. $\quad$ if $Sim(u, v) < \delta$ then
5. $\quad\quad$ Remove (u, v) from S_0;
6. $\quad\quad$ Add (u, v) to S_0_del;
7. $\quad$ for $\%(u, w) \in E_0$ do　// Update E_0
8. $\quad\quad$ if $data$ is an edge and (u,w) is equal to $data$ then
9. $\quad\quad\quad$ Remove edge (u, w) from E_0;
10. $\quad\quad$ else if $\exists (u,v) \in S_0\_del \, \& \, IC_{(u, v)}(u,w) \neq 0$ then
11. $\quad\quad\quad IC_{(u, v)}(u, w) = 0$;
12. $\quad\quad$ if $IC(u, w) = 0$ then
13. $\quad\quad\quad$ Remove edge (u, w) from E_0;
14. $\quad$ for each check-in record ci in C_0 do　// Update C_0
15. $\quad\quad$ if $data$ is a check-in record and ci is $data$ or $data$'s common check-in record then
16. $\quad\quad\quad$ Remove ci from C_0;
17. $\quad\quad$ else if $\exists (u, v) \in S_0\_del \, \& \, IC_{(u, v)}(ci) \neq 0$ then
18. $\quad\quad\quad IC_{(u, v)}(ci) = 0$;
19. $\quad\quad$ if $IC(ci) = 0$ then
20. $\quad\quad\quad$ Remove ci from C_0;

Algorithm 4 first updates each sensitive edge' $Sim(u,v)$ affected by the removed object $data$ through $IC_{(u,v)}(data)$. The set of S_0_del is initialized to all sensitive edges affected by the removed data in S_0 (Line 6). Then, we update the data influenced by S_0_del through the inference contribution. Moreover, if the updated edge (u, w) is no longer sensitive to S_0, it also needs to be removed from E_0 (Line 12, 13). Similarly, the above update operation is also performed for each check-in data. It should be noted that when $data$ represents a check-in data, its common check-in data should be also removed at the same time in C_0 (Line 15, 16).

4.4　Theoretical Analysis of Security, Data Availability and Complexity

In this section, we will give the theoretical analysis and proof of the proposed algorithm's security, availability of the obtained social network graph data, and the complexity of the algorithm.

Theorem 2. *The social network graph G' obtained by Algorithm 1 (STGA) is inference secure.*

Proof. In Algorithm 1, the removable edge set E_0 and the removable check-in set C_0 respectively contain all the edges and check-in data that have contribution on inferring the sensitive edges in S. Obviously, after removing all the data in E_0 and C_0, all the sensitive edge (u, v) in S must be satisfied that $Sim(u, v) = 0$, and make G' become inference secure. Clearly, during the iterative deletion operation through the heuristic, the user similarity of sensitive pairs in S will continue to decrease, and all of them eventually satisfy $Sim(u, v) \leq \delta$. At this time, the obtained social network G' is inference secure. After removing an edge, Algorithm 3 only considers the new added edge that have no contribution to infer sensitive edges. In summary, the social network G' obtained by STGA is inference secure.

Theorem 3. *Given graph $G(V, E, C)$, sensitive edge set S, similarity threshold δ, similarity parameter α and the information loss parameter β, the inference secure graph G' can be obtained by Algorithm 1. Then the maximum number of the removed edges and check-in data is respectively $|S|d_{max}(1-\delta)$ and $|S|f_{max}(1-\delta)$, where d_{max} and f_{max} are the maximum degree and maximum check-in number in $G(V, E, C)$.*

Proof. When Algorithm 1 generates a security graph, we assume that the number of the removed edges and checked-ins is respectively x and y, while ensuring that the removed edges are not connected to a same common neighbor and the removed checked-in data are not co-check-in data. Obviously, x is at most when the similarity parameter $\alpha = 1$ and meets $\frac{|N(u) \cap N(v)| - x}{|N(u) \cup N(v)|} \leq \delta$, that is, $x \geq |N(u) \cap N(v)| - \delta|N(u) \cup N(v)|$. From above, to let $Sim(u, v) \leq \delta$ only need to satisfy $x = |N(u) \cap N(v)| - \delta|N(u) \cup N(v)|$. In this case, when $|N(u) \cap N(v)| = |N(u) \cup N(v)|$, x take maximum, i.e., the maximum of x is $|N(u) \cap N(v)|(1-\delta)$. Because the maximum number of common neighbors and check-in data in any two vertices is d_{max} and f_{max}, and $|N(u) \cap N(v)|$ is always much smaller than d_{max}. The number of removed edges satisfies $x < d_{max}(1-\delta)$. Similarly, the maximum number of removed check-in data is when the similarity parameter $\alpha = 0$. So y is max in the case of all the check-in data is both two vertices' co-check-in data and is equal to f_{max}. In this case, y satisfies $y < f_{max}(1-\delta)$. In summary, when the number of sensitive edges in graph $G(V, E, C)$ is $|S|$, the number of the removed edges and checked-in data will also be much smaller than $|S|d_{max}(1-\delta)$ and $|S|f_{max}(1-\delta)$ with the heuristic.

Now, we analyze the computational complexity of our algorithm. In Algorithm 1, the most time-consuming process is iterative deletion (Lines 10–23). The set of $|S_0|$, $|E_0|$, $|C_0|$ and $|CNode|$ are at most $|S|$, $2d_{max}|S|$, $2f_{max}|S|$ and $2|S|$. Since Algorithm 1 removes up to $|S|d_{max}(1-\delta)$ edges and $|S|f_{max}(1-\delta)$ check-in data. In each delete operation, there are $|E_0| + |C_0|$ calculations of Algorithm 2 and its computational complexity to select the object is at most $O(|S|(d_{max} + f_{max}))$. Algorithm 3 finds and adds a new edge using $|S| + |CNode|$ operations and requires the computational complexity at most $O(|S|)$. The partial update in Algorithm 4 needs to traverse the set S_0, E_0, C_0 separately, where the number of sensitive edges affected by removed *data* in S_0 is at most $|S_0|$ and the sensitive edges reduced in S_0 after deleting *data* is also at most $|S_0|$. Therefore, Algorithm 4 performs $|S_0| + |S_0|(|E_0| + |C_0|)$ operations with the computational complexity of $O(|S|^2 (d_{max} + f_{max}))$. To sum up, the computational complexity of Algorithm 1 is $O(|S|^3(d_{max} + f_{max})^2(1-\delta))$.

5 Experimental Evaluation

In this section, we provide extensive experiments to evaluate our methods. We use two real network datasets **Brightkite** and **Gowalla**, which are published by the Stanford Network Analysis Platform (SNAP).

There are 58,228 vertices, 214,078 edges and 4,491,143 check-ins from April 2008 to October 2010 in **Brightkite**, and 196,591 vertices, 950,327 edges and 6,442,892 check-ins from February 2009 to October 2010 in **Gowalla**.

We implement three versions of Algorithm STGA, which are R-STGA, H-STGA and HA-STGA, where R and H refer to select object to be removed with randomization and heuristic method respectively as well as no edge addition, HA refers to the heuristic deletion method and adopts edge addition strategy. We also implement Algorithm LIP in [3] for comparison, replacing the user similarity and information loss with Definition 1 and 8 in our work.

We evaluate our algorithm from both the runtime and information loss. We set the similarity threshold δ in [0.04, 0.2], both the similarity parameter α and the information loss parameter β range from [0,1]. We randomly select different number k of sensitive relationships for protection, and set $k = 100, 200, 300, 400, 500$. All the programs are implemented in Java. The experiments are performed on a 3.10 GHz Intel Core i5 with 4 GB DRAM running the Windows 7 operating system.

5.1   Runtime

We evaluate runtime with similarity threshold δ, similarity parameter α, information loss parameter β, and the number of sensitive edges k in Figs. 3, 4, 5 and 6. Figure 3 shows that when δ increases, the demand for inference security decreases, making less data be deleted, so the runtime decreases. In Fig. 4 with the increase of α, the inferred sensitive edges become more, causing the runtime of the algorithm be longer. LIP's runtime is basically stable in Fig. 4 because α does not affect it. It can be seen in Fig. 5 that the runtime of all algorithm remains basically the same because runtime is not affected by the information loss parameter β. As depicted in Fig. 6, with the increase of k, the number of the sensitive edge to protect is increased, so all algorithms need more time to protect these edges.

Figures 3, 4, 5 and 6 show the results that both runtime of HA-STGA and LIP higher than others because of the edge-adding strategy. And LIP has an average 100 times higher than HA-STGA. One aspect that produces this result is that LIP only considers the common neighbor on the user similarity, that is, the similarity parameter α is always 1, and most user's common neighbor similarity is higher than spatio-temporal relevance similarity. Thus, the user similarity in LIP is higher and inferring more sensitive edges when other parameters are constant. On the other hand, HA-STGA only needs to find new edges in a small candidate edges set, so its runtime is shorter that LIP. Since heuristic and random deletion do not have much effect on runtime, H-STGA and R-STGA is basically the same runtime and less than 1 s.

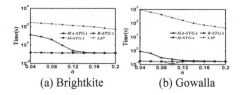

| (a) Brightkite | (b) Gowalla | (a) Brightkite | (b) Gowalla |

Fig. 3. Runtime varies with parameter δ **Fig. 4.** Runtime varies with parameter α

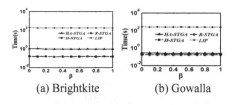

| (a) Brightkite | (b) Gowalla | (a) Brightkite | (b) Gowalla |

Fig. 5. Runtime varies with parameter β **Fig. 6.** Runtime varies with parameter k

5.2 Information Loss

We use the function *IL* (Definition 8) to evaluate information loss and the results are shown in Figs. 7, 8, 9 and 10. The change of *IL* varies with δ is shown in Fig. 7 that the higher δ, the less sensitive edges need to be protected, and the smaller the *IL*. In Fig. 8, when α increased, the user similarity increases, the sensitive edges inferred more, and results in higher information loss. Among them, because of the uncertainty of the randomization method, the information loss of R-STGA fluctuates greatly. As can be seen from Fig. 9, except for the fluctuation of R-STGA, the *IL* of other algorithms is only affected by the information loss parameter β. So when β increases, the overall information loss increases evenly. In Fig. 10, as the number of sensitive edges k increases, data removed more and the information loss *IL* must be increased.

From the experimental results, it can be seen that since HA-STGA and H-STGA both adopt the heuristic, making the total information loss *IL* smaller. And HA-STGA is 80% smaller than H-STGA. This is because that HA-STGA considers the vertex whose neighbors or check-in data is deleted to form the new edge and LIP ignores the additional effects during adding new edge. On the other hand, under the same conditions, the number of sensitive edges deduced by LIP is larger than others. Therefore, the information loss caused by LIP is 3 times higher than that of the H-STGA.

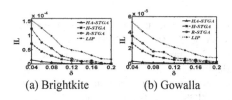

| (a) Brightkite | (b) Gowalla | (a) Brightkite | (b) Gowalla |

Fig. 7. *IL* varies with parameter δ **Fig. 8.** *IL* varies with parameter α

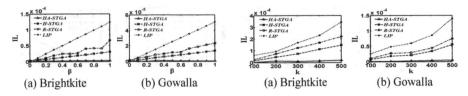

| (a) Brightkite | (b) Gowalla | (a) Brightkite | (b) Gowalla |

Fig. 9. *IL* varies with parameter β **Fig. 10.** *IL* varies with parameter k

6 Conclusions

In this paper, we focused on the issue of protecting sensitive relationships in social network with spatio-temporal features. We solve the problem by proposing a Spatio-Temporal features based Graph Anonymization algorithm (denoted as STGA). In addition, we also design optimizing strategies for edge addition and data updating in order to reduce the information loss and improve the execution efficiency. Our algorithm performs well on real social networks. In the future, we will conduct in-depth research on privacy protection of sensitive relationship in dynamic social networks.

References

1. Cheng, J., Fu, W.C., Liu, J.: K-isomorphism: privacy preserving network publication against structural attacks. In: SIGMOD, pp. 459–470 (2010)
2. Jiang, H., Zhan, Q., Liu, W., Hai, Y.: Clustering-anonymity approach for privacy preservation of graph data-publishing. J. Softw. 2323–2333 (2017)
3. Liu, X., Yang, X.: Protecting sensitive relationships against inference attacks in social networks. In: Lee, S., Peng, Z., Zhou, X., Moon, Y.-S., Unland, R., Yoo, J. (eds.) DASFAA 2012. LNCS, vol. 7238, pp. 335–350. Springer, Heidelberg (2012). https://doi.org/10.1007/978-3-642-29038-1_25
4. Huo, Z., Meng, X., Zhang, R.: Feel free to check-in: privacy alert against hidden location inference attacks in GeoSNs. In: Meng, W., Feng, L., Bressan, S., Winiwarter, W., Song, W. (eds.) DASFAA 2013. LNCS, vol. 7825, pp. 377–391. Springer, Heidelberg (2013). https://doi.org/10.1007/978-3-642-37487-6_29
5. Heatherly, R., Kantarcioglu, M., Thuraisingham, B.: Preventing private information inference attacks on social networks. TKDE **25**, 1849–1862 (2013)
6. Wang, Y., Zheng, B.: Preserving privacy in social networks against connection fingerprint attacks. In: ICDE, pp. 54–65 (2015)
7. Fu, Y., Zhang, M., Feng, D.: Attribute privacy preservation in social networks based on node anatomy. J. Softw. **25**, 768–780 (2014)
8. Huo, Z., Meng, X., Hu, H., Huang, Y.: You can walk alone: trajectory privacy-preserving through significant stays protection. In: Lee, S., Peng, Z., Zhou, X., Moon, Y.-S., Unland, R., Yoo, J. (eds.) DASFAA 2012. LNCS, vol. 7238, pp. 351–366. Springer, Heidelberg (2012). https://doi.org/10.1007/978-3-642-29038-1_26
9. Wang, W., Ying, L., Zhang, J.: The value of privacy: strategic data subjects, incentive mechanisms and fundamental limits. In: SIGMETRICS, pp. 249–260 (2016)

10. Abawajy, J., Ninggal, M., Herawan, T.: Privacy preserving social network data publication. Commun. Surv. Tutor. **18**, 1974–1997 (2016)
11. Karwa, V., Raskhodnikova, S., Yaroslavtsev, G.: Private analysis of graph structure. Trans. Database Syst. 1146–1157 (2014)
12. Wang, W., Ying, L., Zhang, J.: On the relation between identifiability, differential privacy, and mutual-information privacy. IEEE Trans. Inf. Theory **62**, 5018–5029 (2016)

Research on Influence Ranking of Chinese Movie Heterogeneous Network Based on PageRank Algorithm

Yilin Li, Chunfang Li[✉], and Wei Chen

Communication University of China, Beijing 100024, China
{lyler77, LCF}@cuc.edu.cn

Abstract. As the Chinese film industry flourishes, it is of great significance to assess the influence of film and film participants. Based on the theory of complex networks, this paper studies the ranking of influence in the film-participant heterogeneous network. Participants may have multiple identities such as directors, screenwriters, and actors. Referring to the PageRank algorithm of the page ranking algorithm and combining the features of the film industry, a new ranking algorithm, MovieRank, is proposed. The core three rules are as follows: (1) If the movie rank is high, the ranking of the participating players is also high; and if the participants have a high ranking. It also has a high ranking in participating movies; (2) the rankings of films and participating players are influenced by their social attributes; (3) the movie contributes more to their high-position participants, and the participants contribute more to the movie that they play an important role in it. Experimenting with Chinese movie information as experimental data, it is found that the new algorithm MovieRank actually performs better than the original PageRank algorithm. At the same time, through the analysis of the experimental results, it is found that the cooperation between actors from Hong Kong and Taiwanese is very close in the Chinese movie network, and that the directors and screenwriters have higher stability and less change than the actors.

Keywords: PageRank · Heterogeneous information network
Influence assessment · Film · Actor

1 Introduction

In recent years, China's entertainment industry has grown vigorously. Deloitte's "China Culture and Entertainment Industry Preview" report predicts that by 2020, the overall scale of the entertainment industry is expected to approach one trillion yuan. Among them, the film has developed rapidly as an important part of the entertainment industry. The total box office revenue rose 13.45% to $8.5 billion (55.9 billion yuan) in 2017, second only to the United States and ranking second in the world [1]. In the first quarter of 2018, the "Red Sea Action" and "Chinatown Investigative 2" were hotter at the box office, and the China box office beat North American for the first time with $3.17 billion (20.2 billion yuan) [2].

© Springer Nature Switzerland AG 2018
X. Meng et al. (Eds.): WISA 2018, LNCS 11242, pp. 344–356, 2018.
https://doi.org/10.1007/978-3-030-02934-0_32

To evaluate the success of a movie, you can't just look at the box office performance. For most ordinary audiences, who is the director of a film and who participates in it has greatly influenced its judgment of the value of the movie; at the same time, when evaluating the influence of an actor, it is usually also consider the value of the movie he participates in. Therefore, the influence between film and film participants is a complementary process.

How to systematically calculate the ranking of movies and the influence of their participants is a question worth studying. For the IP production and production distribution stage in the early stage of the film industry, the influence rankings of directors, screenwriters and actors can provide valuable references for film staffing. The later stages of film promotion, marketing, and the development of derivatives will also take into account the influence ranking of the participants. According to the movie's influence ranking feedback, it can also serve as an important indicator to support the next movie production decision. For participants, whether it is a director, screenwriter, or actor, effectively improving their influence rankings can help them obtain better resources and develop more effective business strategies. Scientifically predicting their value rankings also provides effective data support for collaboration between filmmakers and participants.

The network is an effective way to explore the interrelationships between things in mathematics and social sciences. Using complex network theory, it is easy to study the relationship between films and participants. In general, the PageRank algorithm can be used to calculate the ranking of each node based on the relationships in the edges of the network. In this article, we built a network of film-participant, where participants included directors, screenwriters and actors. This is a complex heterogeneous network.

To adapt to the characteristics of this type of network, we improve the PageRank algorithm, consider three rules, and propose a MovieRank algorithm that is more suitable for our research context. Through experiments, it is found that the effect of the new algorithm is more significant than the original PageRank algorithm for this network, and it is expected to provide some references for the future development of the Chinese film industry.

2 Related Work

For the movie partnership network, there are many studies. They use complex network methods to model actor collaboration relationships. The nodes in the network correspond to the actors and links between actors were defined by the same movies in which they performed. Initially Watts and Strogatz found the network shows a small-world property, for it has a higher clustering coefficient than a random graph of the same size and average degree [3]. According to Barabási and Albert, the network exhibits a scale-free behavior [4]. Some scholars in China have also conducted statistics and analysis of the Chinese actors network and analyzed the degree of nodes, the shortest path, and the clustering coefficient of the network, and proved that they have small-world characteristics [5].

In addition to the study of network characteristics, the assessment of node influence has gradually become the focus of many scholars. There are many methods to evaluate the influence of nodes using the complex network theory. The basic features of the network are usually used in the early stage [5]. The simplest is to measure the degree of the node, such as the betweenness of nodes. However, this method considers only part of the characteristics of the network and is not completely accurate. For large-scale networks, the computational complexity is very high. Later, many researches applied the PageRank algorithm to calculate the network node's influence ranking. The algorithm has high time efficiency and is suitable for large-scale networks. However, there are also some problems in specific applications, and appropriate improvements are needed.

For the movie partnership network, most of the current researches are based on actors-actors' homogeneous networks, and less research is based on heterogeneous networks. Research based on academic heterogeneous networks may provide some references. In an academic heterogeneous network, publication time has a great influence on the importance of the paper. According to the publication time of different papers, a PageRank algorithm that incorporates the time function is proposed and obtained a good experimental result [6]. Clustering algorithms are also attracting attention. RankClus is a ranking clustering algorithm based on dual-type literature networks [7].

To sum up, there are three problems in the study of the influence of the film cooperation network so far, which have been solved in our experiment. These problems have been solved in our experiments.

1. Most of the research is based on an actor-actor homogenous cooperation network, which contains much less information than heterogeneous networks.
2. When using the PageRank algorithm to calculate the network in a specific area, if you do not make some improvements based on the characteristics of the field, you will not achieve the desired effect.
3. There is not much research on the network of Chinese movie relations, and the data volume is relatively small.

3   Ranking Algorithm Based on Heterogeneous Film - Participants Network

3.1   Heterogeneous Film - Participant Network

In the study of complex networks, we refer to a network with only one type of node and only one type of edge relationship as a homogeneous network. In contrast, a network with two or more nodes or two or more edges is called a heterogeneous network.

The film-participant network discussed in this article is a heterogeneous network. The model is shown in Fig. 1. The network is composed of two types of nodes: participants and movies. The relationship between nodes and edges is based on who is involved in the movie. Among them, the relationship between the edges includes three types: director, screenwriter, and starring, that is, participants may have at most three different identities in a movie: director, screenwriter, and actor.

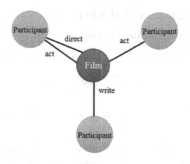

Fig. 1. Film-participant network model.

Compared with homogeneous networks, heterogeneous networks have many different entities and relationships and can contain more abundant information. For example, in a heterogeneous film-participant network, it can be shown which directors made the film, which scriptwriters wrote and which actors participated in the film. It can also show which movies a person has shot, which scripts he wrote, and which films he played. It also implies the cooperation between characters and characters. However, if we construct a homogeneous network based on the cooperative relationship between characters and characters, we can only show that a character and other characters have a cooperative relationship and cannot know the relationship between the character and the movie, and cannot learn more information. Moreover, for the network we are researching, under the same data, the number of edges for constructing the corresponding film-participant heterogeneous network is much smaller than that in the homogenous network of people-people. Using less edges to express more information while reducing computational complexity greatly improves the efficiency of research.

3.2   The PageRank Algorithm

The PageRank [8] algorithm was proposed by Larry Page and Sergey Brin, graduate students of Stanford University at the time, and was first applied to Google's web page ranking algorithm. The PageRank value of a web page is a comprehensive indicator of the importance of a page. The basic idea of the algorithm is: in the Internet, the importance of a page depends not only on the number of pages pointing to the page, but also on the quality of the page pointing to the page. That is, pages pointing from many high-quality pages must be high-quality pages. For example, if a page is linked by another page with a high PR value (PageRank value) such as Google, the PR value of the page will also be high. If it's linked by many of these pages, the PR value is higher. However, if it's linked by a number of pages that don't make much sense, the PR value doesn't change much.

Build directed networks based on the Web's massive link construction. Each page is a node. Build edges by linking between pages to relationships. The basic algorithm of PageRank is as follows:

1. Initialization. Set I represents all nodes in the network. Initializes the PR value for each node. For example, the initial *PR* value of node i is $PR_i^{(0)}$, where the initial PR value of all nodes satisfies Eq. (3.1):

$$\sum_{i \in I} PR_i^{(0)} = 1 \tag{3.1}$$

Usually set the initial PR value of each node is $1/N$, N is the number of nodes in the set I.

2. The following calculation is repeated until the PR value of each node converges, which is the final result. It is now the nth round. Then calculate the *PR* of node i according to Eq. (3.2):

$$PR(i)^{(n)} = \frac{(1-\alpha)}{N} + \alpha \sum_{j \in M_i} \frac{PR(j)^{(n-1)}}{L(j)} \tag{3.2}$$

Among them, M_i is a set of all nodes that have outgoing links to node i, $L(j)$ is the number of outgoing links of node j, and $PR(j)^{(n-1)}$ indicates the previous round of *PR* values on node j. α generally takes 0.85.

However, since PageRank is an algorithm proposed for solving webpage rankings, applying it to the film-participant's network in our study does not necessarily have a good effect. In the network structure, the network constituted by web pages is a directed homogeneous network, and the film-participant network is an undirected heterogeneous network. Therefore, we need to make some improvements, as detailed in the next section.

3.3   MovieRank: An Improved PageRank Algorithm

According to the topic context of the movie, we summarize the following three empirical rules as a theoretical support for improving the PageRank algorithm.

The Movie is Good - The Actors are Good. A good movie is made up of many good actors, good directors, and good screenwriters. A good actor has participated in many good movies, and a good director or screenwriter should also have many good movies. Therefore, in a heterogeneous network, the score of the movie should be calculated using the score of the participant, and the score of the participant is calculated using the score of the movie.

Social Attributes Influence. On the web page we assume that the probability of jumping between different web pages is the same, but in the network we study, it is clear that certain films, actors, directors, and screenwriters themselves will attract more attention. And the influential factors of such a priori accumulation are very complex. We call it social attributes. For the movie, it may include the movie's score, box office, release time, type, and so on. For actors, it may include fan appeal, age, gender, and so on. Therefore, we assign different initial scores according to different social attributes, and at the same time, we also take this impact into account in each iteration.

Position Contribution Value. The position here is the ranking of participants in the promotional materials and films, showing the influence of the participants and the importance of the role played in the movie. For example, a film contributes more to its main role than the supporting role. For an actor, comparing the importance of his position in the movie to all his movies, the contribution value of the movie to him can be known. Therefore, according to the difference in the contribution value, the assigned score should also be different, but the original PageRank algorithm actually considers this contribution value as equal.

According to the above three rules, based on the original PageRank algorithm, we constructed a new algorithm named MovieRank that is suitable for computing heterogenous networks with multi-type nodes and multi-type relationships such as film-participant network. The specific idea of the algorithm is described as follows:

1. Set the initial value. M is a collection of all movies. For a movie m_i, its initial value $MR(m_i)^{(0)}$ is the proportion of the score $Q(m_i)$ calculated according to some comprehensive social attributes in all movie scores.

$$MR(m_i)^{(0)} = \frac{Q(m_i)}{\sum_{m_j \in M} Q(m_j)} \tag{3.3}$$

V is a collection of all participants and k represents different identities such as directors, screenwriters, and actors. Then the initial value of a participant p_i with identity k, which is $MR(p_i, k)^{(0)}$, is calculated as the formula (3.4). $Q(p_i, k)$ is the score of the participant with identity k. $V(k)$ is a set of all participants with identity k.

$$MR(p_i, k)^{(0)} = \frac{Q(p_i, k)}{\sum_{p_{(j,k)} \in V(k)} Q(p_j, k)} \tag{3.4}$$

2. Calculate iteratively until all MR values converge. For the nth round, the calculation method is as follows:
For the movie m_i, $MR(m_i)^n$ is its MR value of the nth round:

$$MR(m_i)^n = (1 - \alpha)MR(m_i)^0 + \alpha \sum_{p_j \in Vm_i} MR(p_j, k)^{n-1} T(p_j, k, m_i) \tag{3.5}$$

In the formula (3.5), Vm_i represents the set of all cast members of the movie m_i, and $T(p_j, k, m_i)$ represents the proportion of the character p_j contributed to this movie by the identity of k. The calculation method is:

$$T(p_j, k, m_i) = \frac{R(p_j, k, m_i)}{\sum_{m_k \in V(m_k, k)} R(p_j, k, m_k)} \tag{3.6}$$

In the formula (3.6), $V(m_k, k)$ represents all movie sets in which the person p_j participates as the identity k. The movie for the collection is sorted by importance (position for the person) from low to high, where the order of the movie m_i is $R(p_j, k, m_i)$.

For the people p_i with the identity of k, $MR(p_i, k)^n$ is its MR value of the nth round:

$$MR(p_i, k)^n = (1 - \alpha)MR(p_i, k)^0 + \alpha \sum_{m_j \in V(p_i, k)} MR(m_j)^{n-1} T(m_j, k, p_i) \tag{3.7}$$

$V(p_i, k)$ represents a set of all movies, the person p_i is the identity k, and $T(m_j, k, p_i)$ represents the contribution ratio of the movie m_j to the person p_i who has identity k, and the calculation method is as formula (3.8):

$$T(m_j, k, p_i) = \frac{R(m_j, k, p_i)}{\sum_{p_k \in V(p_k, k)} R(m_j, k, p_k)} \tag{3.8}$$

$V(p_k, k)$ represents a set of all characters as the identity k in the movie m_j. $R(m_j, k, p_i)$ represents the reverse order of the position of the character p_i in the movie m_j.

After all the MR values have been calculated, the four MR of the movie, director, screenwriter, and actor are normalized and mapped to (0, 1). Then go to the next round of calculations.

4 Experimental Results and Analysis

4.1 Data and Algorithm

We have crawled Mtime, the influential platform of the Chinese movie industry, and only studied films and cast members in the Chinese region. Finally obtained data on a total of 15,156 films and 5090 participants. Among the participants, there were 913 as directors, 774 as screenwriters and 4249 as actors. The film includes information on the director, screenwriter and actor. Based on this, a film-participant network was constructed through the graph database platform Neo4j, according to whether the character participates in a movie by starring, directing, or writing. As shown in Fig. 2 below, it is easy to find all the actors, directors, and screenwriters of a movie.

Based on this network, we used the basic PageRank algorithm and the improved MovieRank algorithm to calculate the rankings of films, directors, screenwriters and actors. The experimental results and analysis are as follows.

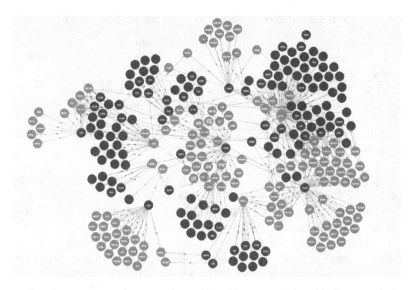

Fig. 2. The film-participant network on the graph database Neo4j.

4.2   Results and Analysis

Movies. Table 1 contains the top 20 movies ranked by the two algorithms. From the comparison, we can see that the PageRank ranking has some anomalies, improved algorithms have led to more of the excellent works on the top of the list, such as *Forever Young*, *The Stolen Years*, *Goodbye Mr. Loser* and *Detective Chinatown*. Analysis shows that the MovieRank corrects some exceptions brought by the PageRank.

The first exception arises from the fact that the actor is too closely related and isolated from the network as a whole. Take the two films with the highest PR calculated by the PageRank as an example. These two films have low status and low scoring in the real world, but the calculated PR is exceptionally high. The reason for this is that their participants are highly coincident, and almost all of the films in which the participants are involved are limited to these two films, so that their PR values are too concentrated. So the MovieRank uses film scoring as a social attribute in rule 2, weakened the impact of the centralized PR value distribution, which ranked 3,976 and 3,977, respectively, in the MovieRank ranking of the two films.

Run For Love, the fifth film in the original algorithm, belongs to another kind of abnormality. The film gets more PRs from participants because of its large number of participants, most of whom are well-known. As a result, even though the quality of the work itself is not high, the film ranks higher. The improved algorithm corrects this type of exception by rule 3: *Run For Love* falls into 90th place in the MovieRank ranking. Similarly, the film The Founding of A Republic, which has a large number of heavyweight participants, has also been downgraded after improvements in algorithms.

Table 1. Top20 movies based on PageRank and MovieRank.

Rank	PageRank			MovieRank		
	Ch. name	Eng. name	PR	Ch. name	Eng. name	PR
1	两个疯子的神经话	The Ramblings of Two Madmen	0.0013486	街头之王	The King of The Streets	0.0006145
2	两只神经的神经话	The Ramblings of Two Lunatics	0.0013486	无问西东	Forever Young	0.0005504
3	夏洛特烦恼	Goodbye Mr. Loser	0.0012010	超级保镖	The Bodyguard	0.0005441
4	西虹市首富	Hello Mr. Millionaire	0.0009802	缝纫机乐队	City of Rock	0.0005437
5	奔爱	Run For Love	0.0009566	猎时者	Time Hunter	0.0005321
6	1:99电影行动	1:99 Dian Ying Xing Dong	0.0009037	夏洛特烦恼	Goodbye Mr. Loser	0.0005268
7	西游降魔篇	Journey to the West: Conquering the Demons	0.0008928	寒战	Cold War	0.0005102
8	寒战2	Cold War II	0.0008323	飞行日志	Air Diary	0.0005078
9	一路惊喜	Crazy New Year's Eve	0.0008257	80'后	Heaven Eternal, Earth Everlasting	0.0005021
10	有一天	One Day	0.0007913	煎饼侠	Jian Bing Man	0.0004996
11	街头之王	The King of The Streets	0.0007701	猪太狼的夏天	Mr. Zhu's Summer	0.0004908
12	爱到底	L-O-V-E	0.0007700	闺蜜	Girls	0.0004862
13	煎饼侠	Jian Bing Man	0.0007654	被偷走的那五年	The Stolen Years	0.0004835
14	超级保镖	The Bodyguard	0.0007523	西游降魔篇	Journey to the West: Conquering the Demons	0.0004812
15	建国大业	The Founding of A Republic	0.0007457	唐人街探案	Detective Chinatown	0.0004781

Directors and Screenwriters. As can be seen from Tables 2 and 3, after the algorithm was improved, only three of the top 25 directors and screenwriters left the list, that is to say, the gap between the two rankings was not significant. This is because the number of directors and screenwriters per film is usually small and stable, and there is not much of a gap, so the added position contribution in the MovieRank has not had a significant impact.

Table 2. Top10 directors based on PageRank and MovieRank.

Rank	PageRank			MovieRank		
	Ch. name	Eng. name	PR	Ch. name	Eng. name	PR
1	王晶	Jing Wong	0.0147912	王晶	Jing Wong	0.0207024
2	杜琪峯	Johnnie To	0.0111251	杜琪峯	Johnnie To	0.0145666
3	邱礼涛	Herman Yau	0.0083543	张彻	Cheh Chang	0.0121104
4	徐克	Hark Tsui	0.0077387	邱礼涛	Herman Yau	0.0115155
5	刘伟强	Andrew Lau	0.0076421	徐克	Hark Tsui	0.0102098
6	陈木胜	Benny Chan	0.0063457	刘伟强	Andrew Lau	0.0092473
7	刘镇伟	Jeffrey Lau	0.0060135	吴宇森	John Woo	0.0077701
8	马伟豪	Joe Ma	0.0058305	刘镇伟	Jeffrey Lau	0.0077084
9	陈嘉上	Gordon Chan	0.0056307	马伟豪	Joe Ma	0.0076944
10	叶伟信	Wilson Yip	0.0056181	陈嘉上	Gordon Chan	0.0073595

Table 3. Top10 screenwriters based on PageRank and MovieRank.

Rank	PageRank			MovieRank		
	Ch. name	Eng. name	PR	Ch. name	Eng. name	PR
1	王晶	Jing Wong	0.0262014	王晶	Jing Wong	0.0364447
2	文隽	Manfred Wong	0.0083685	张彻	Cheh Chang	0.0119743
3	阮世生	James Yuen	0.0080821	黄百鸣	Bak-MinWong	0.0112662
4	刘镇伟	Jeffrey Lau	0.0076310	文隽	Manfred Wong	0.0110283
5	叶念琛	Patrick Kong	0.0075911	刘镇伟	Jeffrey Lau	0.0107969
6	陈庆嘉	Hing-Ka Chan	0.0071825	阮世生	James Yuen	0.0107298
7	黄百鸣	Bak-Ming Wong	0.0067439	徐克	Hark Tsui	0.0099178
8	谷德昭	Vincent Kuk Tak Chiu	0.0063889	陈嘉上	Gordon Chan	0.0090799
9	徐克	Hark Tsui	0.0059534	叶念琛	Patrick Kong	0.0088784
10	陈嘉上	Gordon Chan	0.0059456	马伟豪	Joe Ma	0.0083098

It may be noted that in the two top 10 rankings, directors and screenwriters are all from Taiwan China and Hong Kong. In fact, only Yimou Zhang, Xiaogang Feng and Kaige Chen made the top 25 in both rankings, while only Heng Liu, the mainland screenwriter, made the top 25 in both rankings.

Actors. Similar to the case with directors and screenwriters, Hong Kong and Taiwan actors make up the majority of the top performers. In two rankings, mainland China actors appeared in 37th and 42nd respectively, apart from the exceptionally ranked Matthew Ma in the PageRank: he acted the leading role in the films which are mentioned in the previous analysis of the movie as the example of the first anomaly.

Therefore, Tables 4 and 5 list the top 10 Hong Kong and Taiwan actors and Mainland actors respectively. In contrast to the director and screenwriter, the cast list has changed markedly under the two algorithms. Known as the Golden Supporting Role, Wu Ma, Lin Xue and Xu Shaoxiong, as well as the Mainland actor Liu Hua and Liu Ying, the fall in the ranking of such actors is a reflection of the effect of the position contribution in MovieRank on the PR allocation.

Table 4. Top10 Hong Kong and Taiwan actors based on PageRank and MovieRank.

Rank	PageRank			MovieRank		
	Ch. name	Eng. name	PR	Ch. name	Eng. name	PR
1	曾志伟	Eric Tsang	0.0036753	刘德华	Andy Lau	0.0040148
2	林雪	Suet Lam	0.0035440	曾志伟	Eric Tsang	0.0036894
3	任达华	Simon Yam	0.0029747	任达华	Simon Yam	0.0034510
4	午马	Ma Wu	0.0026597	梁家辉	Tony Leung Ka Fai	0.0030424
5	黄秋生	Anthony Wong	0.0026597	黄秋生	Anthony Wong	0.0028667
6	刘德华	Andy Lau	0.0025566	吴镇宇	Francis Ng	0.0027485
7	梁家辉	Tony Leung Ka Fai	0.0021931	刘青云	Ching Wan Lau	0.0026518
8	吴镇宇	Francis Ng	0.0021554	午马	Ma Wu	0.0026409
9	谷峰	Feng Ku	0.0019873	谷峰	Feng Ku	0.0026306
10	罗兰	Lan Law	0.0019154	洪金宝	Sammo Hung	0.0025500

Table 5. Top10 Mainland China actors based on PageRank and MovieRank.

Rank	PageRank			MovieRank		
	Ch. name	Eng. name	PR	Ch. name	Eng. name	PR
1	马茸原	Matthew Ma	0.0024734	周迅	Xun Zhou	0.0015451
2	黄渤	Bo Huang	0.0014924	葛优	You Ge	0.0015002
3	牛犇	Ben Niu	0.0013483	黄渤	Bo Huang	0.0014668
4	郭涛	Tao Guo	0.0013121	余男	Nan Yu	0.0013532
5	周迅	Xun Zhou	0.0012015	李连杰	Jet Li	0.0013265
6	于荣光	Ringo Yu	0.0012015	刘烨	Ye Liu	0.0012295
7	刘桦	Hua Liu	0.0011645	于荣光	Ringo Yu	0.0012259
8	余男	Nan Yu	0.0011257	范冰冰	Bingbing Fan	0.0012094
9	葛优	You Ge	0.0011232	巩俐	Li Gong	0.0011607
10	吕聿来	Yulai Lu	0.0011154	张静初	Jingchu Zhang	0.0011434

Close Cooperation Between Hong Kong and Taiwan. In all the participants' rankings obtained by the two algorithms, Hong Kong-Taiwanese participants are always at the forefront. The network density of the Hong Kong and Taiwan network is 0.105, which is significantly higher than the 0.005 in the Mainland network and 0.009 in the overall network. This will lead to the fact that during the calculation process, the PR value of Hong Kong and Taiwan participants tends to flow within close-knit circles, and the PR value of Mainland participants with relatively loose cooperation relationship will be dispersed. In the end, the PR value of participants from Hong Kong and Taiwan has become higher.

5   Conclusion

The ranking of influence on films and the evaluation of the value of directors, screenwriters, and actors are of great significance to the development of the entertainment industry. This article builds a heterogeneous network of film-participant in China. Taking full account of the characteristics of the film industry, three rules were proposed to improve the original PageRank ranking algorithm to a new algorithm MovieRank, which is more suitable for the context of the film industry. And through experimental verification, for the film-participant network, the new algorithm does indeed achieve better results than the original PageRank algorithm. This is mainly reflected in the weakening the influence of the aggregation of PR values, highlighting the influence of important players in the film, and balancing the advantages of movies of many participants in the network.

However, there are still many areas in our research can be improved. First of all, it is one-sided that we only assign the initial weight based on the Mtime score as a social attribute. We Can combine more platforms to crawl more stars and movie data, rich data attributes, such as movie box office, fan number, then will get a more comprehensive star ranking and movie rankings. It is also possible to explore the impact of social attributes on your rankings. Second, considering the rapid development of the movie industry, we can calculate rankings at different time periods to analyze the changes of rankings over time. Finally, we can also increase the types of nodes in the network, such as producers, production companies, and so on, and add some important related entities in the movie industry to see more macro industrial information.

References

1. Xinhuanet Homepage. http://www.xinhuanet.com/politics/2018-01/05/c_129783051.htm. Accessed 30 Apr 2018
2. Maoyan Homepage. http://maoyan.com/films/news/36439. Accessed 30 Apr 2018
3. Watts, D.J., Strogatz, S.H.: Collective dynamics of 'small-world' networks. Nature **393** (6684), 440 (1998)
4. Barabási, A.L., Albert, R.: Emergence of scaling in random networks. Science **286**(5439), 509–512 (1999)

5. Nan, H.E., Gan, W.Y., De-Yi, L.I., et al.: The topological analysis of a small actor collaboration network. Complex Syst. Complex. Sci. **3**, 1–10 (2006)
6. 周金梦. 基于学术异构网络的学者影响力评估算法 Evaluating the Scientific Impact of Scholars under Heterogeneous Academic Networks 大连理工大学 (2016)
7. Sun, Y., Han, J., Zhao, P., et al.: RankClus: integrating clustering with ranking for heterogeneous information network analysis. In: Proceedings of the 12th International Conference on Extending Database Technology: Advances in Database Technology, pp. 565–576. ACM (2009)
8. Page, L., Brin, S., Motwani, R., et al.: The PageRank citation ranking: bringing order to the web. Technical report, Stanford InfoLab (1999)

Extraction Algorithm, Visualization and Structure Analysis of Python Software Networks

Ao Shang, Chunfang Li$^{(\boxtimes)}$, Hao Zheng, and Minyong Shi

Computer School, Communication University of China, Beijing 10024, China
LCF@cuc.edu.cn

Abstract. Software complexity brings software developers and learners a series of challenges to face. Automatically analyzing large-scale software systems with complex network provides a new insight into software analysis, design, evolution, reuse, and iterative developing. Nowadays, extracting network models derived from software systems and making it easily comprehensible remains challengeable for software engineers. This paper focus on Python software. We propose a series of algorithms to extract python software networks, and a concept of visual information entropy to visualize network to an optimal statue by D3.js. Then we analyze python software networks in different perspectives by Pajek. A series experiments illustrate that software network can disclose the internal hidden associations to facilitate programmer and learner to understand the software complex structure and business logic through the simplified complexity. Finally we create a synthetic software tool integrated by above three functions, which can assist programmers to understand software macro structure and the hidden backbone associations.

Keywords: Complex networks · Software networks · Python

1 Introduction

Software networks coming from program itself, and they are also a kind of self-introducing documentary to synchronize with software codes structure. With the deepening of the research of software architecture, people gradually pay more attention to the software structure which is the important factor that determines the quality of software. Based on the analysis and research of software engineering, software engineers get guidance to develop high-quality software. Therefore, providing formal mechanism for representation, measurement and visualization of relational structure of complex systems, network analytic methods may arguably paved the way for a software development.

A complex network is a graph (network) with non-trivial topological features—features that do not occur in simple networks such as lattices or random graphs but often occur in graphs modelling of real systems. The small-world network model [1] and the scale-free network model [2] inspired research on complex networks. The abundance of data on interactions within complex systems allows network science to

© Springer Nature Switzerland AG 2018
X. Meng et al. (Eds.): WISA 2018, LNCS 11242, pp. 357–368, 2018.
https://doi.org/10.1007/978-3-030-02934-0_33

describe, model, simulate, and predict behaviors and states of such complex systems. It is thus important to characterize networks in terms of their complexity, in order to adjust analytical methods to particular networks. The measure of network complexity is essential for numerous applications [3]. Network complexity has been successfully used to investigate the structure of software libraries [4], to compute the properties of chemical structures [5], to assess the quality of business processes [6–8], and to provide general characterizations of networks [9–11].

In this case, changing the perspective and using complex networks to study software systems finds a way out for the above ideas. The complex network grasps the system as a whole, pays attention to the mapping relationship between the internal attributes and the external overall characteristics and the new features emerging in the whole system, which provides a powerful tool for controlling the complexity and ensuring the quality of software.

In 2009, Han [12] etc. analyzed the multi power topological properties of software networks. Through investigated software topology features of package, class, and method of an open-source system with undirected graph, and they consider that software networks can instruct the software configuration. In 2011, He and Ma etc. systematically researched software network evolvement and focused on software complexity metrics with complex networks. They systematically discussed the characteristics of "small world" and "scale-free" complex network in networked software from the perspective of networking, service and socialization, and proposed the concept of software network (software grows in the network and can describe software with network) [13–15]. In 2013, Wang and Lee etc. studied the impact of software network nodes on complex software systems and the modeling and analysis of weighted software networks. They analyzed the characteristics of software network nodes, revealing the gap between the conclusions of the research and the actual performance of software. Moreover, a weighted software network model is proposed to describe the dependencies between software nodes more accurately [16, 17]. In 2015, Zhao and Yu used the information content of nodes in the network model as an index for evaluating the importance of classes in software system, which is used to explore more important functional classes in software system operation [18]. Using the information entropy of the software network to measure the complexity of the software system. In 2016, Wang and Yu put forward an evaluation method for software major nodes (Class-HITS). The method uses complex network theory to break the loop in the software network, then obtaining the class integration test sequence according to the reverse topology sorting of the link free network. The experiment shows that the method can not only guarantee the priority of the important nodes to be tested, but also ensure that the total complexity of the constructed test pile is least [19].

There are some mature tools to analysis complex networks such as Pajek, C-Group, and Ucinet, so extracting network data is more important than analyzing network structures in order to improve the engineering application of software complex nework. Meanwhile, as a relatively simple programming language, python is widely used in various fields result from its rich and colorful expansion packages. Python has gradually become the mainstream programming language because of high efficiency in development. This paper aims to extracts network from python software, visualize and analyze software network structures to facilitate simplifying and modification of software. According to research and practice, developing an application which realizes the function of extracting and simplifying software network and visualization.

2 Extracting Algorithm

The first and crucial step to research software systems using networks is extracting software structure. As different programming language has different syntax, the extracting algorithm is different as well. In the past, many researchers extract network from C, C++, Java software, but rarely from python software. We have researched into approach to extract complex network from the structure of Python software, and established a complete extracting algorithm to extract network on .py file layer. Files of .py are extracted as nodes in the network; calling relationships between files are extracted as edges in the network.

2.1 Extracting Nodes

Python software consists of .py files, python packages and other types of files. Python packages can be regarded as a folder composed of a __init__.py file and some other .py files. Therefore, just open the project directory and do the following:

a. Scan the directory and visit all files successively.
b. Record all .py files as nodes of the network and save their path in a path list. Ignore other types of files.
c. Enter into subdirectory and iterate operations stated above when meet python packages.
d. Return to last directory until all files in this directory have been visited. If this directory is the root directory, then finish.

2.2 Extracting Edges

The edge between two nodes in a python software network indicates there is a calling relationship between two .py files, and the direction of the edge indicates the direction of the call relationship. The call relationship between .py files is realized by import statements. We can open and scan the .py file to find out the import statements, and analyze them, so as to determine the call relationships from this file to the others.

In the last step, we have got a path list of all .py files. Traverse the list, then open and scan every file. When we meet import statement, do operations following:

a. Import statements have two patterns, which are import A and from A import B. Transform the second pattern to the first one, for example, transform from A import B to import A.B.
b. Record what follows import as S. S can be regarded as a path of a .py file.
c. If S is a relative path, transform it to an absolute path. Judge whether S is in path list. If so, do g.

d. S also can be regarded as path.module; module refers to a method or an entity in __init__.py file of a package or in other normal files under path. So convert the last item of S to __init__, and judge whether S is in path list. If so, do g.
e. Remove the last term of S. Judge whether S is in path list. If so, do g.
f. It proves that S belongs to exterior package. So there is no new edge.
g. It proves that there is a directed edge from this file to S.

Then, all edges are found out. Figure 1 shows the flow chart of this procedure.

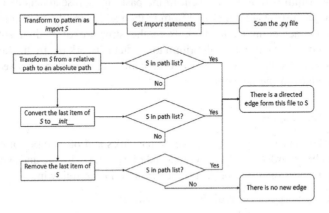

Fig. 1. Flow chat of extracting edges algorithm.

2.3   Output Networks

Output the network into a .net file which is compatible in pajek and a .json file which is commonly used in web pages, which is convenient for subsequent visualization and analysis. Net files are compatible in Pajek coded as Fig. 2. Json files are commonly used in web pages coded as Fig. 3.

```
*Vertices 52
1 "app.main.__init__"
2 "app.models.Drama"
...
51 "app.models.User"
52 "app.models.project_invest_company"
*Arcs :1 "SAMPLK1"
1 27 1
2 46 1
...
51 18 1
52 46 1
```

```
var nodes = [
    { name : "app.models.PassStudents" , type : 1 , num : 4},
    { name : "app.admin.ConfirmerView" , type : 2 , num : 6},
    { name : "app.models.Rollcall" , type : 1 , num : 8},
    ......
    { name : "app.models.Major" , type : 1 , num : 5},
    { name : "app.models.RollcallInfo" , type : 1 , num : 5}
];

var edges = [
    { source : 0 , target: 38 } ,
    { source : 1 , target: 56 } ,
    { source : 1 , target: 73 } ,
    ......
    { source : 75 , target: 38 } ,
    { source : 76 , target: 38 }
];
```

Fig. 2. File of .net format.   **Fig. 3.** File of .json format.

3 Visualization

3.1 Original Visualization with D3.Js

D3.js is a JavaScript library for manipulating documents based on data. It is one of the most popular visualization tools, with simple syntax to achieve colorful visualization effect. Here we visualize the software network with force directed graph realized by D3.js.

Figures 4 and 5 show the force directed graphs painted in the same color and in random color separately. The visualization results can show the network structure of the software, but they're too indifferent to extract useful information.

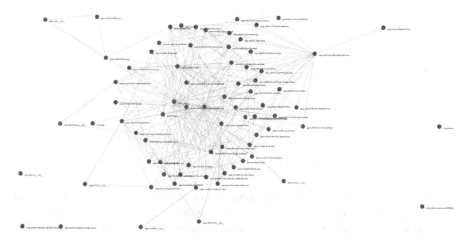

Fig. 4. SSSC (same size same colour).

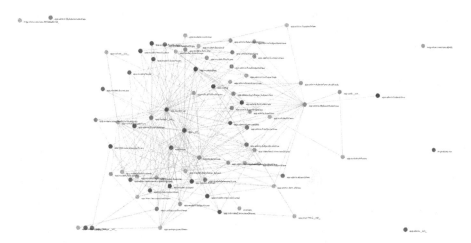

Fig. 5. SSRC (same size random colour).

3.2   Render Graph

In order to increase discrimination and information of graph, we render the graph by adding more features on nodes. Paint the nodes in with same colour, if they are in same package. Figure 6 shows the force directed graph with same size different colours. The internal relationship and regional division are indicated clearly.

Fig. 6. SSDC (same size different colours).

Next, change the size of the nodes according to the number of edges connected to them. Figure 7 shows the force directed graph with different sizes different colors. The macro structure and importance of nodes could be manifested. The internal relationship is also revealed further clearly. The materials and methods section should contain sufficient detail so that all procedures can be repeated. It may be divided into headed subsections if several methods are described.

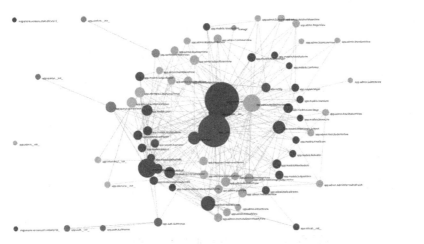

Fig. 7. DSDC (Different Sizes Different Colours).

3.3 Information Entropy Calculation

C. E. Shannon put forward the concept of information entropy in 1948, which solved the problem of quantitative measurement of information. The formula is as Eq. (1). E represents potential energy of certain information source, Z represents summation of potential energy, and H represents information entropy.

$$H = -\sum_{i=1}^{n} \frac{E_i}{Z} log\left(\frac{E_i}{Z}\right) \tag{1}$$

In a graph, the potential energy, or called the amount of information, of a node can be regarded as the probability one notice this node firstly, when he see the graph. We assume the probability relates to its features—color and size. In the Figs. 3 and 4, all nodes have same size, and their colors are indifferent as their colors are all different equal to those are all same. Each node has the same probability to be notice. In the Fig. 4, one tends to firstly notice a certain blue node because it has the most common color in the graph. In the Fig. 5, one tends to notice the biggest blue one firstly. If there are n nodes in a graph, for i-th node, C_i represents the number of nodes painted by same color in the graph, R_i represents the number of edges connecting to it, then the potential energy $E_i = C_i * R_i$, Z represents summation energy of all nodes. The information entropy of a graph can be calculate by Eq. (2).

$$H = -\sum_{i=1}^{n} \frac{C_i * R_i}{Z} log\left(\frac{C_i * R_i}{Z}\right) \tag{2}$$

We researched 10 python software, extracted their software networks and visualized them as different graph like above. Then we calculated their information entropy according to the Eq. 2. Table 1 shows the entropy of these graph. Confirming with the common cognition, we got the conclusion. For a network graph, the more average information is, the greater the entropy is, the less information can be extracted. On the contrary, the more discriminative information is, the smaller the entropy is, the more information can be extracted.

Table 1. Information entropy of visualizations.

Data	SSSC/SSRC	SSDC	DSDC
Ruyi	3.95	3.47	3.25
Exam	4.34	4.12	3.85
Tornado	4.80	4.49	4.07
Keras	5.08	4.89	4.48
Celery	5.79	5.54	5.09
Scrapy	5.88	5.43	5.09
Pandas	6.39	6.26	5.63
Scikt-learn	6.52	5.95	5.71
Tensorflow	7.73	7.17	6.26
Django	7.78	7.18	6.35

Ruyi is a play library system; Exam is an examination management system; Tornado and Django are Python web frameworks; Keras, Scikt-learn and Tensorflow are machine learning frameworks; Celery works as distributed task queue; Scrapy is a crawler framework; Pandas is a data analysis tool. The first two are practical projects; others are popular open-resource tools of Python maintained on GitHub.

4   Experiments

4.1   Parameters of Software Networks

Pajek is a large complex network analysis tool, which has powerful function to compute parameters and disclose the statistical features of networks. A group python software networks which have been mentioned above are analyzed in Pajek listed in Table 2. The statistical parameters include size of vertex N, size of edge E, average length L, cluster coefficients C, average in-degree or out-degree Degree, Diameter, Core.

Table 2. Statistical parameters of software networks.

Data	N	E	L	C	Degree	Diameter	Core
Ruyi	52	134	1.81	0.18	2.58	4	5
Exam	77	269	1.83	0.19	3.49	3	5
Tornado	122	515	1.84	0.21	4.22	6	9
Keras	160	584	2.83	0.18	3.65	7	7
Celery	328	945	2.83	0.19	2.88	10	7
Scrapy	357	896	2.59	0.15	2.51	9	5
Pandas	595	1977	3.46	0.19	3.32	9	12
Scikt-learn	677	2237	3.11	0.08	3.30	11	9
Tensorflow	2282	13183	4.06	0.14	5.78	11	14
Django	2400	7031	3.80	0.11	2.93	11	10

Degree Distribution

Take Scrapy network for example, in-degree, out-degree and external-degree distribution is illustrated in Fig. 8, and in log-log coordinate axis the degree distribution can be well fitted a line which discovers the scale free property of python software networks. In general, the existence of heavy-tailed out-degree distributions suggests a broad spectrum of complexity, and the existence of heavy-tailed in-degree distributions implies a broad spectrum of reuse [20].

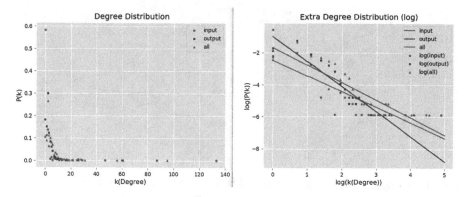

Fig. 8. In-degree and out-degree distribution of scrapy.

4.2 Richer Club Effects

In above 10 python software networks, iteratively delete nodes with degree = 1,2, ..., n each time. Gradually networks structure displays internal stronger association sub-networks, i.e. fewness important nodes with greater degree occupy most associations, and they prefer to connect with each other to support the network core and manipulate the networks backbone [21, 22]. With removing the less important nodes, network node scale and average length are decreased gradually shown in Figs. 9 and 10, network density and clustering coefficient are gradually increased shown in Figs. 11 and 12, which discloses that the rest nodes have closer relevance according with richer club effects.

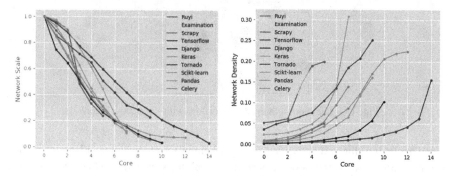

Fig. 9. Network scale and core distribution   **Fig. 10.** Network density and core distribution.

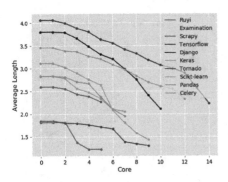

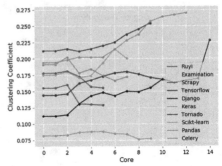

Fig. 11. Average length and core distribution.

Fig. 12. Clustering coefficient and core distribution.

4.3 Backbone Analysis

Tensorflow is one of the most popular tools in machine learning field. Here we extract k-core of its software network until last layer, and finally get its backbone. Figure 13 shows the last layer of this network which has 71 nodes. This sub-network is the foundation of Tensorflow's framework.

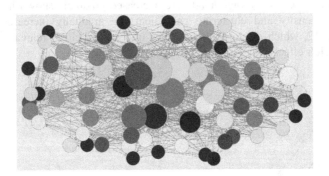

Fig. 13. Backbone graph of Tensorflow with extracting k-core

By analyzing this sub-network, 26 nodes belong to ops package; 13 nodes belong to framework package; 10 nodes belong to kernel_tests package; other nodes respectively belong to other different packages. Ops package implements the function of scientific computation. Framework package implements the function of building graph. Kernel_tests package implement the function of testing kernel. These three packages support the commonest function. Based on these three packages, developers can do advanced researches.

Then the six nodes with most degrees are extracted. Table 3 shows these nodes' information. These nodes built the foundation of Tensorflow's backbone. Software network support a view to comprehend the inner structure of software. The anatomy of

Tensorflow's structure provides guidance to developers to establish specialized machine learning system.

Table 3. The information of six nodes which have most degrees.

Node	Degree	Package	Package
array_ops	64	Ops	Operations of arrays
ops	62	Framework	Operation of framework
math_ops	57	Ops	Scientific computation
dtypes	55	Framework	Type of tensor
constant_op	48	Framework	Operation of constants
control_flow_ops	47	Ops	To control tensor's acceptance

5   Conclusions

This paper systematically studied on the python software network. The extraction algorithm provides an efficient method to extract the network from the python software, and automatically converts the extracted network into .json and .net files which are easy to be visualized by D3.js and be analyzed by Pajek. Both visualization and statistical analysis can help software engineers comprehend the macro structure of software systems. In particular, they provide a perspective in software analyzing, redundancy elimination and software incremental designing. Of course, the analysis of software networks is just the beginning. There are a lot of further work to do, such as to develop plug-ins component to help real-time analysis of Python software and visualization of the internal structure, which can facilitate software design and eliminate redundancy.

Acknowledgements. This paper is partly supported by "Key Cultivation Engineering Project of Communication University of China (Project number: 3132017XNG1606 and 3132017XNG1719)", "the Excellent Young Teachers Training Project (the second level, Project number: YXJS201508)", "Cultural technological innovation project of Ministry of Culture of P. R. China (Project number: 2014–12)". The research work was also supported by "Chaoyang District Science and Technology Project (CYXC1504)".

References

1. Watts, D.J., Strogatz, S.H.: Collective dynamics of "smallworld" networks. Nature **393** (6684), 440–442 (1998)
2. Barabasi, A.-L., Albert, R.: Emergence of scaling in random networks. Science **286**(5439), 509–512 (1999)
3. Morzy, M., Kajdanowicz, T., Kazienko, P.: On measuring the complexity of networks: kolmogorov complexity versus entropy. Complexity **2017**, 12 (2017). Article ID 3250301
4. Veldhuizen, L.T.: Software libraries and their reuse: entropy, kolmogorov complexity, and zipf 's law. In: Library-Centric Software Design (LCSD 2005), p. 11 (2005)

5. Bonchev, D., Buck, G.A.: Quantitative measures of network complexity. In: Bonchev, D., Rouvray, D.H. (eds.) Complexity in Chemistry, Biology, and Ecology, pp. 191–235. Springer, New York (2005)
6. Cardoso, J., Mendling, J., Neumann, G., Reijers, H.A.: A discourse on complexity of process models. In: Eder, J., Dustdar, S. (eds.) BPM 2006. LNCS, vol. 4103, pp. 117–128. Springer, Heidelberg (2006). https://doi.org/10.1007/11837862_13
7. Cardoso, J.: Complexity analysis of BPEL web processes. Softw. Process Improv. Pract. **12**(1), 35–49 (2007)
8. Latva-Koivisto, A.: Finding a complexity measure for business process models (2001)
9. Constantine, G.M.: Graph complexity and the Laplacian matrix in blocked experiments. Linear Multilinear Algebr. **28**(1–2), 49–56 (1990)
10. Neel, D.L., Orrison, M.E.: The linear complexity of a graph. Electron. J. Comb. **13**(1), 19 (2006). ResearchPaper 9
11. Strogatz, S.H.: Exploring complex networks. Nature **410**(6825), 268–276 (2001)
12. Han, Y., Li, D., Chen, G.: Analysis on the topological properties of software network at different levels of granularity and its application. Chin. J. Comput. **32**(9), 1711–1721 (2009)
13. He, K., Ma, Y., Liu, J., Li, B., Peng, R.: Software Networks. Science Press, China (2008)
14. Ma, Y., He, K., Ding, Q., Liu, J.: Research progress of complex networks in software systems. Adv. Mech. **5**, 805–814 (2008). ISSN 1000–0992
15. Ma, Y., He, K., Liu, J., Li, B., Zhou, X.: A hybrid set of complexity metrics for large-scale object-oriented software systems. J. Comput. Sci. Technol. **25**(6), 1184–1201 (2010)
16. Wang, B.: Software system testing based on weighted software network. In: International Conference on Information Technology, Service Science and Engineering Management (2011)
17. Wang, B., Lv, J.: Software network node impact analysis of complex software system. J. Softw. **12**, 1000–9825 (2013)
18. Zhao, Z., Yu, H., Zhu, Z.: The importance of dynamic software network nodes based on the information content. In: Application Research of Computer, no. 7, pp. 1001–3695 (2015)
19. Wang, Y., Yu, H.: An integrated test sequence generation method based on the importance of software nodes. J. Comput. Res. Dev. **3**, 1000–1239 (2016)
20. Myers, C.R.: Software systems as complex networks: structure, function, and evolvability of software collaboration graphs. Phys. Rev. E **68**(4), 046116 (2003)
21. Li, C., Liu, L., Lu, Z.: Extraction algorithms and structure analysis of software complex networks. Int. J. Digital Content Technol. Appl. **6**(13), 333–343 (2012). Binder1, part 36
22. Li, C., Liu, L.: Complex networks with external degree. Chinese J. Electron. **23**(3), 442–447 (2014)

Social Stream Data: Formalism, Properties and Queries

Chengcheng Yu[1], Fan Xia[2], and Weining Qian[2(✉)]

[1] College of Computer and Information Engineering,
Shanghai Polytechnic University, Shanghai 201209, China
ccyu@sspu.edu.cn
[2] School of Data Science and Engineering, East China Normal University,
Shanghai 200062, China
{fanxia,wnqian}@dase.ecnu.edu.cn

Abstract. A social stream, which refers to the data stream that records a series of social stream entities and the dynamic relations between entities, and each entity created by one producer. It is not only can used to model user generate content in online social network services, but also a multitude of systems in which records are combined by graph and stream data. Thus, the research efforts in the area about social stream is one of the hot spots recently. Although the term of "social stream" have appeared frequently, we note there are rarely formal definitions and lacks a unified view on the data. In this paper, we formally define the social stream data model trying to explain the graph stream generating mechanism from the perspective of producers. Then several properties describing social stream data are introduced. Furthermore, we summarize a set of basic operators that are essential to analytic queries based on social stream data, describe their semantics in detail. A classification scheme based on query time window is provided and difficulties lies behind each type are discussed. Finally, three real life datasets are used for the experiment of calculating properties to reveal differences between different datasets and analyze how they may exacerbate hardness of queries.

Keywords: Social stream · Formalism · Properties · Social stream queries

1 Introduction

Social stream is an abstract of a class of data streams, in which each entity created by a producer in the stream are unstructured data with attributes of time, in addition, entities have an interactive relationship. Social stream here is different from that stated in work [1], which is not an architecture model of social network sites, similar to the Twitter or tweet streams stated in works [1–3]. Besides, the changes of entity states in numerous applications can be modeled by social stream, such as email, scientific citation data or even computer network data and so on. Therefore, the research about social stream have attract great attention from both industry and academic. The term of "social stream" has appeared frequently, however, it is just an abbreviation for the sequence of user generate data in most cases.

© Springer Nature Switzerland AG 2018
X. Meng et al. (Eds.): WISA 2018, LNCS 11242, pp. 369–381, 2018.
https://doi.org/10.1007/978-3-030-02934-0_34

In this paper, the social stream data model is formally defined and is a generalization of many real life data. The partial order relationship between social items is defined, which is an important concept related to temporal. We also give several examples to show how different kind of real life data can be represented by our model. Furthermore, the semantic of social stream operators used to construct our social stream query is defined. We describe various constraints that may be imposed on social items and operations that operate on the linkage network. We also give a classification of queries and discuss their difficulties in respect. At last we explore the characteristics of three real life social stream data. One is a large Sina Weibo dataset we collected, and the others are the email and the U.S. patent dataset. We discuss the influence of the high skewness and strong temporal locality found in the dataset, to social stream operators.

2 Related Work

Different from traditional networks and data streams, social stream data is a combination of both. Networks usually can be represented as graphs, which is divided into two classes corresponding to whether edges in the graph have the property of temporal dimension, static networks and temporal networks or dynamic networks. A static network is a graph G:(V, E), in which V is the set of nodes, and E is the set of edges between nodes. A temporal network is modeled as time-varying graph (TVG) [4, 5]. The formalism of evolving graphs [6] is semantically equivalent to TVG. Then the temporal networks can be classified into two types [7]: contact sequence - a set of contacts and intervals graphs. The growing network [8] based on human interactions, such as retweet network in social media, citation network, and Internet are open and continuously growing, which is a specific temporal network. A growing network grows a network over time, new node with edges add to the network at each time step. A graph stream data, is formatted conventionally as a sequence of elements <x,y; t>, which means that the edge (x,y) is encountered at time t. It is similar to the concept about the contact sequence of temporal networks, the time attribute also just be considered as a label or an order. However, the time of attribute just be thought of as a label of the node of a graph in most studies about temporal networks, the properties of the social stream about the producer and temporal attributes cannot be considered for understanding pattern of behavior.

On the other hand, a lot of work have been devoted to problems of social stream or related concepts from different aspects. The home timeline query is one of the fundamental query required by typical OSNs and has been studied extensively [9, 10]. It also suffers from difficulties of traditional data, like high arriving ratio of items and limited resources like CPU and memory. In addition, the interconnection between items makes it hard to find a suitable partition scheme and skewness in the degree distribution of items may cause the performance of the whole system degrade due to activities of popular items. Thus, some work, e.g. [11, 12] etc., attempt to tackle the partition problem by minimization the replications of items or inter-server communications. There are also works [13] that processes streaming edges to detect events in social media data. We hope to abstract a kind of social stream query and discuss the difficulties lying behind.

Now a multitude of research efforts properties or patterns of static and temporal graphs, which together characterizing a real-life graph, and usually different networks have similar patterns, such as power law, small diameter. Power law distributions or heavy-tailed distribution occur in diverse range of phenomena, including the social network [14], email [15], and so on. The diameters are usually much smaller compared to the graph size. The efficient diameter [16] of the AS-level Internet [17] is around 4, and the router-level Internet is about 12. Moreover, there are many other measures to character a graph, including clustering coefficient, expansion, prestige, prestige and so on [18, 19]. A temporal graph can be represented as a sequence of several static graphs, which is the aggregation of interactions that occur in a given interval. The atemporal properties are defined on these static graph. And the temporal ones take into account their temporal nature. For example, the evolution of some measures about citation network over time have been studied [20], such as the exponent of power law degree distribution, diameter, clustering coefficient and so on. And there are two dynamic patterns in many networks, densification power law, and Shrinking diameter [21]. But we want to know how does a social stream look, what are the distinguishing characteristics of social streams, and what rules and patterns hold for them. Several properties or patterns of the social stream summed in this paper will answer these questions. And theses properties may help to analyze how they may exacerbate hardness of social stream queries.

3   Social Stream

In this section, we propose the social stream data model, then introduce the linkship network in a social stream. Finally, we will show many kinds of real-life data can be modeled as social stream.

3.1   Social Stream Data Model

Please note that the first paragraph of a section or subsection is not indented. The first A *social stream* is a series of *social items*. A social item is a tuple,

$$< s_i, t_i, p_i, L_i, C_i >$$

where s_i is the identifier that uniquely identifies a social item, t_i is the timestamp of the emergence of s_i. Note that s_i's are associated with a partial order $<$. For two social items s_i and s'_i, if $t_i < t'_i$, we call s_i is an elder item of s_i'. Moreover, p_i is the producer who post or create the item s_i. The field Li is a set of elder items' identifier. Moreover, Ci is the content. We use $s_i \leftarrow s_i'$ to denote that s'_i points to s_i, if $s'_i \in$ Li. Let us denote by s_i Y s_j the path from s_i to s_j. And $\| s_i \frown s_j \|$ is the number of hops, we call it the length of the path.

For easy of discussion, a naming convention based on the relationship between items is proposed. Social streams within which each item points to at most one elder item are called single-link social streams. Correspondingly, social streams with items that can point to any number of elder item are called multi-link social streams. We have two examples about single-link and multi-link social stream shown in Fig. 1. Two examples on Fig. 1(a) single-link social stream and Fig. 1(b) multi-link social stream, and the final global single-link and multi-link stream we should generate are shown in Fig. 1(c) and (d) respectively. Black lines with arrow are related to producers' sub-timeline in (a) and (b), or global social stream in (c) and (d). Take the link $s_4 \rightarrow s_2$ for example, which producers of endpoints are the same one, we call this link as a self-link. The linkship network formed by all items s_1, s_2, ..., s_7 and links between these items, which has a directed tree graph topology in single-link stream, and has a directed acyclic graph topology in multi-link case.

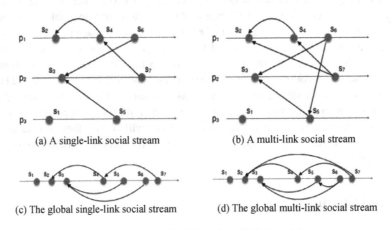

(a) A single-link social stream   (b) A multi-link social stream

(c) The global single-link social stream   (d) The global multi-link social stream

Fig. 1. Examples of single-link and multi-link social streams.

3.2   Social Stream Data Model

The linkship network grows over time, and new node with edges is added to the network in each step. The linkship network can be represented as a directed graph G: (S, E), where S is the set of social items (nodes), for each $s_i \in S$, E is the set of links (edges) between these items. And for each pair of snapshots of graph at time t_1 and t_2, G: (S_1, E_1) and G: (S_2, E_2). If $t_1 < t_2$, then $S_1 \subseteq S_2$ and $E_1 \subseteq E_1$, furthermore, if $S_1 = S_2$, then $E_1 = E_2$.

Lemma 1. *The linkship network is a set of DAGs (directed acyclic graphs). Furthermore, the network formed by single-link stream has a directed tree graph topology.*

Proof. We use reductio ad absurdum to prove that the linkship network G created by a social stream is a set of DAGs. If there is a connected component in G, it must contain a directed cycle $\{s_0 \rightarrow s_1, ..., s_{n-1} \rightarrow s_n, s_n \rightarrow s_0\}$. According to the partial order in the definition of a social item, $s_0 < s_1$ for $s_0 \rightarrow s_1$, and $s_{n-1} < s_n$, then we can get that $s_0 < s_n$, but $s_n < s_0$ for $s_n \rightarrow s_0$, which is inconsistent with the conclusion we deduced $s_0 < s_n$. So G is a set of directed acyclic graphs. And the network formed by single-link stream has a directed tree graph topology, because of each item has one out-degree at most.

3.3   Social Stream Data Model

Social streams can model the spreading dynamic of information in social network. Social stream is a general form of many kinds of real-life data. In this section we illustrate how several real life data can be modeled by the proposed model in detail.

Microblog Retweet: User generated content of social service systems are particularly suitable for modeling to social streams, such as tweet and reply, a term borrowed from terms of Twitter. A tweet can be represented by the tuple $\langle s_a, t_a, p_a, \varnothing, C_a \rangle$, which means user pa post a tweet s_a at the time ta, the content of the tweet is C_a, for example, while a retweet of s_a is of the form $\langle s'_a, t'_a, p'_a, \{s_a\}, C'_a \rangle$. Apparently, $t_a <$ ta', thus $s_a < s'_a$. And the tweet stream is a single-link social stream obviously. Retweet trees created by this stream is the linkship network, following network can be seen as the producer network.

Technical Publications: Many technical publications can be rep-resented by this form, such as patent documents, research papers, e-mail, and so on. For example, a patent document can be modeled to this form $<s_b, t_b, p_b, L_b, C_b>$, which means an inventor or assignee p_b publish a patent s_b, and the grant date is t_b.

Network Traffic: In the internet network, routers exchange message among each other. Now we will demonstrate how social stream model message transmission in the internet of the producer network. A message sent by a router is a social item of the form $\langle s_c, t_c, p_c, \varnothing, C_c \rangle$, where p_c identify the router, and retransmission message of s_c by another router pc' is of the form: $\langle s'_c, t'_c, p'_c, \{s_c\}, C'_c \rangle$. And there is no self-link in the distributed systems. The in-degree of items suggest that whether message is sent by unicast, multicast or broadcast transmission. And the linkship network show a directed tree graph topology.

Besides, the email stream is another typical single-link social stream, an email is related to an item, the sender of the email is the producer of the item, and the relation of reply between emails is related to the linkship between items. Similarly, the other social medias of Patent, Blog, Twitter, Facebook, goods transportation and so on are single-link social stream.

4   Social Stream Properties

In this paper, we focus on properties of social stream structure, the content part of social items is not being considered. Social stream data is the combination of graph and stream data, and the graph is related to the linkship network of a social stream $G: (S, E)$. Some graph patterns or properties are relevant to the social stream, including distributions of degree, connected components diameter and size. Moreover, there are some other important properties of social stream about temporal attribute, such as distributions of productivity, inter-item time, link interval and periodicity.

Degree Distribution: The degree of a node is the number of edges connected with that node. For a directed graph, there are in-degree and out-degree for a node. The in-degree and out-degree distributions are the probability distribution of these degrees over the whole network respectively. $d_{in}(s_i)$ and $d_{out}(s_i)$ are in-degree and out-degree of social item s_i.

$$d_{in}(s_i) = \left|\left\{s_j | s_i \in L_j \wedge s_i, s_j \in S\right\}\right| \tag{1}$$

$$d_{out}(s_i) = \left|L_j\right| \tag{2}$$

A degree distribution follows a power law if $D(k) \propto k^{-\alpha}(-\alpha > 1)$. Such degree distribution have been observed for the Internet graph [17], citation graphs [22] and many other. And most real networks have a power law exponent between 2 and 3.

Connected Components Diameter (CCD) Distribution: It is the probability distribution of these connected components diameter over all weakly connected components in linkship network. We denote by $D(CC_i)$ the diameter of the connected component $CC_i : (S_i, L_i)$. A graph may consist of several connected components $G = \{CC_1 \vee CC_2 \vee \dots \vee CC_N, N > 0\}$.

A connected component (or just a component) is a maximal set of nodes and edges where every pair of nodes in the set are connected to each other by paths. For directed graphs, there are weakly and strongly connected components, a strongly connected component in which exists a directed path connecting any pair of nodes in the component. In this paper, we focus on the weakly connected components.

$$D(CC_i) = \max\left\{\left\|s_j \rightsquigarrow s_k\right\| \, | \, s_j, s_k \in S_i\right\} \tag{3}$$

Connected Components Size (CCS) Distribution: It is the distribution is the probability distribution, which is the fraction of components with a specified size s in all weakly connected components. Let us denote by $S(CC_i)$ the size of connected component $CC_i : (S_i, L_i) \subseteq G$.

$$S(CC_i) = |S_i| \tag{4}$$

Now we will discuss that other properties of social stream about temporal attribute.

Post Rate Distribution: It is the probability distribution of producers' post rate over the number of producers with the specified rate $r < 1$. The post rate of the producer p_i can be calculated by Eq. 5.

$$\lambda(p_i) = \frac{\left|\left\{s_j | p_j = p_i \wedge s_j \in S\right\}\right|}{T} \tag{5}$$

in which $\lambda(p_i)$ is the rate of p_i output social items, T is the time range of the whole social stream. Many streams' items frequency of producers yield a power law distribution or the deviation, such as Twitter [23], citation of papers [24] and so on. The distribution of productivity plays an important role in models of posting items.

Inter-Item Interval Distribution: It is a distribution of the time intervals between two consecutive items a producer published. We denote by T item $(<s_i, s'_i>)$ the time intervals between two consecutive items the producer p_i published.

$$T_{item}\left(<s_i, s'_l>\right) = t'_i - t_i, \; p_i = p'_i, s'_i = max\{s_j | s_j > s_i\} \tag{6}$$

The pattern of inter-item time distribution is of importance to understand the behavior of social items in the stream. The distributions of some systems' follow an exponential distribution. While it is reported that the distribution of inter-item time obey heavy-tailed power-law distribution on Sina Weibo [25], email [15], and other systems.

Link Interval Distribution: Similarly, the distribution of the intervals between an item and its each (direct) linked item is widely interested. We denote by $T_{link}(s_i \rightarrow s_j)$ link interval of $s_i \rightarrow s_j$. It is also reported that retweet intervals of Sina Weibo approximates follow a power-law distribution [25].

$$T_{link}\left(s_i \rightarrow s_j\right) = t_i - t_j \tag{7}$$

Periodicity: The output of a social stream over time may appear the phenomenon of the periodicity and fluctuation. For example, the number of items in the social stream over days within a week may be similar to that within another week, the number of items over hours within a day, and so on. The periodic fluctuations have been observed on Twitter [26], Blogspace [27] and many others.

5 Queries Statement

In the previous section we define a formal model of stream social model and introduce the partial order between items. This section describes several operators that constitute our social stream query.

When designing operators, we mainly consider two aspects, i.e. the temporal and the linkage structure. As mentioned previously, t_i of s_i is used to decide the partial order and we use it to decide its location in the temporal dimension. Unlike the flat structure of traditional streams, social items are connected through the linkage fields, which enriches the semantic of query. The basic operators about social streams contain *merge, join, filter, project* and *aggregation*. In the following we introduce operators one by one in detail. We first describe two relative simple operators, both of which take two social streams as input and produce one as output. The first operator **merge** merges social items in input streams according to the partial order. It behaves similar to the union operator in relation algebra. Duplicates are eliminated and ties are break arbitrarily. The second operator **intersect** takes two streams S_1 and S_2 as input. It selects social item $<s_i, t_i, p_i, ...>$ where p_i contains items in S_2. This operator behaves much like the join operator.

Now we introduce the important operator *filter*, which takes a social stream as input and produce another stream by filtering social items using provided constraints. It takes constraints that are evaluated against different parts of a social item. Based on the constrained part, constraints are categorized into the following three part:

- **Temporal Constraint:** Our temporal constraint is given in the form of time window $[t_s, t_e]$. Let t_n denote the current time and $-\infty$ denote infinite past time. By default, the time window $[-\infty, t_n]$ are imposed on all operators that take a social stream as input and we assume all operators in a social query get the same t_n. A social item s_i may survive at time instance t_i or survive until t_j of item s_j which is semantically related to s_i.
- **Content Constraint:** It is the constraint that is evaluated against the content part of a social item. However, the structure of content depends greatly on specific applications and is essentially opaque to us. It could be simple functions like numeric comparison, keywords containment or some user defined complex functions.
- **Structure Constraint:** This constraint is tested on the linkage set to reflect some structure related conditions. This constraint may be evaluated against the valid linkage set, which only is a subset of the linkage set containing items existing in input stream. The constraint may state the size of the set should satisfy some numeric condition. An example would be select out all root items that don't point to any elder items.

The operator of **project** should be emphasized, which also takes in a social stream and produce another one. In relation algebra, project is used to select out a subset of fields of tuples. In our scenario, it expands input stream with descendants or ancestors of each item in the input stream. Let D_{si} and A_{si} denote descendants and ancestors of s_i respectively. Then $D_{si} = \{s_x \mid s_x \frown s_i\}$ and $A_{si} = \{s_x \mid s_i \curlyvee s_x\}$. It is easy to derive that $D_{si} \cap A_{si} = \varphi$ using Lemma 1.

The last operator we give is **aggregation**, which takes a social stream as input but produces a set of relation tuple. It accepts a function f that is invoked on each social items and returns a real value. This operator transforms stream data to relation data and thus can connects to any relation algebra to achieve functionality like top k ranked items. Based on explicit time window constraints, queries can be categorized into two types of queries.

- **Real Time Query:** The time window of real time query is $[t_n - w, t_n]$, which means the query only cares about items published in recent w time units. It can be ad- hoc queries like home timeline query mentioned previously, or it can be executed periodically to achieve monitoring facility. Similar to data stream problem, this kind of queries also suffer from limited resources, e.g. utilization of CPU, size of memory or bandwidth of network, while still needs to guarantee the evaluation performance and quality of answers. The potential spikes in arriving ratio of social items just make the situation even worse. In addition, the inter-connection among items causes it difficult to find a suitable partition schema of the data.
- **Historical Query:** The time window of historic query could be any past time interval. It is common for analytic jobs to dive into past data. For example, what if we hope to find out what people talk about for different stages when some product was released. To answer those queries efficiently, we may need index tailored for

dynamic graph [28] so that we could rapidly construct a static graph for the query window. Or we should consider that temporal properties when partition the data, thus access to both temporal or structure unrelated data can be avoided.

6 Experiments

In this section, there are three real datasets for experiments of calculating properties to reveal differences between different datasets and analyze how they may exacerbate hardness of queries, including Sina Weibo, Email and patent.

6.1 Real-Life Dataset

Three kinds of real-life datesets' information are listed in Table 1, including Sina Weibo, Email and patent. The real-life data set of tweet stream crawled from Sina Weibo, the most popular microblogging service in China. It includes all tweets from more than 1.4 million real users within 2012 year, is used to determine the parameters of data distributions. The details on the methodology of crawling the data is introduced in [29]. The real dataset of email stream is collected by CALO Project, which contains data from about 150 users. There are actions of reply and forwarding, then the dataset contains 70492 emails sent by 7589 users over eight years actually, although email lists of most users are not incomplete. We assume that all users' emails are complete, and parameters learn from the dataset are used to generate synthetic email stream. We take the patent document stream as an example of multi-link social streams. The data set of patent document stream was originally released by NBER, and includes the US utility patents granted from 1969-01-07 to 1999-12-28.

Table 1. Social stream datasets

Data	Type	#Producer	#Items	Time range
Weibo	Single-link	1.4 M	693.7 M	1 year
Email	Single-link	7.5 K	70 K	8 years
Patent	Multi-link	166 K	2 M	31 years

6.2 Preliminary Results

Firstly, we show results of first two distributions to explore the pattern how items are linked to others. The distributions of items' in-degree and out-degree are shown by Fig. 2(a) and (b). The horizontal axis is items' d_{in} or d_{out}, while vertical axis is the percentage of items having that d_{in} or d_{out} value. The patent stream only has the distribution of items' out-degree, due to the datasets of Weibo and Email are single-link social stream, whose out-degree of items are equal to 0 or 1. We can see that all these distributions obey the power distribution, which agree with previous studies [23, 29]. Moreover, it may consume a lot of resources ranging from disk I/O, CPU utilization to memory for operators like intersect and project if celebrity items appear in input social stream. Systems should be designed elegantly handle those situations, and hence avoid the unavailability of the whole system.

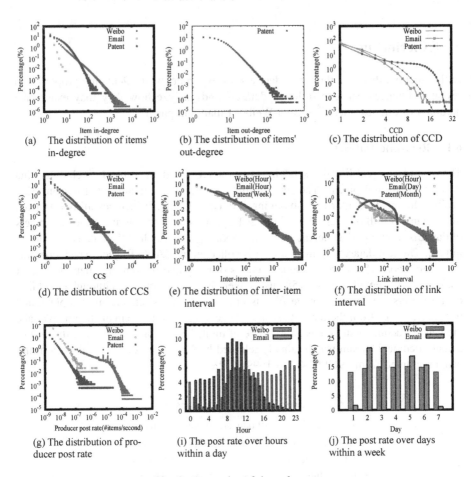

Fig. 2. Properties of three datasets

The diameter and size of connected components, are often treated as indicators of propagation ability of social items. The distributions of CCD and CCS are given by Fig. 2(c) and (d). The horizontal axis is the $D(CC_i)$ or $S(CC_i)$, while the vertical axis is the percentage of connected components having that diameter or size. The connected components whose diameters are smaller and equal to 4 count for 94.7%, 97.7% and 78% of the total for Weibo, Email and Patent. And the obtained power law distributions of CCS show popular items only occupies a small fraction of the total data. However, project operator incurs great performance penalty when those popular items are encountered. Those items may be close to each other due to the temporal locality found previously, which makes the situation worse.

Then we study two interval based distributions which may give us a portrait of arriving pattern of items. Similarity, inter-item interval and link interval distributions shown by Fig. 2(e) and (f) confirm to the power law distribution. The horizontal axis is T_{item} and T_{link}, and vertical axis is the percentage of item pairs with according value. The power law exponents of inter-item interval distribution of Weibo, Patent and Email are 1.4497, 1.8285 and 1.7806, and the link interval exponents of Weibo and Email are 1.4607 and 1.5705, the link interval distribution of patent social stream doesn't follow the common power law distribution. Since items in Sina Weibo are generated rapidly in most of the time, a filter operator with an even small query window can select out huge number of items. Similar conclusion applies to operators like intersect and project as most retweet actions happen immediately after the publish of retweeted tweets.

Figure 2(g) shows the producer post rate distributions, the horizontal axis is post rate λ, and vertical axis is the percentage of producer with the post rate λ. The distributions of Patent and Email follows the power law distribution with exponents 1.815 and 1.579. Weibo's post rate distribution is a combination of two power law distributions, where about $\lambda > 10^{-5}$ fits to a power-law distribution with the exponent of 2.9732, the else part fits' power law exponent of 1.3971. These reveal the activity of producers and the velocities of the whole social stream and different producers' personal social stream. The processing of different producers' items may consider big differences of items' posting rate between producers.

The result of the remaining three distributions of post rate over time within a time period is given by Fig. 2(h), (i) and (j), whose horizontal axis are weeks within a year, hours within a day and days within a week, respectively, while the vertical axis is the percentage of the post rates in according interval respectively. It is easy to find the intriguing phenomena that there exists multiple similar trends in each distribution. This fact reminds us that solutions for social stream query should be aware of such evolution of data over time.

We conduct some supplementary experiments to study whether parameters of those power law distributions changes over time. The time axis is divided into consecutive time slices and distributions of items falling in each time slice are fitted by the power law distribution. The exponents of power law distributions of Weibo, Patent and Email for each time slice are plotted and show in Fig. 3(a), (b) and (c). The power law exponents of each distribution, are basically stable in the datasets of Weibo and Email, increaseing slowly over time in patent stream. We notice that the exponents of producer

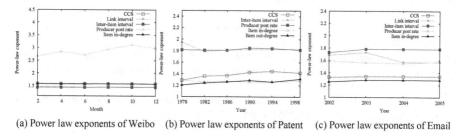

(a) Power law exponents of Weibo   (b) Power law exponents of Patent   (c) Power law exponents of Email

Fig. 3. Power law exponents of three datasets' characteristics.

post rate increase over time in weibo, which can be explained that people tend to be silent and only talk in emerge events. And the power law exponents of producer poster rate decrease over time in Email, because of email data is concentrated in 2003 and 2004, sparse data in other years.

7 Conclusions

In this paper we provide a formal definition of the social stream data. The proposed model can be applied to many real life data and several typical examples are given. Furthermore, we define a set of operators that constitute the social stream query. Their combinations can produce a stream or a set of relation tuples from input streams. We classify queries into two categories and discuss their challenges. At last, exploration of characteristics of three real life data, i.e. Sina Weibo, U.S. Patent and Email, reveals the high skewness and strong temporal locality in social stream data. Operators may incur performance penalty when such kind of data may be confronted. Hence we believe our work will benefit future researches on social streams.

References

1. Tapiador, A., Carrera, D., Salvachua, J.: Social stream, a social network framework. In: First International Conference on Future Generation Communication Technologies (FGST 2012), pp. 52–57 (2012)
2. Sasahara, K., Hirata, Y., Toyoda, M., Kitsuregawa, M., Aihara, K.: Quantifying collective attention from tweet stream. PLoS ONE 8(4), 61823 (2013)
3. Nishida, K., Hoshide, T., Fujimura, K.: Improving tweet stream classification by detecting changes in word probability. In: Proceedings of the 35th International ACM SIGIR Conference, pp. 971–980. ACM (2012)
4. Casteigts, A., Flocchini, P., Quattrociocchi, W., Santoro, N.: Time-varying graphs and dynamic networks. Int. J. Parallel Emerg. Distrib. Syst. 6811(5), 346–359 (2010)
5. Santoro, N., Quattrociocchi, W., Flocchini, P., Casteigts, A., Amblard, F.: Time-varying graphs and social network analysis: temporal indicators and metrics. In: Artificial Intelligence and Simulation of Behaviour, pp. 32–38 (2011)
6. Ferreira, A.: Building a reference combinatorial model for manets. IEEE Netw. Mag. Glob. Internetw. 18(5), 24–29 (2004)
7. Holme, P., Saramki, J.: Temporal networks. Phys. Rep. 519(3), 97–125 (2012)
8. Krapivsky, P.L., Redner, S., Leyvraz, F.: Connectivity of growing random networks. Physics 85(21), 4629–4632 (2000)
9. Wei, J., Xia, F., Sha, C., Xu, C., He, X., Zhou, A.: Web Technologies and Applications. Lecture Notes in Computer Science, vol. 7808, pp. 662–673. Springer, Heidelberg (2013)
10. Gionis, A., Junqueira, F., Leroy, V., Serafini, M., Weber, I.: Social piggybacking: leveraging common friends to generate event streams. In: Proceedings of the Fifth Workshop on Social Network Systems (2012)
11. Pujol, J.M., et al.: The little engine(s) that could: scaling online social networks. In: Proceedings of the ACM SIGCOMM. pp. 375–386 (2010)

12. Chen, H., Jin, H., Jin, N., Gu, T.: Minimizing inter-server communications by exploiting self-similarity in online social networks. In: 20th IEEE International Conference on Network Protocols, ICNP (2012)
13. Angel, A., Koudas, N., Sarkas, N., Srivastava, D., Svendsen, M., Tirthapura, S.: Dense subgraph maintenance under streaming edge weight updates for real-time story identification. VLDB J. 23(2), 175–199 (2014)
14. Kwak, H., Lee, C., Park, H., Moon, S.B.: What is twitter, a social network or a news media? In: WWW, pp. 591–600 (2010)
15. Holger, E., Lutz-Ingo, M., Stefan, B.: Scale-free topology of e-mail networks. Phys. Rev. E: Stat. Nonlinear Soft Matter Phys. 66(3), 035103 (2002)
16. Tauro, S.L., Palmer, C., Siganos, G., Faloutsos, M.: A simple conceptual model for the internet topology. In: Global Telecommunications Conference, 2001. GLOBECOM 2001. IEEE, vol. 3, pp. 1667–1671 (2001)
17. Faloutsos, M., Faloutsos, P., Faloutsos, C.: On power-law relationships of the internet topology. In: SIGCOMM, pp. 251–262 (1999)
18. Newman, M.E.J.: The structure and function of complex networks. In: SIAM Rev, pp. 167–256 (2006)
19. Watts, D.: Collective dynamics of 'small-world' networks. Nature 393, 440–442 (1998)
20. Quattrociocchi, W., Amblard, F., Galeota, E.: Selection in scientific networks. Soc. Netw. Anal. Min. 2(3), 1–9 (2010)
21. Leskovec, J., Kleinberg, J., Faloutsos, C.: Graph evolution: densification and shrinking diameters. Physics 1(1), 2 (2006)
22. Redner, S.: How popular is your paper? an empirical study of the citation distribution. Phys. Condens. Matter 4(2), 131–134 (1998)
23. Welch, M.J., Schonfeld, U., He, D., Cho, J.: Topical semantics of twitter links. In: WSDM, pp. 327–336 (2011)
24. Martin, T., Ball, B., Karrer, B., Newman, M.E.J.: Coauthorship and citation in scientific publishing. CoRR, abs/1304.0473 (2013)
25. Xie, J., Zhang, C., Wu, M.: Modeling microblogging communication based on human dynamics. In: FSKD, pp. 2290–2294 (2011)
26. Bollen, J., Pepe, A., Mao, H.: Modeling public mood and emotion: twitter sentiment and socio-economic phenomena. CoRR, arXiv:0911.1583 (2009)
27. Gruhl, D., Guha, R.V., Liben-Nowell, D., Tomkins, A.: Information diffusion through blogspace. In: WWW, pp. 491–501 (2004)
28. Mondal, J., Deshpande, A.: Managing large dynamic graphs efficiently. In: Proceedings of the ACM SIGMOD International Conference on Management of Data, SIGMOD pp. 145–156 (2012)
29. Ma, H., Qian, W., Xia, F., He, X., Xu, J., Zhou, A.: Towards modeling popularity of microblogs. Front. Comput. Sci. 7(2), 171–184 (2013)

Weibo User Influence Evaluation Method Based on Topic and Node Attributes

Wei Liu[1], Mingxin Zhang[2(✉)], Guofeng Niu[2], and Yongjun Liu[2]

[1] College of Computer Science and Technology, Soochow University,
Suzhou, China
[2] Department of Computer Science and Engineering,
Changshu Institute of Technology, Suhzou, China
mxzhang163@163.com

Abstract. The influential users in social networks contain huge commercial value and social value, and have always been concerned by researchers. For the existing research work, there is a disadvantages in the influence of user node attributes and interest topic between users on the contribution of user influence, ignoring the oldness of the relationship between users. The paper analyzed the user topic similarity from the user's interest similarity and tag similarity, divided the entire time interval of the user interaction relationship into a time window, and selects the four status attributes of the user's fans number, followers number, original blog number, and number of levels. The influence of user status attribute value and user topic similarity on the contribution degree of three kinds of behavior influence on users' forwarding, comment and mention was analyzed, and the influence of users was calculated, and a user influence evaluation method based on topic and node attributes was proposed for Weibo users. Experimenting by crawling the real dataset of Sina Weibo, compared with the typical influence analysis methods WBRank, TwitterRank, and TURank, this method was superior to the other three algorithms in terms of accuracy and recall rate.

Keywords: Social network · Status attributes · Behavior attributes
Topic similarity · User influence

1 Introduction

With the development of the Internet, social networks have become an important part of Internet applications, and social networks are no longer limited to information exchange, but are integrated with business transactions, communication and chat applications. Not only their own development and expansion, with the help of users of other applications, it also forms a more powerful network relationship structure chain. Social networks are increasingly being applied to online marketing. Enterprises can choose influential users to advertise and bring economic benefits to enterprises. Massive users of social networks have brought great challenges in various aspects such as Internet public opinion. In Sina Weibo, the quality of users is uneven, and user

© Springer Nature Switzerland AG 2018
X. Meng et al. (Eds.): WISA 2018, LNCS 11242, pp. 382–391, 2018.
https://doi.org/10.1007/978-3-030-02934-0_35

influence analysis based on Sina Weibo user data can reveal its commercial value and social value.

The influence of social network users has been paid continuous attention by researchers. At present, the influence of the user's node attribute value and the interest topic between the users on the contribution degree of the user influence is insufficient, and the obsolescence of the user relationship is ignored. Weng [1] computed user influence and comprehensive user influence on specific topics by considering Twitter [2] topicality and network structure. This method focuses on the topic-relatedness between users and ignores the user node attributes' impact on users, users with better attribute values are more likely to be experts in this field; Li [3] By improving the random walk model to improve the LeaderRank algorithm, it can find influential communicators and have higher tolerance to noise data. This method focuses on optimizing the random walk model and ignores the oldness of social network user relationships, it has a Disadvantages that cannot identify new influence users with a significant increase in influence; Hong [4] quantifies user influence by analyzing the dynamic interaction between users on the social network and the interaction between users. This method focuses on the analysis of the influence of interactive behavior on the strength of interaction between users, ignoring the influence of the interest topic between users on the user's influence, inability to accurately analyze, recognition from users with the same interests has a higher contribution to influence.

In summary, the existing research insufficiency includes the following: (1) The influence of user attribute and user interest topic on user influence is insufficient; (2) Ignoring oldness of user relationships. This paper selects Sina Weibo data as the research object. For the first point, this paper mined the topic similarity between users based on user blogs and their own tags, and selected the user's four states (number of fans, number of followers, number of original blogs and number of levels) to calculate user state attribute values. At the same time, the PageRank idea is used to calculate the user's three behavioral influences (forwarding influence, commenting influence and referring influence) and the user's comprehensive influence; for second points, the entire time interval of users' forwarding, commenting and mentioning behaviors is divided into time Windows to analyze the contribution of users' forwarding, commenting and mentioning behaviors to users' influence in different time Windows. Capture the changing trend of user influence, so that the old low-influence users with old relationships sink and the new users with high influence rise. A TBUR algorithm is used to evaluate the Weibo user influence based on topic and node attributes. The experiment is conducted by crawling the real dataset of Sina Weibo, and compared with the typical influence analysis methods WBRank, TwitterRank, and TURank. This method is superior to the other three algorithms in terms of accuracy and coverge.

2   User Topic Analysis

Users often express their feelings or views through publishing or forwarding according to their own interests in the social network, which reflects the user's interest. Weibo also provides users with the ability to assign tags to themselves, users can describe their interests, hobbies, and personal information through several short words. Therefor, this article uses Weibo content and tag content to mine user interest and tag similarity.

2.1 Interest Similarity and Tag Similarity

LDA (Latent Dirichlet allocation) [6] is a probabilistic topic model and is also a three-layer variable parameter Bayesian probability model that can be used to identify hidden topic information in large-scale document sets or corpora. The user's Weibo content [7] uses the LDA model to estimate the probability distribution of the user on different topics. The user's interest topic distribution V_u is expressed as follows.

$$V_u = (p_u^1, p_u^2, \ldots, p_u^T) \tag{1}$$

Among them, p_u^i is the probability that user u publishes the content of the weibo to generate topic i, T is the number of all topics.

KL divergence is a method of describing the difference between two probability distributions, which is consistent with the characteristics of the user interest similarity degree calculated in this paper. Therefore, this paper improved the KL method, then uses it to measure the distance between the topic probability distribution vectors of two users. In the actual calculation of user similarity, $ContSim(u, v)$ is as follows:

$$ContSim(u, v) = \frac{2}{D_{KL}(V_u||V_v) + D_{KL}(V_v||V_u)} \tag{2}$$

Among them, $D_{KL}(V_u||V_v)$ is the distance of the topic probability distribution of users u and v, Therefore, the larger the value of $ContSim(u, v)$, the more similar the similarity of interest between user u and user v.

Because of the small number of tags and the short text length, this paper uses the shared tag proportion between user u and user v to measure the similarity of user u and user v in the content of the tag. Calculated as follows:

$$TagSim(u, v) = |S_u \cap S_v| / |S_u \cup S_v| \tag{3}$$

Among them, S_u and S_v represent the tag list of user u and user v, and $TagSim(u, v)$ represents the tag similarity between user u and user v. The formula reflects the similarity of tags between users through the proportion of shared tags, the more shared tag, the higher the similarity of tag between user u and user v.

2.2 User Topic Similarity

This article uses the tags and micro-blog content that best represent the interests of users to calculate the similarity of user topics. Based on the user interest similarity and tag similarity already obtained above, the topic similarity between user u and v is calculated by considering the interest similarity and tag similarity between the user u and the user v, and the calculation formula is as follows:

$$Sim(u, v) = \lambda_1 ContSim(u, v) + \lambda_2 TagSim(u, v) \qquad (4)$$

Among them, λ_1 is the proportion of the similarity of interest in the calculation of the similarity of the user's topic, and λ_2 is the proportion of the similarity of the tag in the calculation of the similarity of the user's topic, set $\lambda_1 = \lambda_2 = 0.5$, representing tag similarity is as important as topical similarity.

3   User Influence Analysis

3.1   Time Window Division

In the specific analysis of the interaction between Weibo users on the impact of user influence, the time factor is very important, because the number of posts, the number of forwards, and the number of comments made by Weibo users are different in a certain time period. Moreover, Weibo exist a very low quality old user who used to be active and has no interaction now. The time window method is that, according to the time window interval, we combine the user's state attribute values and behavior activities in different periods to analyze the user's short-term user influence in this time window, the status attribute of the user changes little in the short term, and the changes impact on user influence is not obvious, so we focus on the influence of user behavior of time windows on user influence. The method can also solve the problem of old data, which is conducive to the sinking of low-quality old users and the floating of high-quality new users, thereby effectively suppressing "zombie users". Finally set the time window to 6 days.

3.2   User Status Attribute Analysis

For ease of reading, the user status attribute list and user behavior attribute list selected in this article are listed as follows, as shown in Table 1:

Table 1. User attribute list and symbol definition

Category	Symbol	Attributes
Number of fans	$n_1(u)$	User fans number
The number of followers	$n_2(u)$	The number of user followers
Original microblogs	$n_3(u)$	User original microblogging number
User level	$l(u)$	User level

The user of Weibo includes a variety of state attributes. This article selects four state attributes (number of fans, number of followers, number of original microblogs, and number of levels) that best represent the characteristics and quantify users, In order to calculate the contribution of users with different state attribute characteristics to users' influence, the value of user state attribute is calculated as follows:

$$M(u) = \sum_{k=1}^{3} \omega_k \frac{n_k(u)}{d_i(u)} + \omega_4 \frac{l(u)}{\sum l(u)} \tag{5}$$

Among them, $\sum l(u)$ represents the sum of user level, ω_1, ω_2, and ω_3 represent the weight of number of fans, number of followers, and the number of original microblogs respectively, set $\omega_1 = \omega_2 = \omega_3 = \omega_4 = 0.25$, $M(u)$ represents the state attribute value of user u.

3.3   User Behavior Influence Analysis

The influence between users is passed through three behaviors (forward, comment, and mention). Therefore, the paper constructs three kinds of behavioral relationship networks. It should be noted that the reason why the article considers three behavioral influences separately is that the forwarding, commentary, and mentioning are generated based on a specific topic. Separate considerations for more granular analysis.

Forward Influence
Aforwarding relationship network is represented as $G_1 = (V_1, E_1, R_1)$, where V_1 represents the user node, E_1 represents the edge formed by the forwarding relationship, and R_1 represents the weight of the edge. In the forwarding relationship network, the definition of the transition probability is $RP_t(v, u)$, which means that the weight distribution proportion of user v forwards user u. The formula is as follows:

$$RP_t(v, u) = \frac{Rn_t(v, u)}{\sum_{x \in o_1(v)} Rn_t(v, x)} \tag{6}$$

Among them, $o_1(v)$ is a set of user v forwarding other users. $Rn_t(V, U)$ is the total number of times that v forwarded u.

The user's forwarding relationship network is very similar to the web page link structure. However, the disadvantage is that since the user state attribute value, the user topic similarity, and the number of forwarding times are all different, the transmission of influence between users is not evenly distributed. Therefore, In the forwarding relationship network, the forwarding influence of the user u is $RR_t(u)$, it is calculated by improving the PageRank [9] algorithm, the formula is:

$$RR_t(u) = (1 - c) + c \times \sum_{v \in o_2(u)} \left(M(v) \times RP_t(v, u) \times Sim(u, v) \times \frac{RR_t(v)}{L_1(v)} \right) \tag{7}$$

Among them, $o_2(u)$ denotes all user sets for forwarding u, and c denotes the damping coefficient, commonly used as an empirical value set to 0.85, $L_1(v)$ represents a set of users who forward user v, $M(v)$ represents the status attribute value of user v.

Comment Influence and Mention Influence

Constructing to the comment and mention relationship network, similar to the forward influence calculation, the comment influence of the user u is $CR_t(u)$

$$CR_t(u) = (1 - c) + c \times \sum_{v \in o_4(u)} \left(M(v) \times CP_t(v, u) \times Sim(u, v) \times \frac{CR_t(v)}{L_2(v)} \right) \quad (8)$$

Among them, $o_4(v)$ is the collection of all users who comment user u, $L_2(v)$ represents a set of users who comment user v, $CP_t(v, u)$ is the comment transition probability

The mention influence of the user u is $AR_t(U)$, it is calculated as followed:

$$AR_t(u) = (1 - c) + c \times \sum_{v \in o_6(u)} \left(M(v) \times AP_t(v, u) \times Sim(u, v) \times \frac{AR_t(v)}{L_3(v)} \right) \quad (9)$$

Among them, $o_6(u)$ is the set of all users who mention the user, $L_3(v)$ represents a set of users who mention user v, $AP_t(v, u)$ is the mention transfer probability.

3.4   User Influence Calculation

Based on the user's forwarding influence, comment influence, and mention influence, then the user's comprehensive influence is calculated by linear fusion. The formula is as follows:

$$F_{TBUR}(u) = \alpha RR_t(u) + \beta AR_t(u) + \gamma CR_t(u) \quad (10)$$

Among them, α, β, γ are the weights of mention influence, forward influence and comment influence in the comprehensive influence of the user respectively. Through experiment set $\alpha = 4/7$, $\beta = 2/7$, $\gamma = 1/7$, shows the order of importance of the three behavioral relationships is the forward influence, mention influence, comment influence, $F_{TBUR}(u)$ is the user's comprehensive influence.

4   Experimental Results and Analysis

4.1   Experimental Data Set

This paper uses crawler technology to obtain Sina Weibo data from August 2017 to October 2017. It also preprocesses the data and filters out user data with a sufficient number of tags and a number of original microblogs greater than 10. The statistics of microblog users' data finally obtained are shown in Table 2.

Table 2. Experimental data

Data name	Value	Data name	Value
Total number of users	18 471	Total number of comments	103 467
Weibo total number	269 327	Total number mentioned	79 213
Total number of forwards	91 659	Total number of friends	1 849 361

The LDA model's solution process uses the Gibbs sampling method [11], and the specific parameter settings take its empirical value: $\Theta = 50/T$ (T is the number of topics), $\varphi = 0.01$. Data classification through data preprocessing results in eight categories. The experimental parameter α, β, γ is determined by multiple experiments, $\alpha = 4/7$, $\beta = 2/7$, $\gamma = 1/7$.

4.2 Experimental Verification Analysis

In order to verify the effectiveness of the proposed algorithm, three algorithms are selected for comparison. They are WBRank [12] algorithm, TwitterRank [1] algorithm and TURank algorithm [13]. First verify the accuracy of the algorithm.

Because the actual influence in the microblog is difficult for the user to identify, the real influence user is determined by the cross validation used by Zhaoyun [14]. The cross-validation method takes the correct results considered by various algorithms (N types) as the final correct result. For example, the correct results obtained by each algorithm are I_A, I_B, I_C, and I_D, when set N = 2, the final reference results I_2 as follows:

$$I_2 = (I_A \cap I_B) \cup (I_A \cap I_C) \cup (I_A \cap I_D) \cup (I_B \cap I_C) \cup (I_B \cap I_D) \cap (I_C \cap I_D) \quad (11)$$

The accuracy of algorithm A mining influences user's definition as shown in the formula (15):

$$I_A = |I_A \cap I_2|/I_A \quad (12)$$

Accuracy Verification

For $N = 2$, 3, 4, the paper verify the accuracy of the four algorithms under top10, top20, top30, top50, top-80, and top-100. The result is shown in Figs. 1 and 2.

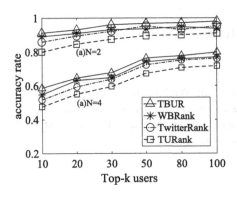

Fig. 1. Comparison of accuracy

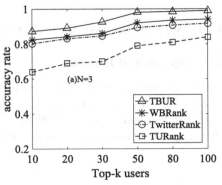

Fig. 2. Comparison of accuracy

The Figs. 1 and 2 show that when the number of reference standards is N = 3, the accuracy rate of each algorithm is significantly different, and the experimental results are the best. If N is set too small (N = 2), the number of reference set elements is too high, the intersection of elements of the algorithm and the reference standard is basically the same, resulting in little discrimination between the algorithms; if N is set too large (N = 4), the number of reference standard set elements is too small, and the intersection of each algorithm and the reference standard set element is also caused. The elements are basically the same, resulting in little differentiation in accuracy.

Correlation Comparison

To further compare the four algorithms, in order to highlight which algorithm is consistent with the actual real value, the paper uses the Spearman rank correlation coefficient to compute the correlation between the predicted ranking and the actual ranking, it can reflect the accuracy of the prediction.

The experiment analyze the accuracy of the ranking results of the top 1%, 10%, and all users, and set N = 3, that is, the three algorithm considers the correct result as a truly influential user ranking, The correlations between the rankings of the four algorithms and the real rankings are analyzed respectively. The experimental results are as follows.

From Table 3, 4 and Fig. 3, we can see that the correlation rate of the prediction method TBUR algorithm is significantly higher than the other three algorithms, and the more the user is predicted by the algorithm, the more the predicted ranking is related to the actual ranking, which is more suitable for high influence user mining.

Table 3. Comparison of user rankings with real user rankings

Actual VS TURank			Actual VS TwitterRank			Actual VS WBRank			Actual VS TBUR		
All	top10%	top1%	All	top10%	top1%	All	top10%	top1%	All	top10%	top1%
0.674	0.786	0.837	0.689	0.812	0.857	0.698	0.822	0.877	0.755	0.853	0.897

Coverage Rate Comparison

This part compares the coverage rate of the number of top-k% users affected by the four algorithms and compares it with the coverage rate of real users to verify the effectiveness of each algorithm. The result is shown in Fig. 4.

From the results of Fig. 4, we can see that the TBUR algorithm's coverage rate is better than the WBRank, TwitterRank and TURank algorithm. This indicates that it is reasonable to analyze the status attribute value and behavior influence of weibo users and the similarity of user topics, so as to more accurately identify influential users, at the same time, better results can be obtained; and, as can be seen from Fig. 4, the user coverage rate of the top 20% of the influence is about 80%, which is consistent with the prevailing Two-eight law.

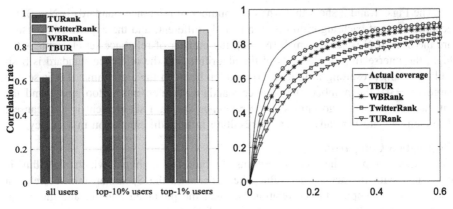

Fig. 3. Correlation rate results   **Fig. 4.** Coverage ratios results

5 Conclusion

This article uses the time window method to overcome the oldness of user relationships, to identify new users with high influence, and to consider the impact of user state attribute values and similarity of user topics on user influence, it uses PageRank to calculate users influence. The experimental results show that the algorithm can more accurately evaluate the user influence. User influence research has guiding significance for network marketing and public opinion monitoring.

References

1. Weng, J., Lim, E.P., Jiang, J., et al.: TwitterRank: finding topic-sensitive influential twitterers. In: WSDM, pp. 261–270 (2010)
2. Shipilov, A., Labianca, G., Kalnysh, V., et al.: Network-building behavioral tendencies, range, and promotion speed. Soc. Netw. **39**(1), 71–83 (2014)
3. Li, Q., Zhou, T., Lü, L., et al.: Identifying influential spreaders by weighted LeaderRank. Physica A **404**(24), 47–55 (2013)
4. Hong, R., He, C., Ge, Y., et al.: User vitality ranking and prediction in social networking services: a dynamic network perspective. IEEE Trans. Knowl. Data Eng. **PP**(99), 1 (2017)
5. Zhu, Z., Su, J., Kong, L.: Measuring influence in online social network based on the user-content bipartite graph. Comput. Hum. Behav. **52**, 184–189 (2015)
6. Blei, D.M., Ng, A.Y., Jordan, M.I.: Latent dirichlet allocation. J. Mach. Learn. Res. Arch. **3**, 993–1022 (2003)
7. Zha, C., Lyu, Y., Yin, H., et al.: UCPR: user classification and influence analysis in social network. In: IEEE International Conference on Distributed Computing Systems Workshops, pp. 311–315. IEEE (2017)
8. Shipilov, A., Labianca, G., Kalnysh, V., et al.: Network-building behavioral tendencies, range, and promotion speed. Soc. Netw. **39**(1), 71–83 (2014)
9. Page, L.: The PageRank citation ranking: bringing order to the web. Stanf. Digit. Librar. Work. Pap. **9**(1), 1–14 (1999)
10. Gleich, D.F.: PageRank beyond the Web. Comput. Sci. **57**(3), 321–363 (2014)

11. Sun, J., Tang, J.: A Survey of Models and Algorithms for Social Influence Analysis. Social Network Data Analytics, pp. 177–214. Springer, New York (2011). https://doi.org/10.1007/978-1-4419-8462-3_7

12. Hu, M., Gao, H., Zhou, J., et al.: A method for measuring social influence of Micro-blog based on user operations. In: Information Technology and Applications (2017)

13. Yamaguchi, Y., Takahashi, T., Amagasa, T., et al.: TURank: Twitter user ranking based on user-tweet graph analysis. In: Web Information Systems Engineering—WISE 2010, pp. 240–253. DBLP (2010)

14. Zhaoyun, D., Yan, J., Bin, Z., et al.: Mining topical influencers based on the multi-relational network in micro-blogging sites. Chin. Commun. **10**(1), 93–104 (2013)

Query Processing

Gscheduler: A Query Scheduler Based on Query Interactions

Muhammad Amjad$^{(\boxtimes)}$ and Jinwen Zhang

Taiyuan University of Technology, 209 University Street, Yuci District,
Jinzhong 030600, China
amjadsadiq786@yahoo.com,
zhangjinwen0039@link.tyut.edu.cn

Abstract. The workload in a database system encompasses cluster of multiple queries running concurrently. The requirement of business is that the workload which consists of different mixes of queries should complete within a short period. We propose a scheduler called Gscheduler, which schedules queries to form good queries mixes in order to finish the workload quickly. The rationale is that a query mix consisting of multiple queries that interact each other and the interactions can significantly delay or accelerate the execution of the mix. We propose a notion called mix rating to measure query interactions in a mix, which is used to differentiate good mixes from bad mixes. Experimental results show the effectiveness of the scheduler.

Keywords: Query interactions · Workload management · Query scheduler
Performance management

1 Introduction

The data analytics for business need to run a batch of queries within a limited time period, which calls for a wise scheduler being able to finish the workload in a short time period by scheduling individual queries. The workload means external client requests to database. The typical workload in such a database system consists of different mixes of queries running concurrently and interacting with each other. The performance of a query that runs concurrently with another query could be impacted in different ways that may be positive or negative. The negative impact means queries interfere at internal resources, such as latches, locks and buffer pools. The positive impact means, for example, one query may bring data into the buffer pool that is then utilized by another query. Both negative and positive impacts are called query interactions which delay or accelerate the execution of a query.

A wise scheduler should consider query interactions, and tries to run mixes with positive query interactions and avoid mixes with negative query interactions. Such a scheduler needs to be able to differentiate good query mixes from bad ones. The first step to do so is to be able to measure the query interactions in a query mix. Zhang et al. [1] propose *query rating* to measure query interactions for two queries. Section 2.1 gives a brief introduction to query rating. Query rating measures the impact on the response time of a query given by the other query, or query interactions of the two

X. Meng et al. (Eds.): WISA 2018, LNCS 11242, pp. 395–403, 2018.
https://doi.org/10.1007/978-3-030-02934-0_36

queries. We need to measure the total impact on the response time of both queries of the query interactions to judge a two-query mix is good or bad. So, we developed a measure, called *mix rating* for such a purpose. For a mix consisting of more than two queries, we can enumerate all the two query mixes, calculate their mix ratings and combine all the two query mix ratings to measure the total impact of query interactions in the mix, or its mix rating. We use mix rating to judge a mix is good, neutral or bad. The proposed scheduler, Gscheduler, tries to form good query mixes for a workload with best effort. The core contribution of this paper is an end to end solution that demonstrates how Gscheduler enables good query interactions to gain huge performance improvements.

The contributions of this work are as follows:

- We introduce a notion called mix rating to measure the total query interactions in a mix. The mix rating allows us to differentiate query mixes into good, neutral and bad.
- We propose Gscheduler based on mix rating, which schedules queries to form good query mixes with best effort, in order to minimize the total running time of a workload.

The rest of paper organized as follow: Sect. 2 discusses the approach to measure query interactions in a mix to differentiate query mixes. Section 3 presents the proposed Gscheduler. Section 4 evaluates the performance of Gscheduler. Section 5 summarizes the related work. Finally, we conclude our work in Sect. 6.

2   Quantifying Query Interactions in a Mix

In order to differentiate query mixes in terms of query interactions, it is essential to measure query interactions in a mix. Before discussing this issue, we first give out the problem statement as follow.

The database system concurrently runs M queries drawn from a workload, $W = \{\langle q_j, w_j \rangle | j = 1, 2, \ldots, N\}$, until all the queries are finished, where q_j is a query and w_j is the number of q_j in the workload. Schedule the queries to minimize the total running time of the workload.

We then briefly introduce query rating in Sect. 2.1, present mix rating for measuring query interactions in a mix in Sect. 2.2

2.1   Query Rating

The query rating is used for characterizing the behaviors of a query in a 2-query mix. Zhang et al. define query rating as follows. "The response time of query q_1 running in isolation denoted as t_{q_1}, may be delayed or accelerated when running concurrently with query q_2, and we denote it as t_{q_1/q_2}. The ratio $\gamma_{q_1/q_2} = \frac{t_{q_1/q_2}}{t_{q_1}}$ can be considered as a measurement of the degree of the interaction between q_1 and q_2 in the point of view of q_1, or the rating of q_1 given by q_2. $\gamma_{q_1/q_2 < 1}$ indicates that query q_2 speeds up query q_1, the smaller γ_{q_1/q_2} is, the greater the acceleration is. $\gamma_{q_1/q_2} > 1$ means that query q_2 slow down query q_1. The larger γ_{q_1/q_2} is, the greater the deceleration is."

Table 1 lists some examples of query rating. These queries are queries from *TPC-H*. According- to our examples, query rating $\gamma_{q_{10}/q_4} > 1$, q_4 is slow down q_{10}. This means bad query interactions. The query rating $\gamma_{q_{10}/q_5} < 1$, q_5 is speed up q_{10}. This means good query interactions. The response time of a query is the total effects of query interactions when it runs concurrently with other queries. It is reasonable to assume that the response time of a query in a fixed mix is stable, though may vary a little. Query rating measures the impact of query interactions of two queries on the response time of a query. We need to measure the total impact of query interactions on the response time of both queries to judge a two-query mix is good or bad. We discuss how the total effects of query interactions in a mix can be measured by using query rating in the next section.

Table 1. Example of query rating

	Q_4	Q_5	Q_6	Q_8
Q_1	1.11451	1.04957	1.35747	1.01932
Q_{10}	1.09167	0.98558	0.99700	0.91156
Q_{12}	1.04763	1.05513	0.88998	0.78040
Q_{14}	1.31390	0.90046	0.88033	4.03535
Q_{18}	1.08223	0.86466	1.28926	0.81513

2.2 Mix Rating

A query mix consists of at least 2-queries. Let us first discuss how to measure the mix rating for a 2-query mix consisting of q_i and q_j. Taking the same approach as query rating, we compare the sum of the response time of 2-queries running in isolation to that of running in the mix to measure the total effects of query interactions in the mix. In order to make them comparable, we concurrently run two queries continuously until they finish at the same time point, as shown in Fig. 1. We count the number of q_i, denote as n_i, and the number of q_j, denote as n_j. We compare $(n_i * t_i) + (n_j * t_j)$ to $\left(t_{q_i/q_j} * n_i + t_{q_j/q_i} * n_j \right)/2$ as shown in Eq. (1).

$$\gamma_{q_i/q_j} = \frac{(n_i * t_i) + (n_j * t_j)}{\left(t_{q_i/q_j} * n_i + t_{q_j/q_i} * n_j \right)/2} \tag{1}$$

In practice, it is almost impossible to see q_i and q_j finishing at the same time. Instead, we concurrently run the two queries for a sufficiently long time and count the number of each query executed, and calculate their average response time, t_{q_i/q_j} and t_{q_j/q_i}. q_i and q_j finishing at the same time mean $q_i/q_j * n_i = t_{q_j/q_i} * n_j$. We find the smallest pair of (n_i, n_j) and apply them to Eq. (1).

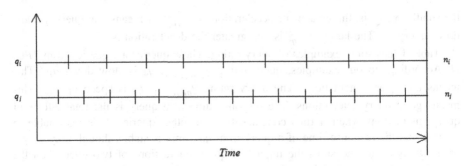

Fig. 1. Two queries concurrently running continuously until finishing at the same time point

After we taking the mix rating of the 2-query mix, we can characterize the mixes according to the mix rating. For example, $\gamma_{q_i/q_j < 1.2}$ indicates the mix being a good mix and, $\gamma_{q_i/q_j > 0.8}$ indicates the mix being a bad mix, and in between defines the mix being a neutral mix. As shown in Table 1, ... are good mixes, ... are bad mixes, and ... are neutral mixes. We assign a score 3, 2, 1 to each good, neutral, and bad mixes, respectively. A good mix means its query interactions are positive and accelerate query execution. A bad mix means its query interactions are negative and delay query execution. A neutral mix means its query interactions have little to do with query execution.

Now, we discuss how to calculate the mix rating for mix consisting of more than two queries. An intuitive approach is to use the mix rating for 2-query mixes. The rationale is that 2-query mixes are the basic form of query interactions and can any query interactions can be formed by combining them. So, we take the most intuitive approach to calculate the mix rating of mixes with more than 2-queries, which simply enumerate all the 2-query mixes in a mix, and average their score as mix rating. A score near to 3, 2 and 1 means the mix is good, neutral, and bad, respectively.

3   Gscheduler

The system architecture of Gscheduler is shown in Fig. 2. The goal of Gscheduler is to schedule appropriate query mixes for a given query workload W in order to minimize W's total completion time.

In database system that workload emerges from clients that generate fixed set of query types. Let $q_i, q_j, \ldots, q_T$ be the T query types in a database system that M represents the number of queries execute concurrently in the system. The proposed Gscheduler is online scheduler. The total number of queries of type q_j in W as follow:

$$|W| = \sum_{j=1}^{T} Nj$$

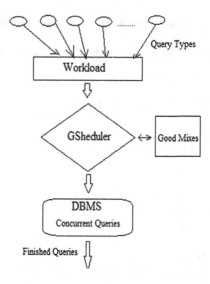

Fig. 2. Query Gscheduler illustration

The clients place their queries in a queue and Gscheduler schedules queries from it. In Gscheduler, algorithm generates a schedule whose total completion time is within an additive constant factor of the total completion time of the optimal schedule. The algorithm considers a large set of mixes $W = \{m_1, m_2, \ldots, m_{|T|}\}$ such that the schedule chosen for W will consist of a subset of mixes selected from W and a specification of how the N_j instances of each query type should be executed using the selected mixes. The W is a very large subset of the full space of query mixes. Given M and T, the total size of the query mixes is the number of ways we can select M objects from T objects types, unordered and with repetition. This is an M-selection from a set of size T and the number of possible selections is given by $S(T, M) = \binom{M + T - 1}{M}$ [15]. The proposed Gscheduler uses a linear program to pick the subset of W used in the chosen schedule [16]. The inputs to the linear program used by the scheduler consist of the set of query mixes $m_i \in W$ and N_j, $1 \leq j \leq T$, to be scheduled. The linear program contains an unknown variable $v_i (v_i \geq 0)$ corresponding to each mix $m_i \in W$. The v_i is the total time for which queries will be scheduled with mix m_i in the chosen schedule. The chosen schedule should perform the work required to complete all N_j input instances of each query type q_j. This requirement can be written in the form of T constraints as follow.

$$\sum_{i=1}^{|T|} v_i \frac{N_{ij}}{A_{ij}} \geq N_j, \forall_j \{1, 2, 3, \ldots, T\} \tag{2}$$

For example, the work needed to complete the execution of one instance of q_j is 1, then the total work required to complete the execution of N_j instances of q_j in the input workload is N_j. The $\frac{N_{ij}}{A_{ij}}$ is the fraction of this work that gets completed per unit time

when mix m_i is scheduled. The $\sum_{i=1}^{|T|} v_i \frac{N_{ij}}{A_{ij}}$ is the total work done for q_j in the chosen

schedule which must not be less than N_j.

```
Algorithm: Propose Gscheduler
Input:
    W = {<q_i, w_i >| i=1,2, 3..., N}
    M // Multiprogramming Level
    GM // good query mix
Output:
    q_i = // Required query to be scheduled.
    If (|w| <M)
        Break
    else
        for (q_i in w)
        if (q_i U runningMix ⊆ GM)
          return q_i
        end
        m← identifyGoodMixes (M, W)
        for (q_i in W)
        if (q_i in U runningMix ⊆ GM)
          return q_i
        end
    end
```

The input workload has been partitioned among the mixes $m_1, m_2, \ldots, m_T$ these mixes are scheduled in decreasing order of v_i values. Each mix m_i we schedule queries from the set of instances assigned m_i until they all complete, and finish workload. A sequence is corresponding to an execution of the workload with M. The problem is that when in query mixes finished exiting one query, the other running two query mixes confusing to split the third query so that it will become good query mixes. The scheduler check when running query will come it will be became good query mix otherwise split this arrival query and take another query for making good query mix as soon as possible.

This is reasonable assumption because we should have to find the maximum good query mixes. The algorithm selects the maximum good query for executing good query mixes.

4   Experiment Evaluation

This section describes the experiments. Section 4.1 gives out the experimental settings. Section 4.2 discusses the results.

4.1 Experimental Settings

Environments. We describe the hardware and software used in our experiments. our experiment was run on a machine with Intel E3500 2.7 GHz CPU and 4 GB RAM. We run *PostgreSQL* database under Window 2007x64. All the statistics and tuning tools are turned off and all configurations of Postgres are default. We use the *TPC-H* database with 10 GB scale factors. We ran the *PostgreSQL* configuration advisor to ensure that the configuration parameters are well tuned.

Workload. We use 9 queries from 10 GB *TPC-H* shown in Table 2, with different parameters values that are chosen according to the *TPC-H*. There are 125 queries in total.

Table 2. Workload of queries

SF	Q_1	Q_4	Q_5	Q_6	Q_8	Q_{10}	Q_{12}	Q_{14}	Q_{18}
10	9	15	14	16	15	16	13	18	9

Scheduling Algorithms. We experimented with three different scheduling algorithms, Gscheduler, First Come First Served (*FCFS*), and Shortest Job First (*SJF*). For *FCFS*, the sequence of queries is randomly generated. For *SJF*, the sequence of queries is generated by order the shortest query with the response time of t_{q_i} in isolation.

Performance Metric. Our performance metric is total completion time for the workload. Figure 3 shows the performance comparison.

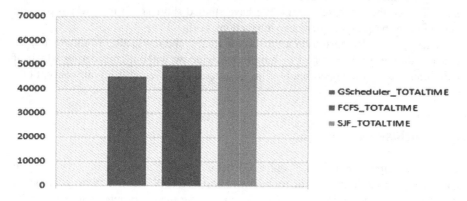

Fig. 3. Accuracy of Gscheduler

4.2 Effectiveness of Gscheduler

Figure 3 clearly demonstrates the performance gap between the propose scheduler algorithm and other two algorithms. The propose Gscheduler is better than *FCFS* and *SJF*.

We examine our proposed Gscheduler and find that the heterogeneous mixes are among the best mixes. The arrival order is very nearly approaching round robin, not too many queries of the same type can arrive together. *FCFS* is able to keep up when the arriving order generates good query mixes, and starts degrading significantly when the arriving order generates bad query mixes. *SJF* is the consistently turns out to be the worst scheduler.

5   Related Work

The presence of query interactions can engender database performance to reduce significantly if interaction unaware query schedulers are used [2]. In multi-tier applications, the transaction mix models for resource provisioning and capacity planning. The author suggests the use of transaction mix models for detecting performance anomalies in multi-tier enterprise applications [3]. The scheduling in database systems has been studied in the context of concurrency control where the focus is on minimizing lock contention [4, 5]. The statistical prediction models predict query response time by training and sampling [6]. The model divided into two categories, linear regression models [7, 8] and nonlinear models [9–11].

The data center query scheduling is an important issue SLQs-achieving [12], balancing and according to cost [13]. Although different scheduling intensions depend on different scheduling algorithms. All the scheduler's aims are based on a performance of response time and utilizations of resources. It is hard to develop an accurate analytical model in practice. Statistical modeling takes the database system as black box to avoid the complexity encountered by analytical modeling and makes the simple models. The workload management is an area of growing importance in database system, the work in this area includes techniques for admission control and setting the multiprogramming level of the database system. We have studied shortest job first, which perform very poor in the presence of query interactions [14].

These papers demonstration a transaction mix as the transaction of different types that run during a time interval or monitoring window without considering which of these transactions run concurrently. This is structurally different from our concept of a concurrent query mix.

6   Conclusions

This paper proposes a novel scheduler, Gscheduler, which always selects the query to enter the system to form good query mix with best efforts. Whether a mix is good or not is judged by its mix rating, our notion for measuring the query interactions in a mix. The experimental results show that a scheduler taking query interactions into account can significantly shorten the total execution time of a workload.

Acknowledgment. This work is supported by the National Natural Science Foundation of China under contract #61572345.

References

1. Zhang, J., Niu, B.: A clustering-based sampling method for building query response time model. Comput. Syst. Sci. Eng. **32**(4), 319–331 (2017)
2. Ahmad, M., Aboulnaga, A., Babu, S., Munagala, K.: Modeling and exploiting query interactions in database systems. In: Proceedings of the 17th ACM Conference on Information and Knowledge Management (CIKM), pp. 183–192. ACM, New York (2008)
3. Kelly, T.: Detecting performance anomalies in global applications. In: Proceedings of Second Workshop on Real, Large Distributed Systems (WORLDS), San Francisco (2005)
4. Ibaraki, T., Kameda, T., Katoh, N.: Cautious transaction schedulers for database concurrency control. IEEE Trans. Softw. Eng. **14**(7), 997–1009 (1988)
5. Katoh, N., Ibaraki, T., Kameda, T.: Cautious transaction schedulers with admission control. Trans. Database Syst. **10**(2), 205–229 (1985)
6. Duggan, J., Çetintemel, U., Papaemmanouil, O., Upfal, E.: Performance prediction for concurrent database workloads. In: Proceedings of the ACM SIGMOD International Conference on Management of Data, pp. 337–348. ACM, New York (2011)
7. Ahmad, M., Aboulnaga, A., Babu, S., Munagala, K.: Interaction-aware scheduling of report generation workload. Int. J. Very Large Data Bases **20**(4), 589–615 (2011)
8. Tozer, B., Brecht, T., Aboulnaga, A.: Q-Cop: avoiding bad query mixes to minimize client timeouts under heavy loads. In: Proceedings of the 26th International Conference on Data Engineering, pp. 397–408. IEEE Computer Society, Long Beach (2010)
9. Mozafari, B., Curino, C., Jindal, A., Madden, S.: Performance and resource modeling in highly-concurrent OLTP workloads. In: Proceedings of the ACM SIGMOD/PODS Conference, pp. 301–312. ACM, New York (2013)
10. Akdere, M., Çetintemel, U., Riondato, M., Upfal, E.: Learning-based query performance modeling and prediction. In: Proceedings of 28th International Conference on Data Engineering, 2012, pp. 390–401. IEEE Computer Society, Washington, DC (2012)
11. Ganapathi, A., Kuno, H.A., Dayal, U., Wiener, J.L.: Predicting multiple metrics for queries: better decisions enabled by machine learning. In: Proceedings of the 25th International Conference on Data Engineering, pp. 592–603. IEEE Computer Society, Shanghai (2009)
12. Baoning, N., Patrick, M., Wendy, P., Paul, B.: Adapting mixed workloads to meet SLOs in autonomic DBMSs. In: Proceeding of the 23rd International Conference on Data Engineering Workshops, pp. 478–484. IEEE Computer Society, Istanbul (2007)
13. Marcus, R., Papaemmanouil, O.: WiSeDB: a learning-based workload management advisor for cloud databases. Proc. VLDB Endow. **9**(10), 780–791 (2016)
14. Elnikety, S., Nahum, E., Tracey, J., Zwaenepoel, W.: A method for transparent admission control and request scheduling in e-commerce web sites. In: Proceedings of the 13th International Conference on World Wide Web, New York, NY, USA, pp. 276–286 (2004)
15. Ryser, H.: Combinatorial Mathematics (14), p. 154. The Mathematical Association of America (1963)
16. Schrijver, A.: Theory of Linear and Integer Programming. Wiley, New York (1986)

An Efficient Algorithm for Computing
k-Average-Regret Minimizing Sets
in Databases

Xianhong Qiu[1] and Jiping Zheng[1,2(✉)]

[1] College of Computer Science and Technology,
Nanjing University of Aeronautics and Astronautics, Nanjing, China
{qiuxianhong,jzh}@nuaa.edu.cn
[2] Collaborative Innovation Center of Novel Software Technology
and Industrialization, Nanjing, China

Abstract. Returning a small set of data points instead of the whole dataset to a user is a major task of a database system which has been studied extensively in recent years. In this paper, we study k-average-regret query, a recently proposed query, which uses "average regret ratio" as a metric to measure users' satisfaction to avoid the biases towards a few dissatisfied users that the best-known k-regret query suffers from. The main challenge of executing a k-average-regret query is the low efficiency of existing algorithms. Fortunately, as the average regret function exhibits the properties of *supermodularity* and *monotonictity*, the computational complexity of k-average-regret query can be significantly reduced exploiting lazy evaluations, thus leading to our accelerated algorithm which we called Lazy-Greedy. Experiments on both synthetic and real datasets confirm the efficiency and quality of output of our proposed algorithm.

Keywords: k-average-regret query · Representative skyline
Lazy evaluation

1 Introduction

Finding a small set of data points from a large dataset to support multi-criteria decision making is an important functionality in many application domains. A number of queries have been proposed in the literature to effectively support such functionality. Top-k [1] and skyline [2] are two representative queries. A top-k query returns k points that have the greatest scores under the utility/score functions specified by a user where k is a positive integer. A skyline query returns points that are not dominated by any other point in the database with no need to ask users to appoint utility functions. Instead, a concept called domination is applied in skyline queries. Specifically, a point p is said to dominate another point q if p is as good or better in all dimensions, and strictly better in at least one dimension. However, both queries suffer from some drawbacks. Top-k queries

© Springer Nature Switzerland AG 2018
X. Meng et al. (Eds.): WISA 2018, LNCS 11242, pp. 404–412, 2018.
https://doi.org/10.1007/978-3-030-02934-0_37

ask for users to specify their exact utility functions, which is difficult to most users. Skyline queries find all points that are not dominated by other points in the database, so the exact number of result set is uncontrollable and cannot be foreseen before the whole database is accessed. In addition, the output size of skyline queries will increase rapidly with the dimensionality.

Recently, the k-regret query [3] was proposed which integrates the merits of top-k and skyline queries. For k-regret queries, any utility function specified by a user is not required and the output size is controllable since only k points that minimize a criterion called maximum regret ratio are returned. However, k-regret queries only consider the regret ratio of the most unhappy user, in other words, there is only one point that satisfies a user among the selected points, so this query suffers from the drawback that it will be skewed towards the least satisfied users only and ignore the other users. Zeighami et $al.$ [4] proposed "average regret ratio" as a metric to measure a user's satisfaction which gives a better impression of how a user in general feels towards the selected points and provided a result set of k points that the average regret ratio of the result set is minimized. Specifically, their method exploits N utility functions sampled from the probability distribution of all utility functions and picks out the point whose removal makes the average regret ratio minimized. Unfortunately, the calculation of the average regret ratio is time-consuming. Motivated by this, in this paper, an efficient algorithm called Lazy-Greedy is proposed with a $(1-1/e)$ approximation guarantee to the optimum solution. Lazy-Greedy is extended from existing greedy algorithm by exploiting lazy evaluations and obtains significant speedups.

The main contributions of this paper are listed as follows.

1. Based on the supermodularity and monotonicity of the average regret ratio, we introduce an efficient approximation algorithm called Lazy-Greedy. The algorithm exploits some lazy evaluations to avoid some unnecessary calculations when picking out the point whose removal increases the average regret ratio the least.
2. Extensive experiments on both synthetic and real datasets are conducted to evaluate our method and the experimental results confirm that our proposed algorithm achieves the same minimum average regret ratio as WO-Greedy proposed in [4] but runs much faster.

The remainder of this paper is organized as follows. In Sect. 2, previous work related to this paper is discussed. Section 3 introduces the problem definition and background techniques which are applied in our algorithm. Followed by the accelerated greedy algorithm in Sect. 4. The performance of our algorithm compared with existing algorithms on synthetic and real datasets is presented in Sect. 5. Finally, Sect. 6 concludes this paper.

2   Related Work

Due to the drawbacks of top-k and skyline queries, a lot of alternatives have been proposed in recent years. First, efforts are put forward to improve top-k

queries. [5,6] asked users to specify some kinds of utilities/preferences to provide alternative ways to users to specify utility functions for top-k queries. Secondly, researchers attempt to reduce the output size of the skyline queries. The representative skyline [7,8] was proposed which returns k skyline points best representing the full skyline. Unfortunately, all these methods are not stable, scale-invariant or with deficiencies of top-k or skyline queries.

Recently, k-regret queries were first proposed in [3] which do not heavily rely on top-k queries and skyline queries and have been studied in [9–11] using different approaches. However, the k-regret queries have the deficiencies that they will be skewed towards the least satisfied users only, ignoring the other users, as they only consider the regret ratio of the most unhappy user. [4] proposed a k-average-regret query which returns a result set that minimizes the average regret ratio and can avoid the drawback that the k-regret query suffers from. But, the efficiency of the algorithm proposed in [4] is very low as it will result in a total running time of $O(dNn^3)$ where d is the dimensionality of the database, n is the size of the database and N is the number of utility functions. Our research aims at answering the k-average-regret query with an efficient algorithm.

3   Problem Definition and Background Techniques

In this section, we first formulate our problem and then point out the background techniques used in our algorithm.

3.1   Problem Definition

Let D be a set of n d-dimensional points over positive real values. Each point in D can be regarded as a tuple in the database. For each point $p \in D$, the value on the i-th dimension is represented as $p[i]$. We assume that smaller values are better. If users prefer large values, we convert them to small values by subtraction with the maximum value. Before we define our problem, definitions of utility function, regret ratio and average regret ratio are given [3,4].

Definition 1 (Utility Function). *A utility function u is a mapping $u: \mathbb{R}_+^d \rightarrow \mathbb{R}_+$. The utility of a user with utility function u is $u(p)$ for any point p and shows how satisfied the user is with the point.*

Definition 2 (Regret Ratio). *Given a dataset D, a subset $S \subseteq D$ and a utility function u. The regret ratio of S, represented as $rr_D(S, u)$, is defined to be*

$$rr_D(S, u) = \frac{\max_{p \in D} u(p) - \max_{p \in S} u(p)}{\max_{p \in D} u(p)}$$

Definition 3 (Average Regret Ratio). *Given a dataset D, a subset $S \subseteq D$, a set U containing all utility functions and the probability distribution function*

$\eta(u)(u \in U)$ of the utility functions for different users. The average regret ratio of S, represented as $arr_D(S, U)$, is defined to be

$$arr_D(S, U) = \int_{u \in U} rr_D(S, u)\eta(u)du$$

Especially, if the utility functions are defined on a discrete space, the average regret ratio of S can be rewritten as follows:

$$arr_D(S, U) = \sum_{u \in U} rr_D(S, u)\eta(u)$$

The calculation of the average regret ratio $arr_D(S, U)$ involves calculating the average of the regret ratios of all utility functions. If the distribution of the utility functions is continuous, the calculation of the average regret ratio requires the calculation of a d-dimensional integral. If the set of the utility functions is defined on a discrete space, we can calculate the average regret ratio by summing for all the utility functions their regret ratios multiplied by their probabilities. In this paper, our utility functions U are sampled from the linear utility space, since linear utility functions are widely used in modeling users' preferences. Specially, a utility function u is linear if there exist non-negative reals $v_1, v_2, \cdots, v_d$ such that $u(p) = \sum_{i=1}^{d} v_i \cdot p[i]$ for any d-dimensional point p. Alternatively, a linear utility function can be represented by a vector $u = (v_1, \cdots, v_d)$, i.e., $u(p)$ is the dot product $u \cdot p$.

Problem Definition: Given a dataset D of size n, a positive integer k, a set U of utility functions of size N and the probability distribution function $\eta(u)(u \in U)$ of the utility functions, the problem of average regret ratio minimizing is trying to find a subset $S \subseteq D$ containing at most k points such that the average regret ratio is minimized while simultaneously keeping the query time as short as possible.

$$S = \arg \min_{s' \subseteq D, |S'|=k} arr_D(S', U)$$

3.2   Background Techniques

We will introduce the concept of supermodular function and describe supermodularity and monotonicity of it as they are the properties used in our Lazy-Greedy.

Definition 4 (Supermodularity). A set function $f : 2^D \rightarrow \mathbb{R}^+$ is supermodular if for every $S_1, S_2 \subseteq D$ it holds that

$$f(S_1) + f(S_2) \leq f(S_1 \cup S_2) - f(S_1 \cap S_2)$$

Definition 5 (Monotonicity). A set function $f : 2^D \rightarrow \mathbb{R}^+$ is monotone and non-increasing if for every $S_1 \subseteq S_2 \subseteq D$, it holds that $f(S_1) \geq f(S_2)$.

Lemma 1. *[4] The average regret ratio* $arr_D(S, U)$ *is a monotone non-increasing supermodular set function, namely, it satisfies the properties of super-modularity and monotonicity.*

Based on the definitions of supermodularity and monotonicity, Lemma 1 can be proved directly. Due to space limitation, we omit proofs here.

4   Lazy-Greedy Algorithm

In the following, we first describe how to determine the appropriate size of utility functions that are required to sample from the linear utility space and then present our Lazy-Greedy whose performance is boosted with lazy evaluations.

4.1   Sampling N Utility Functions

Since we have no idea of user's utility functions, the number of utility functions is infinite in linear utility space. We can select N utility functions based on the probabilities distribution of the utility functions. For this, we need to choose the number of utility functions N that approximates the true value of the average regret ratio with a high confidence and within a reasonable error parameter.

Theorem 1. *Given a confidence parameter* $\delta \in (0, 1]$ *and an error parameter* $\epsilon \in [0, 1]$, *then when the confidence is at least 1-δ and the calculated average regret ratio is within ϵ of its true value, N is at least* $\frac{3\ln(\frac{1}{\delta})}{\epsilon^2}$.

Proof. We need to show that for $0 < \epsilon \le 1$,

$$Pr(\frac{X - \mu}{\mu} \ge \epsilon) = Pr(X \ge (1 + \epsilon)\mu) \le \delta \tag{1}$$

According to Chernoff bounds and let $\mu = N > 0$, we can get

$$Pr(X \ge (1 + \epsilon)\mu) \le e^{-\mu\epsilon^2/3} \le e^{-N\epsilon^2/3} \le \delta$$

Take the logarithm of both sides and rearrange it, we obtain $N \ge \frac{3\ln(\frac{1}{\delta})}{\epsilon^2}$.   □

4.2   Lazy Evaluation for Greedy Algorithm

At each iteration, WO-Greedy [4] must identify the point p whose removal makes the average regret ratio $arr_D(S_i \backslash \{p\}, U)$ minimized then remove it from S_i, where S_i is the result set of the $(i + 1)$-th iteration. Unfortunately, a large number of calculations are needed when we run WO-Greedy algorithm [4] which is time-consuming. The key insight from the supermodularity of $arr_D(S, U)$, the average regret ratio obtained by any fixed point $p \in D$ is monotonically non-decreasing during the iterations of removing points, *i.e.*, $arr_D(S_i \backslash \{p\}, U) \ge arr_D(S_j \backslash \{p\}, U)$, whenever $i \le j$. Instead of recomputing for each point $p \in S_i$, we can use lazy evaluations to maintain a list of lower bounds $\{\triangle(p)\}$ on the average regret ratio sorted in ascending order. Then in each iteration, Lazy-Greedy needs to extract the minimal point $p \in \arg\min_{p':S_{i-1}\backslash\{p\}}\{\triangle(p')\}$ from the ordered list and then updates the bound $\triangle(p) \leftarrow arr_D(S_{i-1}\backslash\{p\}, U)$. After this update, if $\triangle(p) \leq \triangle(p')$, then $arr_D(S_{i-1}\backslash\{p\}, U) \leq arr_D(S_{i-1}\backslash\{p'\}, U)$ for all $p \neq p'$, and therefore we have identified the point with the minimal average regret ratio, without having to compute $arr_D(S_{i-1}\backslash\{p'\}, U)$ for a potentially large number of point p'. We set $S_i \leftarrow S_{i-1}\backslash\{p\}$ and repeat until $i = n - k$. This idea of using lazy evaluations is useful to our algorithm and can lead to orders of magnitude performance speedups. The pseudocodes of lazy-Greedy are shown in Algorithm 1.

Algorithm 1. Lazy-Greedy(D, k, U)

Input: A set of n d-dimensional points $D = \{p_1, p_2, \cdots, p_n\}$ and an integer k. U denotes d-dimensional utility functions whose size is N.
Output: A result set S, $|S| = k$.

1 Initially, let $S = D, p^* = NULL$;
2 **for** $(i = 1; i \leq n - k; i++)$ **do**
3   **if** $i = 1$ **then**
4   **for** each $p \in S$ **do**
5   calculate the value of $arr_S(S\backslash\{p\}, U)$;
6            $\triangle(p) = arr_S(S\backslash\{p\}, U)$;
7   p^*=Lazy-Evaluation$(D, S, i, \{\triangle(p)\}, U)$;
8      $S = S\backslash\{p^*\}$;
9 **return** S;

The calculation of the average regret ratio based on WO-Greedy takes time $O(dnN)$ and there are $O(n^2)$ iterations in the greedy algorithm thus resulting in a total running time of $O(dNn^3)$. For our Lazy-Greedy, the algorithm keeps track of the points which have not been visited and their average regret ratios are the smallest in the list $\{\triangle(p)\}$ instead of calculating the average regret ratio of all points that still inside the current result set S in each iteration. Hence, Lazy-Greedy has some important features. First, it provides a greedy solution identical to the solution provided by WO-Greedy [4]. Secondly, Lazy-Greedy is more efficient compared with WO-Greedy as it can avoid some calculations to minimum average regret ratio by exploiting lazy evaluations. Unfortunately, these cannot be demonstrated theoretically and it is easy to build worst-case examples for which Lazy-Greedy requires the same number of calculations as WO-Greedy. However, subsequent experiments provide an experimental confirmation of the efficiency of Lazy-Greedy when applied to answer the k-average-regret minimizing set. In a sense, Lazy-Greedy is still optimal in terms of number of calculations.

5 Experimental Results

In this section, we show the performance of our proposed algorithm via experiments. The algorithms were implemented in C++ and run on a 64-bit 3.3 GHz Intel Core machine which was running Ubuntu 14.04 LTS operating system. We ran our experiments on both synthetic and real datasets. Unless stated explicitly, for synthetic datasets created adopting the dataset generator of [2], the number of points is set to 10,000 (*i.e.*, $n = 10,000$) and the dimensionality is set to 6 (*i.e.*, $d = 6$) and k is set to 10. The real-world datasets include a 5-dimensional *ElNino*[1] of 178,080 points, a 6-dimensional *Household*[2] of 127,391 points and an 8-dimensional *NBA*[3] of 17,265 points. All experiments are conducted with 964 utility functions sampled from a uniform distribution on the linear class of the utility functions where $\epsilon = 0.0707716$ and $\delta = 0.2$. Moreover, like studies in the literature [3,9], we computed the skyline first and our queries returned anywhere from 5 to 30 points on these datasets except the ElNino dataset (when $k > 10$, the average regret ratio on ElNino dataset is close to 0, so we only show the results when k is small). In our experiments, we consider Lazy-Greedy introduced in this paper. To verify the superiority of our proposed algorithm, we compare it with RDP-Greedy (they called Greedy in [3]) and WO-Greedy [4]. We measure the computational cost in terms of the running time of CPU and the quality of result set by means of the average regret ratio.

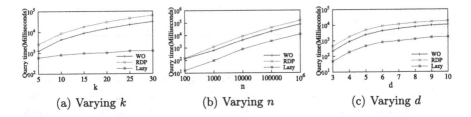

(a) Varying k (b) Varying n (c) Varying d

Fig. 1. Query time on the anti-correlated dataset

Results on Synthetic Datasets: The query time on anti-correlated datasets for different k, n and d are shown in log scale in Fig. 1. In all cases, our Lazy-Greedy has negligible query time as unnecessary calculations to average regret ratio are avoided while RDP-Greedy and WO-Greedy result in a much longer query time. The average regret ratio on anti-correlated datasets with k, d and n varied are presented in Fig. 2. Lazy-Greedy and WO-Greedy have the same and low average regret ratio which are much lower than RDP-Greedy as they share the same greedy skeleton. Besides, the average regret ratio of all algorithms degrade with the increase of k. But, the average regret ratio increases with d for all algorithms due to the curse of dimensionality, and increases with n.

---

[1] http://archive.ics.uci.edu/ml/datasets/El+Nino.
[2] http://www.ipums.org/.
[3] https://www.basketball-reference.com/.

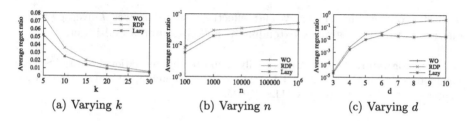

Fig. 2. Average regret ratio on the anti-correlated dataset

Results on Real Datasets: Figure 3 shows the query time on real datasets. The query time of all algorithms increase with k, but our Lazy-Greedy keeps much less query time and maintains a stable level as we analyze in the previous section. The average regret ratio of all algorithms for different k are shown in Fig. 4. We observe similar trends as the experiments on synthetic datasets presented in Fig. 2(a). Besides, similar to the experiments on synthetic datasets, our Lazy-Greedy achieves near-minimal average regret ratio with substantially shorter query time compared with RDP-Greedy and WO-Greedy as it's optimal in terms of numbers of calculations.

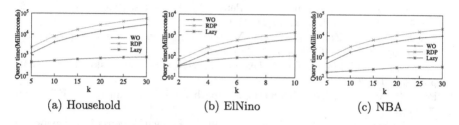

Fig. 3. Query time on the real datasets

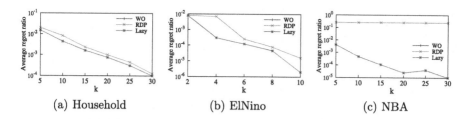

Fig. 4. Average regret ratio on the real datasets

6   Conclusions

This paper studies a k-average-regret query and proposes an accelerated algorithm exploiting some lazy evaluations. Experiments on synthetic and real datasets confirm the efficiency and effectiveness to answer k-average-regret queries. Future work aims at extending a single user into a multi-user.

Acknowledgment. This work is partially supported by the National Natural Science Foundation of China under grants U1733112,61702260, Funding of Graduate Innovation Center in NUAA under grant KFJJ20171605.

References

1. Ilyas, I.F., Beskales, G., Soliman, M.A.: A survey of top-k query processing techniques in relational database systems. ACM Comput. Surv. **40**(4), 11:1–11:58 (2008)
2. Börzsöny, S., Kossmann, D., Stocker, K.: The skyline operator. In: ICDE, pp. 421–430 (2001)
3. Nanongkai, D., Sarma, A.D., Lall, A., Lipton, R.J., Xu, J.: Regret-minimizing representative databases. In: VLDB, pp. 1114–1124 (2010)
4. Zeighami, S., Wong, R.C.W.: Minimizing average regret ratio in database. In: SIGMOD, pp. 2265–2266 (2016)
5. Mindolin, D., Chomicki, J.: Discovering relative importance of skyline attributes. Proc. VLDB Endow. **2**, 610–621 (2009)
6. Lee, J., You, G.W., Hwang, S.W.: Personalized top-k skyline queries in high-dimensional space. Inf. Syst. **34**(1), 45–61 (2009)
7. Lin, X., Yuan, Y., Zhang, Q., Zhang, Y.: Selecting stars: the k most representative skyline operator. In: ICDE, pp. 86–95 (2007)
8. Tao, Y., Ding, L., Lin, X., Pei, J.: Distance-based representative skyline. In: ICDE, pp. 892–903 (2009)
9. Peng, P., Wong, R.C.W.: Geometry approach for k-regret query. In: ICDE, pp. 772–783 (2014)
10. Xie, M., Wong, R.C.W., Li, J., Long, C., Lall, A.: Efficient k-regret query algorithm with restriction-free bound for any dimensionality. In: SIGMOD (2018)
11. Qi, J., Zuo, F., Samet, H., Yao, J.: K-regret queries using multiplicative utility functions. ACM Trans. Database Syst. **43**, 10 (2018)

Efficient Processing of k-regret Queries
via Skyline Priority

Sudong Han[1], Jiping Zheng[1,2(✉)], and Qi Dong[1]

[1] College of Computer Science and Technology,
Nanjing University of Aeronautics and Astronautics, Nanjing, China
{sdhan,jzh,dongqi}@nuaa.edu.cn
[2] Collaborative Innovation Center of Novel Software Technology
and Industrialization, Nanjing, China

Abstract. Extracting interesting points from a large database is an important problem in multi-criteria decision making. The recent proposed k-regret query attracted people's attention because it does not require any complicated information from users and the output size is controlled within k for users easily to choose. However, most existing algorithms for k-regret query suffer from a heavy burden by taking the numerous skyline points as candidate set. In this paper, we define a subset of candidate points from skyline points, called *prior skyline points*, so that the k-regret algorithms can be applied efficiently on the smaller candidate set to improve their performance. A useful metric called skyline priority is proposed to help determine the candidate set and corresponding strategies are applied to accelerate the algorithm. Experiments on synthetic and real datasets show the efficiency and effectiveness of our proposed method.

Keywords: k-regret query · Skyline priority · Prior skyline points
Candidate set determination

1 Introduction

Returning points that users may be interested in is one of the most important goals for multi-criteria decision making. Top-k [1] and skyline [2] queries are two well-studied tools used to return a representative subset of a large database. But these two types of queries suffer in either requiring a predefined utility function to model user's preference over the points, or returning an uncontrollable number of points. To avoid the deficiencies of top-k and skyline queries, Nanongkai *et al.* [3] first proposed regret-based query which returns k points that minimize a criterion called the maximum regret ratio. It quantifies how regretful a user is if s/he gets the best point among the selected k points but not the best point among all the tuples in the database.

Technically, the input of k-regret algorithms is the points in the whole dataset, but existing algorithms [3–5] usually take the skyline points as the candidate points. This is because the points dominated by skyline points have less

© Springer Nature Switzerland AG 2018
X. Meng et al. (Eds.): WISA 2018, LNCS 11242, pp. 413–420, 2018.
https://doi.org/10.1007/978-3-030-02934-0_38

possibility of being k representative points. By removing the non-skyline points, the running time of the algorithms can be largely reduced. However, taking skyline points as candidate points is of low efficiency for computing k representative points because the size of skyline points grows exponentially with dimensionality [6]. Extremely, the size of skyline points is even close to the whole dataset when the dimension is high.

Motivated by these, we devote to finding a small size of candidate set from the entire skyline points so that the k-regret algorithms can be applied efficiently on the smaller candidate set to improve its efficiency. In this paper, we define a set of candidate points called *prior skyline points* based on skyline priority. Skyline priority is a metric that indicates the least dimensionality of subspace in which a point p is a skyline point. Intuitively, a point with a high skyline priority is more interesting as it can be dominated on fewer combinations of dimensions. To avoid the expensive cost of calculating skyline priority given by the naive method, we further propose SP_{SKYCUBE} using sharing and promoting strategies to determine the candidate set. The main contributions of this paper are listed as follows:

- Skyline priority based on subspace skyline is firstly proposed to rank skyline points for candidate set determination of k-regret query.
- We present efficient algorithm based on sharing and promoting strategies to compute skyline priority.
- Results of extensive experiments on both synthetic and real datasets confirm the efficiency of our proposed algorithm.

The rest of this paper is organized as follows. We present related work in Sect. 2. Section 3 contains the required preliminaries and problem definition. In Sect. 4, we present our algorithm. We show our experimental results in Sect. 5. In Sect. 6, we conclude this paper and point out possible future work.

2   Related Work

Motivated by the deficiencies of top-k and skyline queries, various approaches [7, 8] were proposed to find the k skyline points that represented the skyline points. Unfortunately, the two methods are neither stable, nor scale-invariant. The most relevant study to our work is k-regret query was proposed by Nanongkai *et al.* [3]. A number of algorithms were proposed to extend the concept to some extent [4,5,9] However, these studies aim at minimizing the maximum regret ratio of a selected set, ignoring the importance of reducing the size of candidate set to improve the efficiency.

There are also some researches related to subspace skyline. Yuan *et al.* [10] and Pei *et al.* [11] proposed the concept of SKYCUBE, which computes the skylines of all possible subspaces. But the methods mentioned above focus on computing skylines in subspaces efficiently, ignoring to explore the properties, such as skyline priority, and they have not been involved in reducing the number of candidate points for k-regret query.

Table 1. English teacher recruiting example

Candidate	S_1	S_2	S_3	Priority
p_1	8.4	5.1	8.3	2
p_2	5.2	6.2	6.6	3
p_3	9.3	6.1	7.4	1
p_4	4.7	5.2	10	1
p_5	4.2	8.8	9.2	2
p_6	5.6	10	3.1	1
p_7	8.2	9.2	2.5	2
p_8	5.1	9.3	5.3	2

Table 2. Skylines of all subspaces

Subspace	Skyline
S_1	$\{p_3\}$
S_2	$\{p_6\}$
S_3	$\{p_4\}$
S_1, S_2	$\{p_3, p_6, p_7\}$
S_2, S_3	$\{p_4, p_5, p_6, p_8\}$
S_1, S_3	$\{p_1, p_3, p_4\}$
S_1, S_2, S_3	$\{p_1, p_2, p_3, p_4, p_5, p_6, p_7, p_8\}$

3 Preliminaries

To define the problem, we need to introduce the concepts of k-regret query and explain how the size of candidate set influences the performance of k-regret algorithms.

First, we assume that the user's preference to a point can be expressed using a utility function f, where $f(p)$ is the utility of a point p for the user with the utility function f. Then the regret ratio of a user with the utility function f after seeing a subset R instead of a database D is $rr_D(R, f) = \frac{max_{p \in D} f(p) - max_{p \in R} f(p)}{max_{p \in D} f(p)}$.

Since utility functions vary across users, any algorithm for a k-regret query must minimize the maximum regret ratio for a class of utility functions. In this paper, we only consider linear utility functions, denoted by $\mathcal{F}$, because they are very popular in modeling user preferences [3,12,13]. Thus the worst possible regret for any user with a utility function in $\mathcal{F}$ is defined as follows.

Definition 1 (Maximum Regret Ratio). *Define* $rr_D(R, \mathcal{F}) = sup_{f \in \mathcal{F}}$ $\frac{max_{p \in D} f(p) - max_{p \in R} f(p)}{max_{p \in D} f(p)}$.

Now we explain how the size of candidate set influences the performance of k-regret algorithms. One of the classical algorithms for k-regret query is GREEDY [3]. Based on the idea of "greedy", GREEDY iteratively construct the solution by selecting the point that currently contributes to the greatest value of maximum regret ratio. To be specific, the algorithm needs to inspect each of the candidate points (except selected points) by computing the maximum regret ratio to decide whether the point will be included in the result set or not. So the running time of the algorithm is largely dependent on the size of the candidate set. By reducing the size of candidate set, the efficiency of the algorithm can be greatly improved. Now we explain the idea of reducing the size of candidate set via skyline priority by a simple example.

Consider the following example, a Chinese school wants to recruit k teachers. They measure the candidates from d aspects, such as the capability of *speaking, listening, reading, writing, educational background, computer skills* and so on. Table 1 shows a list of candidates which are also the skyline points of all the candidates (only consider three aspects: S_1: *reading*, S_2: *writing*, S_3: *speaking*). The values in Table 1 represent their ability in each corresponding attribute where greater values are better. Now our task is to select k best persons from the candidates. Intuitively, the school may want to hire someone who is the most popular in one aspect among all the candidates. For example, the school may prefer p_3, p_4 and p_6 with the highest priority because they are the best in S_1, S_3 and S_2 respectively. So, when choosing persons from the candidates, p_3, p_4 and p_6 have more probability of being selected than other candidates. When it comes to higher dimensions, the priority of higher dimensional skyline points are lower. For example, we observe that p_2 is only the skyline of the full space and p_2 has the lowest priority. So the possibility of being selected of p_2 is small for it is mediocre in all attributes.

The above example shows that if we consider a smaller subset of skyline points as the candidate set, the points with higher skyline priority are preferred. So we define the problem of candidate set determination based on skyline priority.

Problem Definition. Given a set D of n points in d dimensions, our problem of processing k-regret query via skyline priority is to determine the candidate set of k-regret query by selecting the points with high skyline priority, meanwhile keeping the maximum regret ratio as small as possible.

4   Algorithm

In this section, we first give the formal definition of skyline priority and we concentrate on the computation of skyline priority and develop efficient algorithm to solve our problem.

Given a d-dimensional dataset D, $\mathcal{S}$ is the dimension set consisting of all the d dimensions and S_i represents each dimension. Let p and q be two data points in D, we denote the value of p and q on dimension S_i as $p.S_i$ and $q.S_i$. For any dimension set $\mathcal{B}$, where $\mathcal{B} \subseteq \mathcal{S}$, p dominates q if $\forall S_i \in \mathcal{B}, p.S_i \leq q.S_i$ and $\exists S_j \in \mathcal{B}, p.S_j > q.S_j (i \geq 1, j \leq d)$. The skyline query on $\mathcal{B}$ returns all data points that are not dominated by any other points on $\mathcal{B}$. The result is called subspace skyline points, denoted by $SKY_{\mathcal{B}}(D)$. See the running example in Table 1, $SKY_{\mathcal{B}_1}(D) = \{p_3\}$, $SKY_{\mathcal{B}_2}(D) = \{p_3, p_6, p_7\}$, where $\mathcal{B}_1 = \{S_1\}$ and $\mathcal{B}_2 = \{S_1, S_2\}$.

Given D on $\mathcal{S}$, a SKYCUBE consists of a set of subspace skyline points in $2^d - 1$ non-empty subspaces. The SKYCUBE is shown in Table 2. In SKYCUBE, each $SKY_{\mathcal{B}}(D)$ is called cuboid $\mathcal{B}$. For two cuboids $\mathcal{B}_1$ and $\mathcal{B}_2$ in the SKYCUBE, if $\mathcal{B}_1 \subseteq \mathcal{B}_2$, we call $\mathcal{B}_2$ ($\mathcal{B}_1$) ancestor (descendant) cuboid. If their levels differ by one, we also call $\mathcal{B}_2$ ($\mathcal{B}_1$) parent (child) cuboid.

Definition 2 (Skyline Priority). *Given a skyline point p, skyline priority, denoted by $prt(p)$ is the least dimensionality of subspace in which p is a skyline point. The smaller the value of $prt(p)$ is, the higher priority p has.*

For example, Table 2 shows all the subspace skyline points. p_3 is the skyline point of subspace $\{S_1\}$, $\{S_1, S_2\}$, $\{S_1, S_3\}$ and $\{S_1, S_2, S_3\}$, and the least dimensionality of the subspaces is 1, so the skyline priority of p_3 is 1. Table 1 shows the priority of 8 candidates.

Now we focus on how to compute skyline priority efficiently. The naive method to compute prior skyline points is to enumerate all the $2^d - 1$ subspace skyline points and tag priority of the point with the least dimensionality. The low efficiency of the naive method lies in that the skyline points of each subspace are computed separately, resulting in a great number of redundant computation.

Algorithm 1. SP_{SKYCUBE} algorithm

Input: dataset D, $x\%$, n, k.
Output: the set of prior skyline points SP

1 sort D on every dimension to form d sorted list L_{S_i};
2 initialize SP to be empty;
3 $m = MAX(n * x\%, k)$;
4 **for** *each cuboid $\mathcal{B}$ from bottom to top of the skycube* **do**
5   SKY = the union of all the child cuboids;
6   choose a sorted list L_{S_i};
7   **for** q *in* $D \setminus SKY$ **do**
8   evaluate(q, $SKY_{\mathcal{B}}(D)$);
9   **if** $|SP| \leq m$ **then**
10   remove the point with the largest priority in SP when $|SP| = k$;
11   insert q into SP;

12 return SP;

To cover the problem of repeated computation, we proposed a sharing strategy of subspace skyline points based on the relationship between a parent and child cuboid, as stated in the following Lemma 1.

Lemma 1. *Given a set D of data points on dimension set S. For any two data points p and q, $p.S_i \neq q.S_i (\forall S_i \in S)$, for two sub dimension sets $\mathcal{B}_1, \mathcal{B}_2 (\mathcal{B}_1, \mathcal{B}_2 \subseteq S)$, where $\mathcal{B}_1 \subset \mathcal{B}_2$, $SKY_{\mathcal{B}_1}(D) \subseteq SKY_{\mathcal{B}_2}(D)$.*

The conclusion that the union of child cuboids belongs to the parent cuboid can be easily obtained. Therefore, if we compute the priority from bottom to top, once the priority of a point p in the child cuboid is settled, it is unnecessary to compute the priority in parent cuboids. For example, in Table 2, since p_3 is the skyline of $\{S_1\}$, $prt(p_3) = 1$. So there is no need to compute the priority of p_3 in the skyline of subspace $\{S_1, S_2\}$. A large number of skyline computation are saved due to this strategy.

Besides, a pre-sorting of the data is executed for optimization. This is based on the observation that the efficiency of skyline computation is dependent on the order in which the data points are processed. Furthermore, we also define a value called *Entropy* to promote skyline points without dominating test where $E_{\mathcal{B}}(p) = \sum_{\forall a_i \in \mathcal{B}} p(a_i)$. The procedure using the promoting strategy is shown in Algorithm 2. This is based on the property that for two data points p and q, if

$E_\mathcal{B}(p) \leq E_\mathcal{B}(q)$ and p is already a skyline point, then q is definitely a skyline point (Line 2 to 3), avoiding extra dominating test between p and q. Otherwise, we do a dominating test to decide whether q is a skyline point (Line 4 to 6).

Based on the above techniques, we develop the algorithm SP_{SKYCUBE}, which computes skyline priority and determines the candidate set. The main procedure of our approach is shown in Algorithm 1. At the beginning of the algorithm, the pre-sorting operation of the data is invoked for optimization (Line 1). m, the maximum of the size of candidate points and k, ensures at least k points in candidate set. The tunable parameter x provides a flexible tradeoff between CPU time and maximum regret ratio for users. If a data is a skyline point of child cuboid, there is no necessary to compute its priority. Otherwise it is compared to the current skyline to determine its priority by calling the function *Evaluate* (Line 7 to 8). To get the set of at least k candidate points with high skyline priority, a point p is inserted into SP when the set SP has fewer than m points. Specially, the point with largest priority is replaced by p if p has smaller priority and the size of SP is m.

Algorithm 2. evaluate(p, $SKY_\mathcal{B}(D)$)

Input: a data point q, the computed cuboid $\mathcal{B}$
Output: rank the q with skyline priority if p is a skyline of $\mathcal{B}$

1 **for** *each p in $SKY_\mathcal{B}(D)$* **do**
2 | **if** $f_\mathcal{B}(q) > f_\mathcal{B}(p)$ **then**
3 | | insert q into $SKY_\mathcal{B}(D)$; $prt(p) = |\mathcal{B}|$; return;
4 | **else if** p *dominates* q *on* $\mathcal{B}$ **then**
5 | | discard q; return;
6 | insert q into $SKY_\mathcal{B}(D)$; $prt(p) = |\mathcal{B}|$;

---

5   Experiments

In this section, we show the prior skyline points are efficient to select the k-representative skyline points by comparing them with the skyline points, running on the k-regret algorithm GREEDY [3]. All the experiments were implemented in C++ and the experiments were performed on a 3.3 GHz CPU and 4 G RAM machine running Ubuntu 14.04 LTS.

We ran our experiments on both synthetic and real datasets. For synthetic datasets, we use anti-correlated synthetic data, generated by the generator in [2]. Unless specially stated, the cardinality is ten thousand ($n = 10,000$), the dimensionality is six ($d = 6$), and $k = 10$. The real-world datasets we adopted are three datasets called *Household* (www.ipums.org), *Color* (kdd.ics.uci.edu), and *NBA* (www.basketballreference.com), which are widely used in this area.

We conduct our experiments on two kinds of candidate sets, namely skyline points (S) and prior skyline points ($P - x\%$), where $x\%$ means the candidate set is composed of the top $x\%$ prior skyline points. Then, we show the prior skyline points can be used for the algorithm GREEDY for k-regret query in terms of maximum regret ratio and running time. Different values of x ($x = 2, 4, 6, 8,$ 10) have been chosen.

The effects of P on maximum regret ratio and time on anti-correlated datasets for different k are shown in Figs. 1(a) and 2(a) respectively. When k increases, the maximum regret ratio of all candidate sets decreases which is in accordance with the results in [3]. We observe that the maximum regret ratio of P is close to S or even smaller in some situation while the running time is only $\frac{1}{10}$ of S. This is because the candidate points P with high skyline priority have more possibility to be the representative skyline points. From the above, we have the conclusion that prior skyline points can efficiently speed up the original algorithm with a close maximum regret ratio.

Besides, we evaluate the effect of different cardinalities and dimensionalities on our algorithm. Figure 1(b) shows that when d increases, the maximum regret ratio of different candidate sets increases. Figure 1(c) shows that the maximum regret ratio is independent of n. The same conclusion can be reached that the maximum regret ratio of prior skyline points is close to S and the running time in Fig. 2(b) and (c) is much less than that of S.

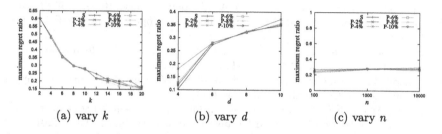

(a) vary k (b) vary d (c) vary n

Fig. 1. Effect on maximum regret ratio of different candidate sets

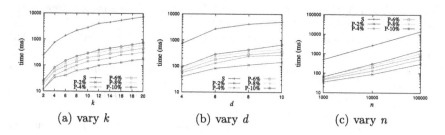

(a) vary k (b) vary d (c) vary n

Fig. 2. Effect on time of different candidate sets.eps

We can observe the similar trends on real datasets. Similar to the experiments on synthetic datasets, with prior skyline points, the algorithm GREEDY achieves a close maximum regret ratio with less running time shown compared with the skyline points. The figures are omitted due to space limitation.

6   Conclusions and Future Work

We studied the problem of reducing the size of candidate set for k-regret query basing on skyline priority. Skyline priority is proposed to rank skyline points and efficient algorithm is provided to determine the candidate set for k-regret query. Extensive experiments verify the efficiency and accuracy of our method. Since our candidate set are the points of "the best - *higher skyline priority* of the best - *skyline points of the whole dataset*", we do think there are certain theoretical results to be found to guarantee our proposed method to be efficient and effective. We leave this as an open problem and also as our future work.

Acknowledgment. This work is partially supported by the National Natural Science Foundation of China under grants U1733112, 61702260, Funding of Graduate Innovation Center in NUAA under grant KFJJ20171601.

References

1. Ilyas, I.F., Beskales, G., Soliman, M.A.: A survey of Top-k query processing techniques in relational database systems. ACM Comput. Surv. **40**(4), 1–58 (2008)
2. Börzsöny, S., Kossmann, D., Stocker, K.: The skyline operator. In: ICDE, pp. 421–430 (2001)
3. Nanongkai, D., Sarma, A.D., Lall, A., Lipton, R.J., Xu, J.: Regret-minimizing representative databases. VLDB **3**, 1114–1124 (2010)
4. Agarwal, P.K., Kumar, N., Sintos, S., Suri, S.: Efficient algorithms for k-regret minimizing sets. CoRR, abs/1702.01446 (2017)
5. Xie, M., Wong, R.C.-W., Li, J., Long, C., Lall, A.: Efficient k-regret query algorithm with restriction-free bound for any dimensionality. In: SIGMOD (2018)
6. Godfrey, P.: Skyline cardinality for relational processing. In: Seipel, D., Turull-Torres, J.M. (eds.) FoIKS 2004. LNCS, vol. 2942, pp. 78–97. Springer, Heidelberg (2004). https://doi.org/10.1007/978-3-540-24627-5_7
7. Lin, X., Yuan, Y., Zhang, Q., Zhang, Y.: Selecting stars: the k most representative skyline operator. In: ICDE, pp. 86–95 (2007)
8. Tao, Y., Ding, L., Lin, X., Pei, J.: Distance-based representative skyline. In: ICDE, pp. 892–903 (2009)
9. Qi, J., Zuo, F., Samet, H., Yao, J.C.: From additive to multiplicative utilities, K-regret queries (2016)
10. Yuan, Y., Lin, X., Liu, Q., Wang, W., Yu, J.X., Zhang, Q.: Efficient computation of the skyline cube. In: VLDB, pp. 241–252 (2005)
11. Pei, J., Jin, W., Ester, M., Tao, Y.: Catching the best views of skyline: a semantic approach based on decisive subspaces. In: VLDB, pp. 253–264 (2005)
12. Nanongkai, D., Lall, A., Das Sarma, A., Makino, K.: Interactive regret minimization. In: SIGMOD, pp. 109–120 (2012)
13. Chester, S., Thomo, A., Venkatesh, S., Whitesides, S.: Computing k-regret minimizing sets. In: VLDB, pp. 389–400 (2014)

Efficient Approximate Algorithms for k-Regret Queries with Binary Constraints

Qi Dong[1], Jiping Zheng[1,2(✉)], Xianhong Qiu[1], and Xingnan Huang[1]

[1] College of Computer Science and Technology,
Nanjing University of Aeronautics and Astronautics, Nanjing, China
{dongqi,jzh,qiuxianhong,huangxingnan}@nuaa.edu.cn
[2] Collaborative Innovation Center of Novel Software Technology and
Industrialization, Nanjing, China

Abstract. Extracting interesting points from a large dataset is an important problem for multi-criteria decision making. Recently, k-regret query was proposed and received attentions from the database community because it does not require any utility function from users and the output size is controllable. In this paper, we consider k-regret queries with binary constraints which doesn't be addressed before. Given a collection of binary constraints, we study the problem of extracting the k representative points with small regret ratio while satisfying the binary constraints. To express the satisfaction of data points by the binary constraints, we propose two models named NBC and LBC in quantitative and qualitative ways respectively. In quantitative way, the satisfaction is expressed by a real number between 0 and 1 which qua_N_tifies the satisfaction of a point by the _B_inary _C_onstraints. While in qualitative way, the satisfaction is modeled qua_L_itatively by a set of _B_inary _C_onstraints which are satisfied by the point. Further, two efficient approximate algorithms called NBC_P-Greedy and NBC_{DN}-Greedy are developed based on NBC while LBC_{DN}-Greedy algorithm is proposed based on LBC. Extensive experiments on synthetic and real datasets confirm the efficiency and effectiveness of our proposed algorithms.

Keywords: k-Regret query · Binary constraints
Multi-criteria decision making · Top-k query · Skyline query

1 Introduction

Extracting interesting points from a large database is an important problem in multi-criteria decision making for end-users. In the literature, top-k queries, skyline queries and k-regret queries were proposed to solve this problem. A top-k query [9–11] requires end-users to specify a utility function (or score function). With the utility function, the top-k query computes the utility for each points in the database, and show the k points with highest utilities to the users. However, the users may not provide their utility functions precisely. A skyline query

X. Meng et al. (Eds.): WISA 2018, LNCS 11242, pp. 421–433, 2018.
https://doi.org/10.1007/978-3-030-02934-0_39

[3,5,13,16] returns a set of interesting data points without the need to appoint a utility function. However, the output size of skyline is uncontrollable, and cannot be foreseen before the whole database is accessed. A k-regret query [15] integrates the merits of top-k and skyline queries, and do not require any utility function from users as well as the output size is controllable since only k points are returned. However, existing k-regret queries [1,2,4,15,18] do not consider user's binary constraints. Consider the following example.

Assume that a real-estate agent maintains a database of available houses along with several attributes, such as *price, number of rooms, number of windows*, and so on. And each available house may satisfy several binary constraints, e.g., *distance from a school less than 5 miles, mountain and lake view, quiet neighborhood* etc. A user might not only be interested in the actual value of an attribute, but also the binary constraints that are satisfied by the house. In order to decide which house to be displayed to the users, we assume that each user has a preference function on the attributes of house called a utility function in his/her mind. Based on the utility function, each point in the database has a utility. A high utility means that this house is favored by the user on the attributes. Meanwhile, if this house satisfies more binary constraints, the user may be more favorite to it.

In the above example, the binary constraints satisfied by houses are also a common and important criterion for users to select houses. Therefore, the real-estate agent considers not only the values of attributes of houses, but also binary constraints satisfied by houses. In the literature of k-regret query, no research addressed this problem. A most related work is the KREP problem proposed by Khan et al. [10], which finds the top-k data points that are representative of the maximum possible number of available options with respect to the given binary constraints. However, they adopted the top-k query which is not convenient for users or real-estate agent because the utility functions are often unavailable. In this paper, we adopt the k-regret query with binary constraints to find k points such that the maximum regret ratio could be as small as possible while maximizing the satisfaction of data points by the binary constraints. Two models to express the satisfaction of data points by the binary constraints namely NBC and LBC are proposed and corresponding efficient approximate algorithms are developed to find k-representatives. The main contributions of this paper are listed as follows.

- We formulate and investigate the problem of selecting the set of k representative points with binary constraints.
- We propose two models NBC and LBC respectively to express the satisfaction of data points by the binary constraints in quantitative and qualitative ways. Then we develop two efficient approximate algorithms with small regret ratio or large satisfaction called NBC_P-Greedy and NBC_{DN}-Greedy respectively based on NBC. Moreover, we provide the efficient algorithm called LBC_{DN}-Greedy based on LBC and it offers a tradeoff between small regret ratio and large satisfaction of data points by the binary constraints.

– Extensive experiments are conducted on both synthetic and real datasets to evaluate our proposed method and the experimental results confirm that our algorithms are able to select the set of k representative points with either small regret ratio or large satisfaction of data points by the binary constraints or a tradeoff between them.

The remainder of this paper is organized as follows. In Sect. 2, previous work related to this paper is described. The formal definitions of our problem are given in Sect. 3. In Sect. 4, we propose two satisfaction models for our problem. We present three efficient algorithms based on the two models in Sect. 5. The performance of our algorithms on real and synthetic data is presented in Sect. 6. Finally, in Sect. 7 we conclude this paper.

2   Related Work

Top-k and skyline queries have received considerable attentions during last two decades as they play an important role in multi-criteria decision making. However, top-k needs to specify the utility functions precisely. Some existing approaches were developed based on top-k queries. Mindolin *et al.* [13] asked users to specify a small number of possible weights each indicating the importance of a dimension. Lee *et al.* [11] asked users to specify some pair-wise comparisons between two dimensions to decide whether a dimension was more important than the other for a comparison. However, these studies require users to specify some kinds of utilities/preferences. For skyline query, the size of skyline is uncontrollable and cannot be foreseen before the whole database is accessed. Therefore, some researchers attempt to reduce the output size of the skyline query, but their approaches are not *stable* like the approach in [12] or not *scale-invariant* like the approach in [19].

To alleviate the burden of top-k for specifying accurate utility functions and skyline query for outputting too many results, regret-based k representative query was first proposed by Nanongkai *et al.* [15] to minimize user's maximum regret ratio. The stable, scale-invariant approach has been extended and generalized to the interactive setting in [14] and the k-RMS problem in [6]. [18] developed efficient algorithms for k-regret queries and [7] generalized the notion of k-regret queries to include nonlinear utility functions. [20] proposed an elegant algorithm which has a restriction-free bound on the maximum regret ratio. In this paper, we follow the definition in [15] and study the k-regret query with binary constraints.

In terms of constrained optimization queries, [21] proposed efficient indexing methods to find an exact solution of the k-constrained optimization query. [17] studied the problem of multi-objective optimization and evaluated the solutions to a combinatorial optimization problem with respect to several cost criteria. [8] proposed OPAC query which considered ϵ-Pareto answers to optimization queries under parametric aggregation constraints over multiple database points. [10] proposed KREP framework to find the top-k data points that are representative of the maximum possible number of available options with respect to the given constraints. However, their optimization function and constraints are aggregated over multiple points in [8], and the scoring function as well as the binary constraints were applied over individual database points in [10]. In our query semantics, the regret ratio is aggregated over multiple points and the binary constraints are applied over individual database points.

3   Problem Definition

Let's denote the set of n d-dimensional data points by D. For each point $p \in D$, the attribute value on the i-th dimension is represented as $p[i]$. We assume that small values in each attribute are preferred by the users. We next start with a few definitions.

Definition 1 (Satiated Constraints by a Data Point). *Given a set of binary constraints $C = \{c_1, c_2, ..., c_r\}$, the satiated constraints $C(p)$ by a data point $p \in D$ are simply the subset of the set C. Formally,*

$$C(p) = \{c_i \in C \colon c_i(p) \text{ is true}\}$$

We say that a data point $p \in D$ is relevant, if it satisfies at least one of the given binary constraints, *i.e.*, $C(p) \neq \emptyset$. We shall use the terms "data points" and "relevant data points" interchangeably, *i.e.*, given the binary constraints, we shall only consider the relevant data points in our algorithms.

Definition 2 (Utility Function). *A utility function f is a mapping $f \colon \mathbb{R}_+^d \to \mathbb{R}_+$. The utility of a user with utility function f is $f(p)$ for any point p and shows how satisfied the user with the point.*

Definition 3 (Regret Ratio). *Given a dataset D, a set of S with points in D and a utility function f. The regret ratio of S on D, represented as $rr_D(S, f)$ is defined to be*

$$rr_D(S, f) = \frac{\max_{p \in D} f(p) - \max_{p \in S} f(p)}{\max_{p \in D} f(p)}$$

Notice that regret ratio is in the range $[0, 1]$. When the regret ratio is close to 0, the user is very happy and when the regret ratio is close to 1 the user is very unhappy.

Since we do not ask users for utility functions, we know nothing about users' preferences. For each user, she/he may have arbitrary utility function. In this paper, we assume that user's utility functions fall in a class of functions, denoted by F and the case of function class F consisting of all liner utility functions is considered since it is widely used in modeling the values of user's preferences.

Definition 4 (Maximum Regret Ratio). *Given a dataset D, a set of S with points in D and a class of utility functions F. The maximum regret ratio of S on D, represented as $rr_D(S, F)$ is defined to be*

$$rr_D(S, F) = \sup_{f \in F} rr_D(S, f) = \sup_{f \in F} \frac{\max_{p \in D} f(p) - \max_{p \in S} f(p)}{\max_{p \in D} f(p)}$$

Problem Definition: Given a dataset D, a positive integer k, our problem is trying to find a subset S of D containing at most k points such that the maximum regret ratio is as small as possible in terms of the attributes of points while maximizing the satisfaction of data points by the binary constraints.

4   Models of Binary Constraints

In this section, we describe two models to express the satisfaction of data points by the binary constraints in detail. Let $C = \{c_1, c_2, ..., c_r\}$ be the set of binary constraints. The satiated constraints $C(p)$ is the subset of binary constraints satisfied by $p \in D$. We aim to find the solution set with small regret ratio while satisfying the binary constraints. Therefore, we need to select the point that makes the greatest contribution to the regret ratio and the satisfaction of data points by the binary constraints.

In this paper, we argue that two data points $p, p' \in D$, we say that p dominates p' in following three cases.

Case 1. p dominates p' with respect to the values of attributes and the set of binary constraints C if one of the following holds true: (1) $\forall i \in d$, $p[i] \leq p'[i]$, at least one $i \in d$, $p[i] < p'[i]$, and $C(p) \supseteq C(p')$, or (2) $\forall i \in d$, $p[i] = p'[i]$ and $C(p) \supset C(p')$.

Case 2. p dominates p' with respect to the regret ratio and the constraints effection on the set S if $rr_{S\cup\{p\}}(S, F) \geq rr_{S\cup\{p'\}}(S, F)$, $\frac{|C(p)|}{r} \geq \frac{|C(p')|}{r}$ and at least $rr_{S\cup\{p\}}(S, F) > rr_{S\cup\{p'\}}(S, F)$ or $\frac{|C(p)|}{r} > \frac{|C(p')|}{r}$.

Case 3. p dominates p' with respect to the regret ratio on the set S and the set of binary constraints C if one of the following holds true: (1) $rr_{S\cup\{p\}}(S, F) \geq rr_{S\cup\{p'\}}(S, F)$ and $C(p) \supseteq C(p')$, or (2) $rr_{S\cup\{p\}}(S, F) = rr_{S\cup\{p'\}}(S, F)$ and $C(p) \supset C(p')$.

We say that a point $p \in D$ is called a skyline point if it is not dominated by any other points in D according to the corresponding case. And the dominance number $N(p)$ of a data point $p \in D$ is defined as the number of data points in D that are dominated by p, i.e., $N(p) = |\{p' : p' \in D, p \text{ dominates } p'\}|$.

Theorem 1. *If we consider the largest dominance number of point p to select the point to the result set, the point p which is selected must be the skyline point.*

Proof. If possible, there is a non-skyline point p' with the largest dominance number $N(p')$. Then at least one of the skyline points, say p, which dominates p', has the larger dominance number $N(p)$ than that of p'. Hence, the non-skyline points do not have the largest dominance number. If we consider the largest dominance number of point to select the point to the result set, the skyline point is selected instead of the non-skyline point.   □

4.1   QuaNtitative Binary Constraints (*NBC*)

We propose model *NBC* in quantitative way to express the satisfaction of data points by the binary constraints. In this way, the satisfaction is expressed by a real number $\frac{|C(p)|}{r}$ between 0 and 1 which quaNtifies the satisfaction of a point p by the Binary Constraints. We argue that if the $\frac{|C(p)|}{r}$ is larger, it means that the point p can make greater contribution to the satisfaction of data points by the binary constraints, since the point p satisfies more binary constraints.

4.2   QuaLitative Binary Constraints (*LBC*)

We propose model *LBC* in qualitative way to express the satisfaction of data points by the binary constraints. In this way, the satisfaction is modeled quaLitatively by a set $C(p)$ of Binary Constraints which satisfied by the point p. We argue that two points can be compared specifically in binary constraints by analyzing binary constraints which satisfied by the two points, and decide which point makes more contribution to the satisfaction of data points by the binary constraints.

5   Efficient Greedy Algorithms

Our algorithms follows the framework of the RDP-Greedy proposed by [15] to select the set of k points with small regret ratio while satisfying the binary constraints based on the two satisfaction models.

5.1   *NBC_P*-Greedy Algorithm

In the RDP-Greedy, it adds the "worst" point that currently contributes to the maximum regret ratio. To be precise, for each round it adds the point p to the solution set S where p is the point such that $rr_D(S, F) = rr_{S \cup \{p\}}(S, F)$. This is done by computing $rr_{S \cup \{p\}}(S, F)$ for each point $p \in D \backslash S$ and keep the point with maximum value. Therefore, we argue that the point is selected whose $rr_{S \cup \{p\}}(S, F)$ and $\frac{|C(p)|}{r}$ are as large as possible in each iteration, since the larger $\frac{|C(p)|}{r}$ can make more contribution to the satisfaction of data points by the binary constraints. Based on the above observations, we design an algorithm which uses the Product of $rr_{S \cup \{p\}}(S, F)$ and $\frac{|C(p)|}{r}$ as a metric to select the point in each iteration, called *NBC_P*-Greedy, based on model *NBC*. We intend to compute the contribution $rr_{S \cup \{p\}}(S, F)$ of the point p above the maximum regret ratio on the set S by using the linear program (LP) used in [15]. This algorithm is similar to the RDP-Greedy [15] and the pseudocodes of *NBC_P*-Greedy are omitted due to space limitation.

Time Complexity: Let's denote the number of data points in D as n, the number of attributes as t and the number of binary constraints as r. An LP solver results in $O(k^2 t)$ running time proposed by [15]. The running time of a $\frac{|C(p)|}{r}$ is at most $O(r)$. The LP and $\frac{|C(p)|}{r}$ are computed at most nk times. Therefore, the overall time complexity of the *NBC_P*-Greedy algorithm is $O(nk(k^2 t + r))$.

5.2 NBC_{DN}-Greedy Algorithm

We know that the result of the product of $rr_{S \cup \{p\}}(S, F)$ and $\frac{|C(p)|}{r}$ is easily affected by the larger value between $rr_{S \cup \{p\}}(S, F)$ and $\frac{|C(p)|}{r}$, and the solution set is only great in regret ratio or the satisfaction. Therefore, based on the model NBC, we propose another algorithm to select the point with the largest Dominance Number by the second dominant case in each iteration, called NBC_{DN}-Greedy, since the point with larger dominance number is more representative and the dominance number of each point p is recomputed in next iteration because of the change of the perselected solution set S in next iteration. The algorithm selects the points only from the skyline set according to Theorem 1.

We maintain a priority queue $\mathcal{L}_1$, which stores all points ids in descending order of their $rr_{S \cup \{p\}}(S, F)$ on the set S, a queue Skl, which stores all skyline points, and a priority queue $\mathcal{L}_2$, which stores all points ids in descending order of their dominance numbers $N(p)$. The pseudocodes of NBC_{DN}-Greedy are shown in Algorithm 1.

Algorithm 1: NBC_{DN}-Greedy$(D, C, \mathcal{L}_1, Skl, \mathcal{L}_2, k)$

Input: A set of n d-dimensional points $D = \{p_1, p_2, \cdots, p_n\}$, a set of binary constraints $C = \{c_1, c_2, ..., c_r\}$, an integer k, the desired output size.
Output: A subset of D of size k, denoted by S.

```
 1  Let S = {p₁*} where p₁* = argmax_{p∈D} p[1];
 2  for i = 1 to k - 1 do
 3      Let rd* = 0, p* = null, L₁ = ∅, Skl = ∅ and L₂ = ∅;
 4      for each p ∈ D\S do
 5          Compute rr_{S∪{p}}(S, F) using Linear Programming;
 6          Compute |C(p)|/r;
 7          Add p to L₁;
 8      while L₁ not empty do
 9          p ← remove top-item from L₁;
10          if Skl is empty then
11              Add p to Skl;
12          else
13              flag = true;
14              while Skl not empty do
15                  p' ← remove top-item from Skl;
16                  if |C(p')|/r ≥ |C(p)|/r then
17                      N(p') + +;
18                      flag = false;
19              if flag == true then
20                  Add p to Skl;
21      remove items from Skl to L₂;
22      p* ← remove top-item from L₂;
23      if rr_{S∪{p*}}(S, F) == 0 && N(p*) == 0 then
24          return S;
25      else
26          S = S ∪ {p*};
27  return S;
```

Time Complexity: The time complexity of computing $rr_{S \cup \{p\}}(S, F)$ and $\frac{|C(p)|}{r}$ is the same as that of NBC_P-Greedy algorithm, *i.e.*, $O(n(k^2t + r))$. The time complexity of skyline queries in all iterations is at most $O((n^2 - n)/2)$ and that of removing items from Skl to $\mathcal{L}_2$ is at most $O(n)$. Therefore, the overall time complexity of the NBC_{DN}-Greedy algorithm is $O(k(n(k^2t + r) + (n^2 - n)/2 + n))$ *i.e.*, $O(\frac{n^2k}{2} + nk(k^2t + r + \frac{1}{2}))$.

5.3 LBC_{DN}-Greedy Algorithm

We know that the number of the integer $\frac{|C(p)|}{r}$ is at most $r < |D|$. Therefore, most points might have the same real number $\frac{|C(p)|}{r}$ in each iteration. The dominance number of a point p is more susceptible to the regret ratio $rr_{S \cup \{p\}}(S, F)$ and the solution set is only good for regret ratio with the increase of k. Hence, we propose an algorithm to select the point with the largest <u>D</u>ominance <u>N</u>umber by the third dominant case in each iteration, called LBC_{DN}-Greedy, based on model LBC. We use the dominance number as the evaluation criterion for the same reason as NBC_{DN}-Greedy algorithm. The pseudocodes of LBC_{DN}-Greedy are shown in Algorithm 2. The definitions of $\mathcal{L}_1$, $\mathcal{L}_2$ and Skl are the same as that in NBC_{DN}-Greedy algorithm.

Time Complexity: The time complexity of LBC_{DN}-Greedy algorithm is almost the same as that of NBC_{DN}-Greedy algorithm. But in LBC_{DN}-Greedy algorithm, we compare the $C(p)$ instead of computing the $\frac{|C(p)|}{r}$. Hence, the time complexity of comparing the $C(p)$ is $O(r)$ and that of skyline queries in all iterators is at most $O((n^2 - n)r/2)$. The overall time complexity of the LBC_{DN}-Greedy algorithm is $O(k(n(k^2t) + (n^2 - n)r/2 + n))$ *i.e.*, $O(\frac{n^2kr}{2} + nk(k^2t + 1 - \frac{r}{2}))$.

6   Experimental Results

In this section, we show the performance of the proposed algorithms via experiments. All algorithms were implemented in C++ and the experiments were all conducted on a 64-bit 2.50 GHz Intel Core machine which was running Ubuntu 16.04 LTS operating system.

We ran our experiments on both synthetic and real datasets. The synthetic dataset includes an anti-correlated dataset, which is generated by the dataset generator in [3]. The cardinality of our synthetic dataset is 1M, the dimensionality of each data point p is 20. The range of each attribute lies between $[0, 100]$. Real dataset we used is the car dataset of 598 different models of cars—each model has a certain number of reviews (range: 11–540) and 9 various ratings (range: 0–10). Moreover, we computed the skyline first and our queries on these datasets returned anywhere from 5 to 30 points and evaluated the maximum regret ratio using Linear Programming implemented in the GUN Linear Programming Kit[1].

---

[1] https://www.gnu.org/software/glpk/.

Algorithm 2: LBC_{DN}-Greedy$(D, C, \mathcal{L}_1, Skl, \mathcal{L}_2, k)$

Input: A set of n d-dimensional points $D = \{p_1, p_2, \cdots, p_n\}$, a set of binary constraints
$C = \{c_1, c_2, ..., c_r\}$, an integer k, the desired output size.
Output: A subset of D of size k, denoted by S.

1 Let $S = \{p_1^*\}$ where $p_1^* = argmax_{p \in D} p[1]$;
2 **for** $i = 1$ to $k - 1$ **do**
3 $\quad$ Let $rd^* = 0$, $p^* = null$, $\mathcal{L}_1 = \emptyset$, $Skl = \emptyset$ and $\mathcal{L}_2 = \emptyset$;
4 $\quad$ **for** each $p \in D \backslash S$ **do**
5 $\quad\quad$ Compute $rr_{S \cup \{p\}}(S, F)$ using Linear Programming;
6 $\quad\quad$ Add p to $\mathcal{L}_1$;

7 $\quad$ **while** $\mathcal{L}_1$ not empty **do**
8 $\quad\quad$ $p \leftarrow$ remove top-item from $\mathcal{L}_1$;
9 $\quad\quad$ **if** Skl is empty **then**
10 $\quad\quad\quad$ Add p to Skl;

11 $\quad\quad$ **else**
12 $\quad\quad\quad$ $flag = true$;
13 $\quad\quad\quad$ **while** Skl not empty **do**
14 $\quad\quad\quad\quad$ $p' \leftarrow$ remove top-item from Skl;
15 $\quad\quad\quad\quad$ **if** $C(p') \supseteq C(p)$ **then**
16 $\quad\quad\quad\quad\quad$ $N(p') + +$;
17 $\quad\quad\quad\quad\quad$ $flag = false$;

18 $\quad\quad\quad$ **if** $flag == true$ **then**
19 $\quad\quad\quad\quad$ Add p to Skl;

20 $\quad$ remove items from Skl to $\mathcal{L}_2$;
21 $\quad$ $p^* \leftarrow$ remove top-item from $\mathcal{L}_2$;
22 $\quad$ **if** $rr_{S \cup \{p^*\}}(S, F) == 0$ && $N(p^*) == 0$ **then**
23 $\quad\quad$ return S;

24 $\quad$ **else**
25 $\quad\quad$ $S = S \cup \{p^*\}$;

26 return S;

Results on Synthetic Datasets: Let's denote the i-th binary constraint by c_i, which is: $p[i + 6] \geq 95, \forall i \in [1, r]$, and each attribute of point p is $p[i], \forall i \in [1, t]$.

The effects on maximum regret ratio and query time on the anti-correlated datasets for different k, t and r are presented in Figs. 1(a–c) and 3(a–c) respectively. We use the number of sets of binary constraints with different size as a metric of the satisfaction of data points by the binary constraints since the set of binary constraints with larger size can satisfy more users and the number of this kind of set is large for the result. The number of sets of binary constraints with different size is presented in Fig. 2(a–c). NBC_{DN}-Greedy and LBC_{DN}-Greedy both have the small maximum regret ratio and their maximum regret ratios decrease with the increase of k as well as increase with the increase of t. Meanwhile, their maximum regret ratio fluctuate with small ranges as r increases since binary constraints do not directly affect the regret ratio. In all cases, the maximum regret ratio of NBC_P-Greedy is always larger than that of NBC_{DN}-Greedy and LBC_{DN}-Greedy. On the contrary, the number of sets of binary constraints with large size in NBC_P-Greedy is always larger than that of NBC_{DN}-Greedy and LBC_{DN}-Greedy. And the number of sets of binary constraints with large size in LBC_{DN}-Greedy is not less than that of NBC_{DN}-

Greedy. That means LBC_{DN}-Greedy provides a great tradeoff between small regret ratio and the large satisfaction of data points by the binary constraints. In terms of query time, the query time of NBC_P-Greedy is less than that of NBC_{DN}-Greedy and LBC_{DN}-Greedy. And the query time of LBC_{DN}-Greedy increases exponentially when binary constraints are more than 10 and is much larger than that of the other two algorithms. This observation can be explained by our time-complexity analysis in Sect. 5.

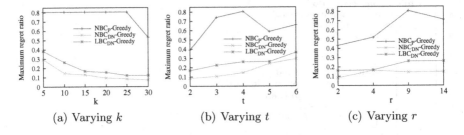

(a) Varying k (b) Varying t (c) Varying r

Fig. 1. Maximum regret ratio for anti-correlated data

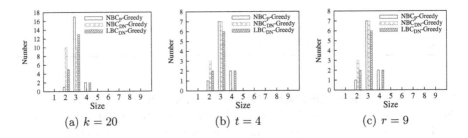

(a) $k = 20$ (b) $t = 4$ (c) $r = 9$

Fig. 2. Number of sets for anti-correlated data

Results on Real Datasets: For the car dataset, we design our query as follows. We use the number of reviews per car-model, the rating of fuel and interior as the three attributes of each point. On the other hand, the rating of the remaining 7 different attributes are utilised as our binary constraints. Specifically, the maximum number of reviews for any car model is 540, and each rating has a value in $(0, 10)$. We say that a car-model is relevant if at least one of the ratings which is utilised as our binary constraints for that car-model is greater than or equal to 9. Out of 598 car-models, we find that 499 of them are relevant ones, while there are also 48 skyline car-models based on the aforementioned query setting. We present 5 representative car-models obtained by our algorithms and respectively list them in Table 1.

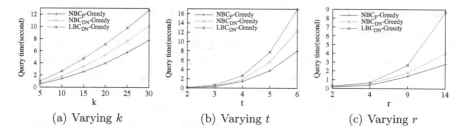

Fig. 3. Query time for anti-correlated data

The first car-model in the first row of the all algorithms is the point whose first dimension is largest in car dataset in all algorithms. And it may not be representative. However, as shown in Table 1, the regret ratios of our algorithms are all small, and each of our representative car-models except the first one satisfies as many as binary constraints. Meanwhile, we only require about 3 ms to retrieve these 5 representatives by each of our algorithm.

Table 1. 5 representative car-models in car dataset on our algorithms

Algorithm	Car-Model(Reviews,Fuel,Interior)	Satiated Constraints	Regret Ratio
$NBC_P - Greedy$	$Honda\ Accord(540, 7.75, 8.85)$	$exterior$	0.1100974
	$Toyota\ Tundra(234, 7.38, 8.56)$	$exterior, performance, comfort, reliability,$ fun	
	$Ford\ Escape$ $Hybrid(25, 9.68, 9.08)$	$exterior, build, performance, comfort,$ $reliability, fun, overall-rating$	
	$Mercedes\ Benz$ $C-class(234, 7.91, 8.79)$	$exterior, build, performance, reliability,$ $fun, overall-rating$	
	$Scion\ Xd(149, 9.13, 8.89)$	$exterior, build, performance, reliability,$ $fun, overall-rating$	
$NBC_{DN} - Greedy$	$Honda\ Accord(540, 7.75, 8.85)$	$exterior$	0.109944
	$Ford\ Escape$ $Hybrid(25, 9.68, 9.08)$	$exterior, build, performance, comfort,$ $reliability, fun, overall-rating$	
	$Audi\ Rs4(26, 8.12, 9.96)$	$exterior, build, performance, comfort,$ $reliability, fun, overall-rating$	
	$Audi\ A8(18, 9.11, 9.94)$	$exterior, build, performance, comfort,$ $reliability, fun, overall-rating$	
	$Toyota\ Camry$ $Hybrid(66, 9.53, 9.17)$	$exterior, build, performance, comfort,$ $reliability, fun, overall-rating$	
$LBC_{DN} - Greedy$	$Honda\ Accord(540, 7.75, 8.85)$	$exterior$	0.109944
	$Ford\ Escape$ $Hybrid(25, 9.68, 9.08)$	$exterior, build, performance, comfort,$ $reliability, fun, overall-rating$	
	$Audi\ Rs4(26, 8.12, 9.96)$	$exterior, build, performance, comfort,$ $reliability, fun, overall-rating$	
	$Audi\ A8(18, 9.11, 9.94)$	$exterior, build, performance, comfort,$ $reliability, fun, overall-rating$	
	$Toyota\ Camry$ $Hybrid(66, 9.53, 9.17)$	$exterior, build, performance, comfort,$ $reliability, fun, overall-rating$	

7   Conclusions

In this paper, we formulate and investigate the problem of selecting the k representative points with the small regret ratio while maximizing the satisfaction of data points by the binary constraints. To express the satisfaction of data points by the binary constraints, we propose two models named NBC and LBC

in quantitative and qualitative ways respectively. And we also design the efficient algorithms called NBC_P-Greedy and NBC_{DN}-Greedy based on the model NBC as well as the efficient algorithm called LBC_{DN}-Greedy based on the model LBC. Experiments on synthetic datasets verified that the NBC_{DN}-Greedy and LBC_{DN}-Greedy achieve the small regret ratio and the large satisfaction of data points by the binary constraints as well as the NBC_P-Greedy obtains large satisfaction of data points by the binary constraints with larger regret ratio. In addition, the three algorithms all achieve small regret ratio with large satisfaction of data points by the binary constraints on real dataset. In future work, we shall consider our framework for finding k representative points with multiple users.

Acknowledgment. This work is partially supported by the National Natural Science Foundation of China under grants U1733112,61702260, the Natural Science Foundation of Jiangsu Province of China under grant BK20140826, the Fundamental Research Funds for the Central Universities under grant NS2015095, Funding of Graduate Innovation Center in NUAA under grant KFJJ20171605.

References

1. Agarwal, P.K., Kumar, N., Sintos, S., Suri, S.: Efficient algorithms for k-regret minimizing sets. In: International Symposium on Experimental Algorithms, pp. 7:1–7:23 (2017)
2. Asudeh, A., Nazi, A., Zhang, N., Das, G.: Efficient computation of regret-ratio minimizing set: a compact maxima representative. In: SIGMOD, pp. 821–834 (2017)
3. Borzsony, S., Kossmann, D., Stocker, K.: The skyline operator. In: ICDE, pp. 421–430 (2001)
4. Cao, W., et al.: k-regret minimizing set: efficient algorithms and hardness. In: ICDT, pp. 11:1–11:19 (2017)
5. Chan, C.-Y., Jagadish, H.V., Tan, K.-L., Tung, A.K.H., Zhang, Z.: Finding k-dominant skylines in high dimensional space. In: SIGMOD, pp. 503–514 (2006)
6. Chester, S., Thomo, A., Venkatesh, S., Whitesides, S.: Computing k-regret minimizing sets. In: VLDB, pp. 389–400 (2014)
7. Faulkner, T.K., Brackenbury, W., Lall, A.: k-regret queries with nonlinear utilities. In: VLDB, pp. 2098–2109 (2015)
8. Guha, S., Gunopoulos, D., Vlachos, M., Koudas, N., Srivastava, D.: Efficient approximation of optimization queries under parametric aggregation constraints. In: VLDB, pp. 778–789 (2003)
9. Ilyas, I.F., Beskales, G., Soliman, M.A.: A survey of top-k query processing techniques in relational database systems. ACM Comput. Surv. **40**(4), 11:1–11:58 (2008)
10. Khan, A., Singh, V.: Top-k representative queries with binary constraints. In: SSDBM, pp. 13:1–13:10 (2015)
11. Lee, J., won You, G., won Hwang, S.: Personalized top-k skyline queries in high-dimensional space. Inf. Syst. **34**(1), 45–61 (2009)
12. Lin, X., Yuan, Y., Zhang, Q., Zhang, Y.: Selecting stars: the k most representative skyline operator. In: ICDE, pp. 86–95 (2007)

13. Mindolin, D., Chomicki, J.: Discovering relative importance of skyline attributes. In: VLDB, pp. 610–621 (2009)
14. Nanongkai, D., Lall, A., Das Sarma, A., Makino, K.: Interactive regret minimization. In: SIGMOD, pp. 109–120 (2012)
15. Nanongkai, D., Sarma, A.D., Lall, A., Lipton, R.J., Xu, J.: Regret-minimizing representative databases. VLDB **3**(1–2), 1114–1124 (2010)
16. Papadias, D., Tao, Y., Fu, G., Seeger, B.: Progressive skyline computation in database systems. TODS **30**(1), 41–82 (2005)
17. Papadimitriou, C.H., Yannakakis, M.: On the approximability of trade-offs and optimal access of web sources. In: FOCS, pp. 86–92 (2000)
18. Peng, P., Wong, R.C.W.: Geometry approach for k-regret query. In: ICDE, pp. 772–783 (2014)
19. Tao, Y., Ding, L., Lin, X., Pei, J.: Distance-based representative skyline. In: ICDE, pp. 892–903 (2009)
20. Xie, M., Wong, R.C.-W., Li, J., Long, C., Lall, A.: Efficient k-regret query algorithm with restriction-free bound for any dimensionality. In: Proceedings of the 2018 International Conference on Management of Data, pp. 959–974 (2018)
21. Zhang, Z., Hwang, S.W., Chang, C.C., Wang, M., Lang, C.A., Chang, Y.C.: Boolean + ranking: querying a database by k-constrained optimization. In: SIGMOD, pp. 359–370 (2006)

Efficient Processing of k-regret Queries via Skyline Frequency

Sudong Han[1], Jiping Zheng[1,2(✉)], and Qi Dong[1]

[1] College of Computer Science and Technology,
Nanjing University of Aeronautics and Astronautics, Nanjing, China
{sdhan,jzh,dongqi}@nuaa.edu.cn
[2] Collaborative Innovation Center of Novel Software Technology
and Industrialization, Nanjing, China

Abstract. Helping end-users to find the most desired points in the database is an important task for database systems to support multi-criteria decision making. The recent proposed k-regret query doesn't ask for elaborate information and can output k points for users easily to choose. However, most existing algorithms for k-regret query suffer from a heavy burden by taking the numerous skyline points as candidate set. In this paper, we aim at decreasing the candidate points from skyline points to a relative small subset of skyline points, called *frequent skyline points*, so that the k-regret algorithms can be applied efficiently on the smaller candidate set to improve their efficiency. A useful metric based on subspace skyline called skyline frequency is adopted to help determine the candidate set and corresponding algorithm is developed. Experiments on synthetic and real datasets show the efficiency and effectiveness of our proposed method.

Keywords: Regret minimization query
Candidate set determination · Skyline frequency
Frequent skyline points

1 Introduction

Extracting a few points to assist end-users to make multi-criteria decisions is an important functionality of database systems. Top-k [1] and skyline [2] queries are two well-studied tools that can effectively reduce the output size. But top-k query requires the utility functions from users and skyline query cannot control the output size. To solve these problems, Nanongkai *et al.* [3] first proposed regret-based query which gives the user a concise, summary result of the entire database and outputs k points that minimize the users' maximum regret ratio.

Technically, the input of k-regret algorithms is the points in the whole dataset, but existing algorithms [3–7] usually take the skyline points as the candidate points. This is because the points dominated by skyline points have less possibility of being k representative points. However, taking skyline points

© Springer Nature Switzerland AG 2018
X. Meng et al. (Eds.): WISA 2018, LNCS 11242, pp. 434–441, 2018.
https://doi.org/10.1007/978-3-030-02934-0_40

as candidate points is of low efficiency for computing k representative points because the size of skyline points grows exponentially with dimensionality [8].

Motivated by these, we devote to finding a small size of candidate set from the entire skyline points so that the k-regret algorithms can be applied efficiently on the smaller candidate set to improve its efficiency.

In this paper, we define a set of candidate points called *frequent skyline points* based on skyline frequency. Skyline frequency is a metric that indicates how often the skyline points have been returned in the skyline when different numbers of dimensions are considered. Intuitively, a point with a high skyline frequency is more interesting as it can be dominated on fewer combinations of dimensions.

To avoid the expensive cost of calculating skyline frequency given by the naive method, we propose SF_{APPROX} to determine the candidate set. The main contributions of this paper are listed as follows:

- We define the concept of frequent skyline points based on skyline frequency which provide candidate points for k-regret query.
- We present efficient algorithm to determine the candidate points and corresponding strategy is provided to make a tradeoff between the maximum regret ratio and time complexity.
- Extensive experiments on both synthetic and real datasets are conducted to evaluate our proposed method and the experimental results confirm the efficiency and effectiveness of our proposed algorithm.

The rest of this paper is organized as follows. We present related work in Sect. 2. Section 3 contains required preliminaries and problem definition for our approach. In Sect. 4, we describe our algorithms to determine the candidate set. We show our experimental results in Sects. 5 and 6 concludes this paper and points out possible future work.

2   Related Work

Motivated by the deficiencies of top-k and skyline queries, the k-regret query was proposed by Nanongkai *et al.* [3]. A number of algorithms were proposed to extend the concept to some extent [4,9]. Recently, a number of algorithms [5–7] were proposed to obtain a solution with a small maximum regret ratio. Specially, the algorithm Sphere proposed by [7] is an elegant algorithm, which is considered as the state-of-art algorithm, because the maximum regret ratio of the result is no restriction bound and can be executed on datasets of any dimensionality. However, these studies aim at minimizing the maximum regret ratio of a selected set, ignoring the importance of reducing the size of candidate set to improve the efficiency.

There are also some researches related to subspace skyline. Yuan *et al.* [10] and Pei *et al.* [11] proposed the concept of SKYCUBE, which computes the skylines of all possible subspaces. But the methods mentioned above can only compute the skylines in subspace, they have not been involved in reducing the number of candidate points for k-regret query.

We adopted the metric for ranking skyline points called skyline frequency proposed by [12]. However, the efficiency of skyline frequency is hard to verify. In this paper, we combine the concept of skyline frequency with our candidate set determination for k-regret query and provide a reasonable verification to the superiority of the points with high skyline frequency.

3   Preliminaries

In this section, the concepts of k-regret query and candidate set are introduced to bring out our problem definition.

Given a d-dimensional dataset D with n points and an output size k, where $d \leq k \leq n$, the k-regret query takes a dataset D as input and outputs a set R of k points from D such that the maximum regret ratio of R is minimized.

A user's preference to a point is represented by the utility function f. Thus the *utility* of a point p for the user with utility function f is $f(p)$. The maximum utility derived from a dataset D is the gain of D, denoted by $gain(D, f)$. If we consider $gain(D, f)$ as the user's satisfaction to the set D, then the *regret* the user seeing a representative set R instead of the whole database is $gain(D, f) - gain(R, f)$ and the regret ratio is $rr_D(R, f) = \frac{gain(D,f) - gain(R,f)}{gain(D,f)}$.

Since utility functions vary across users, any algorithm for a k-regret query must minimize the maximum regret ratio for a class of utility functions. In this paper, we only consider linear utility functions, denoted by $\mathcal{F}$, because they are very popular in modeling user preferences [3,4]. Thus the worst possible regret for any user with a utility function in $\mathcal{F}$ is defined as follows.

Definition 1 (Maximum Regret Ratio). *Define* $rr_D(R, \mathcal{F}) = sup_{f \in \mathcal{F}}$ $\frac{max_{p \in D} f(p) - max_{p \in R} f(p)}{max_{p \in D} f(p)}$.

We take one of the classical k-regret query algorithms RDPGREEDY [3] for example to illustrate how the size of candidate set affects the running time. According to the definition of maximum regret ratio, RDPGREEDY needs to find the point that currently contributes to the maximum regret ratio. To be specific, the algorithm needs to inspect each of the candidate points (except selected points) by computing the maximum regret ratio to decide whether the point will be included in the result set or not. By reducing the size of candidate set, the efficiency of the algorithm can be greatly improved. Now we explain the idea of reducing the size of candidate set via skyline frequency by a simple example.

Considering the following example, a Chinese school wants to recruit k English teachers. They measure the candidates from d aspects, such as the capability of *speaking, listening, reading, writing* and so on, where greater values are better. Table 1 shows a list of candidates which are also the skyline points of all the candidates (only consider the first three aspects). Intuitively, we always want to hire someone who is competent in more aspects among all the candidates. Table 2 lists all the subspace skylines. We observe that p_3 is the skyline of

Table 1. English teacher recruiting example   **Table 2.** Skylines of all subspaces

$Candidate$	S_1	S_2	S_3	$Frequency$
p_1	8.4	5.1	8.3	2
p_2	5.2	8.5	6.6	1
p_3	9.3	6.1	7.4	4
p_4	4.7	5.2	10	4
p_5	4.2	8.8	9.2	2
p_6	5.6	10	3.1	4
p_7	8.2	9.2	2.5	2
p_8	1.1	9.3	5.3	2

$Subspace$	$Skyline$
S_1	$\{p_3\}$
S_2	$\{p_6\}$
S_3	$\{p_4\}$
S_1, S_2	$\{p_3, p_6, p_7\}$
S_2, S_3	$\{p_4, p_5, p_6, p_8\}$
S_1, S_3	$\{p_1, p_3, p_4\}$
S_1, S_2, S_3	$\{p_1, p_2, p_3, p_4, p_5, p_6, p_7, p_8\}$

subspace $\{S_1\}$, $\{S_1, S_2\}$, $\{S_1, S_3\}$ and $\{S_1, S_2, S_3\}$, which means p_3 is better than anyone else when considering these aspects mentioned above. So, if we choose only one person, p_3 has more probability of being selected. On the contrary, we observe that p_2 is not the skyline of any subspace, so the possibility of being selected of p_2 is small.

The above example shows that if we consider a smaller subset of skyline points as the candidate set, the points with high skyline frequency are preferred. So we define the problem of candidate set determination based on skyline frequency.

Problem Definition. Given a set D of n points in d dimensions, our problem of processing k-regret query via skyline frequency is to determine the candidate set of k-regret query by selecting the points with high skyline frequency, meanwhile keeping the maximum regret ratio as small as possible.

4 Candidate Set Determination via Skyline Frequency

In this section, we mainly concentrate on the computation of skyline frequency and develop efficient algorithm to solve our problem.

Given a d-dimensional dataset D, S is the dimension set consisting of all the d dimensions and S_i represents each dimension. Let p and q be two data points in D, we denote the value of p and q on dimension S_i as $p.S_i$ and $q.S_i$. For any dimension set B, where $B \subseteq S$, p dominates q if $\forall S_i \in B$, $p.S_i \leq q.S_i$ and $\exists S_j \in B, p.S_j > q.S_j (i \geq 1, j \leq d)$. The skyline query on B returns all data points that are not dominated by any other points on B. The result is called subspace skyline, denoted by $SKY_B(D)$. See the running example in Table 1, $SKY_{B_1}(D) = \{p_3\}$, $SKY_{B_2}(D) = \{p_3, p_6, p_7\}$, where $B_1 = \{S_1\}$ and $B_2 = \{S_1, S_2\}$.

Given D on S, a SKYCUBE consists of a set of subspace skyline points in $2^d - 1$ non-empty subspaces. The SKYCUBE is shown in Table 2.

Definition 2 (Skyline Frequency). *Given a skyline point $p \in D$, skyline frequency, denoted by $freq(p)$ is the number of subspaces in which p is a skyline point.*

For example, considering the example in Table 1, we can get all the subspace skyline points (Table 2) and count the $freq(p)$ of each point. The skyline frequencies of each skyline point are shown in Table 1.

The naive skycube-based approach, called SF_{SKYCUBE}, calculates the skyline frequency by dividing the full space into all $2^d - 1$ subspaces and counting the number of occurrences of each subspace skyline points. However, it requires the skylines to be computed for each of an exponential number of combinations of the dimensions, and the algorithm does not scale well with dimensionality. Thus, we propose an approximate algorithm named SF_{APPROX} to overcome these two drawbacks. From this point of view, to avoid the high complexity of precise counting, we present an effective approximate counting method that is based on extending a Monte-Carlo counting algorithm [13]. In this section, we illustrate how we adapt the method to combine with the computation of skyline frequency and show the optimization we do to make the algorithm more effective. The main procedure of our approach is shown in Algorithm 1.

Algorithm 1. SF_{APPROX} algorithm

Input: dataset D, n, $x\%$, k, δ, ϵ.
Output: the set of candidate points FP

1 Sort the points in non-descending order of the sum of dimension values;
2 initialize FP to be empty, $\theta = 0$, $m = MAX(n * x\%, k)$, $T = 2nln(2/\delta)/\epsilon^2$;
3 **for** $i = 1$ *to* T **do**
4   randomly choose a subspace $\mathcal{B} \subseteq \mathcal{S}$;
5      $SKY_{\mathcal{B}}(D, \theta) = SkylineWithThreshold(D, \mathcal{B}, \theta, m, |FP|)$;
6   **for** *each point* $p \in SKY_{\mathcal{B}}(D)$ **do**
7          $fre(p) + +$;
8   **if** $|FP| < m$ *or* $fre(p) > \theta$ **then**
9   **if** $|FP| = m$ **then**
10   remove the point with the smallest frequency in FP;
11   insert p into FP and update θ to be the smallest frequency in FP;

12 return FP;

The main difference between SF_{APPROX} and SF_{SKYCUBE} is that SF_{APPROX} returns an approximate rank of skyline frequency within an error of ϵ and a confidence level of at least $1 - \delta$. To obtain the desired error bound, we select a sample of $T = 2nln(2/\delta)/\epsilon^2$ number of subspaces out of $2^d - 1$ subspaces randomly. The proof of the error bound follows from [13].

At the beginning of the algorithm, a preprocessing operation of the data is invoked for optimization (Line 1). This is based on the observation that the efficiency of the algorithm is dependent on the order in which the data points are processed. For sake of simplicity, we pre-sort the points in non-descending order of the sum of dimension values. Intuitively, a smaller sum is likely to have smaller values on more dimensions and is therefore likely to have a higher skyline frequency. Initialization of parameters are done in Line 2. To speed up the algorithm, we maintain a frequency threshold θ to avoid dominating tests of a point p if p is determined not to be among the $x\%$ frequent skylines. For each sample (Line 3), we randomly choose a subspace $\mathcal{B}$ (Line 4 to 11), we compute the skyline of $\mathcal{B}$ and for each point p in the skyline of $\mathcal{B}$, we increase the frequency of p by 1. A point p is inserted into FP when the set FP has fewer than m points or the threshold is larger than θ. Meanwhile, we update θ to be the smallest frequency in FP. Specially, the point with smallest frequency is replaced by p if p has larger frequency and the size of FP is m. Finally, the set of candidate points FP is returned (Line 12).

The $SkylineWithThreshold$ procedure computes the skyline points of D in subspace $\mathcal{B}$ where skyline frequency exceeds θ when the size of candidate points $|FP|$ equals to m. The main idea of computing skyline in subspace is to inspect all pairs of points and returns an object if it is not dominated by any other object (Line 4 to 10). Specially, if there are already m intermediate frequent skylines exceeds θ (Line 3), then the point clearly cannot be among the top $x\%$ frequent skylines. So it is unnecessary to do the dominating test. Therefore, the algorithm can be accelerated by only considering the points with potentiality to be frequent skyline points.

Algorithm 2. SkylineWithThreshold(D, $\mathcal{B}$, θ, m, r)

Input: dataset D, subspace $\mathcal{B}$, threshold θ, size m, size r.
Output: $SKY_{\mathcal{B}}(D, \theta)$

1 initialize $SKY_{\mathcal{B}}(D, \theta)$ with the first point in D, $isSkyline$ = false;
2 **for** each point $q \in D$ **do**
3   **if** $r = m$ and $fre(q) > \theta$ **then**
4   **for** each point $p \in SKY_{\mathcal{B}}(D, \theta)$ **do**
5   **if** q dominates p in subspace $\mathcal{B}$ **then**
6   replace p with q;
7   **else if** p dominates q in subspace $\mathcal{B}$ **then**
8                  $isSkyline$ = false; break;
9   **else**
10                 $isSkyline$ = true; insert q to $SKY_{\mathcal{B}}(D, \theta)$;
11 return $SKY_{\mathcal{B}}(D, \theta)$;

---

5 Experiments

In this section, we show the frequent skyline points are efficient to select the k-representative points by comparing them with the skyline points and the points in the whole dataset, and taking them as input of the state-of-art k-regret algorithm Sphere [7]. All the experiments were implemented in C++ and the experiments were performed on a 3.3GHz CPU and 4G RAM machine running Ubuntu 14.04 LTS.

We ran our experiments on both synthetic and real datasets. For synthetic datasets, we use correlated and anti-correlated synthetic data, generated by the generator in [2]. Unless specially stated, the cardinality is ten thousand (*i.e.* $n = 10,000$), $k = 10$, $d = 6$, $\epsilon = 0.2$, $\delta = 0.05$, $x\% = 0.02$. The real-world datasets we adopted are three datasets, a 6-dimensional *Household* of 127,931 points (www.ipums.org), a 9-dimensional *Color* of 68,040 points (kdd.ics.uci.edu), and an 8-dimensional *NBA* of 17,265 points (www.basketballreference.com), which are widely used in this area.

For the simplicity of describing, we first denote the three candidate sets, the whole dataset points as D, skyline points as S, frequent skyline points $F - x\%$, where $x\%$ means the candidate set is composed of the top $x\%$ frequent skyline points. Different values of x ($x = 1, 10$) have been chosen.

Then, we show the frequent skyline points can be used for the state-of-art algorithm Sphere [7] for k-regret query in terms of maximum regret ratio and running time. The effects of F on maximum regret ratio and time on anti-correlated datasets for different k are shown in Figs. 1(a) and 2(a) respectively. When k increases, the maximum regret ratio of all candidate sets decreases which is in accordance with the results in [3]. We observe that the maximum regret ratio of F is close to D and S or even smaller in some situation while the running time is only $\frac{1}{10}$ of D and S. This is because the candidate points F with high skyline frequency have more possibility to be the representative skyline points. From the above, we have the conclusion that frequent skyline points can efficiently speed up the original algorithm with a close maximum regret ratio.

Besides, we evaluate the effect of different cardinalities and dimensionalities on our algorithm. Figure 1(b) shows that when d increases, the maximum regret ratio of different candidate sets increases. Figure 1(c) shows that the maximum regret ratio is independent of n. The same conclusion can be reached that the maximum regret ratio of frequent skyline points is close to D and S and the running time in Figs. 2(b) and (c) is much less than that of D and S.

We can observe the similar trends on real datasets. Similar to the experiments on synthetic datasets, with frequent skyline points, the algorithm achieves a close

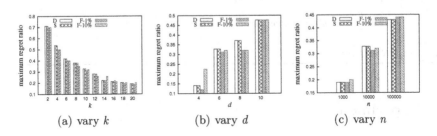

    (a) vary k (b) vary d (c) vary n

Fig. 1. Effect on maximum regret ratio of different candidate sets

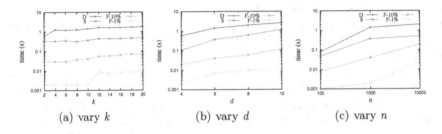

    (a) vary k (b) vary d (c) vary n

Fig. 2. Effect on time of different candidate sets

maximum regret ratio with less running time compared with experiments on the whole dataset and the skyline points. The figures are omitted for the same reason.

6 Conclusions and Future Work

We studied the problem of reducing the size of candidate set for k-regret query basing on skyline frequency. Skyline frequency was adopted to rank skyline points and an algorithm, SF_{APPROX} was provided to determine the candidate set for k-regret query. Our experimental study demonstrated the effectiveness and efficiency of the proposed algorithm. Since our candidate set are the points of "the best - *higher skyline frequency* of the best - *skyline points of the whole dataset*", we do think there are certain theoretical results to be found to guarantee our validity. We leave this as an open problem and also as our future work.

Acknowledgment. This work is partially supported by the National Natural Science Foundation of China under grants U1733112, 61702260, Funding of Graduate Innovation Center in NUAA under grant KFJJ20171601.

References

1. Ilyas, I.F., Beskales, G., Soliman, M.A.: A survey of Top-k query processing techniques in relational database systems. ACM Comput. Surv. **40**(4), 1–58 (2008)
2. Börzsöny, S., Kossmann, D., Stocker, K.: The skyline operator. In: ICDE, pp. 421–430 (2001)
3. Nanongkai, D., Sarma, A.D., Lall, A., Lipton, R.J., Xu, J.: Regret-minimizing representative databases. VLDB **3**, 1114–1124 (2010)
4. Chester, S., Thomo, A., Venkatesh, S., Whitesides, S.: Computing k-regret minimizing sets. VLDB **7**, 389–400 (2014)
5. Asudeh, A., Nazi, A., Zhang, N., Das, G.: Efficient computation of regret-ratio minimizing set: a compact maxima representative. In: SIGMOD, pp. 821–834 (2017)
6. Agarwal, P.K., Kumar, N., Sintos, S., Suri, S.: Efficient algorithms for k-regret minimizing sets. CoRR, abs/1702.01446 (2017)
7. Xie, M., Wong, R.C.-W., Li, J., Long, C., Lall, A.: Efficient k-regret query algorithm with restriction-free bound for any dimensionality. In: SIGMOD (2018)
8. Godfrey, P.: Skyline cardinality for relational processing. In: Seipel, D., Turull-Torres, J.M. (eds.) FoIKS 2004. LNCS, vol. 2942, pp. 78–97. Springer, Heidelberg (2004). https://doi.org/10.1007/978-3-540-24627-5_7
9. Qi, J., Zuo, F., Samet, H., Yao, J.C.: K-regret queries: from additive to multiplicative utilities (2016)
10. Yuan, Y., Lin, X., Liu, Q., Wang, W., Yu, J.X., Zhang, Q.: Efficient computation of the skyline cube. In: VLDB, pp. 241–252 (2005)
11. Pei, J., Jin, W., Ester, M., Tao, Y.: Catching the best views of skyline: a semantic approach based on decisive subspaces. In: VLDB, pp. 253–264 (2005)
12. Chan, C.-Y., Jagadish, H.V., Tan, K.-L., Tung, A.K.H., Zhang, Z.: On high dimensional skylines. In: Ioannidis, Y., et al. (eds.) EDBT 2006. LNCS, vol. 3896, pp. 478–495. Springer, Heidelberg (2006). https://doi.org/10.1007/11687238_30
13. Karp, R.M., Luby, M., Madras, N.: Monte Carlo approximation algorithms for enumeration problems. J. Algorithms **10**(3), 429–448 (1989)

Recommendation Systems

A Multi-factor Recommendation Algorithm for POI Recommendation

Rong Yang$^{(\boxtimes)}$, Xiaofeng Han, and Xingzhong Zhang

Taiyuan University of Technology, 209 University Street, Yuci District,
Jinzhong 030600, China
yangrong@tyut.edu.cn, 632443217@qq.com,
1659898176@qq.com

Abstract. Point-of-Interest (POI) recommendation is an important service in Location-Based Social Networks (LBSNs). There are several approaches, such as collaborating filtering or content-based filtering, to solving the problem, but the quality of recommendation is low because of lack of personalized influencing factors for each user. In LBSNs, users' history check-in data contain rich information, such as the geographic and textual information of the POIs, the time user visiting POIs, the friend relationship between users, etc. Recently, these factors are exploited to further improve the quality of recommendation. The major challenges are which factors to use and how to use them. In this paper, a multi-factor recommendation algorithm (MFRA) is proposed to address these challenges, which initially exploits the locality of user activities to create a candidate set of POIs, and for each candidate POI, the influence of friend relationship and the category and popularity of POI on each user are considered to improve the quality of recommendation. Experiments on the check-in datasets of Foursquare demonstrate a better precision and recall rate of the proposed algorithm.

Keywords: POI recommendation · Locality of POI · Category of POI
Popularity of POI · Friend relationship

1 Introduction

With the rapid development of GPS technology and increased number of POIs, such as malls, parks, attractions, etc., LBSNs have attracted more and more users. LBSNs provide users with the functions of check-in, review, and POI sharing with friends, which have accumulated a large number of data that can be used for user behavior analysis and personalized POI recommendation.

Collaborative filtering [1] is widely used for recommendations. In POI recommendation, users and their friends in social networks often show similar behaviors and common interests on POIs, and suggestions from friends easily affect users [2–5], which are primary factors to be considered. Other factors of POI, such as location, category and popularity, also play an important role when users make decisions, but are not considered in collaborative filtering. We argue that these factors can be used to further improve the quality of recommendations when they are properly used. In this

© Springer Nature Switzerland AG 2018
X. Meng et al. (Eds.): WISA 2018, LNCS 11242, pp. 445–454, 2018.
https://doi.org/10.1007/978-3-030-02934-0_41

paper, a personalized multi-factor recommendation algorithm, denoted as MFRA, is proposed. In MFRA, the location of POI is an excellent factor for filtering unpromising POIs, and the category and popularity of POIs and the friend relationship of users can be used to personalize recommendations.

The rest of the paper is organized as follows. Section 2 briefly discusses the related work. Section 3 formally defines the problem and introduces the multi-factor recommendation algorithm. Section 4 reports the experimental studies. Finally, Sect. 5 concludes the paper.

2   Related Work

POI recommendation methods fall into two categories, content filtering and collaborative filtering.

Content-filtering constructs a recommendation model by extracting features of users and/or POI. Gao [6] combines the features of POI, user interests and user emotions into a unified framework. Bao [5] combines personal preferences obtained from POI features and the expert information extracted from the dataset to score POI. These recommender systems only consider the content characteristics of user and POI, while ignoring various links between users and POI.

Collaborative filtering can be memory-based or model-based [7]. Memory-based collaborative filtering can further be divided into user-based and item-based. Side factors can be incorporated into them to improve the quality of recommendations. One way is to tap the friend relationship between users to improve the quality of recommendation. Konstas [9] uses the LDA model to obtain the similarity from friend relations and seamlessly integrates to the user-based collaborative filtering. Another way is to use temporal information and geographical information to improve recommendation quality. Ye [8] proposed to use the distribution of check-in probability, geographical influence, and social influence for POI recommendations. Yuan [10] defines the time-aware POI recommendation which argues that user behaviors are constrained by the time factor, and can use time similarity to find similar users. To some degree, these approaches can improve the quality of recommendations, but they still use similar users' preferences as users' own preferences, resulting in a recommendation with poor personalization. Introduction of side factors often leads to the noise which is detriment to the quality of recommendations.

The core of the model-based collaborative filtering is to use user-POI score matrix to construct the prediction model. Liu [4] proposed an improved SVD model, which can solve the problem of matrix sparsity, but the matrix still needs to restore after decomposition, which requires a lot of computation. Cao [11] builds the meta-path feature set, and uses the random walk to measure the relevancy between the head and tail nodes in path instances, and the supervised learning method to obtain the weight to infer the check-in probability. However, the meta-path in the collection phase needs to traverse the entire networks, which is very expensive.

3   The Multi-factor Recommendation Algorithm

3.1   Problem Definition

In this subsection, we abstract the problem of POI recommendation and present some related concepts as follows.

Definition 1 (Points of Interest, POI). A POI is represented as $p = (loc, cat)$, where $p.loc$ denotes the geo-location of the POI, and $p.cat$ denotes the category of the POI. The categories of all POIs constitute the category set Cat.

Definition 2 (Intimacy between friends). The intimacy between user u_i and u_j is represented as $Intif(u_i, u_j) = Jarccard(F_i, F_j)$, where F_i and F_j are the set of friends of user u_i and u_j and $Intif(u_i, u_j)$ is the Jarccard distance of their set of friends.

Definition 3 (Check-in Item). A check-in item collected from a social network is modeled as $item = (u, p, t)$, where $item.u$, $item.p$ and $item.t$ respectively denotes the user, POI, and time of check-in.

Problem Statement. Given a set of POIs, a set of users containing the set of friends, and a set of history check-in items, POI recommendations are to recommend unvisited POIs to user by analyzing these check-in history data such that the probability that user visiting these POIs is as large as possible.

3.2   The Influence of Location of POI

Usually people's activities concentrate in certain areas, which means that the user's check-in behavior occurs in a relatively small geographical range, and this phenomenon is called geographical clustering [9]. Figure 1 shows 267,867 check-in items from Fousquare in Los Angles for nearly 2 years involving 31,540 users, 1103 check-in items of which are listed in Fig. 2, which involves 10 users and each color represents a user's check-in locations. As illustrated in Figs. 1 and 2, frequent check-in locations of users in the same city concentrated in certain areas.

地图 – Tip_Veneue
■ (267867)

Fig. 1. 267,867 check-in items from Fousquare in Los Angles for nearly 2 years.

In POI recommendations, compared with further POI, the nearer POI has a higher probability of being recommended. So we first scan the checked-in dataset, cluster the

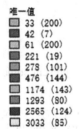

唯一值
☐ 33 (200)
■ 42 (7)
☐ 61 (200)
☐ 221 (19)
☐ 278 (101)
■ 476 (144)
☐ 1174 (143)
■ 1293 (80)
■ 2565 (124)
☐ 3033 (85)

Fig. 2. 1103 check-in items involving 10 users.

visited POIs for each user, for each unvisited POI, we calculate the minimum distance to the nearer visited POI cluster, and if it is less than the distance threshold $Dist_u$, it is added to the candidate set of POIs to recommend.

Example 1. Figure 3 demonstrates the process of creating a candidate set of POIs to recommend for a user. The visited POIs constitute three clusters, and the minimum value from p_1 to the three clusters is 4.3 km, if $Dist_u = 5$ km, then p_1 should be added into the candidate set. In the same way, p_2 is added, and p_3 is abandoned for this user.

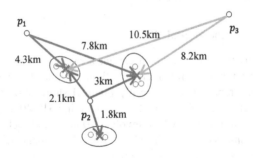

Fig. 3. Create candidate set of POIs to recommend.

3.3   The Influence of Category and Popularity of POI

Usually, different categories imply the different products and services provided by POIs. A user will visit POIs with the same category many times, which can be verified from the real-life check-in datasets. Take the dataset in Fig. 1 as an example, for each user, we statistic the number of check-in items and the times visiting the POIs with same category and illustrate in Figs. 4 and 5. In Fig. 4, 8032 users only have 1 check-in item and the other users have more than 2 check-in items. In the Fig. 5, the number of users visiting POIs with same categories more than 2 times is 21372. Therefore, if a user has more than 2 check-in items, the probability that they are the same category is 51.96%. Therefore, the category information of the POI that the user has visited can be used to derive the user's preferences.

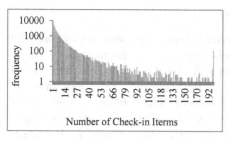 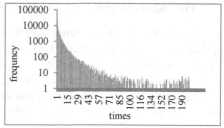

Fig. 4. The frequency with different number of check-in items.

Fig. 5. The frequency with different times visiting POIs with same category.

Definition 4 (Preference of POI Category). Given a user u_i, a POI category cat_k, the number of check-in items of u_i in POIs with cat_k is $NumItem(u_i, p, t)_{\{p \cdot cat = cat_k\}}$, the maximum check-in items of u_i in POIs of all categories is $CMAX(NumItem(u, p, t))_{\{p \cdot cat \in Cat\}}$, the preference of POI category cat_k for u_i, denoted as $cpref(u_i, cat_k)$, is computed in Eq. 1.

$$cpref(u_i, cat_k) = \frac{NumItem(u_i, p, t)_{\{p \cdot cat = cat_k\}}}{CMAX(NumItem(u_i, p, t))_{\{p \cdot cat \in Cat\}}} \tag{1}$$

Category information can partly specify the user preferences, but it is unreasonable to assume that all POIs with the same category are equally important to the user. In order to get the weight of different POIs, we introduce the popularity of POIs, which reflects the quality of the services provided by POIs. We believe that for the same category of POIs, the higher the popularity, the higher the quality of POI.

Definition 5 (Popularity of POI). Given a POI p_k, the number of check-in items in POI is $NumItem(u, p_k, t)$, the maximum check-in items in POIs of the category $p_k.cat$ is $CMAX(NumItem(u, p, t))_{\{p \cdot cat = p_k \cdot cat\}}$, and the maximum check-in item in all POIs is $MAX(NumItem(u, p, t))_{\{p \cdot cat \in Cat\}}$, the popularity of POI, denoted as $pop(p_k)$, is computed in Eq. 2.

$$pop(p_k) = \frac{NumItem(u, p_k, t)}{CMAX(NumItem(u, p, t))_{\{p \cdot cat = p_k \cdot cat\}}} \bigg/ \frac{CMAX(NumItem(u, p, t))_{\{p \cdot cat = p_k \cdot cat\}}}{MAX(NumItem(u, p, t))_{\{p \cdot cat \in Cat\}}} \tag{2}$$

Definition 6 (POI preference). Given a user u_i, a POI p_k, the preference of POI category $p_k.cat$ is $cpref(u_i, p_k)$, the popularity of POI is $pop(p_k)$, the POI preference of p_k for u, denoted as $Ppref(u_i, p_k)$, is computed in Eq. 3.

$$Ppref(u_i, p_k) = cpref(u_i, p_k.cat)pop(p_k) \tag{3}$$

In summary, the POI preference of a candidate POI takes into account both the category and the popularity of POI.

3.4 The Influence of Friend Relationship

Friend recommendation is a fundamental service in both social networks and practical applications, and is influenced by user behaviors such as interactions, interests. A statistic shows that more than 30% stranger POI visited by a user, had been visited by the user's friends [3]. Therefore, in POI recommendations, friend relationship is introduced by considering the friends and similar check-in items of users. Similar to the user-based collaborative filtering algorithm, the similarity between users u_i and u_j is calculated by $sim(u_i, u_j) = \cos sim(Flag\_Item(u_i, p, t),\ Flag\_Item(u_j, p, t))$, where $Flag\_Item(p, u_i, t)$ indicates whether $Item(u_i, p, t)$ exists, if yes, $Flag\_Item(u_i, p, t) = 1$, otherwise $Flag\_Item(u_i, p, t) = 0$. The similarity $sim(u_i, u_j)$ is the cosine distance between $Flag\_Item(u_i, p, t)$ and $Flag\_Item(u_j, p, t)$.

Definition 7 (Influence between Friends). Given two users u_i and u_j, the intimacy between them is $Intif(u_i, u_j)$, the similar between them is $sim(u_i, u_j)$, the influence between them, denoted as $FI(u_i, u_j)$, is computed in Eq. 4.

$$FI(u_i, u_j) = Intif(u_i, u_j)sim(u_i, u_j) \tag{4}$$

In summary, given a user u_i, a POI p_k, the influence between u_i and his friends is $FI(u_i, u)_{\{u \in F_{u_i}\}}$, for user u_i, the preference of visiting POI p_k influenced by friends is calculated in Eq. 5.

$$Fpref(u_i, p_k) = \frac{\sum_{u \in F_{u_i}} FI(u_i, u) \cdot Flag\_Item(u, p_k, t)}{\sum_{u \in F_{u_i}} FI(u_i, u)} \tag{5}$$

3.5 Multi-factor Recommendation Algorithm

This section takes into account the above factors, i.e. location, category and popularity of POI, and friend relationship of users to compute the probability of visiting POI for users. In the first stage, we adopt the operation in Sect. 3.2 and create the candidate POIs. Then for each candidate POI, we compute the POI preference, and the probability of visiting POI by friends to improve the quality of recommendations. But how to balance the two values is also important. In this paper, we utilize the simple linear combination, as shown in Eq. 6, to get the recommended probability of each candidate POI.

$$Rprob(u_i, p_k) = \alpha \cdot Ppref(u_i, p_k) + (1 - \alpha) \cdot Fpref(u_i, p_k) \tag{6}$$

Where $Ppref(u_i, p_k)$ and $Fpref(u_i, p_k)$ respectively denote the preference of POI and Friends, $\alpha \in (0, 1)$ denote a weighting parameter.

4   Experiments

4.1   Experimental Settings

Datasets. Two real-life check-in datasets from February 2009 to July 2011 in New York and Los Angeles are collected from Fousquare for our experimental evaluations. We remove the wrong check-in items and construct the following check-in datasets. In the experiments, 75% of each dataset severs as the training sets and the remaining 25% as the test sets (Table 1).

Table 1. Datasets statistics

Datasets	Check-in item	Number of user	Number of POI
New York (NY)	221,128	4,902	92,018
Los Angeles (LA)	267,867	31,540	144,295

4.2   Baselines and Proposed Algorithms

- User based CF(UCF) Considering the similarity between users to calculate the probability of candidate POIs.
- Friend influence based CF(FCF) Considering the friend influence to calculate the similarity between users, in accordance with the similar recommended POI.
- USG(USG) Considering the influence of location and friends.
- MFRA Considering the location, category and popularity of POI and user's friends.

Evaluation Criteria. As the other works, the precision and recall rate are used as our evaluation criteria. The precision rate is the ratio of the number of POIs that user actually visited to the total number of recommended POIs. The recall rate is the ratio of the number of recommended POIs that the user visited to the total number of POIs that the user visited. The two indexes reflect the accuracy and integrality of the recommendation algorithms.

4.3   Performance Evaluation

Effect of the Weighting Parameter α. As illustrated in Fig. 6, the precision and recall rate of the two datasets are similar when α varies. For the New York dataset, when α is close to 0.21, the precision and recall rate are the highest, and for the Los Angeles dataset, when α is close to 0.19, the precision and recall rate are the highest. Therefore, the α is set to 0.2 by default.

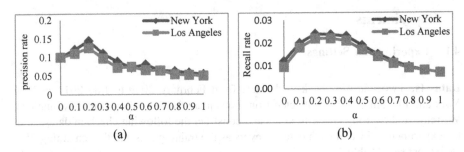

Fig. 6. The precision and recall rate of recommendation in the two datasets with varied α.

Effect of the Number of Recommended POIs. The two datasets are used for this test. In Fig. 7, as the number of recommended POIs increases, (a)–(b) shows the precision and recall rate of the four algorithms for the New York dataset, and (c)–(d) shows the precision and recall rate of the four algorithms for the Los Angles dataset. For the two datasets, UCF algorithm has the lowest precision and recall rate. Compared with the other three recommendation algorithms, the precision and recall rate of the MFRA algorithm are highest.

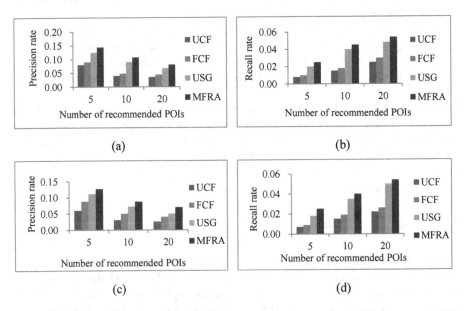

Fig. 7. The precision and recall rate with varied recommended POIs.

5 Conclusions

To solve the POI recommendation, a personalized multi-factor recommendation algorithm is proposed, which first creates a preliminary candidate set by introducing the location factor to remove the unpromising POIs, and consider the category, popularity of POI and friend relationship of the user. Experiments conducted over large-scale Foursquare datasets show the precision and recall rate of our algorithm are improved compared with the other POI recommendation algorithms.

It takes a long time to generate candidate set in the initial stage, so the emphasis of our future work should be laid on reducing this part of time. Meanwhile, the time and sex factors can be introduced into the recommendation algorithm to further improve the quality of the recommendation.

Acknowledgement. This work is financially supported by the National Science and Technology Support Program of China (Grant No. 2015BAH37F01) and the CERNET Next Generation Internet Technology Innovation Project of China (Grant No. NGII20170518).

References

1. Schafer, J.B., Konstan, J.A., Riedl, J.: E-commerce recommendation applications. Data Min. Knowl. Discov. **5**(1–2), 115–153 (2001)
2. Yang, X., Guo, Y., Liu, Y., Steck, H.: A survey of collaborative filtering based social recommender systems. Comput. Commun. **41**(5), 1–10 (2014)
3. Wang, H., Terrovitis, M., Mamoulis, N.: Location recommendation in location based social networks using user check-in data. In: 21st ACM Sigspatial International Conference on Advances in Geographic Information Systems, pp. 374–383. ACM, Orlando (2013)
4. Liu, B., Fu, Y., Yao, Z., Xiong, H.: Learning geographical preferences for point of interest recommendation. In: 19th ACM SIGKDD International Conference on Knowledge Discovery and Data Mining, pp. 1043–1051. ACM, Chicago (2013)
5. Bao, J., Zheng, Y., Mokbel, M.F.: Location-based and preference-aware recommendation using sparse geo-social networking data. In: 20th ACM SIGSPATIAL International Conference on Advances in Geographic Information Systems, pp. 199–208. ACM, California (2012)
6. Gao, H., Tang, J., Hu, X.: Content-aware point of interest recommendation on location-based social networks. In: 29th AAAI Conference on Artificial Intelligence, pp. 1721–1727. AAAI, Texas (2015)
7. Schafer, J.B., Dan, F., Herlocker, J.: Collaborative filtering recommender systems. Adapt. Web **22**(1), 291–324 (2007)
8. Ye, M., Yin, P., Lee, W.C., Lee, D.L.: Exploiting geographical influence for collaborative point-of-interest recommendation. In: 34th International ACM SIGIR Conference on Research and Development in Information Retrieval, pp. 325–334. ACM, Beijing (2011)
9. Konstas, I., Stathopoulos, V., Jose, J.M.: On social networks and collaborative recommendation. In: 32nd International ACM SIGIR Conference on Research and Development in Information Retrieval, pp. 195–202. ACM, Boston (2009)

10. Yuan, Q., Cong, G., Ma, Z.: Time-aware point-of-interest recommendation. In: 36th International ACM SIGIR Conference on Research and Development in Information Retrieval, pp. 363–372. ACM, Dublin (2013)
11. Cao, X., Dong, Y., Yang, P., Zhou, T., Liu, B.: POI recommendation based on meta-path in LBSN. Chin. J. Comput. **39**(04), 675–684 (2016)
12. Ricci, F., Rokach, L., Shapira, B., Kantor, P.B.: Recommender Systems Handbook, 1st edn. Springer, New York (2011). https://doi.org/10.1007/978-0-387-85820-3

Social Network User Recommendation Method Based on Dynamic Influence

Xiaoquan Xiong[1,2], Mingxin Zhang[2(✉)], Jinlon Zheng[2],
and Yongjun Liu[2]

[1] College of Computer Science and Technology, Soochow University,
Suzhou, China
[2] Department of Computer Science and Engineering,
Changshu Institute of Technology, Changshu, China
mxzhang163@163.com

Abstract. The rapid development and wide application of Online Social Network (OSN) has produced a large amount of social data. How to effectively use these data to recommend interesting relationships to users is a hot topic in social network mining. At present, the user relationship recommendation algorithm relies on similarity, and the user's influence is insufficiently considered. Aiming at this problem, this paper proposes a new user influence evaluation model, and on this basis, a new user relationship recommendation algorithm (SIPMF) is proposed by combining similarity and dynamic influence. 2522366 Sina Weibo data were crawled to build an experimental data set for experiment. Compared with the typical relational recommendation algorithms SoRec, PMF, and FOF, the SIPMF algorithm improved 4.9%, 7.9%, and 10.3% in accuracy and recall respectively. And 2.6%, 4.2%, 6.6%, can recommend for users more interested in the relationship.

Keywords: Social network · User similarity · Dynamic influence
User recommendation

1 Introduction

An online social network is a collection of thousands of network users that form a relational connection based on social networking sites in a self-organizing manner [1]. Online social networks are essentially a mapping of real-world social networks in the Internet [2].

Social networking sites allow users to express themselves, display their social networks, and establish or maintain contact with others [3]. Domestic Sina Weibo (hereinafter referred to as Weibo), Tencent, Renren, and foreign countries such as Facebook and Twitter have rapidly emerged. As of September 2017, Weibo monthly active users totaled 376 million, and daily active users reached 165 million [4]. In the first quarter of 2018, the number of Facebook monthly active users in the world's largest social networking site reached 2.20 billion, an increase of 13% over the same period last year [5]. The huge user base in online social networks and the self-organizing relationship structure have spawned a complex social network, resulting in a

© Springer Nature Switzerland AG 2018
X. Meng et al. (Eds.): WISA 2018, LNCS 11242, pp. 455–466, 2018.
https://doi.org/10.1007/978-3-030-02934-0_42

huge amount of social data. These massive social data on the one hand increase user stickiness, but at the same time it also creates information overload problems. Therefore, how to find out the interested relationship for the target user group from the massive Weibo user group becomes a research hotspot. As a method that can effectively relieve information overload, the user recommendation system has received increasing attention [6].

Research on user recommendation algorithms in social networks can be divided into two categories, content recommendation [7] and social relationship recommendation [8, 9]. The content recommendation is to recommend the content of interest to the user from a large amount of data [7]. Social relationships are the link between users in social networks and are important foundations for the dissemination of social network information. Social relationship recommendation is to predict and recommend the unknown relationship between users in social networks [2]. User relationship recommendation algorithms in online social networks can be divided into three types, link-based recommendation algorithms, content-based recommendation algorithms, and hybrid recommendation algorithms. The link-based recommendation algorithm is to transform the relationship recommendation problem into a link prediction problem in a graph composed of a social network topology. For the link-based recommendation algorithm, Lin et al. defined the similarity between nodes and used it to predict new links [10]. Yin et al. used the Non-negtive Matrix Factorization (NMF) method based on social relations to mine the hidden information in microblog social relationships [11]. However, relying solely on the explicit network structure information of microblog users will encounter data sparsity problems and affect the accuracy of the recommendation results. For content-based recommendation algorithms, Xu et al. proposed a variety of user-based attribute information (background information, micro The similarity calculation method of Bo text) and the fusion of these similarity information to achieve recommendation [12]. However, since the Weibo text information is relatively short and mixed with various picture information, it is time-consuming and labor-intensive to handle and the accuracy rate is not high. For the hybrid recommendation algorithm, Chen et al. also use the blog content and social structure information to calculate the similarity between users. Degrees, resulting in a list of mixed recommendations [13]. The hybrid recommendation algorithm can make better use of social relationship information, user attribute information and blog content information, but it also has computational efficiency problems.

Currently proposed social network user recommendation algorithms have problems such as sparse data and insufficient consideration of dynamic changes in user relationships. In response to these deficiencies, this paper builds a user's dynamic impact assessment model based on implicit network information (likes, forwards, comments) among users, and adds explicit network information between users' friends (followers and fans). A matrix decomposition algorithm (SIPMF) that combines similarity and influence is proposed. When recommending new relationships for users, the algorithm considers both the explicit network information between users and the dynamic influence of implicit network information. It obtains similarities and dynamic influences, and uses similarity to constrain dynamic influences. The method of matrix decomposition is introduced to reduce the dynamic influence between users, so as to obtain the potential characteristics of influence among users, and then to predict the relationship between target users.

Based on Scrapy's breadth-first strategy, the Weibo crawler was designed to collect microblog data in a continuous time interval. After preprocessing, an experimental data set composed of user relationship data and interaction behavior data (approximately 2522366) was obtained. Compared with the typical relational recommendation algorithms SoRec, PMF, and FOF, the algorithm improved by 4.9%, 7.9%, 10.3%, and 2.6%, 4.2%, and 6.6% in terms of accuracy and recall. The validity of the SIPMF algorithm is proved and the user's interested relationship can be more accurately recommended.

The Sect. 2 introduces the calculation of user similarity and the evaluation model of dynamic influence. The Sect. 3 is the introduction of data sources and the analysis of the results of the experiment; the last is the summary of the full text.

2   Matrix Decomposition Method Combining Similarity and Influence

2.1   User Similarity

The phenomenon of homogeneity is reflected in many aspects of social life. Nearly all people are red and black is a good example. The same is true in social networks, and the formation of a concern between users is often influenced by whether there is a common interest or hobby among users. In Weibo relationship prediction, a good method is to find users similar to the recommended users, sort them according to similarity, and select TopN users to recommend. Therefore, in the microblog relationship prediction, a primary problem is to find a user similar to the recommended user, that is, how to solve the problem of measuring the similarity of the microblog user. Currently, mainstream microblog user similarity measurement algorithms can be divided into three categories: (1) Krishnamurthy et al. [14] represent users based on attention and fan relationships, and establish network topologies based on user relationships. And to calculate the similarity between users. (2) Considering the number of common neighbors between users, the greater the number of common neighbors, the higher the similarity of users. The CN (Common Neighbors) method, the Jaccard method, and the Cosine method are all based on this idea. The difference is that different methods have different choices for the proportion of common neighbors [15]. (3) Ma et al. [9] viewed Weibo as an undirected weighted graph, and added various user background information, social information, etc. to the nodes in the graph to calculate user similarity.

In Weibo, there are two kinds of relationships between users. If user v_u is follow v_c, then v_c is said to be the follower of v_u and v_u is a follower of v_c. The Weibo network topology structure built on the basis of attention and fan relations is a directed graph. The attention relation is the out-edge of the node in the graph, and the fan relationship is the edge of the node in the graph. Therefore, when calculating the similarity of Weibo users, it is necessary to comprehensively consider the relationship between attention and fans, that is, the outbound and inbound edges in the microblog network topology. The SimRank algorithm considers these two relationships well. The calculation of the two similarities is shown in Eqs. (1) and (2).

$$Sim_I(v_u, v_c) = \frac{c \sum\limits_{v_i \in I(v_u)} \sum\limits_{v_j \in I(v_c)} Sim_I(v_i, v_j)}{|I(v_u)||I(v_c)|} \tag{1}$$

$$Sim_O(v_u, v_c) = \frac{c \sum\limits_{v_i \in O(v_u)} \sum\limits_{v_j \in O(v_c)} Sim_O(v_i, v_j)}{|O(v_u)||O(v_c)|} \tag{2}$$

Among them, $c \in (0,1)$ is the damping coefficient, usually taking $0.6 \sim 0.8$, $I(v_u)$ is the set of all v_u neighbors (fans) and $O(v_u)$ is the set of v_u neighbors (follow). In Weibo network topology, the contributions of outbound and inbound to the similarity are not consistent. The number of fans of Weibo V is much larger than the number of concerns. Relatively speaking, the importance of outbound similarity is more important than the similarity of incoming edges. Sex, so a linear weighted combination of two similarities. The formula for calculating the similarity is shown in (3).

$$Sim(v_u, v_c) = \lambda Sim_I(v_u, v_c) + (1 - \lambda)Sim_O(v_u, v_c) \tag{3}$$

Among them, λ is the weight of fan similarity.

2.2 Dynamic User Influence

In Weibo, users do not exist in isolation. Yin et al. [16] studied the formation mechanism of the attention relationship in Weibo. The experimental results show that about 90% of new links are formed by two-hop relations. This shows that the formation of the vast majority of new concerns in Weibo is formed between friends (friends who have concerns or fan relationships). For example, v_u and v_i are related to each other, and v_i is focused on v_c. If the impact of v_i is large enough, v_u is likely to focus on v_c. In addition, if v_u, v_i, and vj are related to each other and v_i and v_j are focused on v_c, v_u may be affected by v_i and v_j. Follow v_c. For a microblog of v_c, the user's v_u likes, comments, or forwards represent different degrees of recognition of v_u for the microblog v_c, and further, v_u The behavior of v_c Weibo is also a manifestation of the influence of v_c. Therefore, the higher the likes, comments, and frequency of v_u posts on the v_c blog post, the stronger the influence of v_c on x. The following definitions are made for the three impact factors.

$$I_{like}(v_u, v_c) = In \frac{Like(v_u, v_c)}{\sqrt{\sum\limits_{v_i \in U_{v_u}^l} Like(v_u, v_i) \sum\limits_{v_j \in U_{v_c}^l} Like(v_c, v_j)}} \tag{4}$$

Among them, Like (v_u, v_c) is the number of likes between v_u and v_c, and $U_{v_c}^l$ and $U_{v_c}^l$ are shown with v_u and v_c respectively is a bit like the collection of users where the behavior occurred.

The influence between user pairs is constantly changing, but within a relatively short period of time, the change is small. Therefore, the interval can be divided according to the time sequence of microblog posting, and the influence of the current interval can be considered as a local influence.

$$I_{like_i}(v_u, v_c) = In \frac{Like_i(v_u, v_c)}{\sqrt{\sum\limits_{v_i \in U_{v_u}^I} Like_i(v_u, v_i) \sum\limits_{v_j \in U_{v_c}^O} Like_i(v_j, v_c)}} \tag{5}$$

Among them, $Like_i(v_u, v_c)$ is the number of v_c in $[(i-1)*k, i*k)$ like v_u, $U_{v_u}^I$ represents the point A collection of users who like v_u, $U_{v_c}^O$ represents a collection of users who like v_c.

Define the user's forwarding influence and comment influence in the same way. Finally, the user's local influence on the ith interval is obtained.

$$I_i(v_u, v_c) = \alpha_1 I_{like_i}(v_u, v_c) + \alpha_2 I_{comment_i}(v_u, v_c) + \alpha_3 I_{retweet_i}(v_u, v_c) \tag{6}$$

Among them, the parameters α_1, α_2, α_3 are the weighting factors for the likes, comments, and forwarding behaviors, respectively, and $0 \leq \alpha_1 \leq 1$, $0 \leq \alpha_2 \leq 1$, $0 \leq \alpha_3 \leq 1 \square \alpha_1 + \alpha_2 + \alpha_3 = 1 \square$ Construct a weight function.

$$F(t) = 1/t, t = 1, 2, \cdots, n \tag{7}$$

Dynamic Influence of v_u on v_c:

$$I(v_u, v_c) = F(1)I_1(v_u, v_c) + F(2)I_2(v_u, v_c) + \cdots + F(n)I_n(v_u, v_c) \tag{8}$$

$I_i(v_u, v_c)$ is the local influence of v_u on v_c in the ith interval.

2.3   Matrix Factorization

In Weibo, building an adjacency matrix based on user concerns and fan relationships is the simplest binary matrix. The user v_u follows v_c, then $R(v_u, v_c) = 1$, otherwise $R(v_u, v_c) = 0$. In such a scoring matrix, it is necessary to predict whether the relationship of $R(v_u, v_c) = 0$ is based on existing relationships and social information. Changes occur in the later period. For m users, the rating matrix $R^{m \times n}$ for n recommended objects, each element of the matrix r_{v_i, v_j} represents vj for vj Scoring, this value is 0 or 1 in the binary score matrix. $W \in R^{m \times k}$ and $H \in R^{k \times n}$ means that the decomposed user is associated with the recommended user k-dimensional feature matrix, its column vector W_{v_u} and the row vector H_{v_j} represents the corresponding potential feature vector respectively. However, because of system noise, $R^{m \times n}$ cannot be perfectly decomposed into $W^T H$. So get an approximation $\hat{R}$ of R.

$$\hat{R} = W^T H \tag{9}$$

At the same time, in order to prevent over-fitting, you need to constrain W and H. In the Probabilistic Matrix Factorization (PMF), it is assumed that the observation noise $\delta = |R - \hat{R}|$ obeys the mean $W^T H$ and the variance is a Gaussian distribution of σ. The probability density function of the scoring matrix is shown in Eq. (10).

$$p(R|W,H) = N(\hat{R}, \sigma^2) \tag{10}$$

Assume that both W and H obey a mean of 0 and the variances are Gaussian distributions of σ_W and σ_H, respectively. As shown in Eqs. (11) and (12).

$$p(W) = N(0, \sigma_W^2) \tag{11}$$

$$p(H) = N(0, \sigma_H^2) \tag{12}$$

Formula (13) can be obtained by using the posterior probability formula.

$$p(W,H|R) \propto p(R|W,H)p(W)p(H) \tag{13}$$

2.4   SIPMF

In the scoring matrix, not only need to consider the explicit structure information between users, but also need to consider the user's similarity and mutual influence between users, take these two together to get a new scoring matrix.

$$R_{v_i,v_j} = \beta Sim(v_i, v_j) + (1 - \beta)I(v_i, v_j) \tag{14}$$

Among them, β is the weight coefficient of the user's similarity.

A SIPMF model was proposed after the similarity and influence were combined. After the fusion, the logarithm of the posterior probability formula satisfies formula (15).

$$\begin{aligned} In\, p(W,H|R) = &-\frac{1}{2\sigma^2}\sum_{i=1}^{m}\sum_{j=1}^{n}(R_{ij} - W_i^T V_j)^2 - \frac{1}{2\sigma_W^2}\sum_{i=1}^{m} W_i^T W_i \\ &-\frac{1}{2\sigma_H^2}\sum_{i=1}^{n} H_i^T H_i - \frac{1}{2}(In\,\sigma^2 + mIn\,\sigma_W^2 + nIn\,\sigma_H^2) + C \end{aligned} \tag{15}$$

To simplify the calculation, let $\sigma_W = \sigma_H$. Maximizing the posterior probability is equal to minimizing the objective function E(W, H).

$$E(W,H) = \frac{1}{\sigma^2}\sum_{i=1}^{m}\sum_{j=1}^{n}(R_{ij} - W_i^T V_j)^2 + \frac{1}{\sigma_W^2}(\sum_{i=1}^{m} W_i^T W_i + \sum_{i=1}^{n} H_i^T H_i) \tag{16}$$

SIPMF algorithm design: First, user similarity is calculated based on the user relationship; then, an evaluation model of dynamic influence is established based on the interaction behavior; the two are integrated into the PMF model through adaptive weight wij; and the user relationship is obtained by gradient descent algorithm Feature vector.

Input: initial influence matrix I, similarity matrix S, number of recommend user m, number of recommended user n, learnrate, Output: precision, recall, F1-measure

```
{
  for i=1 to m do // first step:calculate matrix S and I
    for j=1 to n do
      {
        S1 = formula(1)
        S2 = formula(2)
        // calculate similarity matrix S
```
$$S[i][j] = \lambda S_2 + (1-\lambda)S_1$$
```
      }
  for i=1 to m do
    for j=1 to n do
    // calculate influence matrix I
        I[i][j] = formula(8)
  for i=1 to m do
    for j=1 to n do
//choose final dynamic self-adapting weight by formula(9)
        w = w[i][j]
// second step:gradient descent
```
initialize latent factor: $W, H \sim N(x \mid 0, \sigma^2))$
```
  for (i,j) TrainData do
  {
```
$$W_i = W_i - learnrate \times \partial_E / \partial_{W_i}$$
$$H_j = H_j - learnrate \times \partial_E / \partial_{H_j}$$
```
  }
//third step:compute rating score
predict(i,j)
Compute precision, recall and F1-measure
}
```

3   Experimental Results and Analysis

3.1   Experimental Data Set

For the Weibo user recommendation issue, there is currently no publicly available Chinese standard data set. The experimental data in this article comes from a set of Sina Weibo user data collected by a self-developed Weibo crawler (acquisition started in

October 2017). Weibo acquisition adopts a breadth-first strategy, first select 10 users, crawl 10 users' attention and fan list, get user set U (total 464), and then use user set U as a basis to expand one layer outward. Catch these user's attention and fan list to get more users (27362 total). At the same time, collect the microblog information of these users, as well as users who like, comment, and forward these microblogs (Table 1).

Table 1. Data set statistics.

Statistical content	Number
User	111516
Weibo	60615
Like	1093821
Comment	645966
Retweet	610448

On the basis of these already obtained data, a preliminary screening of the data was conducted. The screening was mainly based on the user's activity on Weibo, and the number of followers and the number of followers was greater than 10 for screening. At the same time, users who have fewer than 10 microblogs are further screened. The purpose of these screenings is to remove inactive users.

3.2 Experimental Evaluation Index

In this paper, Precision rate, Recall rate and F1 measure are used as evaluation indicators of experiment results.

The accuracy rate refers to the ratio between the predicted number of relations Nt and the total number of predictions Na in the predicted relationship. The accuracy rate is proportional to the ratio, and the higher the ratio, the better the prediction result. The accuracy formula is:

$$Precision = \frac{N_t}{N_a} \tag{17}$$

The recall rate refers to the ratio between the predicted relationship number Nt and the total relationship number Nb in the predicted relationship. The recall rate is proportional to the ratio, and the larger the ratio, the better the prediction result. The formula for the recall rate is as follows:

$$Recall = \frac{N_t}{N_b} \tag{18}$$

The F1 measure is the evaluation index of the comprehensive accuracy and recall rate. The larger the value, the more accurate the prediction result. The F1 measure calculation results are as follows:

$$F1 = \frac{2 * Precision * Recall}{Precision + Recall} \tag{19}$$

3.3 Experimental Parameters

For the user's attention similarity and the user's fan similarity weight parameter λ, rely on fans and attention relations for Top-5 recommendation, parameters λ value from 0.1 to 1, according to the similarity recommendation, get the recommendation accurately Rate line chart for the parameter λ.

The Fig. 1 shows that the recommended accuracy is the highest when the parameter $\lambda = 0.7$.

The parameters α_1, α_2, and α_3 are calculated using the analytic hierarchy process. The hierarchical weight matrix of the three types of interaction information for commenting, forwarding, and praising is shown in Table 2. The final parameters α_1, α_2, and α_3 are set to 0.14, 0.33, and 0.53, respectively.

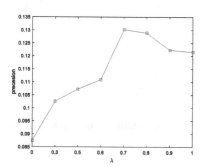

Table 2. Discriminant matrix for calculating weights

Interactive	Like	Comment	Retweet
Like	1	1/3	1/3
Comment	3	1	1/2
Retweet	3	2	1

Fig. 1. Parameter λ line chart of accuracy

3.4 Analysis of Results

In order to evaluate the effectiveness of the proposed algorithm proposed in this paper, the SIPMF algorithm is based on a friend-based recommendation algorithm FOF [17], PMF algorithm (Probabilistic Matrix Factorization, PMF) [18], and SoRec [19] algorithm (Social Recommendation Using Probabilistic Matrix). Factorization, SoRec compares the captured microblogging dataset. Among them, the PMF algorithm only considers the relationship matrix between users, and considers the relationship between all users to be equal. SoRec integrates the user's static influence on the basis of the user relationship matrix.

From the real microblogging data crawled, 30% were randomly selected as the test set and 70% were used as the training set. The important parameter settings are shown in Sect. 3.3. The friend recommendation performances of Top-2, Top-4, Top-6, and

Top-8 are respectively compared in the feature dimension k = 5. The accuracy, recall rate, and F1 value are shown in Figs. 2, 3, and 4, respectively. As can be seen from the figure, the SIPMF algorithm is more effective than other algorithms in terms of recommendation accuracy, recall rate, and F1 metrics. This shows that the user's dynamic influence has a significant impact on the recommendation effect. This also proves the effectiveness of the method of fusion dynamic influence and similarity under matrix decomposition.

From Figs. 2, 3 and 4, it can be seen that as the length of the recommendation list continues to increase, the recommendation accuracy rate and the recommendation recall rate of these four recommendation algorithms are affected to varying degrees, but the SIPMF algorithm is recommended in different recommended lengths. On the other hand, relatively good recommendations can be obtained.

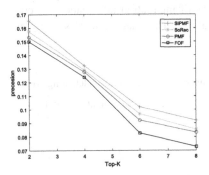

Fig. 2. Precesion of Top-k

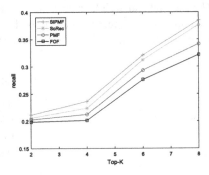

Fig. 3. Recall rate of Top-k

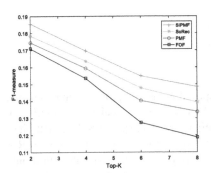

Fig. 4. F1 measure of Top-k

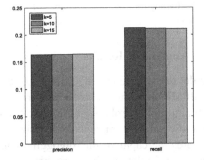

Fig. 5. The influence of feature dimension on recommendation precision

3.5   Influence of Parameters on Recommendation Accuracy

Under the premise that the other parameters are not changed, the impact on the accuracy of top-4 recommendation is affected when the dimension of the hidden feature $k = 5$, $k = 10$, and $k = 15$, respectively.

From the Fig. 5, we can find that with the increase of feature dimensions, the algorithm's recommendation accuracy has been improved, but the increase is not large, so the experimental dimension of the experiment is 5 in this paper.

4   Conclusion

The existing social network recommendation algorithm based on user influence evaluates the user's static influence according to historical records, and does not fully consider the dynamic influence of the user. This article aims at this issue, put forward the user's dynamic influence evaluation model. At the same time, the fusion of user similarity restricts the user's dynamic influence. A social network user recommendation method that combines dynamic influence model and user similarity is proposed. The effectiveness of the algorithm is verified by experiments.

In order to highlight the problem itself, this article only considers the user's network structure information and interactive information between the recommended objects, while ignoring the impact of other background information between the recommended objects on the recommendation results, such as the user's blog content, user tags, etc. In the future work, we will study how to better integrate the user's background information into the recommendation method in order to further improve the effectiveness of the recommendation.

References

1. Fang, P.: Research of community detection's algorithm based on friends similarity from online social networks. HuaZhong University, Wuhan (2013)
2. Zhao, S., Liu, X., Duan, Z., et al.: A survey on social ties mining. Chin. J. Comput. **40**(3), 535–555 (2017)
3. Ellison, N.B., Steinfield, C., Lampe, C.: The benefits of Facebook "friends": social capital and college students' use of online social network sites. J. Comput. Med. Commun. **12**(4), 1143–1168 (2007)
4. Sina Weibo: 2017 Weibo User Development Report [EB/OL]. (2018-00-00). http://data. weibo.com/report/reportDetail?id=404. Accessed 27 May 2018
5. Facebook: Facebook 2018 first quarter earnings [EB/OL]. (2018-00-00). https://www.sec. gov/Archives/edgar/data. Accessed 27 May 2018
6. Liu, H.F., Jing. L.P., Jian, Y.U.: Survey of matrix factorization based recommendation methods by integrating social information. J. Soft. (2018)
7. Chen, J., Liu, X., Li, B., et al.: Personalized microblogging recommendation based on dynamic interests and social networking of users. Acta Electron. Sin. **45**(4), 898–905 (2017)
8. Wang, R., An, W., Fen, Y., et al.: Important micro-blog user recommendation algorithm based on label and pagerank. Comput. Sci. **45**(2), 276–279 (2018)

9. Ma, H., Jia, M., Zhang, D., et al.: Microblog recommendation based on tag correlation and user social relation. Acta Electron. Sin. **1**, 112–118 (2017)
10. Lin, D.: An information-theoretic definition of similarity. In: Fifteenth International Conference on Machine Learning, pp. 296–304. Morgan Kaufmann Publishers Inc. (1998)
11. Yin, D., Hong, L., Davison, B.D.: Structural link analysis and prediction in microblogs, pp. 163–1168 (2011)
12. Xu, Z., Li, D., Liu, T., et al.: Measuring similarity between microblog users and its application. Chin. J. Comput. **37**(1), 207–218 (2014)
13. Chen, H., Jin, H., Cui, X.: Hybrid followee recommendation in microblogging systems. Sci. China Inf. Sci. **60**(1), 012102 (2016)
14. Krishnamurthy, B., Phillipa, G. et al.: A few chirps about twitter. In: Proceedings of the First Workshop on Online Social Networks, pp. 19–24. ACM (2008)
15. Liben-Nowell, D., Kleinberg, J.: The Link-Prediction Problem for Social Networks. Wiley, New York (2007)
16. Yin, D., Hong, L., Xiong, X., et al.: Link formation analysis in microblogs, pp. 1235–1236 (2011)
17. Chen, J., Geyer, W., Dugan, C., et al.: Make new friends, but keep the old: recommending people on social networking sites. In: Sigchi Conference on Human Factors in Computing Systems, pp. 201–210. ACM(2009)
18. Salakhutdinov, R., Mnih, A.: Probabilistic matrix factorization. In: International Conference on Neural Information Processing Systems, pp. 1257–1264. Curran Associates Inc. (2007)
19. Ma, H., Yang, H., Lyu, M.R., et al.: SoRec: social recommendation using probabilistic matrix factorization. Comput. Intell. **28**(3), 931–940 (2008)

Computing User Similarity by Combining Item Ratings and Background Knowledge from Linked Open Data

Wei Xu$^{(\boxtimes)}$, Zhuoming Xu, and Lifeng Ye

College of Computer and Information, Hohai University, Nanjing 210098, China
{xwmr, zmxu, lfye}@hhu.edu.cn

Abstract. User similarity is one of the core issues in recommender systems. The boom in Linked Open Data (LOD) has recently stimulated the research of LOD-enabled recommender systems. Although the hybrid user similarity model recently proposed by the academic community is suitable for a sparse user-item matrix and can effectively improve recommendation accuracy, it relies solely on the historical data (item ratings). This work addresses the problem of computing user similarity by combining item ratings and background knowledge from LOD. We propose a computation method for the user similarity based on feature relevance (USFR), which is an improvement on the user similarity based on item ratings (USIR) in the hybrid user similarity model. The core idea of our improvement is replacing the item ratings in the original model with feature relevance, thereby forming our hybrid user similarity model. Experiments on benchmark datasets demonstrate the effectiveness of the proposed method and its strengths of rating prediction accuracy compared to the USIR measure. Our work also shows that the incorporation of background knowledge from LOD into a hybrid user similarity model can improve recommendation accuracy.

Keywords: User similarity · Hybrid model · User-feature matrix
Knowledge graph · Linked open data
Linked open data enabled recommender systems

1 Introduction

As a personalized information service, recommender systems play an increasingly critical role in alleviating the information/choice overload problem that people face today. User similarity is one of the core issues in recommender systems. For instance, the user-based collaborative filtering (CF) model [1] employs user similarity to make rating predictions from the historical data (the item ratings stored in the user-item matrix) by leveraging the collaborative power of the ratings provided by multiple users. Traditional computation methods for user similarity rely solely on the item ratings, which usually faces the sparsity problem of the user-item matrix, resulting in low accuracy of rating predictions. Recently, Wang et al. [2] proposed a hybrid user similarity model comprehensively considering multiple influence factors of user similarity. Experimental results show that the hybrid model is suitable for the sparse data and effectively improves prediction accuracy [2]. However, according to our observations,

X. Meng et al. (Eds.): WISA 2018, LNCS 11242, pp. 467–478, 2018.
https://doi.org/10.1007/978-3-030-02934-0_43

this hybrid model still relies solely on the item ratings and does not take full advantage of the background knowledge provided by Linked Open Data (LOD) [3].

The LOD cloud provides a wealth of background knowledge for various domains. For example, DBpedia [4], as a central interlinking hub for the LOD cloud, is a huge RDF dataset [5] extracted from Wikipedia; it is also a knowledge base (graph) describing millions of entities that are classified in a consistent ontology. The boom in LOD has recently stimulated the research of a new generation of recommender systems —LOD-enabled recommender systems [6, 7]. The ultimate goal of such a system is to provide the system with background knowledge about the domain of interest in the form of an application-specific knowledge graph [8], which is extracted from LOD. For instance, recent studies (such as [9] and [10]) on features-based user similarity have exploited the background knowledge extracted from LOD to reflect users' preferences. Specifically, the feature rating and relevance method proposed by Anelli et al. [9] can transform a user-item matrix into a user-feature matrix which can be used to compute features-based user similarity.

In the above context, this work addresses the problem of computing user similarity by combining item ratings and background knowledge from LOD. We propose a computation method for the *user similarity based on feature relevance* (abbreviated as USFR), which is an improvement on the *user similarity based on item ratings* (abbreviated as USIR) in the hybrid user similarity model proposed by Wang et al. [2]. The core idea of our improvement is replacing the item ratings (stored in the user-item matrix) in the original hybrid model with the transformed feature relevance (stored in the user-feature matrix), thereby forming our hybrid user similarity model. We also propose an *aggregated similarity* (abbreviated as aggrSim) that is a weighted aggregation of the USFR measure and the USIR measure. Our experimental results demonstrate the effectiveness of the proposed USFR and aggrSim measures and their strengths of rating prediction accuracy compared to the USIR measure. Our work also shows that the incorporation of background knowledge from LOD into a hybrid user similarity model can improve recommendation accuracy.

The remainder of this paper is organized as follows. In Sect. 2 we elaborate on the computation methods for our proposed USFR measure and aggregated similarity measure. Section 3 presents our experimental evaluation results and discussions. Finally, we conclude our work in Sect. 4.

2 User Similarity Computation Methods

2.1 Basic Idea of Our Proposed USFR Method

Our USFR method is based on feature relevance and is an improvement on the user similarity based on item ratings (USIR) proposed by Wang et al. [2]. The basic ideas of the USIR method and our USFR method are shown in Fig. 1.

More specifically, in the USIR method the user similarities are computed by means of the hybrid user similarity model [2], as shown in the lower left portion of Fig. 1. In our USFR method, the user-item bipartite graph is built by relying on users' ratings in the user-item matrix. The knowledge graph containing the items involved in the system is built by extracting relevant knowledge from the LOD cloud. The user profiles are then built by using the mappings between the items in both the user-item bipartite graph and the knowledge graph, as shown in the middle of Fig. 1. According to the built user profiles, we can create a user-feature matrix via the feature rating and relevance method [9] as described later. As shown in the lower right portion of Fig. 1, we can then compute the user similarity based on feature relevance (USFR) by means of our hybrid user similarity model as described later. Moreover, in order to further improve the accuracy of rating prediction, we propose an aggregated similarity by combining the USFR method and the USIR method, as shown in the bottom of Fig. 1.

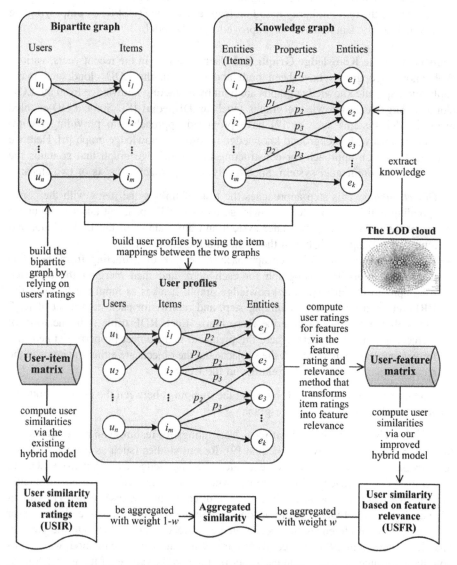

Fig. 1. Overview of the computation methods of USIR, USFR and the aggregated similarity.

2.2   The USFR Method

In order to compute user similarities by combining users' ratings and background knowledge from LOD, our USFR method needs to complete the following three major steps, as shown in Fig. 1:

- Constructing the knowledge graph by extracting relevant knowledge from the LOD cloud and then building the user profiles by using the mappings between the items in both the bipartite graph and the knowledge graph.
- Computing user ratings for features via the feature rating and relevance method [9] that can transform the item ratings into the feature relevance.
- Computing the user similarity based on feature relevance (USFR) by employing the user ratings for features via our improved hybrid model.

Constructing the Knowledge Graph and User Profiles. In the recent years, various Web knowledge graphs have been built and stored in the LOD cloud, and many application-specific knowledge graphs are built by extracting knowledge from the LOD cloud or large-scale knowledge graphs (such as DBpedia) [8]. Some LOD-enabled recommender systems such as [7] have proposed approaches to providing recommender systems with background knowledge in form of knowledge graph [6]. Here we can adapt such an approach for constructing the knowledge graph that contains the items in the recommender system. Specifically, the approach consists of two steps:

- *Object linking:* This step approaches the task of linking the items with the corresponding entities in the LOD knowledge bases like DBpedia. It takes as input the list of items in the recommender system and any dataset in the LOD cloud, and returns the mappings between the items and the entity URIs.
- *Subgraph extraction:* This step approaches the task of extracting from the LOD dataset an informative subgraph for each item, and then merging the extracted subgraphs to obtain a specific knowledge graph. It takes as input the list of entity URIs returned by the object linking step, and returns for each item a set of RDF triples that ontologically describe these items. Each RDF triple is in the form of (subject, predicate, object), where the predicate is an object property defined in a formal ontology, whereas both the subjects and the objects are either the items in the recommender system or other entities in the LOD.

The user profiles can be built by using the mappings between the items in both the bipartite graph and the knowledge graph.

Computing User Ratings for Features. User ratings for features can be computed via the feature rating and relevance method [9]. Recent studies (such as [9] and [10]) on features-based user similarity have exploited the knowledge extracted from LOD to reflect users' preferences. For example, when rating movies, the users implicitly evaluate the features of a movie (i.e., genres, directors and actors), which are knowledge extracted from LOD. Therefore, some movies can then be recommended to a user according to his preferences for specific genres, directors and actors. On the basis of this idea, Anelli et al. [9] proposed the feature rating and relevance method, which uses both the relevance of a feature in each user profile and the ratings of items that contain the feature to rate each feature in each user profile, thus moving from a user-item matrix to a user-feature matrix. The data model and the calculation method of the feature rating and relevance method are as follows.

Data Model of User Profiles. A user profile contains not only the information about items and ratings but also the information about items, properties and entities. For a user u, the items, properties and entities are defined as a RDF triple set $\langle I, P, E \rangle$, where I is a set of items (i.e., subjects), P a set of properties (i.e., predicates), and E a set of entities (i.e., objects). A feature is represented by a property-entity pair, denoted as pe, in the profile of a user. The function $\rho^{uf}(\cdot)$ is used to represent the relevance of feature f in the profile of user u. The relevance of item i rated by user u is defined as $\rho^{ui}(\cdot)$, which is used to estimate the importance of the item to the user. Finally, in the profile of user u, the ratings for feature f is computed via the function $r^{uf}(\cdot)$ considering the ratings of all the items containing f.

Calculation Method. The relevance of feature pe is computed via the function $\rho^{uf}(\cdot)$, denoted as $\rho^{uf}(pe)$, which is built by the proportion of the number of the items that contain the feature in the profile of user u to that of all items rated by the user. The computation formula is defined as Eq. (1) [9].

$$\rho^{uf}(pe) = \frac{\sum_{i \in I_u} |\{\langle i, p, e \rangle | \langle i, p, e \rangle \in \langle I, P, E \rangle\}|}{|I_u|} \tag{1}$$

After completing the calculation of the relevance of all the features in the profile of user u, the relevance of items i rated by the user, denoted as $\rho^{ui}(\cdot)$, can be computed as the normalized summation of the relevance for all the features in the user profile, as defined by Eq. (2) [9].

$$\rho^{ui}(i) = \frac{\sum_{\langle i,p,e \rangle \in \langle I,P,E \rangle} \rho^{uf}(pe)}{|\{\langle i, p, e \rangle | \langle i, p, e \rangle \in \langle I, P, E \rangle\}|} \tag{2}$$

The rating of user u for feature pe, denoted as $r^{uf}(pe)$, is then computed by using both the ratings and the relevance of each item i containing the feature in the user profile, as defined by Eq. (3) [9].

$$r^{uf}(pe) = \frac{\sum_{\langle i,p,e \rangle \in \langle I,P,E \rangle} r_{ui} \cdot \rho^{ui}(i)}{\sum_{\langle i,p,e \rangle \in \langle I,P,E \rangle} \rho^{ui}(i)} \tag{3}$$

Computing the User Similarities Based on the User-Feature Matrix. We can use the constructed user-feature matrix to compute user similarities by means of our hybrid user similarity model, which is an improvement of the hybrid user similarity model proposed by Wang et al. [2] with the goal of making the model suitable for the computation of feature relevance-based user similarity (i.e., USFR). As mentioned earlier, the feature rating and relevance method [9] can be used to transform the item ratings provided by each user into the feature relevance implicitly expressed by the user. Therefore, the core idea of our improvement is replacing the item ratings in the hybrid model with the transformed feature relevance, thereby forming our hybrid user similarity model.

On the basis of the influence factors of all possible rated items [2], our hybrid user similarity model mainly considers the following four influence factors of all possible relevant features:

- *The non-linear relationship between user ratings for features:* User ratings for features kept in the user-feature matrix can be calculated from the feature relevance. Similar to the non-linear relationship in the original hybrid model [2], the relationship between user ratings for features should be non-linear. This factor is reflected in function $S_1^f(\cdot,\cdot)$, which will be described later.
- *The non-co-rated features:* Similar to the non-co-rated items in the original hybrid model [2], the co-rated features are very rare in the user-feature matrix. Therefore, the non-co-rated features need to be considered in our hybrid model. This factor is reflected in function $S_{feature}(\cdot,\cdot) \cdot S_1^f(\cdot,\cdot)$, which will be described later.
- *The asymmetry between users:* Similar to the asymmetry of user similarity in the original hybrid model [2], the user similarity in our hybrid model should also be asymmetric. This factor is reflected in function $S_2^f(\cdot,\cdot)$, which will be described later.
- *The implicit rating preference of users:* Similar to the user preference factor in the original hybrid model [2], the implicit rating preference of users (i.e., the user ratings for features) should also be considered in our model to distinguish the preference between difference users. This factor is reflected in function $S_3^f(\cdot,\cdot)$, which will be described later.

On the basis of the original hybrid model [2], the feature relevance-based user similarity between users u and v, denoted $S^f(u,v)$, is then defined as Eq. (4).

$$S^f(u,v) = S_2^f(u,v) \cdot S_3^f(u,v) \cdot \sum_{i \in F_u} \sum_{j \in F_v} S_{feature}(i,j) \cdot S_1^f(r_{ui}, r_{vj}) \tag{4}$$

where F_u and F_v represent the sets of the features associated to user u and user v, respectively. In the following, the four functions are described in detail, respectively.

Function S_1^f. On the basis of an improved Proximity-Significance-Singularity (PSS) model [2], the non-linear relationship between user ratings for features, denoted $S_1^f(r_{ui}, r_{vj})$, is calculated using Eq. (5).

$$S_1^f(r_{ui}, r_{vj}) = Proximity(r_{ui}, r_{vj}) \times Significance(r_{ui}, r_{vj}) \times Singularity(r_{ui}, r_{vj}) \tag{5}$$

where r_{ui} represents user u's ratings for feature i and r_{vj} represents user v's ratings for feature j. Function *Proximity* is used to compute a similarity value according to the absolute value of the difference between r_{ui} and r_{vj}. Function *Significance* is used to evaluate the impact of rating pair (r_{ui}, r_{vj}) on the final result. That is, the rating pair is more important if the two ratings are more distant from median rating r_{med}. Function *Singularity* represents the difference between the rating pair and other ratings. The three functions are defined as Eqs. (6), (7) and (8) [2], respectively.

$$Proximity(r_{ui}, r_{vj}) = 1 - \frac{1}{1 + \exp(-|r_{ui} - r_{vj}|)} \tag{6}$$

$$Significance(r_{ui}, r_{vj}) = \frac{1}{1 + \exp(-|r_{ui} - r_{med}| \cdot |r_{vj} - r_{med}|)} \tag{7}$$

$$Singularity(r_{ui}, r_{vj}) = 1 - \frac{1}{1 + \exp\left(-\left|\frac{r_{ui} + r_{vj}}{2} - \frac{\mu_i + \mu_j}{2}\right|\right)} \tag{8}$$

where μ_i and μ_j are the average rating values of features i and j, respectively.

It is noteworthy that function $S_1^f(r_{ui}, r_{vj})$ cannot be used to compute independently the similarity between two users u and v. Therefore, function $S_{feature}(i,j)$ is used as a weight factor to adjust the value of this function, that is, $S_{feature}(i,j) \cdot S_1^f(r_{ui}, r_{vj})$ are used to evaluate the user similarity.

Function $S_{feature}$. Function $S_{feature}(i,j)$ is defined as a feature similarity measure based on Kullback-Leibler (KL) divergence (a.k.a. KL distance) to fully use all ratings on features i and j. The KL divergence $D(i,j)$ between two features i and j is defined as Eq. (9) [2].

$$D(i,j) = D(p_i \| p_j) = \sum_{r=1}^{max} p_{ir} \log_2 \frac{p_{ir}}{p_{jr}} \tag{9}$$

where max is the maximum value in rating scale; $p_{ir} = \frac{\#r}{\#i}$ is the probability of rating value r on feature i; $\#i$ represents the number of all users rated the feature i and $\#r$ the number of rating value k.

As explained in [2], $D(i,j)$ is asymmetric. In order to balance contribution between two features i and j, a symmetric distance is defined as Eq. (10) [2].

$$D_s(i,j) = \frac{D(i,j) + D(j,i)}{2} \tag{10}$$

The KL divergence-based feature similarity measure $S_{feature}(i,j)$ is thus defined as Eq. (11).

$$S_{feature}(i,j) = \frac{1}{1 + D_s(i,j)} \tag{11}$$

To ensure p_{ir} and p_{jr} are not 0, an adjusted probability formula is defined as Eq. (12) [2] to replace the original one in Eq. (9).

$$\hat{p}_\chi = \frac{\delta + p_\chi}{1 + \delta |D|} \tag{12}$$

where p_χ and $\hat{p}_\chi$ are the original probability value and the adjusted one of the rating χ, respectively; $|D|$ is the number of all possible value in the rating scale; $\delta \in (0, 1)$ is a smooth parameter.

Function S_2^f. Function $S_2^f(u, v)$ is built by the proportion of the number of the co-rated features to that of all features rated by an active user. Specifically, for the active user u, the function is defined as Eq. (13).

$$S_2^f(u, v) = \frac{1}{1 + \exp\left(-\frac{|F_u \cap F_v|}{|F_u|}\right)} \tag{13}$$

Function S_3^f. Similar to the rating preference of users in the original hybrid model [2], function $S_3^f(u, v)$ is defined as Eq. (14) by using the mean value and standard variance of ratings.

$$S_3^f(u, v) = 1 - \frac{1}{1 + \exp(-|\mu_u - \mu_v| \cdot |\sigma_u - \sigma_v|)} \tag{14}$$

where μ_u and μ_v are the mean rating of user u and v, respectively, and $\mu_u = \frac{\sum_{p \in F_u} r_{u,p}}{|F_u|}$. The σ_u and σ_v represents the standard variance of user u and v, respectively, and $\sigma_u = \sqrt{\frac{\sum_{p \in F_u}(r_{u,p} - \bar{r}_u)^2}{|F_u|}}$.

2.3   Computation of the Aggregated Similarity

As shown in Fig. 1, an aggregated similarity (abbreviated as aggrSim) is also proposed in this work to further improve the accuracy of rating prediction provided by the user-based collaborative filtering method. The aggregated similarity $aggrSim(u, v)$ is a weighted aggregation of the feature relevance-based user similarity (USFR) and the item ratings-based user similarity (USIR), as defined by Eq. (15).

$$aggrSim(u, v) = w \cdot S^f(u, v) + (1 - w) \cdot S(u, v) \tag{15}$$

where parameter w is a weight with values between 0 and 1, which is used to adjust the importance of the two similarities; function $S(u, v)$ represents the item ratings-based user similarity (USIR), as defined by Eq. (16) in the original hybrid model [2].

$$S(u, v) = S_2(u, v) \cdot S_3(u, v) \cdot \sum_{i \in I_u} \sum_{j \in I_v} S_{item}(i, j) \cdot S_1(r_{ui}, r_{vj}) \tag{16}$$

3   Experimental Evaluation

3.1   Experimental Design

To validate the effectiveness of our proposed feature relevance-based user similarity (USFR) and aggregated similarity (aggrSim) and the performance strengths of rating predictions provided by the user-based collaborative filtering (CF) method using these similarity measures, we have conducted experiments on benchmark datasets. Our experimental design is described below.

User-Based CF Models and Accuracy Metrics. The original USIR measure [2], our USFR measure and aggrSim measure were applied to the user-based collaborative filtering model as defined by Eq. (17) [1] by substituting the measures for $S(u, v)$.

$$\hat{r}_{ui} = \bar{r}_u + \frac{\sum_{v \in P_u(i)} S(u, v)(r_{vi} - \bar{r}_v)}{\sum_{v \in P_u(i)} |S(u, v)|} \tag{17}$$

where $\hat{r}_{ui}$ represents the predicted rating, $P_u(i)$ refers to the set of the nearest neighbors to target user u, and $\bar{r}_u$ is the mean rating for user u. The calculation $r_{vi} - \bar{r}_v$ represents a mean-centered rating of user v for item i.

When evaluating the prediction accuracy of the CF models, we used two popular accuracy metrics: Root Mean Squared Error (RMSE) and Mean Absolute Error (MAE) [11–13].

Experimental Datasets. The experiment used three benchmark datasets: the Movie-Lens 100k dataset, the DBpedia 2016-04 release, and the DBpedia-MovieLens 100k dataset, as described below.

The MovieLens 100k Dataset. This benchmark dataset [12] (cf. https://grouplens.org/datasets/movielens/) contains 100,000 ratings from 943 users on 1,682 movies. All ratings are in the scale 1 to 5. During the experiment, this dataset was divided into three parts:

– *Training data (50%):* This part of the data was used to build the training model for predicting the missing ratings in the user-item matrix.
– *Validation data (25%):* This part of the data was used for parameter tuning in order to determine the optimal value for parameter w and thus the best model.
– *Testing data (25%):* This part of the data was used to test the accuracies of rating predictions provided by the user-based CF models.

The DBpedia 2016-04 Release. The English version of DBpedia dataset [4] (cf. http://wiki.dbpedia.org/dbpedia-version-2016-04) provides most of the movies in the MovieLens 100k dataset with a wealth of background knowledge in the form of RDF triples described with ontological properties (object properties).

The DBpedia-MovieLens 100k Datasets. This dataset [13] contains the mappings from MovieLens item (movie) identifiers to DBpedia entity URIs. In fact, 1,569 movies out of the 1,682 movies in the MovieLens 100k dataset have been mapped to DBpedia entity URIs. These mappings were used directly (i.e., there's no need to perform the object linking step) to construct the knowledge graph from the *DBpedia dataset*.

3.2 Parameter Tuning Results

The optimal value for the parameter w was determined in the parameter tuning experiment by testing the rating prediction accuracy over the validation data. The range of the parameter is [0, 1]. During the experiment, the value of w started from 0, with the step size being 0.1.

The experiment finally found that for $w = 0.9$, the smallest RMSE value is 0.9815. This optimal parameter value was used for the performance comparison experiment.

3.3 Prediction Accuracy Results

The rating prediction accuracies in terms of RMSE and MAE were yielded using the testing data. Figures 2 and 3 show the accuracy comparisons in terms of RMSE and MAE, respectively, between the three user-based CF models using the similarity measures (USIR, USFR, and aggrSim), abbreviated as CF-USIR, CF-USFR, and CF-aggrSim, respectively.

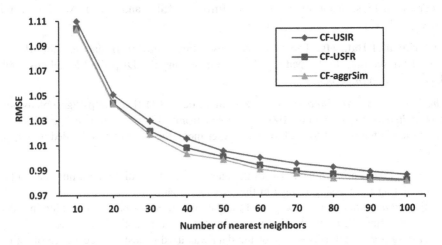

Fig. 2. Comparison between the three CF methods in terms of RMSE.

Figures 2 and 3 indicate that the USIR measure is worse than the USFR and aggrSim measures in terms of both RMSE and MAE. In addition, the aggrSim measure slightly outperforms the USFR measure in terms of both RMSE and MAE. The experimental results show that our feature relevance-based user similarity (USFR) and aggregated similarity (aggrSim) measures have performance strength of rating predictions compared to the original item ratings-based user similarity (USIR) measure.

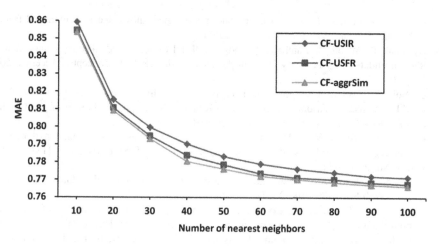

Fig. 3. Comparison between the three CF methods in terms of MAE.

4 Conclusions

Studies on the LOD-enabled recommender systems have shown that LOD can improve recommendation performance. Although the original hybrid user similarity model, as a state-of-the-art method, is suitable for the sparse data and can effectively improve the prediction accuracy through comprehensive consideration of multiple influence factors of user similarity in the model, it relies solely on the historical data (item ratings) and does not take full advantage of background knowledge from LOD. This work addresses the problem of computing user similarity by combining item ratings and background knowledge from Linked Open Data. We propose a computation method for the user similarity based on feature relevance (USFR), which is an improvement on the user similarity based on item ratings (USIR). We have elaborated on the computation methods for the USFR measure and the aggregated similarity measure and demonstrated through experiments the effectiveness of our methods and their strengths of recommendation accuracy compared to the USIR measure. Our work also shows that the incorporation of background knowledge from LOD into a hybrid user similarity model can improve recommendation accuracy. Our future work focuses on evaluating our methods on more datasets and using more metrics.

References

1. Aggarwal, C.C.: Neighborhood-based collaborative filtering. In: Aggarwal, C.C. (ed.) Recommender Systems: The Textbook, pp. 29–69. Springer, Cham (2016). https://doi.org/10.1007/978-3-319-29659-3_2
2. Wang, Y., Deng, J., Gao, J., Zhang, P.: A hybrid user similarity model for collaborative filtering. Inf. Sci. **418**, 102–118 (2017)
3. Heath, T.: Linked data - welcome to the data network. IEEE Internet Comput. **15**(6), 70–73 (2011)

4. Lehmann, J., et al.: DBpedia - a large-scale, multilingual knowledge base extracted from Wikipedia. Semant. Web **6**(2), 167–195 (2015)

5. Cyganiak, R., Wood, D., Lanthaler, M. (eds.): RDF 1.1 Concepts and Abstract Syntax. W3C Recommendation, 25 February 2014. http://www.w3.org/TR/rdf11-concepts/. Accessed 20 May 2018

6. Di Noia, T., Ostuni, V.C.: Recommender systems and linked open data. In: Faber, W., Paschke, A. (eds.) Reasoning Web 2015. LNCS, vol. 9203, pp. 88–113. Springer, Cham (2015). https://doi.org/10.1007/978-3-319-21768-0_4

7. Oramas, S., Ostuni, V.C., Di Noia, T., Serra, X., Di Sciascio, E.: Sound and music recommendation with knowledge graphs. ACM Trans. Intell. Syst. Technol. (TIST) **8**(2), Article No. 21 (2017)

8. Paulheim, H.: Knowledge graph refinement: a survey of approaches and evaluation methods. Semant. Web **8**(3), 489–508 (2017)

9. Anelli, V.W., Di Noia, T., Lops, P., Sciascio, E.D.: Feature factorization for top-N recommendation: from item rating to features relevance. In: Proceedings of the 1st Workshop on Intelligent Recommender Systems by knowledge Transfer and Learning co-located with ACM Conference on Recommender Systems (RecSys 2017), pp. 16–21. ACM (2017)

10. Tomeo, P., Fernández-Tobías, I., Di Noia, T., Cantador, I.: Exploiting linked open data in cold-start recommendations with positive-only feedback. In: Proceedings of the 4th Spanish Conference on Information Retrieval, CERI 2016, Article no. 11. ACM (2016)

11. Aggarwal, C.C.: Evaluating recommender systems. In: Aggarwal, C.C. (ed.) Recommender Systems: The Textbook, pp. 225–254. Springer, Cham (2016). https://doi.org/10.1007/978-3-319-29659-3_7

12. Herlocker, J.L., Konstan, J.A., Terveen, L.G., Riedl, J.T.: Evaluating collaborative filtering recommender systems. ACM Trans. Inf. Syst. **22**(1), 5–53 (2004)

13. Meymandpour, R., Davis, J.G.: A semantic similarity measure for linked data: an information content-based approach. Knowl. Based Syst. **109**, 276–293 (2016)

Reasearch on User Profile Based on User2vec

Ying Wang[1,2], Feng Jin[1(✉)], Haixia Su[2], Jian Wang[2],
and Guigang Zhang[2(✉)]

[1] Beijing Institute of Technology, 100081 Beijing, China
jinfeng226@163.com
[2] Institute of Automation, Chinese Academy of Science, 100081 Beijing, China
guigang.zhang@ia.ac.cn

Abstract. Personalized services for information overload are becoming more common with the arrival of the era of big data. Massive information also makes the Internet platform pay more attention to the accuracy and efficiency of personalized recommendations. The user's profile is constructed to describe the user information of the relevant platform more accurately and build virtual user features online through user behavior preference information accumulated on the platform. In this paper we propose a new user mode named user2vec for personalized recommendation. The construction of user2vec relies on platform and extremely targeted. At the same time, user profile is dynamically changing and need to be constantly updated according to the data and date, therefore we define a new time decay function to track time changes. Dynamic description of user behavior and preference information through user vectorization combined with time decay function can provide reference information for the platform more effectively. Finally, we using a layered structure to build an overall user profile system. And the experiment adapts content-based recommendation algorithm to indirectly prove effectiveness of user profile model. After many sets of experiments proved, it can be found that the proposed algorithm is effective and has certain guiding significance.

Keywords: User profile · User2vec · Time decay function · Personalized recommendation

1 Related Work

User portraits have been studied for some time and research focuses mainly on personalized search [2–9] and folksonomy [10–18]. Among them, Chen et al. [2] is based on fuzzy constraint, Hawalah and Fasli [3] and Kassak et al. [8] use interest vectors to model user profiles, Liang [7] applies clustering algorithm to model user profile and Chen et al. [9] introduces the importance to user profile. Xie et al. [10] which use user profile for sentiment analysis, proposed a new framework to incorporate various sentiment information to various sentiment-based information for personalized search by user profiles and resource profiles, Cai et al. [11] proposes an automatic preference elicitation method based on mining techniques to construct user profile and put forward three ranking strategies to assess the quality of user profiles. Tchuente et al. [12] proposes a approach to resolve data sparsity problems. Du et al. [13] further improve

© Springer Nature Switzerland AG 2018
X. Meng et al. (Eds.): WISA 2018, LNCS 11242, pp. 479–487, 2018.
https://doi.org/10.1007/978-3-030-02934-0_44

the framework to settle some detailed consideration. Wu et al. [14] proposes a new multilevel user profiling model by integrating tags and ratings to achieve personalized search. Ouaftouh et al. [18] uses traditional machine learning algorithms to model user portraits, etc.

For the above models and algorithms, there are several limitations. First, most of the models are aimed at the field of personalized search and research results have been mostly related to folksonomy in recent years, causes the algorithm to be less portable. Second, very little time decay information is taken into account. To deal with these constrictions, we propose a new framework including user2vec, hierarchy labeled system and time decay function to model user profile which as the user changes, the use of paintings is continually updated to ensure the timeliness and accuracy of the user profile.

2 User2vec

Word2vec is a approach to generate word embedding which is a dense vector that can be computed by a computer that translates natural language [19, 20].According to this feature, this paper proposes a method to convert users into vectors. There are three reasons for converting platform users to vectors:

- According to the platform's own characteristics, the vector label structure can effectively measure each user's portrait relative to the platform.
- The user's vector is to project the user into a certain dimension of space, then it is easier to compare the similarity between users. The construction of user vectors can more conveniently measure the similarity of interest preferences between users, and on the other hand provide more comprehensive user preference information for personalized recommendations.
- User portraits are easier to manage, and the size of the label for painting images should not be too fine, because the massive data will cause heavy computational burden.
  Here we describe the user2vec implementation method using the movie platform Douban as a representative:
- Defining specific vector labels and vector dimensions based on the platform itself. Denote X as user vector:

$$X = (x_1, x_2, x_3 \ldots x_k \ldots, x_n)$$

And x_k (k = 1,2, ..., n) is the user preference value corresponding to each tag. It is worth noting that the length of each user's vector is exactly the same, and each dimension represents what the label is also fixed. For example, in Douban (Fig. 1) research and data analysis processing, the defined user label dimension is 1000-dimensional, and the label of each dimension corresponds to below:

(gender, age range... comedy, action, love, plot, sci-fi, suspense... China, America, Japan... classic, youth, magic, inspiration... Marvel, country...)

Fig. 1. Douban film platform

The characteristics of these labels are obvious. The amount of information covered by labels is not very high. It is very important to include all relevant areas of the Douban Film Platform because the concern for a movie platform is definitely the user's preference and recommendation of the movie, and whether the user likes tomatoes or whether they like the sports car is not a platform to care about.

- Labels the tags in each user vector and forms a secondary tag (Fig. 2). Because the tags in the user vector are developed by the platform and have high abstraction, the probability of completely overlapping with the tags in the user vector during the process of tagging and text processing by the user is not high, and therefore, the user vector needs to be generalized. For example, for the comedy tag, some tags and texts with high relevance may include: Mr. Bean, Jackie Chan, Stephen Chow and other relevant tags, then these tags can be included in the next level of the comedy tag. However, note that the next-level labels of different user vector labels are not mutually exclusive, that is, it is possible that a certain label will appear multiple times. To this point, it must be clear that each subordinate label contributes to the above label. It may not be the same.

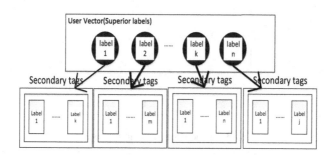

Fig. 2. First level and second level labels

- Calculate the word frequency of the subordinate tag and combine the time decay function (Part 3) to calculate the value of the corresponding superior user tag.

3   Time Decay Function

The time decay function is used to attenuate user preference values based on time variations. Because the user's preferences are highly affected by time, and the processing through time decay can filter out the chance of predicting user preferences. There are few studies on the application of time decay function to user portraits and it has been proposed in literature [21] that Newton's cooling law is used as a time decay function to update the label's heat, but Newton's cooling law is used as a time decay function with two problems. One is that the function is more complex, the computational overhead for large amounts of data is larger, and the other is that the characteristics of time decay and human preference forgetting are not very consistent. In general, there is no external stimulus after passing the heat detector. The capacity for forgetting under the circumstances will increase substantially and eventually reach a more stable stage. And for this feature, this paper proposes to use the time decay function based on the sigmoid function.

$$T(t) = \text{sigmoid}(-wx + \varphi) \tag{1}$$

$$\text{sigmoid}(-wx + \varphi) = \frac{1}{1 + e^{-(-wx + \varphi)}} \tag{2}$$

Defining $T(t)$ as user preference hotness value, $w > 0$ and w, φ are hyperparameters to adjust the median slope of the sigmoid function. $T(t)$ function image as shown in Fig. 3. and Newton's cooling law function image as shown in Fig. 4.

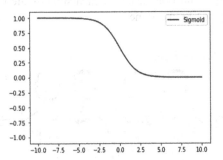

Fig. 3. Time decay function image

4   User Profile Construction

User portrait is a concept based on users, so the process of constructing a user's portrait is guided by a user's image-building process from the initial entry into a platform to the creation of a highly-active user portrait. Note that because all tags are the same for users, when constructing a user's portrait, the heat values of all dimensions need to be sorted, and the top k hotlines, or tags with higher heat than a certain value, will be used as the personality of the user's portrait and label display.

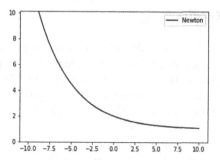

Fig. 4. Newton cooling function image

- The user who logs in to the platform for the first time does not have any preference information. At this time, the user will browse according to novelty, etc., or may click or browse the platform item according to the current popularity. All clicks, browses, and other information that generates user preferences must be associated with the user's tag vector and updated from 0 to θ.

$$x_k = 0 \rightarrow \theta \tag{3}$$

θ is a hyperparameter and is related to the weight of the secondary tag that starts to change

- For all non-zero preference hotness values, daytime or hourly time decay tracking is started on the day, but once the main tag or reselection from the tag's preference is taken, then the hotness update is performed and the time decay is also reinstated. Note that there will be simultaneous changes to the primary and secondary tags, but the hyperparameter settings for the two tags are different. Because the secondary tags contribute different degrees to the primary tags, the degree of update caused by different secondary tags is different.

- When the user information is accumulated to a certain degree or more, the user's preference about the platform will be saturated, and then the user's personalized information may need to be adjusted more meticulously. However, it should be noted that the user's portrait is adjusted instead of the user's vector, because the user's vector is always the same, all users' dimensions are the same, but the corresponding degree of preference is different. For example, in the watercress movie, there is a user's preference is very obvious, the most favorite comedy film Stephen Chow, the probability of this tag appears in the secondary label of the comedy film more than, for example, 50% probability, then the system will be against this obvious features make adjustments to the user's portrait visible properties.

5 Experiments Results

The experiment uses the movielens dataset and indirectly proves the effectiveness of the algorithm through a recommendation algorithm. First of all, it is necessary to declare that the user's portrait is very dependent on the user's log information, because the log information contains very important behavior information about the user's click, browse, browsing time and so on. However, given the data problem, we only use the movielens dataset to build it to a certain extent. However, the reader needs to make clear that under the more comprehensive data conditions, the construction of user portraits will be more comprehensive and accurate, and the effect will be better. At this point we only qualitatively determine the effectiveness of the modeling scheme. In this experiment, the 10 M data in the movielens dataset was used, and the experimental results were verified by the content-based recommendation algorithm [22, 23]. The verification index was Mean Absolute Error (MAE).

First we need to experiment to find the most suitable hyperparameter for this experiment, that is, the optimal selection of w, φ and θ. The selection indicator is content-based recommendation algorithm of the recommended rating and the actual rating of the MAE value. Experimental data is shown in Table 1. The statement made a lot of parametric experiments on the choice of hyperparameters, and the data is given as part of it.

Table 1. Hyper-parameter selection process.

θ	w	φ	MAE
0.4	1	0	0.895
0.5	1	0	0.791
0.6	1	0	0.824
0.8	1	0	0.884
0.4	2.5	5	0.704
0.5	2.5	5	0.691
0.6	2.5	5	0.711
0.8	2.5	5	0.782
0.4	8	10	0.855
0.5	8	10	0.839
0.6	8	10	0.847
0.8	8	10	0.852

From the data in the table and all the hyperparameter test experiments can be roughly determined that the hyperparameter with better effect is $\theta = 0$, $w = 2.5$, $\varphi = 5$.

Next, we compare the algorithm extracted by this method with the TF-IDF method and word2vec which uses word vector similarity to select tags and apply them to content-based recommendation algorithm results, where the abscissa is the number of tags. Experimental data shown in Fig. 5.

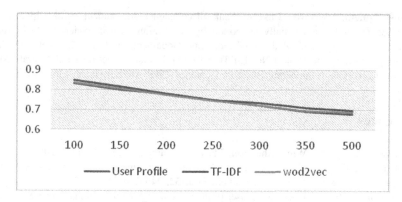

Fig. 5. Comparison of three extraction tags

Next, we compare the content-based recommendation algorithm that combines user portraits with multiple recommendation algorithms. Experimental data shown in Table 2.

Table 2. Multiple recommendation algorithms experiment data

	Average MAE	Minimum MAE
Content-based	0.833	0.801
UserCF	0.758	0.741
ItemCF	0.719	0.739
FunkSVD	0.699	0.686
User profile + content-based	0.702	0.691

From the figure we can see that the MAE of the first experiment is not much different, but we need to pay attention to the fact that the data of this experiment does not contain many of the user's behavioral log data, especially the part based on time is very different. By the same token. In the second experiment, the algorithm results of this paper are slightly inferior to the FunkSVD algorithm. This part of the data has a great impact on the generation and effectiveness of the user's portrait, so the experimental data as a whole can reflect the effectiveness of the algorithm.

6 Summary

This paper proposes a method for constructing a user portrait model based on user vectorization combined with a sigmoid-based time decay function. User vectorization implements the mapping of the user from abstract to high-dimensional space vectors, while the time decay function introduces time information into the user portrait, which is more accurate and more timely. Experiments show that the modeling method presented in this paper can effectively construct user vectors and present accurate user portraits, which are both available and valuable.

Acknowledgments. We would like to thank all colleagues and students who helped for our work. This research was partially supported by (1) National Social Science Fund Project: Research on Principles and Methods of Electronic Document Credential Guarantee in Cloud Computing Environment (Project No. 15BTQ079); (2) Special Project for Civil Aircraft, MIIT; (3) Fund of Shanghai Engineering Research Center of Civil Aircraft Health Monitoring.

References

1. Mulder, S., Yarr, Z.: Win in the User. Machinery Industry Press, Beijing (2007)
2. Chen, S.T., Yu, T.J., Chen, L.C., et al.: A novel user profile learning approach with fuzzy constraint for news retrieval. Int. J. Intell. Syst. **32**, 249–265 (2017)
3. Hawalah, A., Fasli, M.: Dynamic user profiles for web personalisation. Expert Syst. Appl. **42**, 2547–2569 (2015)
4. De Amo, S., Diallo, M.S., et al.: Contextual preference mining for user profile construction. Inf. Syst. **49**, 182–199 (2015)
5. Al-Shamri, M.Y.H.: User profiling approaches for demographic recommender systems. Knowl. Based Syst. **100**, 175–187 (2016)
6. Peng, J., Choo, K.K.R., et al.: User profiling in intrusion detection: a review. J. Netw. Comput. Appl. **72**, 14–27 (2016)
7. Liang, C.: User profile for personalized web search. In: International Conference on Fuzzy Systems and Knowledge Discovery, Shanghai, IEEE, pp. 1847–1850 (2011)
8. Kassak, O., Kompan, M., Bielikova, M.: User preference modeling by global and individual weights for personalized recommendation. Acta Polytech. Hung. **12**(8), 27–41 (2015)
9. Chen, Y., Yu, Y., Zhang, W., et al: Analyzing user behavior history for constructing user profile. In: Proceeding of 2008 IEEE international symposium on IT in medicine and education. IEEE, pp. 343–348 (2008)
10. Xie, H., Li, Q., Mao, X., et al.: Community-aware user profile enrichment in folksonomy. Neural Netw. **58**, 111–121 (2014)
11. Cai, Y., Li, Q., Xie, H., et al.: Exploring personalized searches using tag-based user profiles and resource profiles in folksonomy. Neural Netw. **58**, 98–110 (2014)
12. Tchuente, D., Canut, M., Jessel, N.P.: Visualizing the relevance of social ties in user profile modeling. Web Intell. Agent Syst. Int. J. **10**, 261–274 (2012)
13. Du, Q., Xie, H., Cai, Y., et al.: Folksonomy-based personalized search by hybrid user profiles in multiple levels. Neurocomputing **204**, 142–152 (2016)
14. Wu, Z., Zeng, Q., Hu, X.: Mining personalized user profile based on interesting points and interesting vectors. Inf. Technol. J. **8**(6), 830–838 (2009)
15. Xie, H., Li, X., Wang, T., et al.: Incorporating sentiment into tag-based user profiles and resource profiles for personalized search in folksonomy. Inf. Process. Manag. **52**, 61–72 (2016)
16. Amoretti, M., Belli, L., Zanichelli, F.: UTravel: smart mobility with a novel user profiling and recommendation approach. Pervasive Mobile Comput. **38**, 474–489 (2017)
17. Sahijwani, H., Dasgupta, S.: User Profile Based Research Paper Recommendation. https://uspmes.daiict.ac.in/btpsite. Last accessed 2018/6/5
18. Ouaftouh, S., Zellou, A., Idri, A.: UPCAR: user profile clustering based approach for recommendation. In: International Conference on Education Technology and Computers, Association for Computing Machinery, Barcelona, pp. 17–21 (2017)
19. Mikolov, T., Chen, K., Corrado, G., et al: Efficient estimation of word representations in vector space. 1–12 (2013)

20. Le, Q., Mikolov, T.: Distributed representations of sentences and documents. In: Proceedings of the 31st International Conference on Machine Learning, Beijing, China, JMLR W&CP, vol. 32 (2014)
21. Ranking Algorithm Based on User Voting (4): Newton's Cooling La. https://blog.csdn.net/zhuhengv/article/details/50476306. Last accessed 2018/6/5
22. Aggarwal, C.C.: Recommender Systems: The Textbook. Springer, Basel (2016)
23. Gasparic, M.: Context-based IDE command recommender system. In: RecSys 2016, September 15–19, Boston, MA, USA, pp. 435–438 (2016)

Product Recommendation Method Based on Sentiment Analysis

Jian Yu[1,2,3], Yongli An[1,2,3], Tianyi Xu[1,2,3], Jie Gao[1,2,3], Mankun Zhao[1,2,3], and Mei Yu[1,2,3(✉)]

[1] School of Computer Science and Technology, Tianjin University,
No. 92 Weijin Road, Nankai District, Tianjin, China
{yujian,anyongli,tianyi.xu,gaojie,zmk,yumei}@tju.edu.cn
[2] Tianjin Key Laboratory of Advanced Networking, No. 92 Weijin Road,
Nankai District, Tianjin, China
[3] Tianjin Key Laboratory of Cognitive Computing and Application,
No. 92 Weijin Road, Nankai District, Tianjin, China

Abstract. With the rise of online shopping, massive product information continues to emerge, and it becomes a challenge for users to select their favorite things accurately from millions of products. The collaborative filtering algorithm which is widely used could effectively recommend product to users. However, collaborative filtering algorithm only analyzes the relationship between product and user's evaluation without the analysis of comments. The content contains a lot of useful information and implicates user's pass judgment about product, so collaborative filtering algorithm with no content analysis would reduce the accuracy of the recommendation results. In this paper, we propose a recommendation algorithm based on the content sentiment analysis and the proposed algorithm improves the performance of the traditional product recommendation algorithm based on collaborative filtering. Experimental results demonstrate that the accuracy of the proposed recommendation algorithm based on sentiment analysis is slightly higher than the recommendation algorithm based on collaborative filtering.

Keywords: Collaborative filtering algorithm · Sentiment analysis
Product recommendation algorithm

1 Introduction

With the advent of social networking and shopping, millions of users provide the mass of the data about the user's feelings and views, which contains shopping, entertainment and other activities. Existing widely in areas of movies, music and e-commerce sites, these data aroused attention of domestic and foreign researchers, especially expert scholar in text sentiment analysis and product recommendation.

Sentiment analysis is now a mature machine learning research topic, as illustrated with this review. And the machine learning methods used in this paper is supervised learning which fine-grain is.

© Springer Nature Switzerland AG 2018
X. Meng et al. (Eds.): WISA 2018, LNCS 11242, pp. 488–495, 2018.
https://doi.org/10.1007/978-3-030-02934-0_45

In particular, aiming to mine more connection from user's interest and user's sentiment, we proposes the collaborative filtering recommendation algorithm based on sentiment analysis (CFRBS). The CFRBS makes full use of user shopping comment information of e-commerce website, and calculates the comprehensive similarity between the users through combining the sentiment similarity and the score similarity to receive a score as predict score for product. At the same time, by changing the weight of the sentiment similarity, we explored the level of the sentiment similarity contributes towards the comprehensive similarity.

This remainder of the paper is organized as follows: Sect. 2 provides a reviews of the related work and the research status of the collaborative filtering algorithm. The design details for the collaborative filtering recommendation algorithm based on sentiment analysis algorithm are presented in Sect. 3. We then describes and analyzes the experiment results in Sect. 4, which is followed by the conclusion of the advantage of the algorithm and future directions outlines in Sect. 5.

2   Related Work

In e-commerce, collaborative filtering algorithm is divided into product-based collaborative filtering algorithm and user-based collaborative filtering algorithm [1].

Adomavicius and others [2] have researched various recommendation algorithms and introduced the limitations and possible extensions of the current recommendation algorithm. Faridani [3] proposes Canonical Correlation Analysis (CCA) to perform an offline learning on corpses that have similar structures to Zappos and TripAdvisor. Through the review of the product description and learning from user behavior, Wiet-sma et al. [4,5] believe that the user's recommendation system combined with user reviews will bring more accurate recommendation. Aciar et al. [6] propose a kind of priority mechanism of the recommendation system, which is sorted by the evaluation of the user's degree of expertise and professional degree of user reviews quality to improve user similarity, so that the results are more accurate.

In addition, some researchers propose a lot of improvement suggestions for collaborative filtering. Tohomaru et al. [7] propose a user online shopping behavior model, and the model combines the maximum entropy principle. Chen et al. [8] introduce a recommendation system based on profitability, which integrates the profitability of the traditional recommendation system while meeting the needs of users and seller. Lee et al. [9] analyse the dataset collected from MyStarbucksIdea (MSI), while considering both term-based features and other features of each idea which can be valuable enough for their innovation in initial ideation stage. Dong et al. [10] present a recommendation ranking strategy that combines similarity and sentiment to suggest products that are similar but superior to a query product according to the opinion of reviewers. Cunico et al. [11] extract comments on products from the similar product information and determines an adjustment to the first product rating based on an analysis of the comments and references to the product in the similar product information. Jing et al. [12] propose an enhanced collaborative filtering approach based on sentiment assessment to discover the potential preferences of customers, and to predict customers future requirements for business services or products.

The aforementioned CF algorithms typically only use the relations between users and may lead to recommend the product which users are not interested in rather than the interested product. And sentiment in the data is able to determine the direction of the user's interest, improve the accuracy of the CF algorithm. In view of this, we propose a collaborative filtering recommendation algorithm based on sentiment analysis algorithm which allows us to combine the sentiment and users rating score from the users review.

3   Product Recommendation Method Based on Sentiment Analysis

3.1   Problem Description

When computing the similarity between users, the collaborative filtering algorithm only considers all the common rating items of two users, and user's text sentiment similarity is not considered. So there is a possibility that some similar users are not suitable for the recommendation.

In addition, much of the collaborative filtering algorithms are based on the user's score records to calculate the user's similarity, and does not consider user's sentimental information contained in the online shopping comments. At the same time, user's reviews contain a large number of interest and concern of users and the sentiment tendencies for these concerns and interests, and through analysis of these worthy data which contains user's sentiment.

The main process is looking for suitable shopping website user reviews data sets, getting the user's sentiment tendency and further getting positive on the user's preference and sentiment concerns, and combining user comments set scoring records to calculating the similarity between users, final predict the score of goods and recommend products to users.

3.2   Technological Process of Algorithm

The data is gained from the online shopping site, and the user's sentimental information about product is mainly included in the comment text. Therefore, sentiment analysis of these comment data is one of the key point in experiments. The training data set after the data segmentation is used to classify sentiment polar by the Naive Bayes classification algorithm, then calculates the sentiment orientation and the sentiment value. In this paper, collaborative filtering algorithm based on user (User-based) is used as the basis for the calculation of user's score similarity. According to the method of calculating the similarity of the user's score, the calculation method of the user's sentiment similarity is proposed. Then the comprehensive similarity is calculated by weighted sum of the user's score similarity and the sentiment similarity. Use comprehensive similarity and users' predict score to get users' commodity forecast scores. The pseudocode of CFRBS algorithm is shown in Table 1.

Table 1. The pseudocode of CFRBS algorithm

Algorithm 1: The CFRBS algorithm
//input: $commenttext$, $users'score$,
//output: predict score: S
1. data processing:use SnowNLP to segment text;
2. classify sentiment by Naive Bayes and calculates the sentiment orientation and the sentiment value;
3. calculate the similarity of the users score s^r;
4. calculate the sentiment similarity s^P;
5. the final similarity is $sim(i,j) = \alpha * sim^P(i,j) + (1 - \alpha) * sim^r(i,j)$;
6. calculate the commodity predicted score.

3.3   Sentiment Orientation and Sentiment Value Calculation

The sentiment orientation and the sentiment value of the text can be obtained by the Naive Bayesian classification algorithm. Naive Bayesian classification algorithm is one of the frequently used machine learning method, and the introduction of Native Bayesian classification algorithm is as follows.

Assume that the input of the Naive Bayesian classification algorithm is a set of n-dimensional vectors X, and the classifier output is a class label set $Y = c1, c2, ..., ck$. Given an input vector x which is belongs to X, classifier will outputs an class label y which belongs to Y. $P(X, Y)$ represents the joint probability distribution of input variables X and output variable Y, and the training set $T = (x_1, y_1)(x_2, y_2), ..., (x_n, y_n)$ is generated by the joint probability distribution $P(X, Y)$.

The Naive Bayesian is a very simple probabilistic model that tends to work well on text classifications and usually takes orders of magnitude less time to train when compared to models like support vector machines.

The calculation of the sentiment tendency of the commodity review data is mainly classified into two processes: the judgment of the label of the sentiment body and the calculation of the sentiment tendency. The judgment of the label of the sentiment body is based on the calculation of the similarity between the feature words and the theme of the review data, and by the introduction of the polarity of sentiment words and the intensity of the expression of sentiment, it is possible to accurately calculate the corresponding sentimental tendency values, then combine the sentiment orientation values to analyze the commodity reviews and sentiment analysis. The value contains the user's sentimental polarity and the user's sentimental intensity information. The greater the value is, the more positive the users' sentiment is.

3.4 Comprehensive Similarity Calculation

In order to obtain better recommendation results, the collaborative filtering algorithm based on user (User-based) is used as the basis of user similarity computing. The reason for choosing this algorithm is that the User-based algorithm is more accurate than the content-based recommendation, and it is not based on knowledge which requires professional and statistical information of the domain. The main idea of this algorithm is that the users who have similar interests have a similar score to similar product, so the degree of similarity between users can be calculated by the score of the common rated product. In this paper, the method for calculating the score similarity of users is defined as the Eq. (1).

$$sim^r(i,j) = \frac{\sum_{u \in U}(R_{u,i} - \overline{R_i})(R_{u,j} - \overline{(R_{(j)})}))}{\sqrt{\sum_{u \in U}(R_{u,i} - \overline{R_i})^2}\sqrt{\sum_{u \in U}(R_{u,j} - \overline{R_j})^2}} \tag{1}$$

In Eq. (1), there is modified by cosine similarity algorithm. U represents a set of products which the user i and the user j had both evaluated, $R_{u,i}$ and $R_{u,j}$, respectively, means the user $i's$ and the user $j's$ score for product u, $\bar{R_i}$ and $\bar{R_j}$, respectively, represents the mean value of the score to the products which both of the user i and user j had evaluated.

According to Eq. (1), we can find that the sentiment similarity of the two users is not included. Therefore, the calculation method of the users' sentiment similarity is defined as the Eq. (2).

$$sim^p(i,j) = \frac{\sum_{u \in U}(S_{u,i} - \overline{S_i})(S_{u,j} - \overline{S_j}))}{\sqrt{\sum_{u \in U}(S_{u,i} - \overline{S_i})^2}\sqrt{\sum_{u \in U}(S_{u,j} - \overline{S_j})^2}} \tag{2}$$

In Eq. (2), $S_{u,i}$ and $S_{u,j}$, respectively, said the user $i's$ and the user $j's$ sentiment score for product u which both of them had evaluated, $\bar{S_i}$ and $\bar{S_j}$, respectively, represents the mean value of the sentiment score to the products which both of the user i and user j had evaluated.

However, Eq. (2) in the calculation of user similarity only concern about the sentiment similarity and the results are more one-sided, and the users' score similarity is not included, so Eq. (5) is not enough to completely show the user's similarity. Therefore, the calculation method of the user's comprehensive similarity is proposed. In the paper, comprehensive similarity which is weighted sum between sentiment similarity and score similarity is defined as the Eq. (3).

$$sim(i,j) = \alpha \cdot sim^p(i,j) + (1 - \alpha) \cdot sim^r(i,j) \tag{3}$$

In Eq. (3), $sim^r(i,j)$ is score similarity of user i and user j, $sim^p(i,j)$ is sentiment similarity of user i and user j. And the range of the value of weight factor which is determined by experiments is specified in the $[0, 1]$ range.

3.5   Calculation of Commodity Predicted Scores

The method for calculating the prediction score of the product is shown in the formula (4).

$$Score(a) = \frac{\sum_{j=1}^{K} sim(i,j) \times (R_{i,a} - \bar{R}_j)}{\sum_{j=1}^{K} sim(i,j)} + \bar{R}_i \tag{4}$$

Score(a) is the user $i's$ predict score for the product, K represents the number of users that are most similar to the user i, $sim(i,j)$ is comprehensive similarity of user i and user j, $R_{j,a}$ is the score of user j to product a, $\bar{R}_i$ and $\bar{R}_j$, respectively, is mean score of the user$i$ and user j for all products.

From the formula (4), this paper can get the predict score of the user i. Then the accuracy of the prediction can be obtained by comparing with the actual score of user i to measure the accuracy of the proposed algorithm.

4   Experimental Analysis

4.1   Data Introduction and Data Processing

We used a publicly available dataset of review data of nearly 1000 kinds of products from Jingdong, Taobao and other shopping sites and the dataset including product ID, user ID, topic, user area, user categories, user areas, product ratings, product reviews, and purchase times and so on. It is a set of 50 thousand polar movie reviews and 80% of the data set is used as the training set, which is used to calculate the comprehensive similarity between users, and the remaining 20% as target data.

We chose review data as our data set because it covers a wide range of human sentiment and captures most of the product information relevant to sentiment classification. Also, most existing research on sentiment classification uses product review data for benchmarking. We used the 40,000 documents in the training set to build our supervised learning model. The other 10,000 were used for evaluating the accuracy of our classifier. The average absolute error (MAE) is used as the evaluation criterion of the result.

4.2   Experiment 1: Comparison of CFRBS Algorithm and CF Algorithm

In the Experiment 1, we ran CFRBS algorithm in comparison of CF algorithm. In order to avoid the effect of the number of projections K on the experimental results, we conducted experiments by changing the value of the K.

Figure 1 shows the results for comparison of the product recommendation algorithm based on sentiment analysis (CFRBS) and the testbed. By comparing the accuracy of the CFRBS algorithm and the CF algorithm, we could see from Fig. 1 that regardless of the K value, the MAE value of the CFRBS algorithm is lower than that of CF algorithm, which demonstrate the ability of the CFRBS algorithm garner enhanced performance when using sentiment information. And it was found that the K value had no effect on the experimental results.

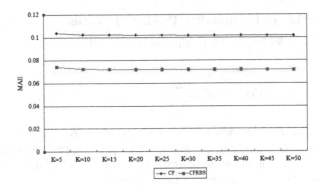

Fig. 1. Comparison of two algorithms

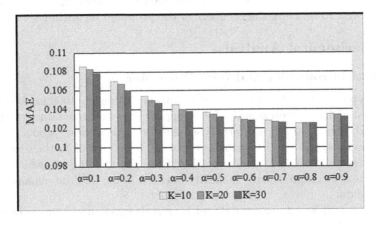

Fig. 2. K values varied with the change of the MAE value at different time

4.3   Experiment 2: Comparison of Different Weight Factors α

In the second experiment, we changed value of α to discover the importance of sentiment information. Experimental results show that with the increase of the weight factor α from 0.1 to 0.9, the curve of the MAE value of the prediction score is in line with the law of the first decline and then rise. And the results can be seen from the experimental results, when α is equal to 0.8, the user's score similarity is 80%, and the sentiment similarity was 20%, the accuracy of the prediction is the highest. As shown in Fig. 2. We can come to a conclusion that the comprehensive similarity is still determined by the user's score similarity, however, after adding sentiment similarity to the user's comments, we can get more accurate prediction score.

5   Conclusion and Future Work

Through the analysis of the advantages and disadvantages of the traditional collaborative filtering algorithm, this paper makes full use of the sentiment analysis technology and proposes the comprehensive similarity which combines the user's sentiment similarity and the user's score similarity. And the CFRBS effectively improve the accuracy of product recommendation, which is based on the traditional collaborative filtering algorithm. The key point of this paper is the comprehensive similarity which is combination of sentiment similarity and score similarity.

The further study will get more detailed data, and consider the time and interest in the product recommendation which will affect the accuracy of the recommendation results. In order to design more reliable collaborative filtering algorithm, these factors would be the focus in further research.

References

1. Lin, C., Tsai, C.: Applying social bookmarking to collective information searching (CIS): an analysis of behavioral pattern and peer interaction for co-exploring quality online resources. Comput. Hum. Behav. **27**, 1249–1257 (2011)
2. Adomavicius, G., Tuzhilin, A.: Toward the next generation of recommender systems: a survey of the state-of-the-art and possible extensions. IEEE Trans. Knowl. Data Eng. **17**, 734–749 (2005)
3. Faridani, S.: Using canonical correlation analysis for generalized sentiment analysis, product recommendation and search. In: Proceedings of the fifth ACM conference on Recommender systems. ACM, pp. 355–358 (2011)
4. Wietsma, R.T.A., Ricci, F.: Product reviews in mobile decision aid systems. In: PERMID, vol. 15, p. 18 (2005)
5. Ricci, F., Wietsma, R.T.A.: Product reviews in travel decision making. In: Hitz, M., Sigala, M., Murphy, J. (eds.) Information and Communication Technologies in Tourism 2006. Springer, Vienna (2006). https://doi.org/10.1007/3-211-32710-X_41
6. Aciar, S., Zhang, D., Simoff, S.: Recommender system based on consumer product reviews. In: Proceedings of the 2006 IEEE/WIC/ACM International Conference on Web Intelligence. IEEE Computer Society, pp. 719–723 (2006)
7. Iwata, T., Saito, K., Yamada, T.: Modeling user behavior in recommender systems based on maximum entropy. In: Proceedings of the 16th International Conference on World Wide Web. ACM, pp. 1281–1282 (2007)
8. Chen, M.C., Chen, L.S., Hsu, F.: HPRS: a profitability based recommender system. In: 2007 IEEE International Conference on Industrial Engineering and Engineering Management, pp. 219–223 (2007)
9. Lee, H., Choi, K., Yoo, D.: The more the worse? Mining valuable ideas with sentiment analysis for idea recommendation. In: PACIS, p. 30 (2013)
10. Dong, R., O'Mahony, M.P., Schaal, M., McCarthy, K., Smyth, B.: Combining similarity and sentiment in opinion mining for produce recommendation. J. Intell. Inf. Syst. **46**, 285–312 (2016)
11. Cunico, H.A., Silva, A.: Product recommendation using sentiment and semantic analysis (2017)
12. Jing, N., Jiang, T., Du, J., Sufumaran, V.: Personalized recommendation based on customer preference mining and sentiment assessment from a Chinese e-commerce website. Electron. Comm. Res. **18**, 1–21 (2018)

Author Index

Printed in the United States
By Bookmasters